COLOUR PHYSICS FOR INDUSTRY
Second Edition

Colour Physics for Industry

Second Edition

Edited by Roderick McDonald

1997

Society of Dyers and Colourists

Copyright © 1997 Society of Dyers and Colourists. All rights reserved. No part of this publication may be reproduced, stored in a retrieval system or transmitted in any form or by any means without the prior permission of the copyright owners.

Published by the Society of Dyers and Colourists, PO Box 244, Perkin House, 82 Grattan Road, Bradford, West Yorkshire BD1 2JB, England, on behalf of the Dyers' Company Publications Trust.

This book was produced under the auspices of the Dyers' Company Publications Trust. The Trust was instituted by the Worshipful Company of Dyers of the City of London in 1971 to encourage the publication of textbooks and other aids to learning in the science and technology of colour and coloration and related fields. The Society of Dyers and Colourists acts as trustee to the fund, its Textbooks Committee being the Trust's technical subcommittee.

Typeset by the Society of Dyers and Colourists and printed by Staples Printers Rochester Ltd.

ISBN 0 901956 70 8

Contributors

David R Battle
Instrument development manager, Datacolor International, Lawrenceville, USA

Adrian R Hill
Principal optometrist, Oxford Eye Hospital, Radcliffe Infirmary, Oxford, UK

Lindsay W MacDonald
Professor of multimedia systems, Cheltenham & Gloucester College of Higher Education, Cheltenham, UK

Roderick McDonald
Formerly principal scientist, Coats Viyella PLC, Glasgow, UK

James H Nobbs
Senior lecturer, Department of Colour Chemistry, University of Leeds, Leeds, UK

Bryan Rigg
Senior lecturer, Department of Computing, University of Bradford, Bradford, UK

Roy S Sinclair
Formerly professor, Department of Chemistry & Chemical Engineering, University of Paisley, Paisley, UK

Kenneth J Smith
Colour consultant, Manchester, UK

Preface

Since the first edition of this book was published in 1987 there has been a revolution in information technology and computing power, and included in this has been the enormous expansion of scientific colour control in the colour-using industries. In today's industry, there are very few organisations dealing in coloration who do not use instrumental colour measurement and instrumental colorant recipe formulation systems. Since the first edition of *Colour physics for industry* some aspects of the science have changed rapidly and others more slowly. To this end some chapters have been extensively revised and new authors have joined the editorial team. A new chapter on the colour science behind the generation of colour in visual displays has been added. The chapters on light, light sources and light interactions, colour-match prediction for pigmented materials, colour-order systems, colour spaces, colour difference and colour scales, and the measurement of colour have been rewritten and brought up to date by new authors. Other chapters have been revised to incorporate the latest technological and scientific developments.

The book is intended to provide in a single volume a comprehensive text on those aspects of colour science that are important to the colour technologist in the day-to-day manufacture and control of coloured products. In particular it is intended as a source book for students preparing for the Society of Dyers and Colourists' Associateship examinations and for university courses involving the technological aspects of coloration.

The present book is the latest in a series produced with the support of funds provided by the Dyers' Company Publications Trust, which was generously established by the Worshipful Company of Dyers. Grateful thanks are expressed to authors, referees, members of the Society's Textbooks Committee and the Society's staff, particularly to Paul Dinsdale, the Society's editor, whose expertise in technical supervision has been invaluable, and to our copy editor Jean Macqueen, whose eagle eye has assisted greatly in achieving consistency of presentation throughout the various chapters of the book. I also wish to thank Carol Davies for her work on the production of the finished pages.

Some of the figures and appendices used in this book have been previously published elsewhere. We are grateful to the owners of the copyright for granting permission to reproduce such material. In addition to permissions recorded in several of the figure captions, we also wish to acknowledge the authors and publishers of the following works:

A H Taylor and G P Kerr, *J. Opt. Soc. Amer.*, **31** (1941) (Figure 1.8)

R S Hunter, *The measurement of appearance* (New York: John Wiley and Sons, 1975) (Figures 1.7 and 1.19)

M B Halstead in *Color 77* (Bristol: Adam Hilger, 1978) (Figures 1.11 and 1.12)

D B Judd and G Wyszecki, *Color in business, science and industry*, 3rd Edn (New York: John Wiley and Sons 1975) (Figures 1.10, 3.1, 3.4 and 3.6 and Appendix 2)

R A Funk, PhD thesis (Clemson University, 1980) (Figures 5.14 and 5.22–5.25)

D McKay, PhD thesis (University of Bradford, 1976) (Figures 5.16–5.21)

H J A Dartnall, J K Bowmaker and J D Mollon, *Proc. Royal Soc.*, **220B** (1983) 115 (Figure 8.6)

S Zeki, *Nature*, **284** (1980) 412 (Figure 8.12)

G Wyszecki and W S Stiles, *Color science* (New York: John Wiley and Sons, 1982) (Appendix 4)

RODERICK McDONALD

Contents

Contributors v
Preface vi

CHAPTER 1   **Light, light sources and light interactions**
by Roy S Sinclair   1
1.1 Light, colour physics and colour technology 1
1.2 Light, colour and the electromagnetic spectrum 2
1.3 Planckian radiators and colour temperature 8
1.4 Daylight and the CIE standard illuminants 11
1.5 Sources of artificial light 15
1.6 Properties of artificial light sources 23
1.7 Colour-matching booths and visual colour matching 27
1.8 The interactions of light with matter 29
1.9 Light absorption, reflection and colour 38
1.10 Light interaction with atoms and molecules 43
1.11 Photophysics, photochemistry and light fastness 51

CHAPTER 2   **The measurement of colour**
by David R Battle   57
2.1 Introduction 57
2.2 The tristimulus colorimeter 58
2.3 The spectrophotometer 59
2.4 Reflectance measurement 61
2.5 Spectrophotometer light sources 63
2.6 Instrument geometry 65
2.7 The dual-beam spectrophotometer 68
2.8 The spectral analyser 68
2.9 The CE3000 spectrophotometer 71
2.10 The Microflash spectrophotometer 72
2.11 Fluorescence 72
2.12 Application of transmission spectrophotometry to dyes 76
2.13 Comparing results from different designs of spectrophotometer 79
2.14 Conclusion 80

CHAPTER 3   **Colorimetry and the CIE system**
by Bryan Rigg   81
3.1 Introduction 81
3.2 Basic principles 83

3.3   Additive and subtractive mixing 84
3.4   Properties of additive mixtures of light 88
3.5   A possible colour specification system 90
3.6   Standard observer: colour-matching functions 94
3.7   Calculation of tristimulus values from measured reflectance values 95
3.8   The 1931 CIE system 98
3.9   Additions to the CIE system 102
3.10 Calculation of tristimulus values from R values measured at 20 nm intervals 104
3.11 Relationships between tristimulus values and colour appearance 105
3.12 Chromaticity diagrams 105
3.13 Usefulness and limitations of the CIE system 108
3.14 Non-uniformity of the CIE system: colour differences 110
3.15 Metamerism 112
3.16 Colour constancy and chromatic adaptation 116
3.17 Colour rendering of light sources 119

CHAPTER 4   **Colour-order systems, colour spaces, colour difference and colour scales**
by Kenneth J Smith   121

4.1   Colour-order systems and colour spaces 121
4.2   Colour-difference evaluation 140
4.3   Colourists' components of colour and colour difference 162
4.4   Evaluation of depth and relative depth 169
4.5   Evaluation of fastness-test results 185
4.6   Evaluation of whiteness and yellowness 195

CHAPTER 5   **Recipe prediction for textiles**
by Roderick McDonald   209

5.1   Introduction 209
5.2   Spectrophotometric curve matching 214
5.3   Colorimetric matching 217
5.4   The starting recipe 223
5.5   Reducing cost and metamerism by computer formulation 226
5.6   Deviations from linearity in recipe prediction equations 233
5.7   Recipe prediction using fluorescent dyes 249
5.8   Recipe correction 266
5.9   Using historical data to improve accuracy of prediction 270
5.10 Selection of optimum dye combinations for recipe preparation 271
5.11 Recipe formulation on fibre blends 277
5.12 Use of neural networks in recipe formulation 281

5.13 Effect of extraneous factors on recipe formulation accuracy 285
5.14 Hardware for computer colorant formulation 286
5.15 Cost-effectiveness of computer prediction systems 288

CHAPTER 6 Colour-match prediction for pigmented materials
by James H Nobbs — 292
6.1 Introduction 292
6.2 Semitransparent layers 299
6.3 Partial reflection at air/coating interface 304
6.4 Database calibration of opaque layers 307
6.5 Preparation of calibration panels (paint) 328
6.6 Database calibration for semitransparent layers 334
6.7 Match prediction of an opaque layer 340
6.8 Match prediction of a semitransparent layer 367

CHAPTER 7 Colour in visual displays
by Lindsay W MacDonald — 373
7.1 Introduction 373
7.2 CRT displays 373
7.3 Liquid-crystal displays 389
7.4 Colorimetry of displays 399
7.5 Measurement and calibration 414

CHAPTER 8 How we see colour
by Adrian R Hill — 426
8.1 The eye and the brain 426
8.2 Structure and function of the eye 427
8.3 The mechanism of vision 430
8.4 Modelling the colour vision process 452
8.5 Attributes of colour 462
8.6 Methods of investigating the perception of colour 463
8.7 Discrimination of colour attributes 467
8.8 Some colour appearance phenomena 479
8.9 Individual differences in colour vision 486
8.10 Defective colour vision 489
8.11 Tests for defective colour vision 500

Appendices — 514

Subject index — 525

CHAPTER 1

Light, light sources and light interactions

Roy S Sinclair

1.1 LIGHT, COLOUR PHYSICS AND COLOUR TECHNOLOGY

Colour in a manufactured object is normally obtained by applying a colorant (dye or pigment) to a polymer substrate, such as textile, paper or paint medium. The appearance of such surface colours depends on three factors [1,2]:
(a) the nature of the prevailing illumination under which the coloured surface is viewed
(b) the interaction of the illuminating radiation with the coloured species in the surface layers, particularly within the visible region of the electromagnetic spectrum
(c) the ability of the radiation that is transmitted, reflected and scattered from the coloured surface to induce the sensation of colour in the human eye/brain system.

In order to understand these dependencies the colour technologist needs to have an appreciation of the nature of light itself, along with some understanding of the physical, chemical, physiological and even psychological aspects of the interactions involving the light, the coloured medium and the human eye. Physically, for example, the production of colour by a dyed or pigmented surface is most simply explained in terms of the process of selective absorption of certain wavelength portions of the incident white light coupled with transmission, reflection and scattering of the non-absorbed radiation (Figure 1.1). Colour can also be produced, however, by other types of light interaction: by interference and by diffraction, for example. The process of light interference produces coloured effects in peacocks' feathers and the wing-cases (elytra) of beetles, and is increasingly being used to produce interesting forms of coloured materials such as liquid crystal displays and polarisation colours.

Many modern light sources such as lasers, coloured light tubes (such as sodium vapour street lamps) and video or television screens emit coloured radiation directly, rather than producing the colour by selective absorption. Lasers and sodium vapour lamps produce their coloured radiation by electronic excitation and deactivation processes, the latter resulting in direct emission of radiation in appropriate narrow

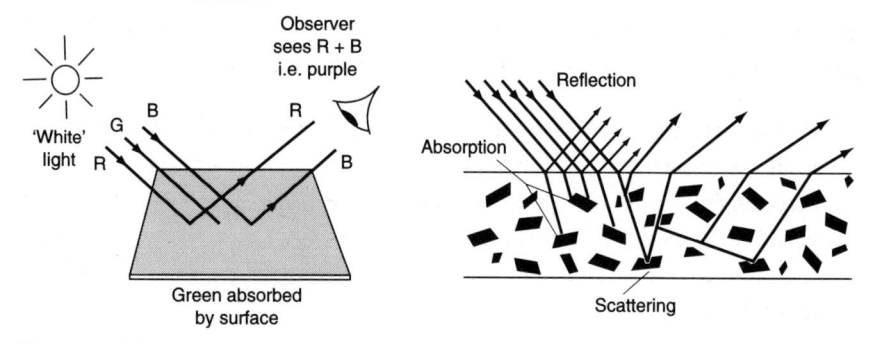

Figure 1.1 (a) Light source, object and eye; (b) light interactions within a coloured surface

wavebands within the visible spectrum, giving rise to a particular hue which is characteristic of the wavelengths or combinations of wavelengths in the emitted radiation. The coloured emissions (red, green and blue) of the video or television viewing screen are stimulated by the cathode rays, via the process known as *luminescence* in the phosphor dot material laid down on the inside of the screen.

It has been claimed that there are 15 different ways of producing colour [3], and some of these processes are considered in more detail in this and later chapters. The dye and pigment technologist may use light:

(a) to illuminate a coloured surface for purposes of assessing colour appearance
(b) to irradiate a coloured surface to assess its light stability
(c) in optical and spectroscopic instruments, to assess optical properties of a surface or to undertake chemical and other scientific studies of the colorants used.

The information provided in this chapter will help in developing the understanding of light and its interaction with matter needed by any colour technologist, whether involved in the manufacture of colorants, the design of colour devices, the application of colorants or the assessment of the characteristics of the coloured material. The main emphases are on the physics and chemistry of the processes of interaction.

1.2 LIGHT, COLOUR AND THE ELECTROMAGNETIC SPECTRUM

Visible light is a form of electromagnetic radiation, distinguished from other forms by its ability to be detected in the retina of the human eye. The first real progress in understanding the relationship between light and colour came in 1672, with the publication of Newton's descriptions and explanations of the effect of passing white light from the sun through combinations of prisms [4]. In describing his experiments, carried out in 1666, Newton used the word 'spectrum' to describe the range of colours (hues) produced, which he noted were in the same order as that observed in a rainbow. He also appreciated that the colours were not a special property of the light but merely a sequence of sensations perceived by the human eye. In his own words:

> And if at any time I speak of light and rays as coloured or endued with colours, I would be understood to speak not philosophically and properly, but grossly, and accordingly to such conceptions as vulgar people in seeing all these experiments would be apt to frame. For the rays to speak properly are not coloured. In them is nothing else than a certain power and disposition to stir up a sensation of this or that colour.

Further insight into the nature of electromagnetic radiation occurred at the beginning of the 19th century, with the discovery by the astronomer Herschel of the heating effect of the rays beyond the red end of the visible spectrum, a region now referred to as the infrared (IR). At about the same time the German physicist J W Ritter, who was exploring the other end of the spectrum, found that silver nitrate was darkened more rapidly by rays beyond the violet than by blue and violet radiation from the visible spectrum; he is credited with discovering ultraviolet (UV) radiation.

In 1860 the Scottish physicist and mathematician James Clerk Maxwell developed a theory that predicted a whole family of wave radiations having associated electrical and magnetic fields, distinguished by having a common velocity in space but differing in wavelength and frequency (as defined in section 1.2.1). It was more than 25 years before his prediction was confirmed. Proof was provided in 1887 by the German physicist Heinrich Hertz who, generating an oscillating current from the spark of an induction coil, produced and detected radiation of extremely long wavelengths, now known as radio waves. Subsequently other forms of electromagnetic radiation were discovered, leading to the realisation that the visible spectrum (from violet to red) forms only a small part of the total electromagnetic spectrum known today.

1.2.1 Wave characterisation of electromagnetic radiation

All electromagnetic radiation, including light, travels through a vacuum with a velocity c of $2.997\,925 \times 10^8$ m s^{-1} (about 186 000 miles per second). This value is constant for all types of radiation, and whatever its intensity. When light has to travel through a medium of refractive index n, however, its speed is given by Eqn 1.1:

$$\text{Speed of light in medium} = \frac{c}{n} \qquad (1.1)$$

Treating electromagnetic radiation as a transverse wave motion, the different forms can be distinguished by their wavelength λ, defined in Figure 1.2 by the distance AB. An alternative parameter is the frequency ν, defined as the number of complete waves

4   LIGHT, LIGHT SOURCES AND LIGHT INTERACTIONS

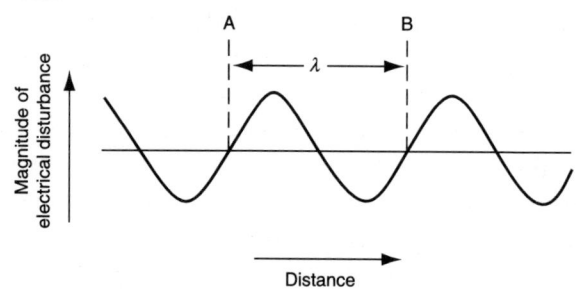

Figure 1.2 Wavelength

from a single wave train passing a given point in space in one second (the unit of frequency is known as the hertz and has dimensions of s^{-1}). Frequency is also measured as the reciprocal of the time taken for one wave to pass the given point in space (the time period T of the radiation, Figure 1.3) (Eqn 1.2):

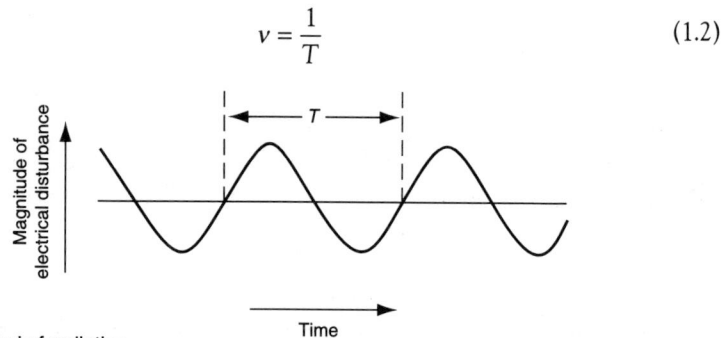

$$v = \frac{1}{T} \tag{1.2}$$

Figure 1.3 Period of radiation

The wavelength and frequency of radiation are related to its velocity by Eqn 1.3:

$$c = v\lambda \tag{1.3}$$

Another common measure of radiation character is the wavenumber $\bar{v}$, which is the number of wavelengths per metre, i.e. the reciprocal of the wavelength (Eqn 1.4):

$$\bar{v} = \frac{1}{\lambda} \tag{1.4}$$

Radiation in the UV and the visible regions of the electromagnetic spectrum is usually described in terms of its wavelength, whereas in the IR both wavelength units and wavenumber units are in common use.

The unit of length that is chosen for wavelength depends largely on the region of the spectrum that is being studied. Thus yellow-green light near the centre of the visible spectrum is said to have a wavelength of 550 nanometres (1 nm = 1×10^{-9} m).

This is preferable to describing the wavelength as 0.000 000 550 m, which is cumbersome. Similarly we might describe radiation in the mid-IR range as having a wavelength of 10 μm (or a wavenumber of 1000 cm^{-1}) rather than a wavelength of 10 000 nm or even 0.000 01 m. The usual rule is, where possible, to choose a unit of wavelength which keeps the numbers in the range 1 to 1000.

The visible region of the electromagnetic spectrum makes up a very small part of the total spectrum (Figure 1.4). The names given to the individual regions are mainly those allocated historically by their discoverers. The wavelength limits of the different regions shown are only approximate, as each region forms a continuum with the next region with ill-defined boundaries. For example, the wavelength range of visible light is often taken as from approximately 400 nm (violet) to 700 nm (red), but occasionally from 380 to 770 nm (as in many CIE tabulations) or even from 350 to 850 nm in the more extended tabulations. These differences are a recognition that the sensitivity range of the human eye depends on the intensity of the radiation source, on the conditions under which the observations are made, and to some extent on the individual making the observations.

Again, within the visible region the main hues are shown in their commonly accepted positions for observers with normal colour vision. Averaged visual assessments have placed a pure (or unitary) blue hue at 436 nm, pure green at 517 nm, pure yellow at 577 nm, with no true red appearing in the spectrum (long-wave red has a yellowish cast).

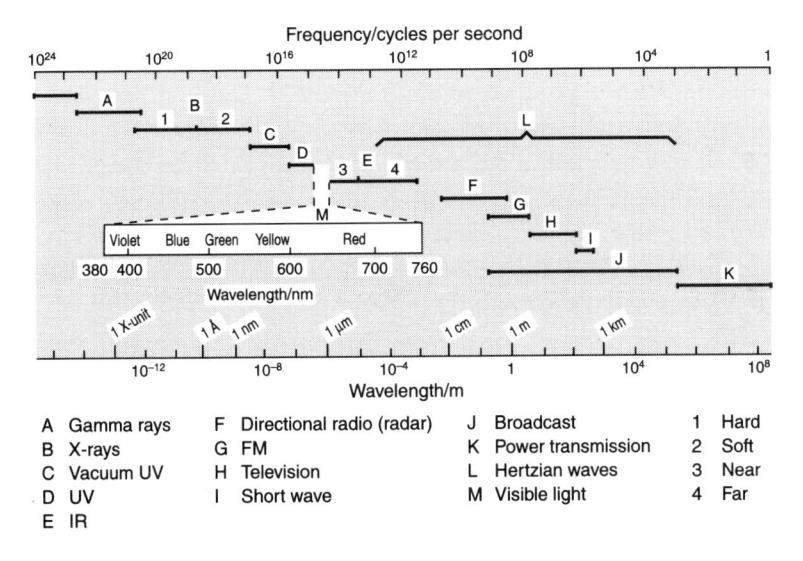

A	Gamma rays	F	Directional radio (radar)	J	Broadcast	1 Hard
B	X-rays	G	FM	K	Power transmission	2 Soft
C	Vacuum UV	H	Television	L	Hertzian waves	3 Near
D	UV	I	Short wave	M	Visible light	4 Far
E	IR					

Figure 1.4 The radiant-energy (electromagnetic) spectrum; the visible portion of the spectrum is expanded to show the hues associated with different wavelengths of light (reproduced from the *IES Lighting Handbook*, courtesy of the Illuminating Engineering Society)

1.2.2 Energy content of radiation

Many properties of light, particularly those relating to absorption and emission of energy by atoms and molecules, cannot be fully explained by the wave theory of light. The alternative (quantum) view that electromagnetic radiation exists as a series of energy packets called photons is accepted as the best model for understanding the energetics of the processes of light absorption and emission. The photon energy depends directly on the frequency of the radiation involved, as given by Planck's relationship (Eqn 1.5):

$$E = h\nu = \frac{hc}{\lambda} \qquad (1.5)$$

where h = Planck's constant = $6.626\,176 \times 10^{-34}$ J s.

The energy given by Eqn 1.5 represents the small amount of energy in a quantum of radiation, and is typically the amount of energy absorbed by a single atom or molecule. The equivalent amount of energy absorbed by a mole of chemical material (assuming complete absorption at the frequency or wavelength concerned) is obtained by multiplying the value of E in Eqn 1.5 by Avogadro's number ($N = 6.023 \times 10^{23}$).

Suppose, for example, that a blue dye molecule absorbs strongly at 600 nm in the orange/red region. This corresponds to energy absorption E_{mol} given by Eqn 1.6:

$$E_{mol} = \frac{6.023 \times 10^{23} \times 6.626\,176 \times 10^{-34} \times 2.997\,925 \times 10^{8}}{600 \times 10^{-9}} \qquad (1.6)$$

or E_{mol} = 199 409 joules per mole (about 200 kJ mol^{-1}). This is a large amount of energy and can, in principle, lead to breakdown of chemical bonds and destruction of the dye molecule. UV radiation at 300 nm wavelength has twice the energy, and its absorption is even more likely to lead to bond scission and hence colorant destruction. The intrinsic highly energetic character of low-wavelength radiation, which can lead to damage to the biological molecules in living cells, is the reason why human beings must be screened from UV and other high-energy radiation.

We will return to this topic briefly in the description of light-fastness testing in section 1.11.3. The quantum concept of radiation is discussed further in section 1.10, which deals with the nature of the processes of light absorption in atoms and molecules.

1.2.3 Measurements of light intensity

In order to quantify the amount of radiation emitted by a source and travelling through space to fall on a surface, we need to specifiy the radiation amount in different units.

Our choice of unit depends on whether we consider all the radiation emitted by the source, only that emitted in a certain direction (or over a certain solid angle), or that falling on a given area of the surface.

The familiar slide projector and screen set-up is a good illustration of the quantities that need to be considered. We would first of all be interested in the total power emitted by the projector, which is measured in watts. We can imagine the radiation as a series of particles or photons travelling between the light source and the screen and hitting the screen at a certain rate, that is, with a definite number of collisions per second (in wave terminology, we are considering the amplitude of the wave here). The total radiant flux hitting the screen will be measured in joules per second or in watts. The projector is designed, using mirrors and lenses, to throw the radiation in a certain direction so that what is important is that the radiant intensity or the number of photons per unit solid angle per second (i.e. the flux per unit solid angle) is reasonably constant over the illuminated screen area. If, however, the screen is moved closer to the projector the radiance or intensity per unit area increases, whilst if it is moved further away the radiance decreases, even although the radiant intensity is constant by definition. For a given distance of the screen we can measure the irradiance or flux per unit area (in watts per square metre, for example), and if we allow the radiation to fall on the screen for a known time we can then compute the total radiant exposure in joules per square metre.

In this discussion we have considered the total radiant emission which, for a typical projector, will include both visible (light) and IR (heat) emissions. In terms of viewing a projector screen what is important is the overall effect on the human eye, which means we should concern ourselves with only the luminous radiation and define appropriate quantities in terms of the visual effect of the radiation. Such terms and the relevant units are indicated in Table 1.1.

The fundamental relationship between radiant and luminous quantities is incorporated in the definition of the lumen, which is defined as the luminous flux of a beam of yellow-green monochromatic radiation whose frequency is 540×10^{12} Hz (equiva-

Table 1.1 Terms and units used in describing light intensity

Radiant energy term	Unit	Luminous energy term	Unit
Radiant flux	W	Luminous flux	lm
Radiant intensity	W sr^{-1}	Luminous intensity	cd (lm sr^{-1})
Radiance	W sr^{-1} m^{-2}	Luminance	cd m^{-2}
Irradiance	W m^{-2}	Illuminance	lm m^{-2} (lux)
Radiant exposure	J m^{-2}	Light exposure	lux s

lent to a wavelength of about 555 nm) and whose radiant flux is 1/683 W. The relationship is of course wavelength-dependent, since radiation outside the visible spectrum has zero luminous contribution. The variation with wavelength is defined by the so-called V_λ curve, which has a maximum at 555 nm and decreases to zero at the ends of the visible spectrum. The use of V_λ values and their variation with wavelength are discussed in section 1.6.

Table 1.2 indicates some typical illumination levels for common sources of light.

Table 1.2 Typical illumination levels (approximate)

Environment	Illuminance/lux
Bright summer sunshine in UK	44 000
Average overcast daylight in UK	5 500
North skylight (open shade) in UK	4 400
Work bench – watch manufacture, etc.	3 300
Textile industry – colour matching of blacks and navies	3 300
Shop windows	1 100–2 200
Football field (floodlight)	1 100
Textile industry – colour matching of bright or pale colours	770
Reading room	550
Screen for slide projection (darkened room)	11–110
Living room, general tungsten illumination	55–165
Dim street lighting	11

1.3 PLANCKIAN RADIATORS AND COLOUR TEMPERATURE

When a coal fire is lit (or when the bar of an electric fire is switched on), it first of all glows a dull red, then orange-red, then yellow; eventually it approaches the 'white-hot' stage as the temperature rises. At the same time the total amount of energy emitted rises (the fire gets steadily hotter). The radiative power emitted by a heated body is best described by a plot showing the variation across the electromagnetic spectrum of the emittance (for example, in watts per square metre) per unit wavelength. Such curves are known as the *spectral power distribution* (SPD) curves of the heat/light source, and Figure 1.5 illustrates how these curves change in the visible region as the temperature of the heated body rises.

A Planckian or black body radiator is an idealised radiation source consisting of a heated enclosure from which radiation escapes through an opening whose area is small compared to the total internal surface area of the enclosure (in practice approximated to by a small hole in the side of a large furnace). The term 'black body' was originally used in recognition that such a model source would radiate energy perfectly and con-

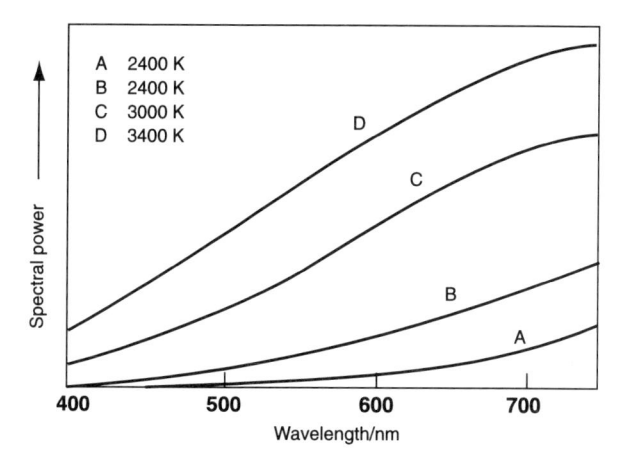

Figure 1.5 Spectral power distribution (SPD) changes during heating

versely would absorb light perfectly, without reflecting any of it away, in the manner of an ideal black object. Nowadays such a model source is referred to as an ideal, full or Planckian radiator.

The Austrian physicist Josef Stefan showed in 1879 that the total radiation emitted by such a heated body depended only on its temperature and was independent of the nature of the material from which it was constructed. Considerable debate about the spectral distribution from these so-called black bodies ensued, in which many of the world's leading theoretical and practical physicists joined: these included Wien, Jeans and Lord Rayleigh. In 1900, however, the German physicist Max Planck developed a theoretical treatment that correctly predicted the form of the spectral power distribution curves for different temperatures (it took the support of Einstein in 1905 to convince the sceptics). Planck's breakthrough came through the assumption that radiation was not emitted continuously but only in small packets or quanta, with the energy of the quantum being directly proportional to the frequency of the radiation involved (as discussed in section 1.2.2).

Planck used his now famous Eqn 1.5 to derive an expression for the spectral emittance from which the SPD curve of the source can be calculated. The Planckian radiation expression has the form of Eqn 1.7:

$$M_e = \frac{c_1}{\lambda^5 [\exp(c_2/\lambda T) - 1]} \tag{1.7}$$

where M_e = emittance per wavelength interval (W m^{-3})
T = absolute temperature of the source (K)
λ = wavelength of the radiation band considered (m)

$$c_1 = 2\pi c^2 h = 3.741\,83 \times 10^{-16} \text{ W m}^2$$
$$c_2 = 1.4388 \times 10^{-2} \text{ m K.}$$

Some examples of the SPD curves for Planckian radiators at different temperatures based on Eqn 1.7 are shown in Figure 1.6. To accommodate the large ranges of values involved, Figure 1.6 shows the power on a logarithmic scale (note the units used) plotted against the wavelength in nm, also on a logarithmic scale, and illustrates how at temperatures below 6000 K most of the energy is concentrated in the long-wavelength IR or heat region of the electromagnetic spectrum. In fact the emission over the visible region is only a small part of the total emission for any of the curves shown.

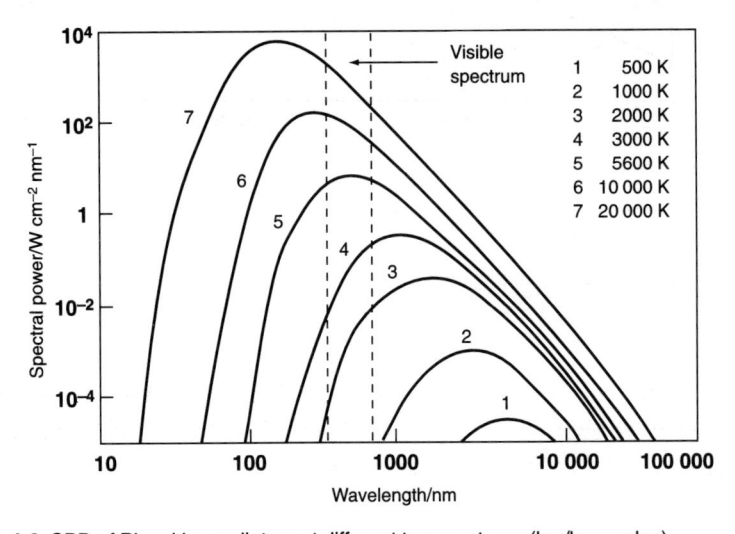

Figure 1.6 SPD of Planckian radiators at different temperatures (log/log scales)

The shape of the SPD curve across the visible region changes significantly, however, from about 1000 K at which the colour appearance of the emitted radiation is predominantly red to 10 000 K, at which it is bluish-white (Figure 1.7). Between these two limits the colour changes from red, through orange-red to yellowish-white and eventually to bluish-white, as discussed above. The closest approach to the ideal equi-energy (ideal white light) source with constant emittance across the visible spectrum occurs somewhere between 5000 and 6000 K.

Thus we can associate the colour appearance of the source with the temperature at which a Planckian radiator will give approximately the same colour appearance. The precise connection between colour temperature and Planckian radiator temperature (and that of correlated colour temperature) is best discussed through a plot of the colour coordinates of the Planckian radiators on a suitable CIE chromaticity diagram (section 3.12).

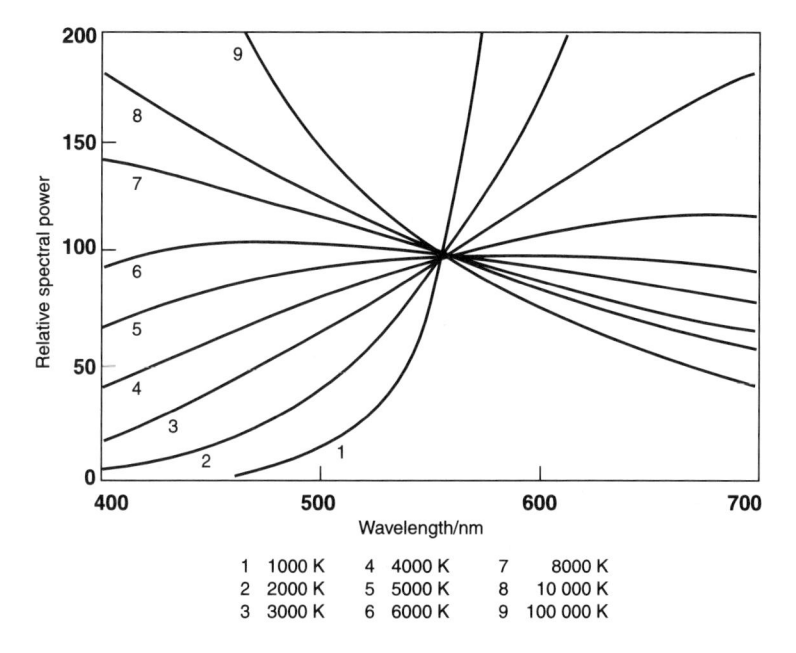

Figure 1.7 Relative SPD in visible region

The typical 100 W domestic tungsten light bulb has a colour temperature of about 2800 K. That of a tungsten–halogen projector bulb is about 3100 K, whilst that of average daylight from an overcast sky is about 6500 K.

1.4 DAYLIGHT AND THE CIE STANDARD ILLUMINANTS

1.4.1 Daylight and its spectrum

Sunlight reaches the earth's surface only after a significant amount of it has been removed by absorption and scattering processes as a result of encounters in the atmosphere with ozone, water vapour, liquid water droplets, ice and dust particles, as well as with an increasing variety of pollutant species. The selective absorptions (mainly by ozone and water vapour) produce the undulations in the SPD curves of daylight shown in Figure 1.8. Scattering is most pronounced at the low wavelengths; this leads to the blue appearance of a clear sky and the rapid fall-off of the energy in the UV (400–300 nm region).

Figure 1.8 shows that the nature of daylight depends on the part of the sky that is viewed, and on the viewing conditions. Variations are also observed which depend on:

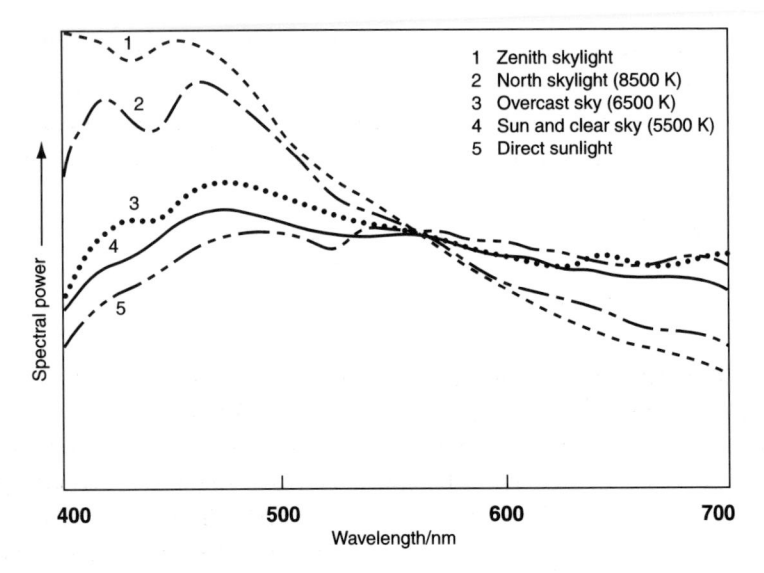

Figure 1.8 SPDs of different forms of daylight (correlated colour temperatures)

- the latitude of the place of measurement
- the season of the year
- the local weather conditions
- the time of day.

Despite these variations it is possible to assign colour temperatures to the different phases of daylight as shown in the legend to Figure 1.8. Thus light from a clear north skylight is distinctly blue and can have a colour temperature from 7500 to 10 000 K, or even higher, whilst that of average daylight (particularly when viewed from indoors) is in the region 5800 to 6500 K. When sunlight is included with clear skylight the influence of the blue scattered light is less and the colour temperature drops to about 5500 K.

In many colour technology applications daylight was traditionally the preferred illuminant for colour matching and colour assessment operations. However, with recognition of the above variability and the trend towards tighter colour tolerances, most colour matching is now done under artificial daylight illumination derived either from fluorescent tubes or from a filtered xenon arc lamp conforming to the appropriate national or international standards.

Although the variability of daylight in the visible region is significant, the much greater variability of the UV portion of daylight (from 300 to 400 nm) is a more significant problem in the visual assessment of fluorescent whitening materials and in the light-fastness testing of coloured materials. Light-fastness testing is increasingly being

carried out using xenon arc sources. Additionally, the characteristics of daylight, and particularly the effect of ozone levels on the proportion of UV in direct sunlight, has been of recent concern as it is the UV radiations that cause skin damage, including sunburn and skin cancers, in humans. For such studies the UV region is split into three subregions:
– UVA: 400–315 nm
– UVB: 315–280 nm
– UVC: less than 280 nm.

Glass, depending on its thickness, generally cuts off UV radiation below 310 nm, so that exposure to daylight behind glass involves mainly UVA radiation. The DNA and the proteins in living cells absorb below 300 nm, and therefore it is exposure to UVB and UVC which cause damage to the eyes and give rise to the most serious types of sunburn damage (and in the long term to skin cancer). The current concerns over ozone 'holes' arise from the increased transmission of UVB and particularly UVC through the atmosphere under conditions of depleted upper atmosphere ozone.

In the standard method of test for light fastness to daylight (discussed in section 1.11.3) samples are exposed behind glass, so under these conditions only the visible and UVA radiation are involved in light fading of colorants. In weathering tests, however, the protective effect of glass is removed.

1.4.2 Standard illuminants and standard sources

When the international system of colour measurement and specification was set up by the Commission Internationale de l'Éclairage (CIE) in 1931 it was recognised that standardised sources of illumination would have to be defined, and three such sources (CIE Standard sources A, B and C) were adopted at that time as approximations to three common illumination conditions. Although these sources were defined in such a way that they could be physically realised (a standardised tungsten lamp in combination with suitable blue-coloured solution filters) the opportunity was also taken to define a set of numerical values representing the relative SPD of the appropriate standard illuminant at 10 nm intervals across the visible spectrum (380–770 nm).

It is thus important to distinguish between standard illuminants, which are defined in terms of spectral power distributions, and standard sources, which are defined as physically realisable emitters of radiant energy and have SPDs that only approximate to those of the corresponding illuminants.

Standard illuminant A was designed in 1931 to be representative of indoor artificial (tungsten lamp) illumination and is defined as an illuminant having the same SPD as a Planckian radiator at a temperature of about 2856 K. An actual source corresponding

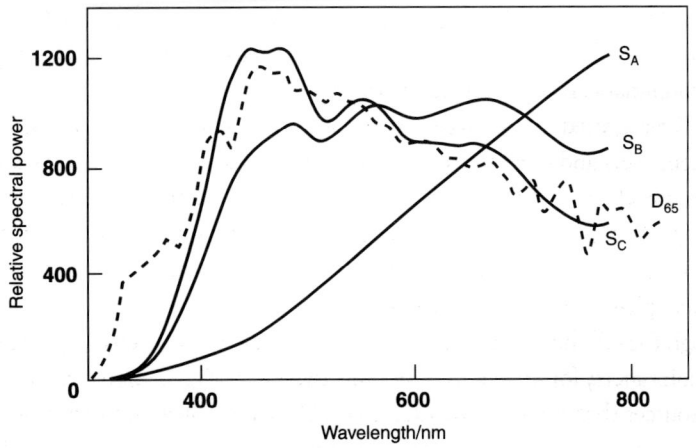

Figure 1.9 SPDs of CIE standard illuminants

to this illuminant is readily achieved, and calibrated standard tungsten lamps are available from standardising bodies in each country; in the UK this is the National Physical Laboratory (NPL). Such an illuminant is relatively yellowish in colour as it is deficient in power in the blue end of the visible spectrum and rich in the red wavelengths (Figure 1.9).

Standard illuminant B, with a correlated colour temperature (CCT) of about 4870 K, was supposed to represent daylight plus sunlight, and standard illuminant C (CCT = 6770 K) was intended to represent average daylight; both are now largely redundant in favour of the D illuminants introduced subsequently.

Studies of daylight and its spectrum in the 1950s and early 1960s confirmed that standard illuminants B and C had too little power in the UV region to be of value in assessing fluorescent brightening agents (or optical brighteners as they were then termed), and the SPDs and the colour coordinates also deviated from those of the natural daylight conditions they were supposed to represent [5]. In 1963, therefore, the CIE recommended several new standard illuminants (the D illuminants) by defining spectral distributions across the UV, visible and near-IR (300–830 nm) to represent various phases of daylight. CIE illuminant D_{65}, with an approximate CCT of 6500 K, is now accepted by the CIE as a standard illuminant (CIE 1986). Its SPD is a good approximation of average daylight, taking into account the following types of variation:
– from early morning to late evening
– from a blue sky to completely overcast conditions
– at different latitudes.

The SPDs of the standard D illuminants were originally defined at 10 nm intervals but values at 5 nm and 1 nm intervals have been obtained by interpolation and are now available (CIE 1971 and 1986). Figure 1.9 compares the SPD of standard illuminant D_{65} with the CIE illuminants A, B and C; the higher UV content of D_{65} compared with CIE illuminants B and C is clearly evident.

The CIE recognised that a single distribution such as D_{65} would be unlikely to satisfy all colour users and suggested others, such as D_{50} and D_{55} with CCTs of 5000 and 5500 K for use where yellower phases of daylight than average were desirable (D_{50} is favoured by the graphic art trade for illumination of colour prints and photographs). D_{75}, with a CCT of 7500 K, is popular in some parts of America for colour assessment where a bluer phase of daylight is preferred.

Although the CIE has published a method for assessing the quality of daylight simulators for colorimetry [6], problems have been encountered in attempts to manufacture practical sources that simulate the illuminant D curves, particularly the undulations present naturally and accentuated by the interpolation procedures. Recently Hunt has suggested that the CIE needs to accept that only approximations to the standard D illuminant curves are ever going to be possible. He advocates that practical D sources be carefully specified and adopted for use as the best approximations to the D illuminants achievable [7] and suggests these could be distinguished as:

– source DT: a tungsten–halogen lamp with a blue glass filter
– source DX: a filtered xenon arc
– source DF: a fluorescent lamp with a suitable CCT.

The demand for a practical and standard D source is now substantial, both in connection with the comparison of the quality of instrumentally measured and visually assessed colour matches and also, more particularly, in the measurement and assessment of fluorescent samples (where the UV content of the source is critical). The situation has been conveniently summarised recently by McCamy [8].

1.5 SOURCES OF ARTIFICIAL LIGHT

Artificial light is produced in many ways. The most important method (and historically the earliest) is to heat or burn matter so that the constituent atoms or molecules of the source are excited to such an extent that they vibrate and collide vigorously, causing them to be constantly activated and as a result to emit radiation over the UV, visible and near-IR regions of the electromagnetic spectrum (similar to the Planckian or black body radiator). This phenomenon, referred to as *incandescence*, produces a continuous spectrum over quite a wide range of wavelengths (dependent mainly on the temperature of the source).

Common incandescent sources range from the sun, through tungsten and tungsten–halogen sources to burning gas mantles, wood, coal or other types of fires and candles

(the last mentioned have colour temperatures in the region of 1800 K). Other methods of producing light, probably in decreasing order of importance, are:
(a) electrical discharges through gases (e.g. sodium and xenon arcs)
(b) photoluminescent sources such as the fluorescent tube, long-lived phosphorescent materials and certain types of laser
(c) cathodoluminescent sources based on phosphors, as used in television and VDU screens
(d) electroluminescent sources based on certain semiconductor solids and phosphors, as in light-emitting diodes (LEDs)
(e) chemiluminescent sources as used in light sticks.

Many of these other sources emit over selected regions of the electromagnetic spectrum giving line and band spectra, and these may be inherently coloured as a consequence of selected emission in the visible region. For example, the sodium-vapour lamp is orange-yellow due to a concentration of emission around 589.3 nm (the sodium D line), although an almost equally intense band of radiation is emitted near 800 nm in the near-IR. The mechanisms of light emission are mentioned only briefly below, but fuller details are given in Nassau [3].

1.5.1 The tungsten-filament lamp

Some light sources show only minor deviations from Planckian distribution: of these, the tungsten-filament lamp is a prime example. The radiation is derived from the heating effect of passing an electric current through the filament while it is held inside a bulb which either contains an inert gas or is evacuated or at a low pressure to keep oxidation of the filament to a minimum. The character of the emitted radiation (and therefore the colour temperature) is controlled to a large extent by the filament thickness (resistance) and the applied voltage. For a given filament, increasing the voltage increases the light output but decreases the lamp lifetime. In practice tungsten lamps are produced with a variety of colour temperatures, ranging from the common light bulb at 2800 K to the photographic flood at 3400 K (which has quite a short lifetime). Temperatures must be kept well below 3680 K, which is the melting point of tungsten.

1.5.2 Tungsten–halogen lamps

Tungsten filaments can be heated to higher temperatures with longer lamp lifetimes if some halogen (iodine or bromine vapour) is present in the bulb. When tungsten evaporates from the lamp filament of an ordinary light bulb it forms a dark deposit on the glass envelope. In the presence of halogen gas, however, it reacts to form a gaseous tungsten halide, which then migrates back to the hot filament. At the hot filament the halide decomposes, depositing some tungsten back on to the filament and releasing halogen back into the bulb atmosphere, where it is available to continue the cycle.

With the envelope constructed from fused silica or quartz, tungsten–halogen lamps can be made very compact with higher gas pressures. They can then be run at higher temperatures (up to 3300 K) with higher efficacy (lumens per watt). Such lamps are commonly used in slide and overhead projectors and in visible-region spectrometers and other optical instruments, and in a low-voltage version in car headlamps. Mains-voltage lamps are used for floodlighting and in studio lighting in the film and television industry.

1.5.3 Xenon lamps and gas discharge tubes

An electric current can be made to pass through xenon gas by using a high-voltage pulse to cause ionisation. Both pulsed xenon flash tubes and continuously operated lamps operating at high gas pressures (up to 10 atm) are available, the latter giving almost continuous emission over the UV and visible region.

Largely because of its spectral distribution, which when suitably filtered resembles that of average daylight (Figure 1.10), the high-pressure xenon arc has become very important for applications in colour technology. It is now an international standard source for light-fastness testing, and is increasingly being used as a daylight simulator for colorimetry, and in spectroscopic instrumentation (flash xenon tubes in diode array spectrometers), as well as in general scientific work involving photobiological and photochemical studies and in cinematography.

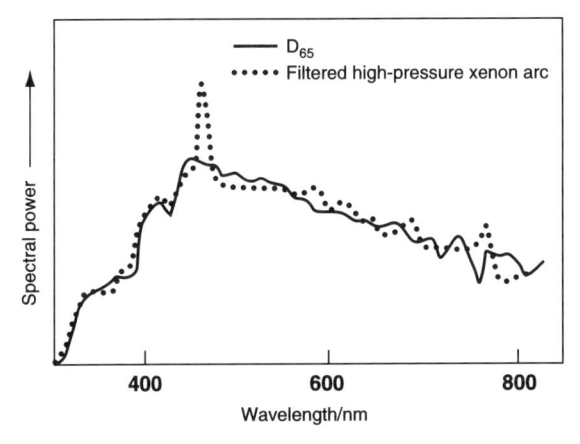

Figure 1.10 SPD of a filtered high-pressure xenon-arc lamp compared with an 'average daylight', CIE illuminant D_{65}

Electrical discharges through gases at low pressure generally produce line spectra. These emissions arise when the electrically excited atoms jump between quantised energy levels of the atom (see section 1.10). The mercury discharge lamp was one of the earliest commercially important sources of this type (Figure 1.11), its blue-green colour being due to line emissions at 405, 436, 546 and 577 nm. There is a high-intensity 366 nm line emission in the UV, which makes it necessary for the user of an unfiltered mercury lamp to wear protective UV-absorbing goggles. When mercury arcs with clear quartz or silica envelopes are used, protection is also required from generated ozone.

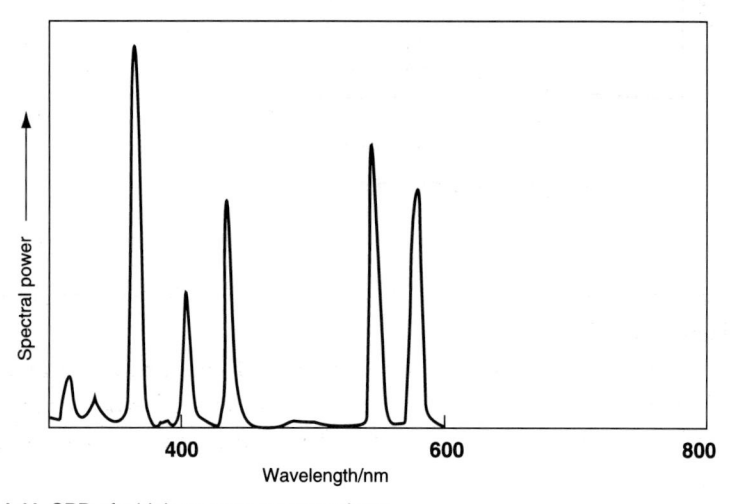

Figure 1.11 SPD of a high-pressure mercury lamp

The intensity and width (wavelength ranges) of the line emissions depend to a large extent on the size of the applied current and the vapour pressure within the tube. By adding metal halides to the mercury vapour, extra lines are produced in the spectrum and the source effectively becomes a white light source (HMI lamp).

Mercury light sources are used extensively in the surface coating industry (UV curing), in the microelectronics industry (photolithography), as the basic element in fluorescent lamps and tubes (section 1.5.4), as an aid to assessment of fluorescent materials in colour-matching light booths and, to a limited extent, for assessing the stability of coloured materials to UV irradiation (section 1.11). The metal halide lamps are used in floodlighting applications, while the special HMI lamp was developed as a supplement to daylight in outdoor television productions.

Another well-known light source of this type is the sodium-vapour lamp which, in its high-pressure form, was developed in the 1960s particularly for street lighting and floodlighting applications. The spectral emission lines in this case are considerably broadened, with the gas pressures being sufficiently high to produce a significant

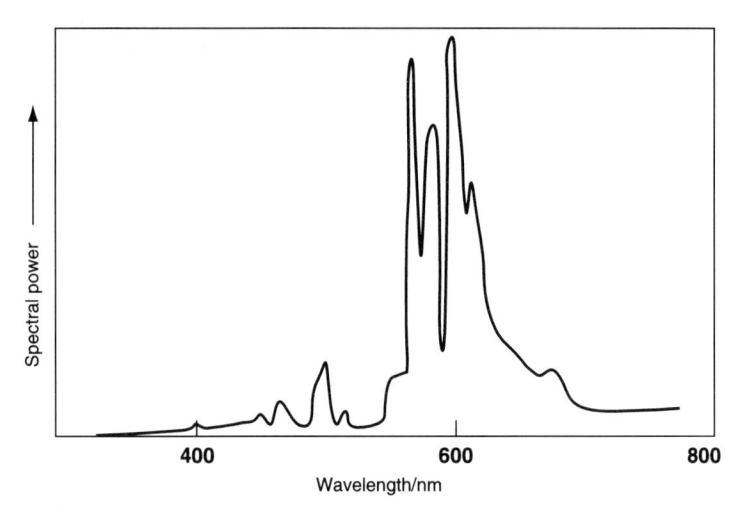

Figure 1.12 SPD of a high-pressure sodium lamp

absorption at the D line wavelength (589.3 nm). A typical SPD curve for a high-pressure sodium lamp is shown in Figure 1.12.

The main value of the sodium-vapour lamp lies in its relatively high efficacy (100–150 lm W^{-1}). Cited refractive index values for liquids and transparent materials are usually based on measurements using the D line radiation from a low-pressure sodium lamp.

1.5.4 Fluorescent lamps and tubes

The ubiquitous fluorescent tube consists of a long glass vessel containing mercury vapour at low pressure sealed at each end with metal electrodes between which an electrical discharge is produced. The inside of the tube is coated with phosphors that are excited by the high-energy UV lines from the mercury spectrum (mainly 254, 313 and 366 nm lines), which by photoluminescence (or a mixture of fluorescence and phosphorescence) are converted to radiation above 400 nm. The spectrum that is produced is dependent on the type of phosphor mixture used; thus the lamps vary from the red-deficient 'cool white' lamp, which uses halophosphate phosphors, to the broad-band type in which long-wavelength phosphors are incorporated to enhance the colour-rendering properties (section 1.6.2).

A third type, known as the three-band fluorescent or prime colour lamp, uses narrow-line phosphors to give emissions at approximately 435 nm (blue), 545 nm (green) and 610 nm (red) and an overall white light colour of surprisingly good colour-rendering properties. The characteristics of these lamps have been extensively studied

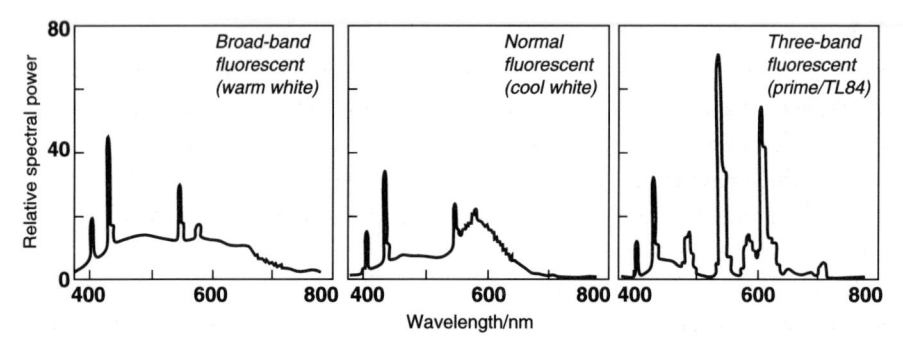

Figure 1.13 Comparative SPDs of three types of fluorescent lamp

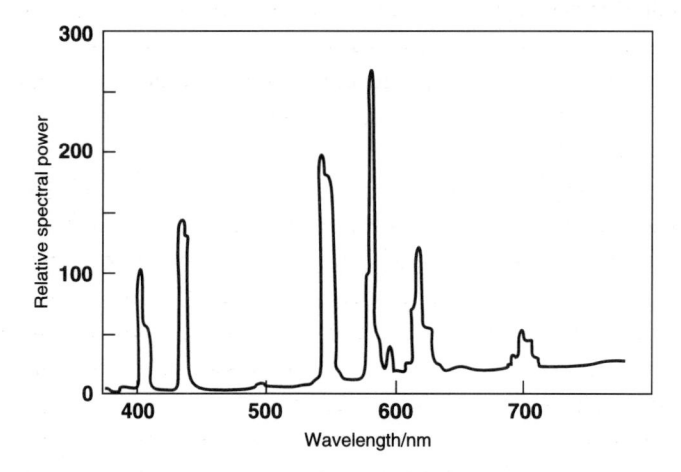

Figure 1.14 SPD of mercury lamp with tungsten ballast (MBTF lamp)

by Thornton [9], and they have been marketed as Ultralume (Westinghouse) in the USA and TL84 (Philips) in the UK.

The characteristics of the three types of fluorescent tubes are compared in Figure 1.13. The first two lamps show prominent line emissions at the mercury wavelengths of 404, 436, 546 and 577 nm. The much higher efficacy of the three-band fluorescent (TL84) lamps over other types has resulted in their use in store lighting, but this has aggravated the incidence of colour mismatches (metamerism) caused by changing illuminants (see sections 1.9.4 and 3.15).

High-pressure mercury lamps have also been designed with red-emitting phosphors coated on the inside of the lamp envelope to improve colour rendering; these include the MBF and MBTF lamps. The latter have a tungsten-filament ballast which raises the background emission in the higher-wavelength regions (Figure 1.14).

1.5.5 Laser light sources and LEDs

Laser sources are increasingly being used in optical measuring equipment, certain types of spectrometers and monitoring equipment of many different types. The red-emitting He–Ne gas laser was one of the earliest lasers developed, but it is the red-emitting diode laser which has become familiar in its application to barcode reading devices in supermarkets and elsewhere. Yet another type emits in the IR region, and is widely used in compact disc (CD) players.

The term 'laser' is an acronym for the process in which light amplification occurs by stimulated emission of radiation. In order to explain laser action we have to appreciate some of the aspects of atomic and molecular excitation covered in section 1.10 below.

In the gas discharge tubes mentioned in section 1.5.4, light emissions arise from electrical excitation of electrons from their normal ground state to a series of excited states and ions, and it is the subsequent loss of energy from these excited states which results in spontaneous emission at specific wavelengths according to the Planck relation given in Eqn 1.5. (The precise electronic levels involved in the He–Ne laser are illustrated in section 1.10.2.)

In a laser means are provided to hold a large number of atoms or molecules in their metastable excited states, usually by careful optical design in which the radiation is reflected many times between accurately parallel end mirrors. The system shown in Figure 1.15 is said to exist with 'an inverted population' allowing stimulated rather than spontaneous emission. Thus if a quantum of light of exactly the same wavelength as the spontaneous emission interacts with the excited state before spontaneous emission has occurred, then stimulated emission can occur immediately (Figure 1.16). It is one of the characteristics of laser light that it is emitted in precisely the same direction as the stimulating light, and it will be coherent with it, i.e. all the crests and troughs occur exactly in step, as indicated in Figure 1.15.

Because of the optical design of the laser cavity and the consequent coherence of laser light, it is emitted in a highly directional manner and can be focused on to very small areas giving a high irradiance capability. The use of Brewster angle windows in the discharge tube section of a gas laser also results in the emitted radiation being highly polarised (Figure 1.17). Certain types of laser can also be operated to give high-power short-lived light pulses, nowadays reaching down to femtosecond (1 fs = 1×10^{-15} s) timescales, which can be used to study the extremely rapid chemical and physical processes that take place immediately after light is absorbed.

Semiconductor materials are used in the manufacture of light-emitting diodes (LEDs) and in diode lasers, the wavelength of emission being determined by the chemical composition of the semiconductor materials. The mechanism of light production in the LED arises from the phenomenon of *electroluminescence*, where the electrical excitation between the conduction band in the n-type semiconductor and the

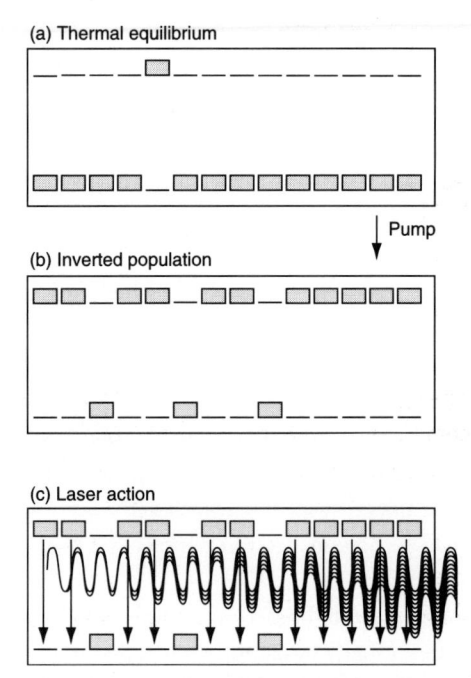

Figure 1.15 A schematic illustration of the steps leading to laser action: (a) the Boltzmann population of states, with more atoms in the ground state; (b) when the initial state absorbs, the populations are inverted (the atoms are pumped to the excited state); (c) a cascade of radiation then occurs, as one emitted photon stimulates another atom to emit, and so on: the radiation is coherent (phases in step)

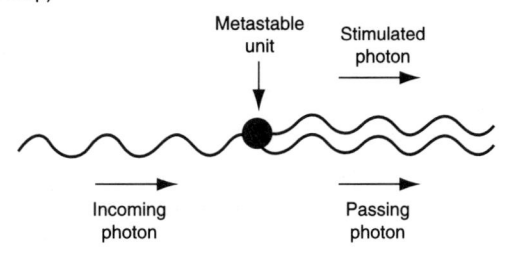

Figure 1.16 Stimulated emission of a photon

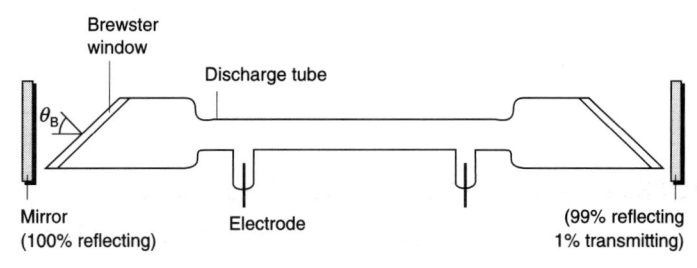

Figure 1.17 Optical design of the He–Ne gas laser (Brewster window to induce polarisation)

Table 1.3 Semiconductor materials used for LEDs

Colour of LED	Materials
Red	Ga, As, P; Ga, Al, As; GaP; GaP:N
Orange	Ga, As, P; In, Ga, P; In, Al, P
Yellow	Ga, As, P; In, Ga, As
Green	AlP; Cd, S, Se
Blue	SiC; ZnS; Cd, Zn, S
Violet	GaN; Cd, Zn, S

valence band in the p-type material results in an energy gap and hence light emission by electron hole recombination across the p–n semiconductor junction [3]. Table 1.3 shows the materials used to make LEDs to produce light of different colours.

The commonest LEDs are manufactured from gallium combined with arsenic and phosphorus in different ratios to give variation in colour and wavelength of the emitted light. For example, with an As : P ratio of 60 : 40 a red emission (690 nm) is produced, a ratio of 40 : 60 gives orange (610 nm) and a ratio of 14 : 86 gives yellow (580 nm).

Similar materials can be used to form a diode laser, where the end faces of the semiconductor double layer are polished to give the necessary multireflection; these materials have a high refractive index, so readily produce the required internal reflections at their surfaces. Figure 1.18 shows diagrammatically the construction of a semiconductor junction laser.

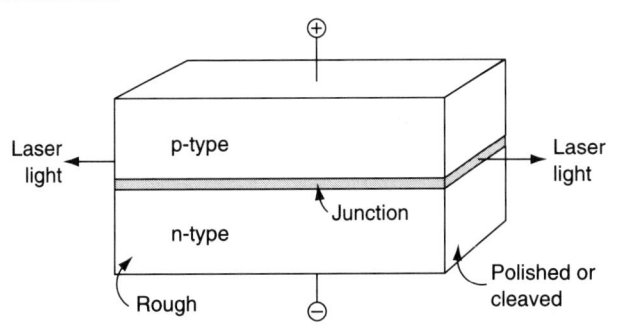

Figure 1.18 Semiconductor junction laser

1.6 PROPERTIES OF ARTIFICIAL LIGHT SOURCES

There are two aspects of artificial light sources that are of particular interest to colour scientists:

(a) the luminous efficacy of the lamp in lumens per watt (lm W^{-1}), which is a measure of the amount of radiation emitted for a given input of electrical power, weighted by the ease by which that radiation is detected by the human observer
(b) the colour-rendering characteristics of the lamp, which is a measure of how good the lamp is at developing the accepted 'true' hues of a set of colour standards.

1.6.1 Lamp efficacy

The human eye is stimulated more strongly by light of some wavelength regions of the visible spectrum than by others; thus yellow-green light at 555 nm is the most readily seen, while blue and red light of the same radiant flux appear quite dim by comparison. The wavelength-dependent factor that converts radiant energy measures to luminous or photometric measures is known as the V_λ function. It varies with wavelength across the visible spectrum (Figure 1.19).

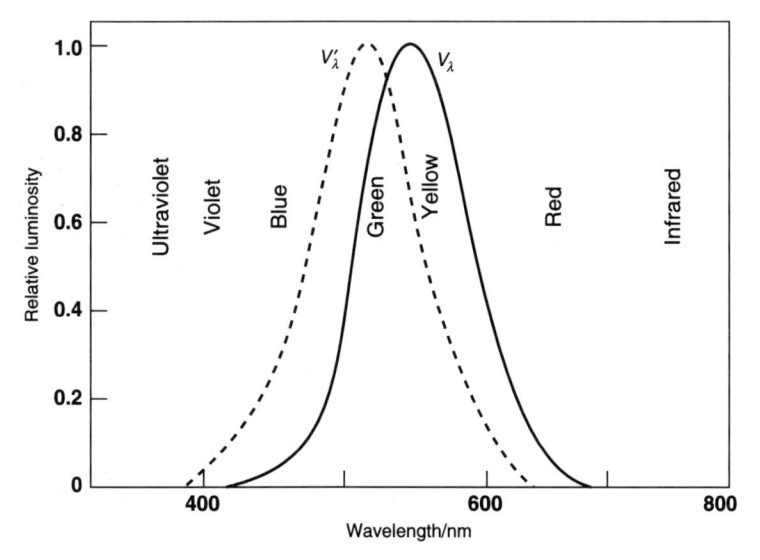

Figure 1.19 The luminosity curves V_λ (photopic) and V'_λ (scotopic)

The photopic V_λ curve describes the relative sensitivity of the average human eye to visible radiation under normal levels of illumination and is an essential quantity for calculations by illuminating engineers; it was first standardised by the CIE in 1924. If a lamp's variation of radiant flux with wavelength P_λ is known, then the total luminous flux F in lumens can be calculated from Eqn 1.8:

$$F = K_m \int P_\lambda V_\lambda \, d\lambda \tag{1.8}$$

where K_m = luminous efficacy of radiation at 555 nm (about 683 lm W^{-1}), at which wavelength the V_λ function has a maximum value of 1.000. The limits of the integral in Eqn 1.7 are effectively those of the visible spectrum, i.e. 380–770 nm.

A lamp emitting radiation only at 555 nm would have this maximum efficacy of 683 lm W^{-1}. The nearest practical approach, however, is the sodium lamp emitting at 589 nm where $V_\lambda = 0.76$, with a maximum efficacy near 150 lm W^{-1}. Some energy is dispersed in nonvisible emission and some by heat loss and other inefficiencies. The poor colour-rendering properties of the sodium lamp have been mentioned already; there is a general compromise between luminous efficacy and colour-rendering properties (see section 1.6.2).

Figure 1.19 also includes the V'_λ curve, effective at scotopic or low light levels (under twilight conditions, for instance); this curve has a maximum at 510 nm and is relatively higher in the blue but becomes effectively zero above 630 nm (many red objects appear black under these conditions).

1.6.2 Colour-rendering properties

A traditional red letter box or red bus illuminated by sodium-vapour street lighting appears a dullish brown; similarly, the human face takes on a sickly greenish hue when viewed in the light from a vandalised fluorescent street lamp (where the phosphor-coated glass envelope has been removed and the light is from the unmodified mercury spectrum). Both these lamps would be recognised as having poor colour-rendering properties. The colour-rendering properties of fluorescent tubes are better, but can span an appreciable range (Table 1.4).

Table 1.4 Properties of some common light sources

Lamp	Luminous efficacy /lm W^{-1}	Correlated colour temperature/K	Colour-rendering index/R_a
Tungsten (100 W)	14	2800	100
Tungsten–halogen	21	3000	100
High-pressure sodium	100	2000	25
Mercury phosphor (MBTF)	20	3800	45
Xenon	25	5500	93
Fluorescent			
Warm white	74	3000	54
Cool white	74	4200	63
Northlight	48	6500	94
Artificial daylight	38	6500	94
Three-band (TL84)	93	4000	85

Colour-rendering properties can be measured by two indices: the CIE special colour-rendering index R_i, and the CIE general colour-rendering index R_a, defined by Eqns 1.9 and 1.10 respectively:

$$R_i = 100 - 4.6 d_i \qquad (1.9)$$

$$R_a = 100 - \left[\frac{4.6}{8} \times (d_1 + d_2 + d_3 + d_4 + d_5 + d_6 + d_7 + d_8) \right] \qquad (1.10)$$

where d_i is the distance in the CIE $U^*V^*W^*$ colour space (section 4.1.12) between the coordinates of the colour concerned when illuminated by the test source compared with those of the sample when illuminated by the nearest D source, and d_1 to d_8 are the R_i values for a series of Munsell colours round the colour circle at value level 6 [10].

No source exists that combines the colour-rendering properties of the black body radiators with the high efficacy of the sodium lamp. For fluorescent lamps there is in general a trade-off between efficacy and colour rendering (Figure 1.20). The exceptional nature of the three-band (TL84) lamp in this respect is evident from Figure 1.20, the high efficacy rating being one of the major reasons for its adoption for general lighting in large stores. The unusually high efficacy of the high-pressure sodium lamp makes it economically valuable for street lighting.

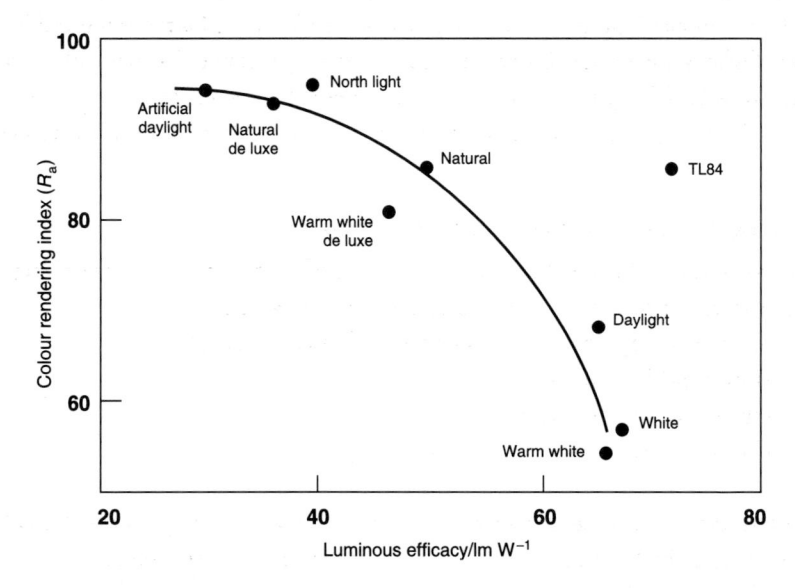

Figure 1.20 Relationship between colour-rendering index and luminous efficacy for some common fluorescent lamps

In general, lamps need to be selected carefully with a view to their end use. Those with good colour-rendering characteristics are chosen for applications where colour comparisons are being made. For critical colour-matching applications, for example in the colour-using industries, the artificial daylight lamp is used since it satisfies the requirements of BS950 : Part 1 : 1967 for colour-matching work, including having the necessary UV content for assessing fluorescent white materials.

1.7 COLOUR-MATCHING BOOTHS AND VISUAL COLOUR MATCHING

Visual assessment of coloured samples for purposes of colour control and specification requires careful specification of several factors, including:
(a) the nature and intensity of the light source
(b) the angle of illumination and viewing of the samples
(c) the colour of the surrounds and sample background
(d) the size and distance apart of the samples
(e) the state of adaptation of the observer
(f) the observer's colour vision characteristics
(g) an agreement on the colour-difference terms to be used.

The modern colour-matching booth takes into account the first three of these factors, whilst recommendations can be laid down for standardising (as far as is possible) the state of adaptation of the observer and for pre-testing assessors for their colour vision characteristics (Chapter 8). For the consistent use of colour-difference terms the observers must be appropriately trained and experienced in colour-difference assessments.

1.7.1 Light sources in colour-matching booths

Most colour-matching booths provide a selection of light sources available, with press-button accessibility. Typical sources that might be available include:
(a) an 'artificial daylight' source, often labelled (incorrectly) D_{65}
(b) a tungsten-filament source
(c) a three-band fluorescent source (TL84)
(d) a UV source (for enhancing fluorescent whites).

Artificial daylight sources can be filtered tungsten-filament lamps but in the UK are more likely to be 'artificial daylight' fluorescent tubes conforming to BS950 : Part 1 : 1967 (artificial daylight for the assessment of colour). Such lamps have a specified UV content that is useful for assessing fluorescent whitening agents on textiles, paper and plastics. No internationally acceptable daylight simulator conforming closely to the D_{65} specification has yet been developed [8]. Users of matching booths should be aware of the SPD of the lamps provided, as many different combinations of sources are marketed.

The illumination levels are usually specified and can vary typically from 700 lux to over 3000 lux, the latter value being preferred when matching dark colours. The illumination level should be checked routinely and the lamps replaced when the output drops by a specified value, or after sources have been run for a specified number of hours. With artificial daylight sources it is usually the UV content that declines most rapidly. It is useful to provide a check of the 'colour balance' characteristics, i.e. a test for the maintenance of the SPD within acceptable limits (which test should be in addition to the illumination level test).

The provision of other sources, such as the tungsten-filament and the TL84 lamps, allows for checks of colour constancy and metamerism of sample pairs, the latter being assessed by the extent of the change in the quality of colour match when the illuminant is changed (from daylight to tungsten-filament lamp, for example).

1.7.2 Matching booth design and sample viewing arrangements

In the typical matching booth the lamps are placed in a recess directly above the matching surface, preferably behind a frosted glass cover and/or filter glass which provides reasonably diffuse illumination (Figure 1.21). The matching surface and the interior of the matching booth should be painted a light to medium grey (about 30% reflectance), and the booth placed at a height such that the viewing angle for the observer of average stature is approximately 45°.

Specimens to be compared should be placed side by side on the same plane. The closer together they are, the more the dividing line between the specimen tends to disappear, so assisting the precision of colour-difference assessment. The position of the sample pairs being compared (sample and standard) should be interchanged to avoid any bias in viewing them in only one order. Samples with surface texture or gloss may need to be viewed at slightly different angles to eliminate surface texture effects or to give some assessment of the gloss, but this should not be a noticeable effect if the viewing illumination is truly diffuse. Some matching booths have a directional light source for assessing gloss, but the assessment of surface gloss requires a matching booth of different and specific design.

When sample and standard are of different sizes it can be helpful to prepare appropriate aperture masks, by cutting rectangular holes in a sample of a grey card. Such masks should always be used when comparing a large sample with a small chip from a colour atlas (when attempting to obtain a Munsell colour specification, for example).

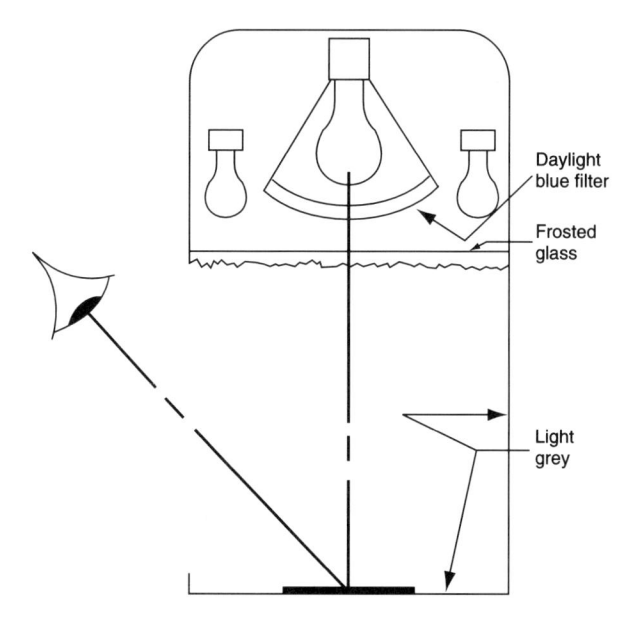

Figure 1.21 Artificial daylight booth

According to Hunter, the general rules for visual examination are as follows [2]:
(a) Place the specimens in immediate juxtaposition; as the dividing line separating objects becomes thinner, visual differences become easier to see.
(b) Keep the intensity of illumination high; only at light levels approaching those of an outdoor overcast sky does the eye make comparisons with its maximum precision.
(c) Have the background similar to the specimens (if anything, greyer) so that it offers no distracting contrast with the visual task.
(d) Have the illumination spectrally representative of that normally employed in critical commercial studies, usually actual or artificial daylight.

1.8 THE INTERACTIONS OF LIGHT WITH MATTER

Consider a beam of white light incident on the surface of a coloured paint film. As soon as the light meets the paint surface the beam undergoes refraction, and some of the light is reflected. The refracted beam entering the paint layer then undergoes absorption and scattering, and it is the combination of these two processes which gives rise to the underlying colour of the paint layer.

In order to have some appreciation of the optical factors which give the surface its overall appearance (including colour and gloss or texture) we need to outline the laws that affect the interactions of the light beam with the surface. In doing so we have to recognise that the white light beam, considered as a bundle of waves with wavelengths covering the range 400–700 nm, can also be considered as a wavebundle in which the waves have components which vibrate in planes mutually at right angles to one another along the line of transmission. If the wave vibrations are confined to one plane we describe the radiation as being plane polarised. Polarisation effects are important when we consider reflections from glossy surfaces and mirrors, as discussed below.

1.8.1 Refraction of light

Refraction into the interior of the film takes place according to Snell's law, which states that when light travelling through a medium of refractive index n_1 encounters and enters a medium of refractive index n_2 then the light beam is bent through an angle according to Eqn 1.11:

$$\frac{\sin i}{\sin r} = \frac{n_1}{n_2} \tag{1.11}$$

where i is the angle of incidence and r is the angle of refraction (Figure 1.22).

A typical paint resin has a refractive index similar to that of ordinary glass ($n = 1.5$) and so a beam of radiation incident on the surface at 45° will be bent towards the normal by 17° to a refraction angle of approximately 28°.

The refraction angle depends on the wavelength; the ability of glass to refract blue

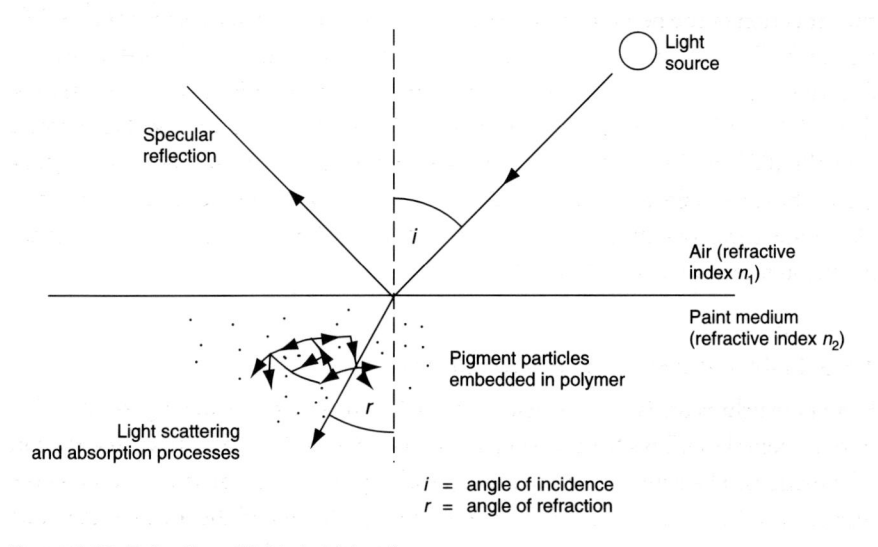

Figure 1.22 Refraction of light at paint surface

radiation more than red radiation is apparent in the production of a visible spectrum when white light is passed through a glass prism. Refractive indices are therefore normally measured using radiation of a standard wavelength – in practice, sodium D line radiation (yellow-orange light of wavelength 589.3 nm).

1.8.2 Surface reflection of light

A light beam incident normally (vertically) on a surface or any boundary between two phases of differing refractive index will suffer partial back-reflection according to Fresnel's law (Eqn 1.12):

$$\rho = \frac{(n-1)^2}{(n+1)^2} \quad (1.12)$$

where ρ is the reflection factor for unpolarised light and n is n_2/n_1.

If the incident light beam is white then the light reflected from the surface will also be white (white light needs to undergo selective absorption before it appears coloured). This small percentage of white light reflected from the surface affects the visually perceived colour, and instrumentally measured reflectance values should indicate whether the specular reflection is included (SPIN) or excluded (SPEX).

For the air ($n = 1$) and resin layer ($n = 1.5$) interface the total surface reflection at normal angles is about 4% ($\rho = 0.04$). At angles away from the normal, however, this surface or *specular* (mirror-like) reflection varies depending on the polarisation of the beam relative to the surface plane (Figure 1.23). The curves in this diagram show that the reflection of the perpendicularly polarised component becomes zero at a certain angle (the Brewster angle), and the reflected light at this angle is polarised in the one direction. The reflection of both polarised components becomes equal at normal incidence (0°), and again at the grazing angle (90°), at which point the surface reflects virtually 100% of the incident light (surfaces always look glossy at high or grazing angles). Thus light reflected from most surfaces is partially polarised. This is why Polaroid glasses are useful for cutting out glare from wet roads when driving, and for seeing under the surface of water on a bright day.

1.8.3 Light scattering and diffuse reflection

Part of the light beam is not specularly reflected at the surface but undergoes refraction into the paint layer. This light will encounter pigment particles, which will scatter it in all directions. The extent of this scattering will depend on the particle size and on the refractive index difference between the pigment particles and the medium in which they are dispersed, again according to Fresnel's laws. With white pigments like

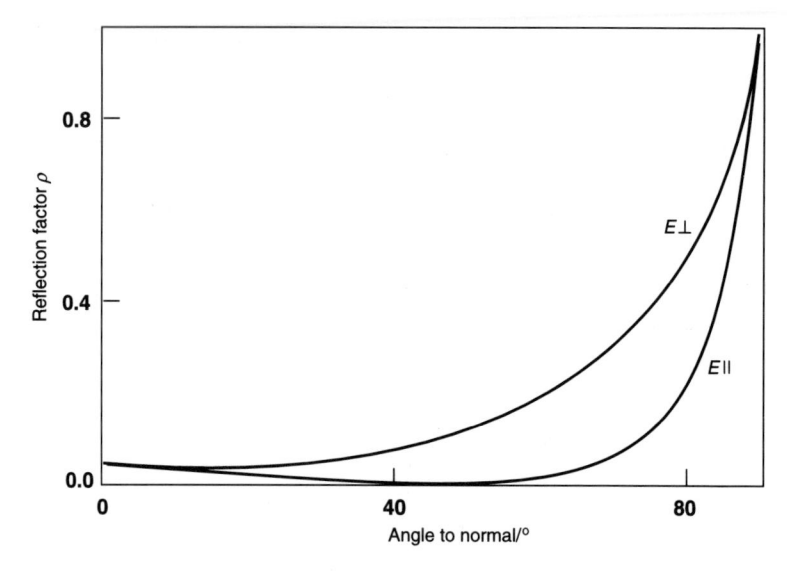

Figure 1.23 Polarised Fresnel reflection at air/glass interface (n =1.5); Brewster angle at 56° (only the perpendicularly polarised component reflected)

Light is diffusely reflected to give a matt finish

Some light is specularly reflected to give, e.g., an eggshell finish

Most of the light is specularly reflected to give a gloss finish

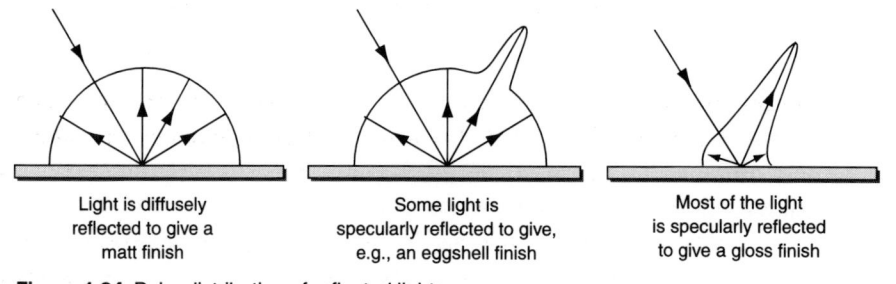

Figure 1.24 Polar distribution of reflected light

titanium dioxide ($n > 2$) the scattering will be independent of wavelength, and most of the incident light will be scattered in random directions. A high proportion will reappear at the surface and give rise to the diffuse reflected component; with a good matt white the diffuse reflection can approach 90% of the incident light. White textile fibres and fabrics produce a high proportion of diffusely reflected light, either because of the scattering at the numerous interfaces in the microfibrillar structure of natural fibres like cotton, wool and silk or, in the case of synthetic fibres, from the presence of titanium dioxide pigment in the fibres.

In practice there will be a balance between specular and diffuse reflected light which can be described by the polar reflection or goniophotometric reflection curve (Figure 1.24). To assess the gloss, determined by the proportion of the specular

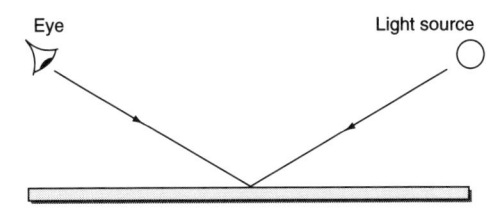

Figure 1.25 Viewing arrangement for the assessment of gloss

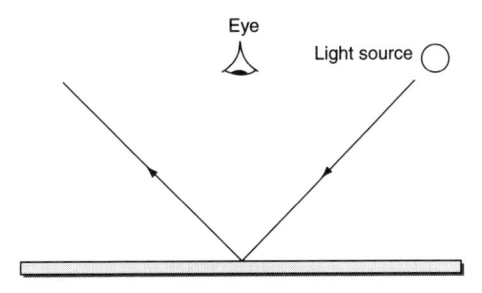

Figure 1.26 Viewing arrangement for the assessment of coloristic properties

component, the sample should be viewed at an angle equal to the incident, i.e. at 60° for the case illustrated in Figure 1.25. The extent of the diffuse component (and any colour contribution) is then assessed by viewing at right angles to the surface (that is, at an incident angle of 0°, Figure 1.26).

Thus the direction of reflected light plays a large part in the appearance of a surface coating. If it is concentrated within a narrow region at an angle equal to the angle of incidence the surface will appear glossy, i.e. it will have a high specular reflection. Conversely if it is reflected indiscriminately at all angles it will have a high diffuse reflection and will appear matt. Gloss is usually assessed instrumentally at high angles (60 or 85°) as the specular component is more important at such high angles (even a 'matt' paint surface shows some gloss at high or grazing angles).

1.8.4 Absorption of light (Beer–Lambert law)

If the paint layer contains coloured pigment particles (usually 0.1–1 µm in size) then the light beam travelling through the medium will be partly absorbed and partly scattered (Figure 1.1). Some particles are so small (< 0.2 µm) that they can be considered to be effectively in solution, and their light-absorption properties can be treated in the same way as those of dye solutions which absorb but do not scatter light.

34 LIGHT, LIGHT SOURCES AND LIGHT INTERACTIONS

The transmission of light of a single wavelength (monochromatic radiation) through dye solutions or dispersions of very small particles is governed by two laws:

(a) Lambert's or Bouguer's law (1760), which states that layers of equal thickness of the same substance transmit the same fraction of the incident monochromatic radiation, whatever its intensity

(b) Beer's law (1832), which states that the absorption of light is proportional to the number of absorbing entities (molecules) in its path; that is, for a given path length, the proportion of light transmitted decreases with the concentration of the light-absorbing solute.

Mathematically Lambert's law can be expressed in differential form (Eqn 1.13):

$$-\frac{dI}{dl} = kI \quad (1.13)$$

which on integrating suggests that the intensity decreases exponentially with thickness or path length l (Eqn 1.14):

$$I = I_o \exp(-kl) \quad (1.14)$$

In practice we use the logarithmic form (Eqn 1.15):

$$\ln(I/I_o) = -kl \quad (1.15)$$

or in decadic logarithms (Eqn 1.16):

$$2.303 \log(I/I_o) = -kl \quad (1.16)$$

where k is known as an absorption coefficient.

By defining the transmittance $T = I/I_o$ and inverting the quantity inside the logarithmic expression to remove the negative sign, we obtain the normal form of the Lambert's law expression (Eqn 1.17):

$$\log(1/T) = k'l \quad (1.17)$$

Suppose that we were to measure the absorption of green light by a purple dye solution contained in a spectrophotometer cell (cuvette) of total path length 1 cm, and that the solution absorbed 50% of the incident radiation over the first 0.2 cm; then the light transmittance through the cell would vary as shown in Table 1.5. Each 0.2 cm layer of solution decreases the light intensity by 50%, as required by the Lambert–Bouguer law. The quantity $\log(1/T)$, known as the *absorbance*, increases linearly with thickness or path length, whilst the intensity decreases exponentially (Figure 1.27).

Table 1.5 Lambert's law variation

Path length/cm	Transmittance ($T = I/I_o$)	$1/T$	$\log(1/T)$
0	1	1.0	0.0
0.2	0.5	2.0	0.301
0.4	0.25	4.0	0.602
0.6	0.125	8.0	0.903
0.8	0.0625	16.0	1.204
1.0	0.0312	32.0	1.505

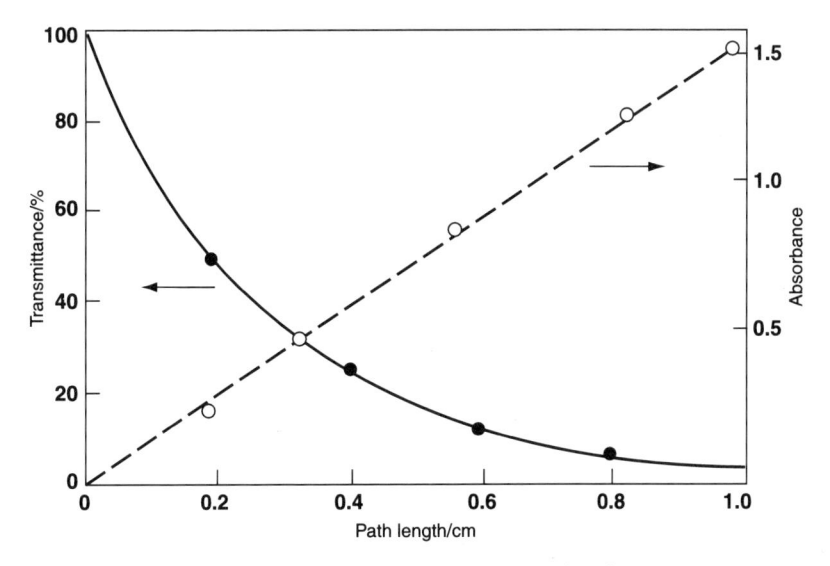

Figure 1.27 Change of transmittance and absorbance with path length

A plot of Beer's law behaviour at fixed path length would show a similar linear dependence of absorbance A with concentration. In fact the combined Beer–Lambert law is often written as Eqn 1.18:

$$A = \log(1/T) = \varepsilon c l \qquad (1.18)$$

where the proportionality constant ε is known as the *absorptivity*; if the concentration is in units of moles per unit volume (litre), it is known as the molar absorptivity.

The combined Beer–Lambert law can alternatively be written as Eqn 1.19:

$$I = I_o \times 10^{-A} \quad \text{or} \quad I = I_o \times 10^{-\varepsilon c l} \qquad (1.19)$$

Measurements of absorbance are widely used, through the application of the Beer–Lambert law, for determining the amount of coloured materials in solution, including measurements of the strengths of dyes (section 1.9). In practice deviations from these laws can arise from both instrumental and solution (chemical) factors, but discussion of these deviations is outside the scope of the present treatment [11].

1.8.5 Combined absorption and scattering (Kubelka–Munk analysis)

Most opaque coloured objects illuminated by white light produce diffusely reflected coloured radiation by the combined processes of light absorption and light scattering. Consider the simple case of a light beam passing vertically through a very thin pigmented layer of thickness dx in a paint film (Figure 1.28). We consider separately the downward (incident) and upward (reflected) components of the incident light beam, assuming that the absorption coefficient is represented by K and the scattering coefficient by S.

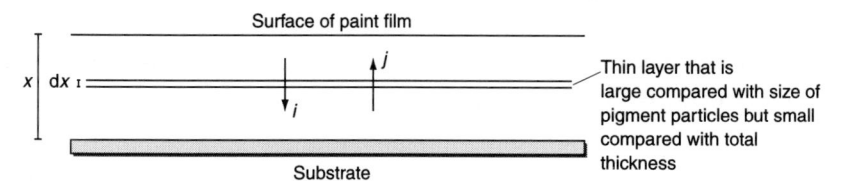

Figure 1.28 Kubelka–Munk analysis

The downward flux (intensity I) is:
- decreased by absorption = $-KI\,dx$
- decreased by scattering = $-SI\,dx$
- increased by backscatter = $+SJ\,dx$ (from the radiation proceeding upwards, of which J is the intensity), which is summarised by Eqn 1.20:

$$dI = -KI\,dx - SI\,dx + SJ\,dx \\ = -(K+S)I\,dx + SJ\,dx \quad (1.20)$$

At the same time, the upward flux (intensity J) is:
- decreased by absorption = $-kJ\,dx$
- decreased by scattering = $-SJ\,dx$
- increased by backscatter = $+SI\,dx$ (from the radiation proceeding downward), summarised by Eqn 1.21:

$$dJ = -KJ\,dx - SJ\,dx + SI\,dx \\ = -(K+S)J\,dx + SI\,dx \quad (1.21)$$

Solution of these differential equations depends on the boundary conditions applied, but in the absence of scattering ($S = 0$) leads to the Lambert–Bouguer law for the downward flux. For an isotropically absorbing and scattering layer of infinite thickness (or at least so thick that the background layer reflection is negligible), it leads to the widely used Kubelka–Munk expression (Eqn 1.22):

$$K/S = \frac{(1-R_\infty)^2}{2R_\infty} \qquad (1.22)$$

where $R_\infty = J_0/I_0$ is the reflection factor at the surface for a sample of infinite thickness.

A comprehensive survey of the solutions to these differential equations, together with examples of their use, is given by Judd and Wyszecki [12]. The particular solution given here has been expanded to illustrate the dependence of reflectance on dye and pigment concentration, as indicated in Chapters 5 and 6. The K, S and K/S values provide the colour technologist with functions which, in principle, are additive and linearly related to concentration of dyes and pigments in solid substrates. For example, for a dyed substrate where the scattering is attributed entirely to the textile substrate and therefore does not vary with dye concentration [D], we have a particularly simple form of concentration dependence (Eqn 1.23):

$$K/S = \frac{K_f}{S_f} + \frac{K_d}{S_f}[D] \qquad (1.23)$$

where K_f and K_d are the light absorption coefficients for the fibre and dye respectively, at the wavelength of measurement, and S_f is the scattering coefficient of the fibre at the same wavelength.

Although this relationship has certain limitations (for example, when dealing with highly exhausting acid dyes on wool and when taking measurements near the wavelength of maximum absorption), good linearity is observed (Figure 1.29; the raw data from which this plot is derived is shown in Figure 1.31).

Some of the major limitations to the Kubelka–Munk type of analysis are that it deals with diffuse monochromatic radiation and handles only two fluxes (diffuse light travelling upwards or downwards) through a homogeneous absorbing and scattering medium. The light loss through edges is thus neglected, as are the surface and the totally internally reflected components of the incident light beam. Other assumptions such as the uniform distribution of the dyes or pigments, and the lack of interactions between them, are also not realised. Such factors lead to a nonlinearity of the Kubelka–Munk function when measured over wide concentration ranges. Further discussion of the limitations of the Kubelka–Munk theory and methods of dealing with nonlinearity are to be found in Chapter 5, whilst Chapter 6 deals with typical

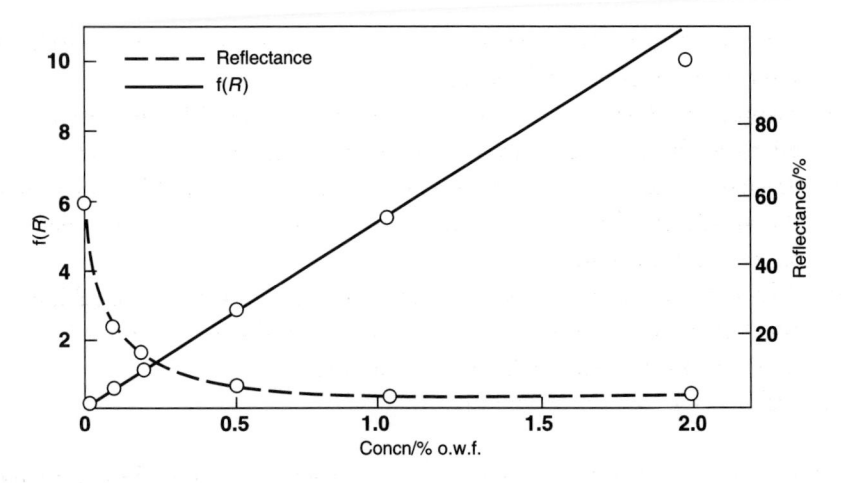

Figure 1.29 Reflectance and Kubelka–Munk plots for CI Acid Red 57 at 525 nm

deviations found with pigment systems and outlines the extension of the light absorption/scattering interactions to multiflux analysis.

1.9 LIGHT ABSORPTION, REFLECTION AND COLOUR

As we have seen, colour arises in dyed or pigmented material as a result of the selective absorption of radiation within the visible region of the electromagnetic spectrum. It has long been recognised that a relation exists between the hue of a coloured sample and the wavelength regions over which light absorption is strong, although the colour is actually determined (at least under normal conditions of illumination and viewing) mainly by the spectral energy distribution of the radiation reflected from the coloured opaque sample.

In this section we shall look at the relationship of the characteristics of spectral absorption and reflection curves to the general colour characteristics of dyed and pigmented samples. First, however, it is necessary to recognise that colours are described by colour technologists in terms of three visual characteristics:

(a) hue
(b) strength or depth
(c) brightness or dullness.

These terms are recognised within the colour industry as a whole, although variants are used in the different colour-using industries. The most recent edition of *Colour terms and definitions*, published by the Society of Dyers and Colourists, gives the following definitions [13]:

(a) *hue*: that attribute of colour whereby it is recognised as being predominantly red, green, blue, yellow, violet, brown, etc.
(b) *strength* (of a dye): the colour yield of a given quantity of dye in relation to an arbitrarily chosen standard; (of a dyeing or print) synonymous with depth
(c) *depth*: that colour quality an increase in which is associated with an increase in the quantity of colorant present, all other conditions (such as viewing conditions, for instance) remaining the same
(d) *dullness* (of a colour): that colour quality an increase of which is comparable to the effect of the addition of a small quantity of neutral grey colorant, whereby a match cannot be made by adjusting the strength
(e) *brightness*: the converse of dullness.

The relationships between these terms and spectral absorption and reflection characteristics are discussed in the following sections.

1.9.1 Hue and wavelength position of light absorption

Basing measurements on the Beer–Lambert law as discussed in section 1.8.4, Figure 1.30 shows the variation of the absorption coefficients in solution of three acid dyes of different hue, compared with the corresponding Kubelka–Munk coefficients (K_d/S_f) derived from reflectance measurements of wool material dyed with the same three dyes. The solution absorption curves are surprisingly similar to the absorption curves derived from the Kubelka–Munk analysis. (Such close similarity may not always be found.) The yellow dye absorbs over the near-UV and blue wavelength regions of the visible region with a maximum absorption λ_{max} near 400 nm, the red dye absorbs in the green region (λ_{max} about 510 nm) and the blue dye absorbs in the orange-red region with λ_{max} about 610 nm.

These absorption curves have half-band widths (range of wavelengths at half the maximum absorption intensity) of about 100 nm, the blue dye showing some absorption over the whole visible spectrum. The general relationship between observed hue and wavelength region in which the maximum value lies is illustrated in Table 1.6. The precise hue description will depend mainly on the wavelength position of the absorption band and partly on the band width and the overall shape of the absorption curve, but also on the observer's personal interpretation of the meaning of the hue terms used. Moreover, the wavelength ranges and associated hue descriptions given elsewhere may vary from that given in Table 1.6 (in section 1.2.1 we noted that the 'true' hues of blue, green and yellow have been observed to occur with monochromatic lights of wavelengths 436, 517 and 577 nm respectively and strictly these should lie near the middle of the appropriate radiation hue regions).

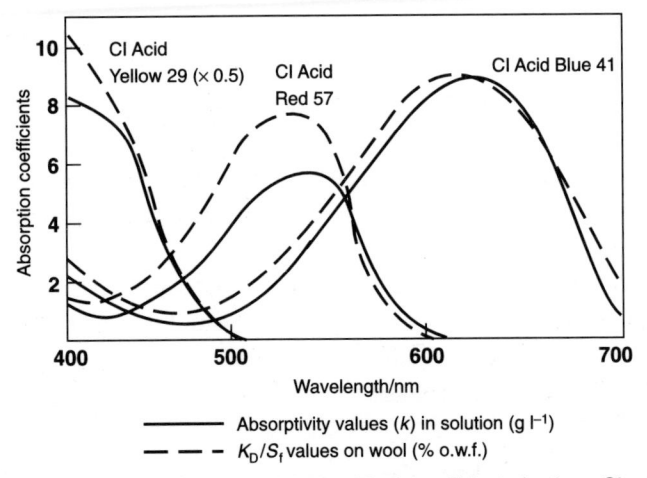

Figure 1.30 Absorption coefficients and Kubelka–Munk coefficients for three CI acid dyes

Table 1.6 Absorption band/hue relationships

Wavelength region of light absorption/nm	Principal hue observed
400–430 (violet-blue)	Yellow
430–470 (blue)	Orange
470–500 (blue-green)	Orange-red
500–540 (green)	Red-purple
540–570 (yellow-green)	Purple
570–590 (yellow)	Violet
590–610 (orange)	Blue
610–700 (orange-red)	Greenish-blue

The observed hues given in Table 1.6 are associated with the sensations arising from the mixture of wavelengths transmitted or reflected into the eye, i.e. from the range of wavelengths *not* absorbed. Thus the yellow hue observed to arise from strong absorption across the near-UV, violet and blue regions results from the mixing of the green and red wavelengths not absorbed and therefore transmitted by the yellow dye solution (or reflected by the yellow-dyed wool). Further discussion on the way in which the human eye/brain system reacts to stimulation from different wavelength distributions of visible radiation is to be found in Chapters 3 and 8.

1.9.2 Measurement of dye and pigment strength

Addition of a dye to an initially undyed or white substrate results in a decrease in

LIGHT ABSORPTION, REFLECTION AND COLOUR   41

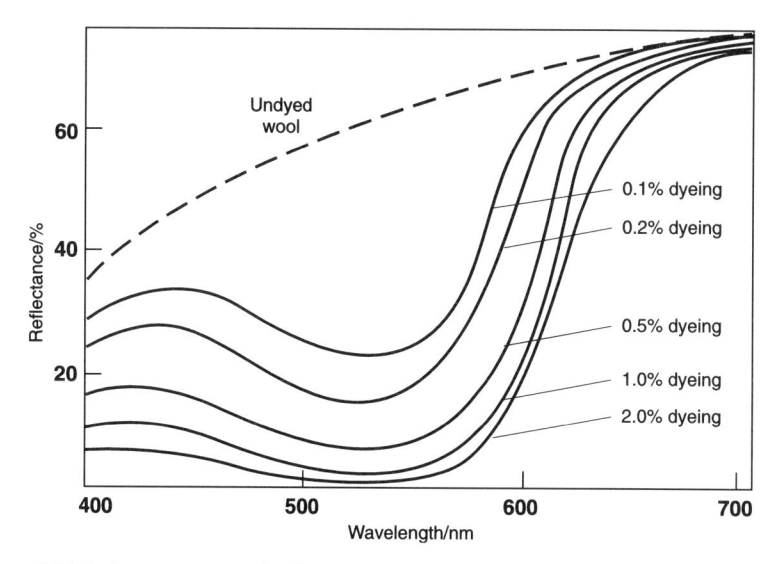

Figure 1.31 Reflectance curves for CI Acid Red 57 on wool

reflectance which is greatest in the region in which the dye absorbs light. For the typical red acid dye considered in Figure 1.29, the changes in reflectance with increasing concentration of dye are illustrated in Figure 1.31. These show that for this dye the absorption maximum (reflectance minimum) occurs in the region of 510 nm, with the decrease in reflectance falling off rapidly as the depth increases. The undyed wool has a distinctly yellowish cast, as suggested by the steeply sloping reflectance curve with minimum reflectance at 400 nm.

The reflectance data at λ_{max} from these curves were used to produce the linear Kubelka–Munk plot shown in Figure 1.29. The actual quoted concentration (expressed as a percentage mass of dye on fibre) is fixed arbitrarily by the dye manufacturer in terms of a so-called *standard depth* defined by samples of pigmented card produced by the Society of Dyers and Colourists, or other standardising body, and defined as international standard depths in DIN 53.235 and BS1006 : A01 : 1978. These standard depths are a series of arbitrarily chosen depths, each judged to be equal for all hues, which enable dyeing, fastness or other properties to be compared on a uniform basis.

Once the standard depth of a particular dye (or pigment) has been defined, relative strength measurements on subsequent batches are evaluated by preparing the dyed or pigmented sample under defined dyeing or application conditions, and testing strength variations such as 80, 90, 100, 110 and 120% for the sample being assessed. The resulting strength series of the sample colorant is then compared visually in a colour-matching booth with the standard sample (prepared simultaneously) accepted as being representative of the 100% strength of the colorant.

In many cases dye manufacturers have gradually replaced the dyeing test for strength determination with solution spectrophotometry based on a simple ratio determination of the absorbance values at λ_{max}, as expressed implicitly by the Beer–Lambert law. The full experimental details of the standard procedures for carrying out such solution strength tests have been published [14]. Such relative strength tests based on optical measurements on the dye solutions are, however, valid only if the chemical composition of the dye can be consistently reproduced, and hence if the dye can be produced with reproducible affinity or uptake characteristics on the appropriate fibres and a reproducible absorption spectrum in solution.

1.9.3 Dullness and brightness characteristics

The dullness/brightness variation is best illustrated in terms of the reflectance curves for two dyes of similar hue and strength which differ mainly in terms of their brightnesses. Thus we may compare CI Acid Red 57 (mentioned above) with a duller metal-complex red dye, Neolan Red BRE, both applied to wool (Figure 1.32). The spectral reflectance curve of the latter dye shows greater background absorption across the absorption spectrum, an effect which is akin to adding a uniformly absorbing grey dye to the brighter acid dye sample. The absorption peak in bright dyes tends to be sharper or more pronounced than in their duller counterparts.

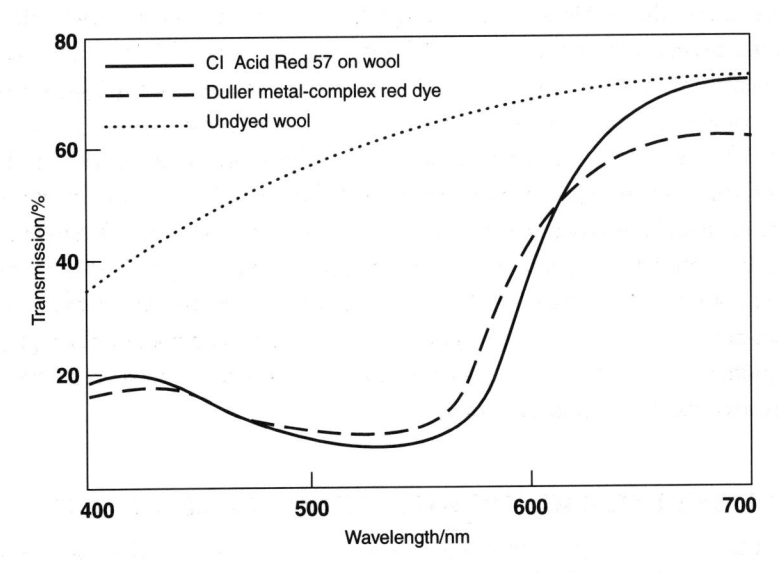

Figure 1.32 Comparing the characteristics of bright and dull red dyes on wool

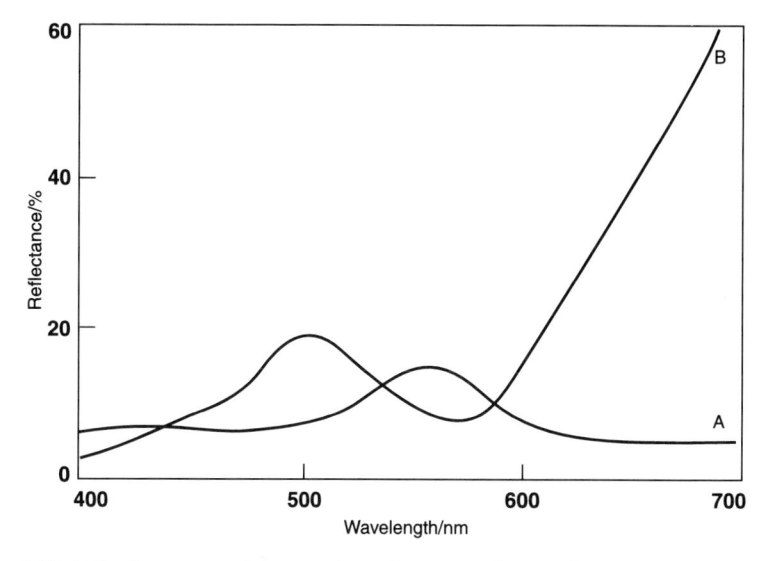

Figure 1.33 Reflection curves of green metameric patterns (see text)

1.9.4 Metamerism

It has been implied in the foregoing sections that different reflectance curves result in different observed colours and that is usually the case; the eye is said to be capable of distinguishing 7 million different surface colours, some of which show minimal difference in their reflectance curves. The eye, however, integrates the colour response across parts of the visible spectrum (Chapter 8), and so certain colours with differing reflectance curves can appear similar in a particular illuminant. The two green samples represented by the curves in Figure 1.33 are both green in artificial daylight and a fairly good colour match to most observers. When viewed under tungsten light sample B appears brown, however, while A remains green. The brown colour arises from the excess of orange and red light in the illumination (cf. Figure 1.9) and the high reflectance of the material for this part of the spectrum. This loss of colour match between a sample pair on changing illuminant is known as *metamerism*. The greens shown here represent an extreme example, but metamerism is a common problem in the colour-producing industries, particularly when the batch and the standard are prepared using different colorants, or if two different materials are being compared. Metamerism is discussed more fully in section 3.15.

1.10 LIGHT INTERACTION WITH ATOMS AND MOLECULES

The interactions of light with atoms and molecules are discussed in most texts on chemical spectroscopy [15]. Here we will only discuss the basic features of these interactions in the context of light sources, with a brief introduction to the origin of colour in organic pigments and dyes [16].

1.10.1 Atomic spectra

The simplest atomic spectrum is that obtained by examining the light emission from a low-pressure hydrogen arc by means of a visual spectrometer. A characteristic series of coloured lines (the Balmer series) is observed (Figure 1.34); these arise from the fall of electrons down the quantum levels of the hydrogen atom, each level being adequately characterised for the present discussion by the relevant principal quantum number (n). The electrons are initially promoted to the excited levels ($n > 1$) by the electrical discharge, and the Balmer series of lines is produced by spontaneous emission of light energy of very characteristic frequencies or wavelengths as the electrons return from the higher excited states to the second energy state ($n = 2$). Observations of the emissions outside the visible range show other line series in the UV (the Lyman series) and in the

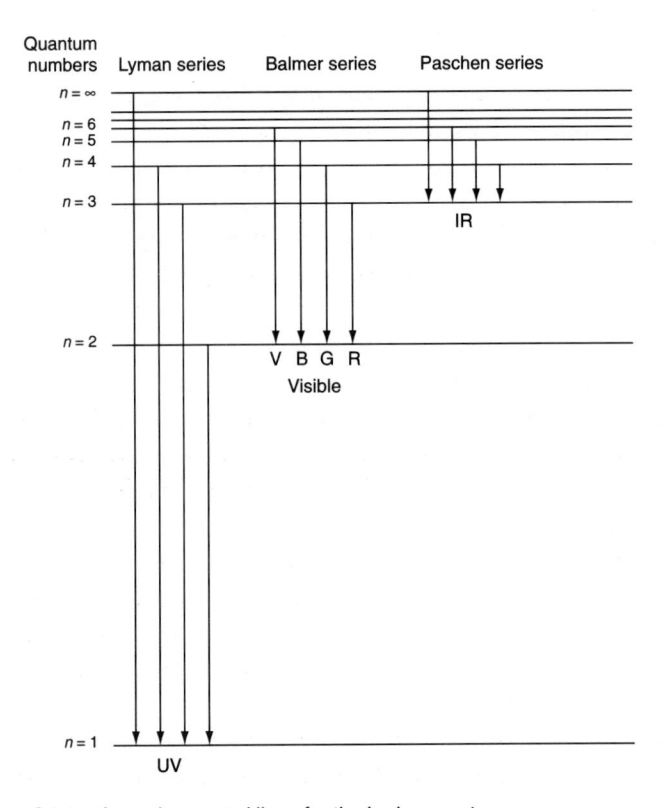

Figure 1.34 Origin of atomic spectral lines for the hydrogen atom

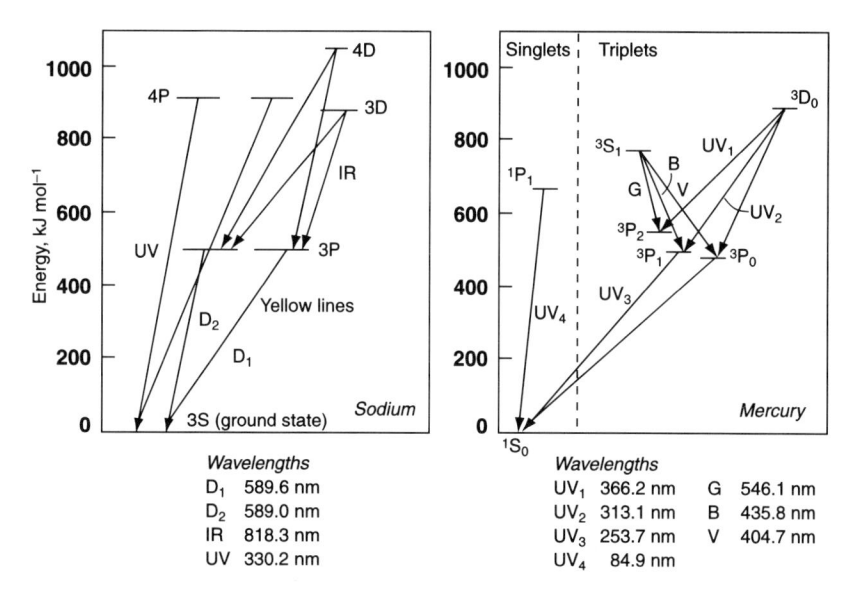

Figure 1.35 Spectral energy-level diagrams showing emission transitions for atomic sodium and mercury vapours (simplified term symbols used for sodium)

near-IR (Paschen series) and far-IR (Pfund series). The energy transitions giving rise to these spectral emissions are also illustrated in Figure 1.34.

To explain the atomic emission spectra of more complex atoms such as sodium and mercury it is necessary to label the states using symbols representative of three of the four quantum numbers which characterise the electrons in an atom. Thus the inclusion of the secondary quantum number l defines s, p, d and f electrons (l = 0, 1, 2 and 3 respectively) while the inclusion of the spin quantum number s (= ±1/2) gives the overall resultant spin indicated by the superscripts in the term symbols used to define the ground and excited states of the atom. These concepts are incorporated in the atomic energy level diagrams for sodium and mercury (Figure 1.35), in which the wavelengths of the characteristic lines in the emission spectra of these atoms are shown (cf. lamp SPDs in section 1.5). The ground state of the sodium atom (electronic configuration 2, 8, 1) arises from the electron in the outer 3s atomic orbital, whilst that of mercury (electronic configuration 2, 8, 18, 32, 18, 2) arises from the spin-paired electrons in the outer 6s atomic orbital. Excited states are formed by promoting one of the ground state electrons into one of the available higher-energy s, p, d atomic orbitals (cf. Figures 1.35 and 1.36). In these figures the energy levels are distinguished by term symbols (capital S, P, D etc.).

Atomic absorption spectroscopy results from the reverse transitions in atoms, in which the absorption of a quantum of radiation absorbed results in the promotion of the electron in the atom from the ground-state energy level to an upper energy level.

Thus atomic sodium shows strong absorption at 589.3 nm due to the reverse 3s to 3p transition (and at 330 nm due to 3s to 4p transition). Atomic absorption spectroscopy has become one of the major analytical tools for determining trace amounts of metals in solution. Atomic absorption is also responsible for the dark lines (the Fraunhofer lines) seen in the spectrum of the sun. The sodium atomic absorption line was the fourth in the dominant series of lines first observed by Fraunhofer and was labelled as line D; to this day the orange-yellow 589.3 nm line of sodium (actually a pair of lines at 589.0 and 589.6 nm due to electron spin differences) is known as the sodium D line.

1.10.2 Electronic transitions in the He–Ne laser

The principles involved in laser action were described in section 1.5.5, the important characteristic being the formation of a relatively long-lived excited state (the metastable state), which allows stimulated emission to be generated before spontaneous emission takes place.

In the He–Ne laser electrical excitation 'pumps' one of the 1s outer electrons in the helium atom to the higher-energy 1s 2s excited state, which then transfers the energy (by collision) to the approximately equi-energy metastable He (2p 5s) state from which the characteristic red 632.8 nm laser radiation is produced by the transition shown in Figure 1.36. Fast deactivation processes from the terminal 3p level of the laser transition ensures that sufficient helium atoms are restored to the ground state

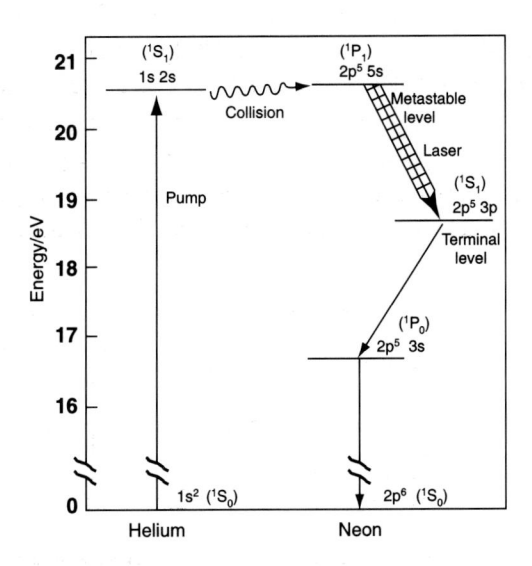

Figure 1.36 The energy-level scheme of the electrically excited helium–neon laser (term symbols shown in brackets)

ready to undergo excitation by energy transfer and hence maintain the laser beam to give a continuous output (possible with this particular type of laser).

Other transitions are possible with the neon atom, but the design of the laser cavity ensures that only the 632.8 nm radiation appears in the output beam (through one of the end mirrors, which is partially transmitting to the extent of about 1%).

1.10.3 UV absorption in simple molecules

In the hydrogen molecule, the simplest of all molecules, the two atoms are held together by a single bond formed by the two atomic electrons combining (with their spins paired) to form a ground-state σ molecular orbital. The promotion of one of the electrons into the nearest excited state can be induced by absorption of radiation very low down in the vacuum UV, at about 108 nm (Figure 1.37).

The absorption occurs so low in the UV because of the significant energy difference between the highest occupied molecular orbital (HOMO) and the lowest unoccupied molecular orbital (LUMO). To obtain absorption in a more accessible region of the UV (i.e. above 200 nm) it is necessary to use organic molecules with double bonds or containing heteroatoms such as oxygen, nitrogen or sulphur. For example, ethene with its single double bond absorbs at about 180 nm, but 1,3-butadiene and 1,3,5-hexatriene absorb at longer wavelengths with increasing strength of absorption as indicated by the values of their molar absorptivities, ε_{max} (Table 1.7).

Molecular orbitals for 1,3-butadiene involving the π-electron double bonds are shown in Figure 1.38, along with a simple energy diagram of the possible electronic transitions that produce absorption in the UV. The HOMO to LUMO (π → π*) transition leads to the longest-wavelength absorption band for butadiene quoted in Table 1.7. Extension of the conjugated (alternate single- and double-bonded) system to four double bonds leads to absorption just above 400 nm and a yellow colour; β-carotene, with eleven conjugated double bonds, is the major orange component in carrots and other vegetables, and one of the most important of the carotenoid plant pig-

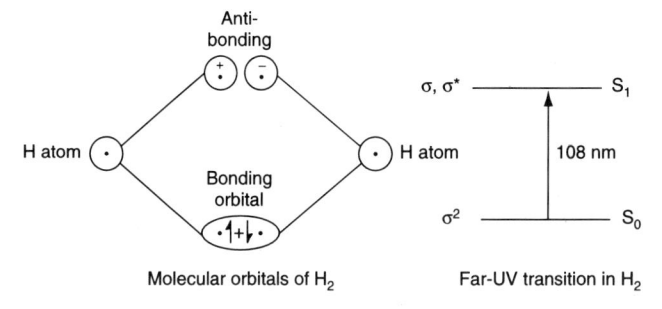

Figure 1.37 Electronic states and spectroscopic transition in hydrogen

48 LIGHT, LIGHT SOURCES AND LIGHT INTERACTIONS

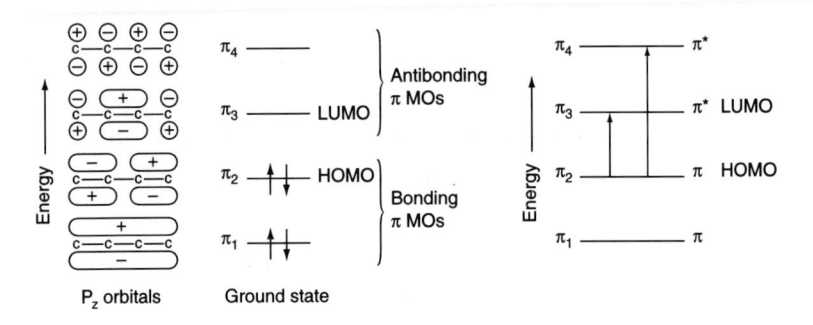

Figure 1.38 Molecular orbitals and electronic transitions in 1,3-butadiene

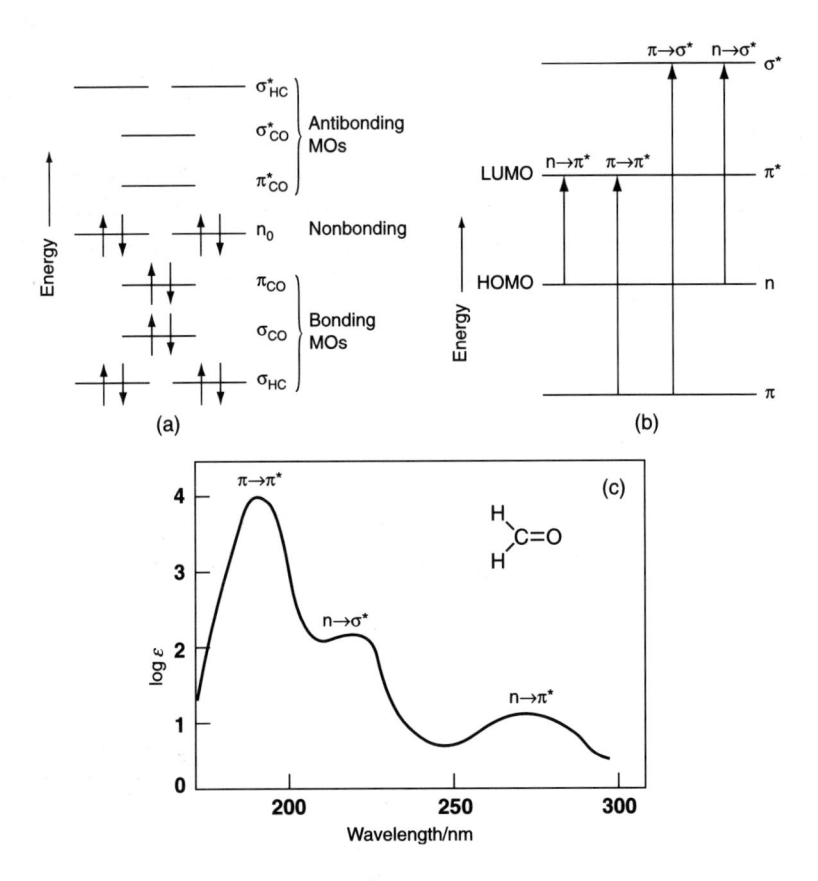

Figure 1.39 Ground-state molecular orbitals, transitions and schematic UV spectrum of methanal (formaldehyde)

Table 1.7 UV absorption of simple polyenes

Molecule	λ_{max}/nm	ε_{max}/l mol^{-1} cm^{-1}
Ethene	180	10 000
Butadiene	217	18 000
Hexatriene	258	25 000

ments. Lycopene, which gives tomatoes their red colour, is another example of a natural carotenoid colouring matter.

The UV absorption characteristics of methanal (formaldehyde) illustrates the important influence of the oxygen heteroatom. In the methanal molecule bonding and nonbonding electrons are both involved in the ground state (Figure 1.39), with the lowest-energy transition arising from a weak absorption band at about 270 nm due to excitation of one of the nonbonding electrons into an antibonding π^* orbital. The schematic UV absorption spectrum shows two bands of significantly different absorption intensities (note the logarithmic absorptivity scale), which is typical of simple carbonyl compounds. In the vapour phase or in solution in a nonpolar solvent, the 270 nm band of methanal shows sub-band fine structure which is due to the simultaneous changes in electronic and vibrational structure. Such vibrational structure in UV and visible absorption bands can be represented schematically in energy level diagrams (Figure 1.40).

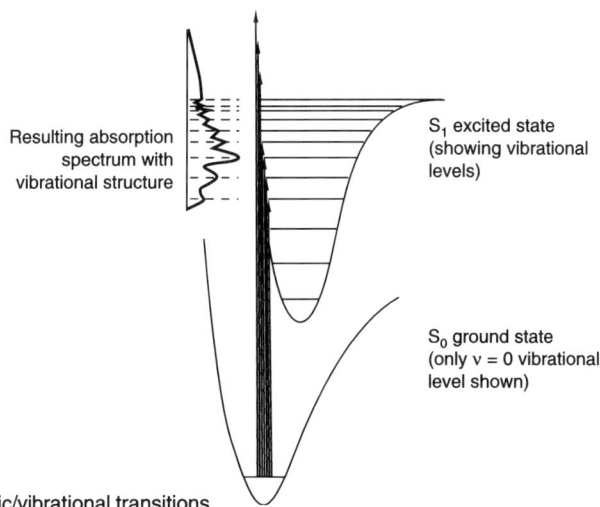

Figure 1.40 Electronic/vibrational transitions

1.10.4 Absorption spectra of aromatic compounds and simple colorants

The structure of benzene is often represented as three pairs of conjugated π-bonds in the hexagonal ring structure, with three of the six π-orbital states available being occupied in the ground state by spin-paired electrons. The UV spectrum of benzene shows an intense absorption band near 200 nm with a weaker but characteristic band near 255 nm. This 'benzenoid' absorption band shows highly characteristic vibrational structure, but this is absent in the phenol spectrum, in which the band appears at longer wavelengths (bathochromic shift) and is of greater intensity. This effect is enhanced if the phenol is made alkaline so that the OH group ionises to O^- (Figure 1.41).

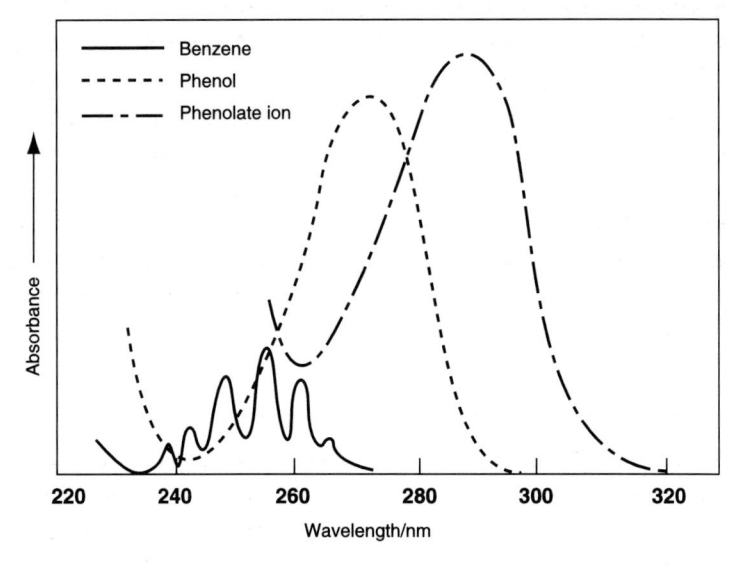

Figure 1.41 UV absorption spectra of benzene and its derivatives

The bathochromic shift and enhanced intensity has been attributed to the electron-donating capabilities of the OH and O^- groups. Such electron-donating effects of so-called auxochromic groups have long been used in the synthesis of dye and pigment molecules, which by definition have to absorb strongly in the visible region.

Azobenzene absorbs weakly just below 400 nm, but substitution with an electron-donating OH or NH_2 group in the *para* position gives a simple disperse dye. Incorporation of both electron-donating and electron-accepting groups (NO_2 groups, for instance) at opposite ends of the azobenzene structure gives an intense orange disperse dye. The principle of incorporating donor–acceptor groups in the synthesis of dyes and pigments is widely applied and is well illustrated in the anthraquinone series [16].

Interaction with radiation during photon absorption causes electron movement and creates excited states with significantly higher dipoles than those in the ground-state molecule. It is presumed that the donor–acceptor groups in dye and pigment molecules help to stabilise the formation of the polar excited states and hence result in strong light absorption.

1.11 PHOTOPHYSICS, PHOTOCHEMISTRY AND LIGHT FASTNESS

The energy content of visible radiation is in excess of 200 kJ mol^{-1}, depending on wavelength (section 1.2.2), which is sufficient in principle to break most chemical bonds. It is therefore remarkable that commercially important dyes and pigments have high light stability despite being designed to absorb light strongly. To appreciate why this is so requires understanding of the processes by which molecules dissipate the initially absorbed photon energy, in particular the fast photophysical and photochemical processes which the initially formed excited states undergo immediately after the absorption of light. It is possible in this section to give only the briefest introduction to this major field of study, but there are now many excellent textbooks on the subject of light-induced physics and chemistry [17]. Several textbooks and review articles have also been published on the relationships between chemical structure of dyes and pigments and their photochemistry and light fastness [18,19].

1.11.1 Excited states and energy deactivation processes

In section 1.10 the light-absorption process was explained in terms of electron excitation within the available molecular orbitals or molecular energy levels of the molecule concerned. The concept of the energy level diagram can be extended to illustrate some of the energy deactivation and reaction pathways open to an electronically excited molecule. Such a diagram is known as a Jablonski diagram (Figure 1.42).

An important consideration is the timescales over which the various photophysical processes take place, the light-absorption process itself occurring within a femtosecond. Most light-stable molecules return to the ground state within a few picoseconds (1 ps = 1 × 10^{-12} s) by efficient deactivation processes as discussed below. Some molecules, however, resist collisional deactivation processes and remain in the excited singlet state for up to a few nanoseconds, after which they either emit fluorescent radiation or undergo an electron spin change and cross over to the longer-lived metastable triplet state (an excited state having two electrons with parallel spin in different orbitals). The lifetime of the triplet state can range from microseconds to milliseconds or longer, and in certain systems can lead to delayed phosphorescent emission.

The energy changes that lead to fluorescence, usually on the long-wavelength side

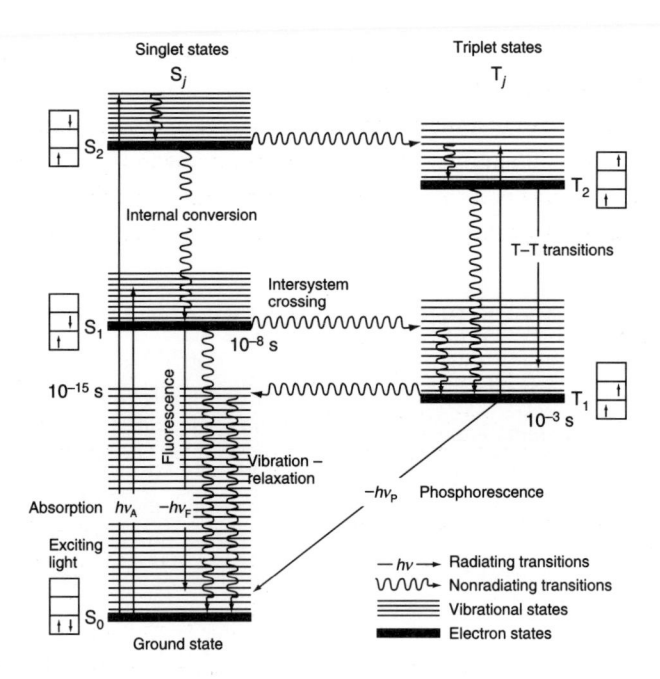

Figure 1.42 Jablonski diagram: electronic states and transitions

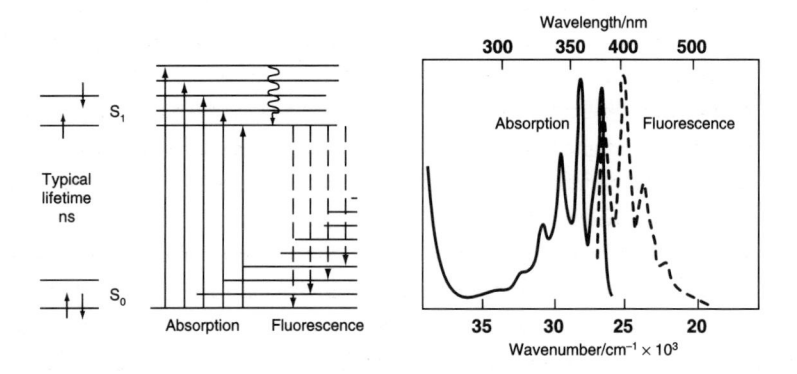

Figure 1.43 Absorption and fluorescence transitions in anthracene

of the lowest-energy ($S_0 \to S_1$) absorption band, are illustrated in some detail in Figure 1.43, which also shows the typical UV absorption and violet-blue fluorescence emission spectrum of anthracene in solution.

The light-absorption process leads to the formation of a series of vibrationally excited levels in the S_1 state, but the vibrational energy is quickly dissipated into thermal energy (by collision with the surrounding solvent or substrate molecules), and the reverse emission transition takes place from the zero vibrational level to a range of vibrational levels in the ground (S_0) state.

Since the emission transitions are of lower energies, the emission spectrum is shifted to lower frequencies or longer wavelengths. The mirror-image appearance of the anthracene absorption and fluorescence emission spectra arises from the similarity of the vibrational energy level spacing in the ground (S_0) and first excited (S_1) states.

1.11.2 Photoreactions from the excited states

As indicated above, it was formerly believed that light-stable molecules passed on their excitation energy by simple energy-transfer processes during collision with surrounding solvent or substrate molecules. Recent work with picosecond and femtosecond pulsed laser spectroscopy suggests, however, that molecules immediately change shape on light absorption and that chemical isomerisations or fast reversible hydrogen atom transfers (reduction/oxidation processes) are involved in processes leading to a return to the ground state with no overall change in the light-absorbing molecule [19]. Very occasionally side reactions in this process may lead to the destruction of the dye molecule, but with dyes of light fastness greater than 5 the chance of a molecule undergoing such destruction is no more than about one in a million (the photochemical process leading to dye destruction is said to have a *quantum yield* of about 10^{-6} or less).

Certain dye systems are susceptible to light degradation with much higher quantum yields. The reactions of CI Disperse Blue 14 (1,4-dimethylaminoanthraquinone) will be used to illustrate this aspect of dye photochemistry. Irradiation of the dye with UV/visible light in solution or in nylon film, in the absence of oxygen, leads to significant photoreduction (partially reversible when oxygen is admitted) accompanied by the spectral changes shown in Figure 1.44(a).

It is believed that the dye reacts via the triplet state in which the lone-pair electrons in the carbonyl chromophore become relatively less electronegative (the long-wavelength $n \rightarrow \pi^*$ transition moves electrons towards the aromatic ring system) and pick up hydrogen atoms from reducible solvent or substrate species, resulting in the formation of the fully reduced quinol ring structure.

If dyed polymer film is irradiated in the presence of oxygen the photoreaction observed is quite different (Figure 1.44(b)) and is initiated with the UV portion of the irradiating light. The principal reaction is a dealkylation of alkylamino groups, leading to a reduction in the electron-donating power of the auxochromic amino groups and hence a blue (or hypsochromic) shift in the absorption band. The light fastness of the dye on polyester substrate has been shown to be 1–2. The requirement of the photoreaction for UV irradiation suggests that it is initiated through one of the higher-energy singlet excited states, such as S_2 or S_3.

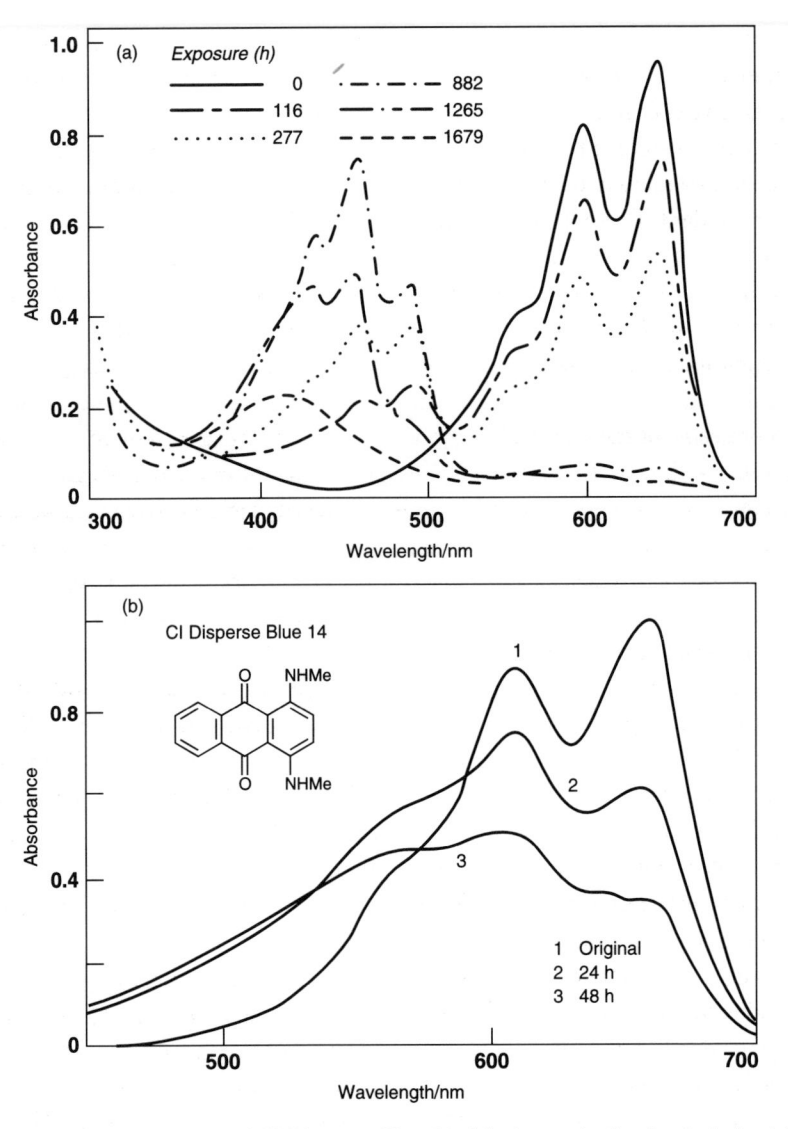

Figure 1.44 Photoreactions of CI Disperse Blue 14: (a) photoreduction in alcohol solution (oxygen-free); (b) photodealkylation in polyester film (oxygen present) [18]

1.11.3 Light-fastness measurements

Measurement of the light stability or light fastness of dyed and pigmented systems is a prerequisite in assessing the overall quality of coloured materials. The international specification for light-fastness testing (ISO 105 : B01 and B02 : 1988, BS EN20105 : 1993) details the exposure conditions for daylight behind glass (B01) and artificial light (xenon arc fading lamp test) (B02). In both methods the samples to be tested are exposed alongside a set of blue-dyed wool standards used to define light fastness on a scale from 1 (very low) to 8 (very high).

The dyes specified for the production of the blue wool standards were chosen so that each standard in daylight tests requires roughly twice the exposure of the next lower standard. This approximation does not hold for some of the standards, which show varying rates of fading. The low-fastness standards (1 and 2) are anomalous in that they are bleached by visible light whereas the others show their maximum sensitivity over the UV region. A different set of blue wool standards is produced in America and the light-fastness values derived using that series are prefixed with the letter L.

Since some of the specified dyes are no longer being manufactured (and anyway batch-to-batch reproducibility has proved a problem), tests are currently under way using a set of blue pigmented samples printed on card as replacements for the blue wool standards. The first set of trial pigment standards is based on varying the ratio of two pigments of low and high fastness along with titanium dioxide in a printing ink formulation to cover the 3–7 fastness range (the range into which most dyed textiles fall). Recent developments have been summarised in an interim report [20].

The specification for the xenon arc used for fading tests under standard B02 indicates that the lamp should have a correlated colour temperature of 5500 to 6500 K and that it should contain a light filter transmitting at least 90% in the visible region above 380 nm and falling to zero between 310 and 320 nm. In this way the UV radiation is steadily reduced over the near-UV region. IR heat filters are also used to minimise the heating effect of the IR region (cf. Figure 1.10).

Existing light-fastness lamps are either water-cooled or air-cooled, and the humidity and temperature conditions have to be adjusted to values laid down in the appropriate standard. This is specified in terms of the maximum temperature recorded in a black panel incorporated in the sample position racks, with humidity control being determined by the fastness rating of a sample of cotton dyed with an azoic red combination whose humidity sensitivity in light-fastness testing has been calibrated.

The two basic light-fastness standards are supplemented by standards B03 to B08, which cover:

- B03 colour fastness to weathering: outdoor exposure
- B04 colour fastness to weathering: xenon arc
- B05 detection and assessment of photochromism
- B06 colour fastness to artificial light at high temperatures: xenon arc fading lamp test
- B07 colour fastness to light of wet textiles
- B08 quality control of light-fastness reference materials.

The related standard BS1006 includes a UK-only test, specifying the use of mercury-vapour fading lamps.

The test B05 for photochromism is a test for change of colour (usually at least partially reversible) caused by irradiation. Photochromism is usually dependent on some reversible change in the chemical structure of the colorant induced through the first excited state.

Light-fastness testing is discussed further in section 4.5.

REFERENCES

1. F W Billmeyer and M Saltzman, *Principles of color technology*, 2nd Edn (New York: John Wiley, 1981).
2. R S Hunter and R W Harold, *The measurement of appearance*, 2nd Edn (New York: John Wiley, 1987).
3. K Nassau, *The physics and chemistry of color* (New York: John Wiley, 1983).
4. W D Wright, *The rays are not coloured* (London: Adam Hilger, 1967).
5. S T Henderson, *Daylight and its spectrum*, 2nd Edn (Bristol: Adam Hilger, 1977).
6. CIE Publication No. 51 (TC 1-3) (Paris: CIE, 1981).
7. R W G Hunt, *Col. Res. Appl.*, **17** (1992) 293.
8. C S McCamy, *Col. Res. Appl.*, **19** (1994) 437.
9. W A Thornton, *J. Illum. Eng. Soc.*, **3** (1973) 61.
10. R W G Hunt, *Measuring colour*, 2nd Edn (Chichester, UK: Ellis Horwood, 1991).
11. E I Stearns, *The practice of absorption spectrophotometry* (New York: Wiley Interscience, 1969).
12. D B Judd and G Wyszecki, *Color in business, science and industry* (New York: John Wiley, 1975).
13. *Colour terms and definitions* (Bradford: SDC, 1988).
14. R Brossman *et al.*, *J.S.D.C.*, **103** (1987) 38, 100, 138.
15. C N Banwell and E M McCash, *Fundamentals of molecular spectroscopy* (London: McGraw Hill, 1995).
16. J Griffiths, *Rev. Prog. Col.*, **14** (1984) 21.
17. P Suppan, *Chemistry and light* (Cambridge: Royal Society of Chemistry, 1994).
18. N S Allen and J F McKellar, *Photochemistry of dyed and pigmented polymers* (London: Allied Science, 1980).
19. H E A Kramer, *Chimia*, **40** (1986) 160.
20. J Guthrie, N Tayan and L Wilson, *J.S.D.C.*, **111** (1995) 220.

FURTHER READING

M Longair in *Colour: art and science*, Ed. T Lamb and J Bourriau (Cambridge: CUP, 1995).

CHAPTER 2

The measurement of colour

David R Battle

2.1 INTRODUCTION

Colour is a physical experience in the mind of an observer and as such is impossible to measure. We can, however, measure some of the physical parameters that create the experience. The colour perceived by an observer results from the interaction of a light source, a sample and the observer. The light source and the observer are covered elsewhere in the book. This chapter will concentrate on measurements of the sample.

An observer perceives colour by detecting light reflected from an object that is imaged on to the retina at the back of the eye. The retina is covered with light-sensitive rod and cone cells that convert light into electrical signals which are passed on to the brain. Rod cells are responsible for low-intensity night vision, while three types of cone receptor cell have maximum sensitivity to light at three different parts of the visible spectrum. (The visible spectrum extends, roughly speaking, from 400 to 700 nm.)

The observer then perceives colour as a ratio of intensity from the R, G and B cones. Colour perception starts with the spectral characteristics of the light source, which are then modified by the reflectance of the object (Figure 2.1). The resulting light then stimulates the eye to generate the R, G and B signals according to the standard observer functions.

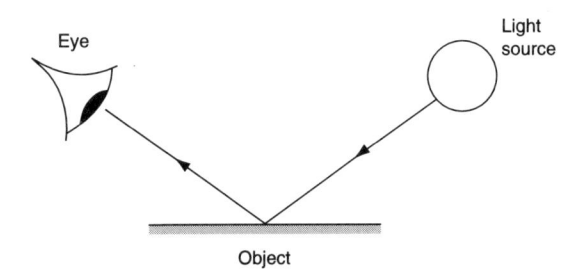

Figure 2.1 Principle of viewing colour

58 THE MEASUREMENT OF COLOUR

Thus the perceived colour depends on the spectral power distribution (SPD) of the light source, the reflectance of the object and the spectral response of the eye. The CIE defines colour mathematically using the XYZ tristimulus values (Eqn 2.1):

$$X = \sum_{\lambda=400}^{700} \bar{x}_\lambda R_\lambda S_\lambda \quad Y = \sum_{\lambda=400}^{700} \bar{y}_\lambda R_\lambda S_\lambda \quad Z = \sum_{\lambda=400}^{700} \bar{z}_\lambda R_\lambda S_\lambda \quad (2.1)$$

where $\bar{x}, \bar{y}$ and $\bar{z}$ = colour-matching functions (i.e. the spectral response of the eye)
R = reflectance of the object
S = SPD of the light source.

This definition of colour is examined in more detail in Chapter 3.

2.2 THE TRISTIMULUS COLORIMETER

A tristimulus colorimeter is the simplest form of instrument for the measurement of colour (Figure 2.2). Like the eye it has red, green and blue photodetectors, and measures tristimulus values. A light source, usually a quartz halogen bulb, illuminates the sample at 45° to the normal. Light reflected from the sample along the normal is then collected and passed to a detector, which consists of three filters each in front of its own light-sensitive diode. The response of the filter/diode combinations is tailored to match the differential spectral response of the eye.

A colorimeter is a very cost-effective colour-measuring instrument and is useful for measuring colour difference in quality-control situations, but it has some serious limitations. Its absolute accuracy is restricted, since the practical realisations of the light source and filter/diode detector response are usually only approximate fits to the CIE

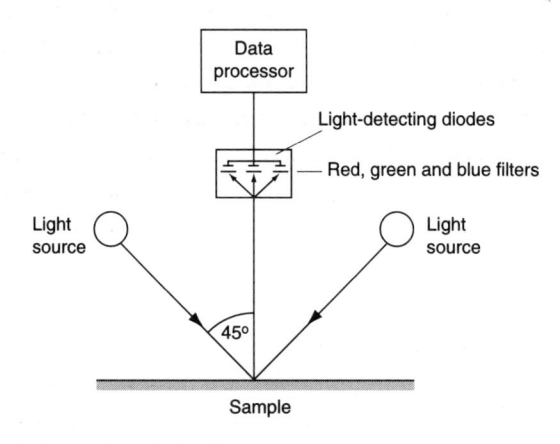

Figure 2.2 Basic features of a colorimeter

definitions. This restriction on absolute accuracy is unimportant when measuring colour difference as here the instrument is being used to compare similar shades in differential mode, and it is the relative difference that is important. The colorimetric values measured are valid for just one illuminant, the instrument light source, but it is often desirable to know the tristimulus values or colour difference for several illuminants. Thus a colorimeter cannot give any indication of metamerism (section 1.9.4). Metamerism occurs when a pair of samples are a close, though not exact spectral match and have tristimulus values that are similar under one illuminant but very different under another (Figure 2.3). The phenomenon is caused by the different SPDs of the two light sources either magnifying or diminishing the reflectance curve differences.

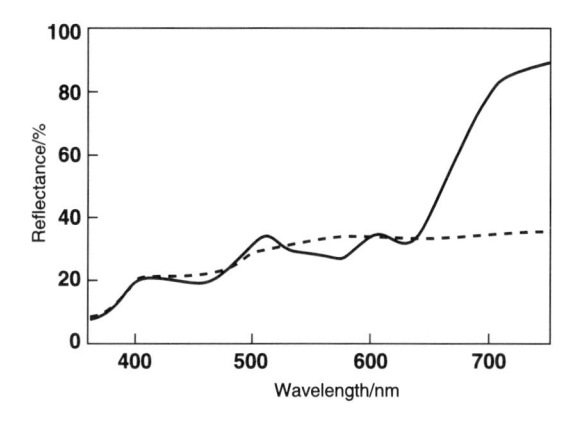

Figure 2.3 Reflectance curves from a pair of metameric samples

Where the requirement is for a simple quality-control instrument to determine if a batch is within a specified colour tolerance of a standard test sample, then a colorimeter can be a satisfactory solution. For match-prediction work, measurement of colour under different illuminants, detection of metamerism, evaluating colour difference against numeric standards and absolute colour measurement, a better and more flexible way of measuring colour is to use a spectrophotometer. This is an instrument that makes accurate measurements at many points across the visible spectrum, thus allowing tristimulus colour coordinates to be calculated under any illuminant of interest by using Eqn 2.1.

2.3 THE SPECTROPHOTOMETER

A colour-measurement spectrophotometer measures the ratio of reflected to incident light (the reflectance) from a sample at many points across the visible spectrum (Eqn 2.2):

$$\text{Reflectance} = \frac{\text{Reflected light}}{\text{Incident light}} \qquad (2.2)$$

Reflectance values are generally expressed as percentages, with the perfect reflecting diffuser having a reflectance of 100%. When used in colorimetric calculations, however, as in Eqn 2.1, they are given as decimal fractions, with the perfect reflecting diffuser having a reflectance of unity.

A colour-measurement spectrophotometer again follows our simple theory of colour vision. A light source is used to illuminate the sample using a specific illumination and viewing geometry. Reflected or transmitted light is then passed on to the spectral analyser, where the light is split into its spectral components. This allows the light detector and control electronics to make measurements at many points across the visible spectrum. It is the spectral analyser that gives the spectrophotometer its advantage over a simple colorimeter.

2.3.1 Principles of operation of the Spectraflash 500

The Datacolor SF500 is a classical dual-beam reference spectrophotometer (Figure 2.4). It was designed in the late 1980s and can be found in many of the standardising laboratories throughout the world.

The SF500 uses a pulsed xenon lamp filtered to approximate to illuminant D_{65} to illuminate a 15 cm diameter sphere coated with barium sulphate. The UV content of this light source is controlled by a motorised filter wheel, which is useful when making measurements on fluorescent materials. The UV part of the illuminating SPD can be cut off below 400, 420 or 460 nm, thus allowing measurements where the fluorescent

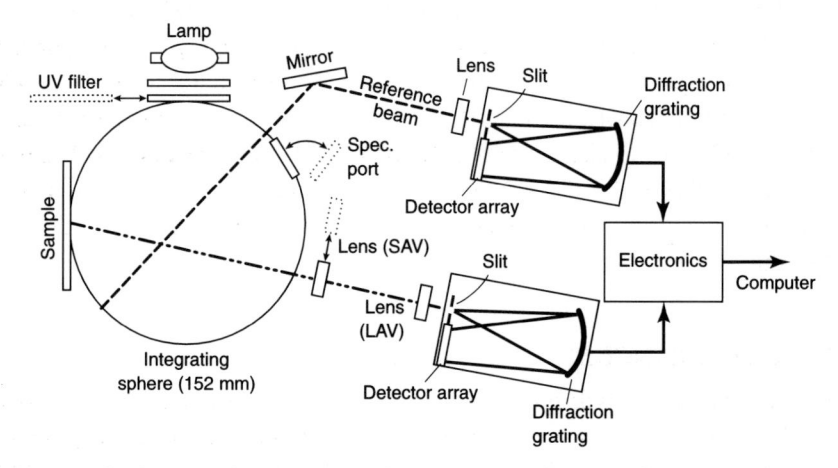

Figure 2.4 Basic features of a dual-beam spectrophotometer (Datacolor Spectraflash 500)

component has been quenched; this enables the effect of fluorescent brightening agents to be measured. It is also possible to adjust and calibrate the UV content to allow for ageing of the light source and sphere; section 2.5 gives further details.

The sample to be measured is placed at the measurement port of the sphere where it is diffusely illuminated (i.e. from all angles). Light reflected from the sample at 8° from the normal is collected and passes through the sample port of the sphere to the spectral analyser. Collection at 8° allows a specular port or gloss trap to be placed at 8° from the other side of the normal to the sample beam. The specular port is a portion of the sphere, a plug coated with barium sulphate, which can be removed by a motorised mechanism to reveal a light trap. The specular component of a measurement results from mirror-like reflection from a shiny surface. Thus if no light is incident on a sample at 8° (due to the light or gloss trap being in place) then no light can be reflected at 8°, and the reflected beam must then consist of diffusely reflected light; section 2.6.4 gives a fuller explanation. Thus the instrument can switch on or off the gloss component of the measurement. The use of a large gloss port ensures good specular exclusion from materials that are glossy but not optically flat.

A second beam (the reference beam), consisting of light reflected from the sphere wall, exits the sphere via the reference port. A steering mirror is then used to direct the beam to a second spectral analyser, which is identical to the sample analyser. The use of a reference beam allows direct measurement of reflectance, as described in Eqn 2.1, between the ratio of reflected and incident light. Dual-beam measurements are preferred due to their inherent stability, as the measurement is of a ratio rather than an absolute value. Errors due to drift of the measurement electronics or variation in the light source are thus effectively eliminated.

Modern spectral analysers use concave holographic diffraction gratings to split the light into its spectral components, which are then focused on to a 40-element diode array. The diode spacing is nonlinear and is set to match the spectral output of the grating. A complex process involving the use of three laser beams (red, green and blue) is used to align the analyser. This is the key to the excellent inter-instrumental agreement provided by this instrument of only 0.15 CIELAB units maximum colour difference on the BCRA tiles. A solid cast aluminium optical bed and the use of thermally compensated materials in the analyser ensure the long-term and thermal stability of the instrument [1].

2.4 REFLECTANCE MEASUREMENT

The visible spectrum is defined by the CIE for the colour-matching functions of the standard colorimetric observer as extending over the range 360 to 830 nm, with data recorded at 1 nm intervals. The data are also published at 5 nm intervals from 380 to

780 nm. The eye is not very sensitive at either end of the spectrum, however, and in practice it is difficult (and rarely necessary) to measure such a wide spectral range at such a small bandwidth. For many years in industrial colorimetry it was considered sufficient to characterise reflectance by measuring from 400 to 700 nm at 20 nm intervals (i.e. 16 data points or channels). Modern instruments have expanded this measuring from about 360 to 750 nm at 10 nm intervals (i.e. 40 data points). It is usual for the channel bandwidth to be the same as, or similar to, the measurement interval.

Instruments with a channel separation of greater than 5 nm are called abridged spectrophotometers due to the abridgement of data caused by the use of summation intervals (channel spacing) greater that 5 nm. Most commercial colour-measurement spectrophotometers fall into this abridged class; commercial instruments that measure to the full CIE specification are unusual nowadays.

Reflectance data usually follow a smooth curve with few hills and valleys. In contrast, transmission curves can be very complex with many sharp peaks and troughs. Thus many more measurement points are required to define a transmission curve adequately, particularly when it is to be used for chemical analysis. For colorimetric purposes, bearing in mind the eye uses just three broad-band filters, 16 or 40 data points are sufficient even though some of the detail within a transmission curve will be lost.

Evaluation of the data obtained from an abridged spectrophotometer requires the use of XYZ weighting function tables that have been calculated for the appropriate channel bandwidth. The CIE only publishes 1 and 5 nm tables [2], whilst the ASTM publishes both 10 and 20 nm tables [3]. It is important to use the correct weighting function tables for the instrument concerned. Care is needed when comparing colorimetric data from instruments from different manufacturers, since certain published sets of weighting functions do not take bandwidth into account. For example, if a reflectance curve is compared with itself when converted from 10 to 20 nm bandwidth, the curves can differ by up to 2 CIELAB units entirely because of weighting function table bandwidth errors. Many of the published tables of XYZ weighting functions are in the process of being revised and republished, together with details for calculating them at any desired channel spacing or bandwidth. The use of these new tables will remove a major source of error when comparing data between different colour systems.

When calculating XYZ tristimulus values it is likely that the tables will cover a wider spectral range than does the measuring instrument. If so, the unused weighting function values at either end of the spectrum should be added together and lumped in with the table value at the appropriate end of the spectrum.

All spectrophotometric measurements are made relative to a reference standard: this is the perfect reflecting diffuser, which is defined as the ideal isotropic diffuser having a reflectance of unity at all points across the spectrum. It replaced magnesium oxide as the reference standard in 1969. In practice the perfect reflecting diffuser cannot be physically realised; it is a concept used to design experimental equipment.

The national standardising laboratories in various countries can calibrate secondary standards against the perfect reflecting diffuser. Secondary standards are usually white ceramic tiles or opal glass. Barium sulphate is sometimes used, as it has the advantage of being diffuse with no specular component, but its surface is easily contaminated and cannot be cleaned. Ceramic tiles and opal glass are preferred as they are more robust and are easy to clean, but they can suffer from translucency (see section 2.6.3).

2.5 SPECTROPHOTOMETER LIGHT SOURCES

In theory the type of light source used in a spectrophotometer should not matter, as we are measuring the ratio of incident to reflected light. Certain materials are fluorescent, however – they absorb energy at one part of the spectrum and re-emit at another. Thus the SPD of the light source can have a dramatic effect on the measured colour.

It is common practice in the paper, textile and plastics industries to use fluorescent whitening agents, to make samples appear whiter. These chemicals absorb UV energy at around 350 nm and re-emit energy in the blue part of the visible spectrum at around 450 nm to counter the natural yellowing of the untreated substrate. There is also increasing use of coloured fluorescent dyes and pigments, such as 'day-glo' colours, which both absorb and emit energy in the visible part of the spectrum. Thus it is important to know the spectral characteristics of the light source in a spectrophotometer when measuring fluorescent samples.

The two light sources most widely used today are quartz halogen and pulsed xenon. Quartz halogen lamps emit very little energy in the UV and so will not excite a fluorescent whitening agent to the same extent as daylight; thus the measured reflectance curve and calculated tristimulus values will be correct only for illuminant A, whereas illuminant D_{65} is often of more interest. Unfortunately quartz halogen lamps produce too much energy in the IR and can thus heat up the sample. This can lead to errors in measurements on thermochromic samples; for instance, the red, orange and yellow BCRA ceramic tile standards change by approximately 0.1 CIELAB unit for a 1 degC change in temperature. A quartz halogen lamp has the advantage of requiring a fairly simple power supply, but although replacement bulbs are quite cheap they often last only a few months. Some of these problems have been overcome in recent years by pulsing the lamp to increase its life and reduce power consumption in portable equipment, a measure that will also reduce sample heating.

Pulsed xenon light sources have become popular for use in spectrophotometers as their physical characteristics are almost ideal. They have been used for many years in photographic flash lamps. The SPD of pulsed xenon is easily filtered to approximate to illuminant D_{65} from the UV to the IR end of the spectrum. Thus they can be used to measure fluorescent colours; moreover, as the light source is pulsed the sample is not heated. Typical pulse durations are of the order of 1 ms. They provide a very intense illumination that is useful for measuring dark colours. As can be seen from the SPD (Figure 2.5), they have several high-intensity lines that can cause problems if the sample and reference analysers do not have identical spectral alignment. Because there are small fluctuations in flash-to-flash intensity and SPD, they have to be used in dual-beam instruments.

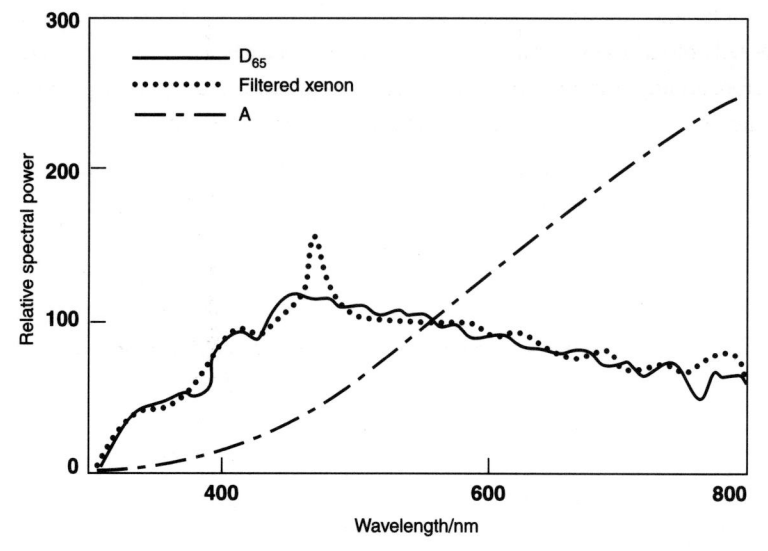

Figure 2.5 Spectral power distributions of illuminants A and D_{65}, and a pulsed xenon source

Care has to be taken in the design of a pulsed xenon light source to limit the peak energy in order to avoid measurement errors caused by triplet absorption. This arises where the sample is illuminated with much higher light intensity (many thousands times greater) than that of natural daylight. This causes electrons to be excited to a triplet rather than a singlet state, which modifies both the absorption and emission curves and hence the measured reflectance of the sample. Triplet absorption has been observed in both fluorescent whitening agents and pale shades of some dyes (see also section 4.6.9) [4].

Where accurate and repeatable measurements of fluorescent samples are required some form of UV adjuster or calibrator is required to compensate for yellowing of the sphere coating and variation in SPD between bulbs. The UV calibrator usually takes the form of a UV cutoff filter that is moved across the bulb to throttle the UV, rather as a valve adjusts the flow of water. The UV is adjusted to the correct level using a stable fluorescent plastic standard of known spectral characteristics.

The use of a UV calibrator was first recommended by Ganz and is essential for the correct use of the Ganz–Griesser whiteness method [4]. Several modern instruments are now supplied with UV calibrators; some are manual devices, while others are motorised and computer-controlled to automate the calibration procedure.

2.6 INSTRUMENT GEOMETRY

The angles and method by which a sample is illuminated and viewed can affect the observed colour dramatically. The geometry of a colour-measurement instrument is therefore an important factor in its design. The CIE has recommended four different instrument geometries for the measurement of colour (Figure 2.6).

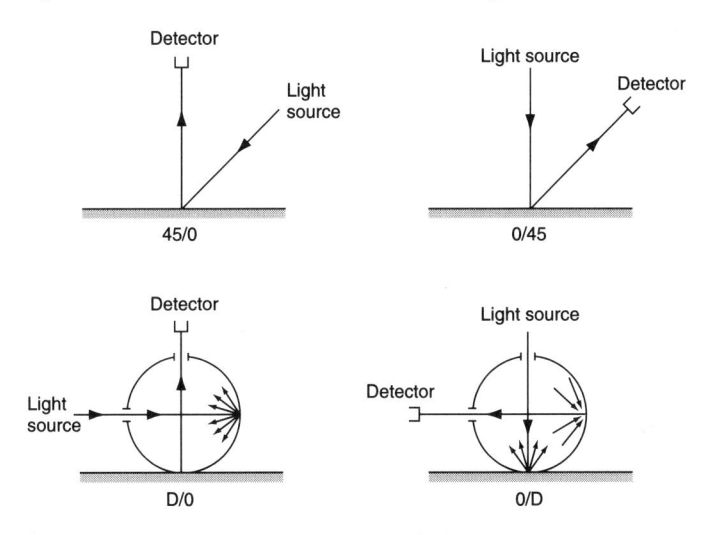

Figure 2.6 CIE-recommended illuminating and viewing geometries

2.6.1 45/normal (45/0)

The specimen is illuminated by one or more beams whose effective axes are at an angle of 45 ± 2° from the normal to the specimen surface. The angle between the direction of viewing and the normal to the specimen should not exceed 10°. The angle between the axis and any ray of an illuminating beam should not exceed 8°. The same restriction should be observed in the viewing beam.

2.6.2 Normal/45 (0/45)

The specimen is illuminated by a beam whose effective axis is at an angle not exceeding 10° from the normal to the specimen. The specimen is viewed at an angle of 45 ± 2° from the normal. The angle between the axis and any ray of the illuminating beam should not exceed 8°. The same restriction should be observed in the viewing beam.

2.6.3 Diffuse/normal (D/0)

The specimen is illuminated diffusely by an integrating sphere. The angle between the normal to the specimen and the axis of the viewing beam should not exceed 10°. The integrating sphere may be of any diameter provided the total area of the ports does not exceed 10% of the internal reflecting area of the sphere. The angle between the axis and any ray of the viewing beam should not exceed 5°.

A typical integrating sphere is 150 mm in diameter, though spheres as small as 50 mm are used in portable instruments. It is usually coated with barium sulphate, as this is one of the whitest substances known. Spheres coated with barium sulphate can be made with very high efficiencies due to its high reflectance of around 98.5% in the visible region. Its powdery surface makes it an almost ideal Lambertian diffuse reflector. While it is an improvement over the magnesium oxide that was used 20 years ago, it is not however an ideal coating. It is difficult to apply, and up to ten coats are required. Its surface is powdery, and easily damaged and contaminated. The natural ageing process of barium sulphate causes spheres to yellow; while the dual-beam design of modern spectrophotometers tends to cancel this effect eventually the sphere will need to be recoated – typically every three to five years, depending on the environment.

As a result of these problems several other substances have been tried. Halon, a form of powdered PTFE, is a little whiter than barium sulphate but suffers from translucency. It can be used as a sphere coating but has to be pressed to form a layer several millimetres thick, rather than applied as a conventional paint. Ceramic spheres have also been used by at least one manufacturer. But despite its problems, barium sulphate remains the most widely used sphere coating.

The illumination port of the sphere must be baffled to prevent direct illumination of the sample from the source, as this would destroy the effect of diffuse illumination.

Usually the diameter of the area that is illuminated is about 4 mm larger than that which is viewed. This is to reduce measurement errors due to translucency, that is, a slight transparency of the surface layers of a material which allows light to be reflected from inside the material beneath the surface (Figure 2.7). The resultant error is sometimes referred to as *lateral diffusion error*. A surprising number of materials are translucent; they include ceramic tiles (including the BCRA colour standard tiles used to test instruments) and, notoriously, opal glass.

INSTRUMENT GEOMETRY 67

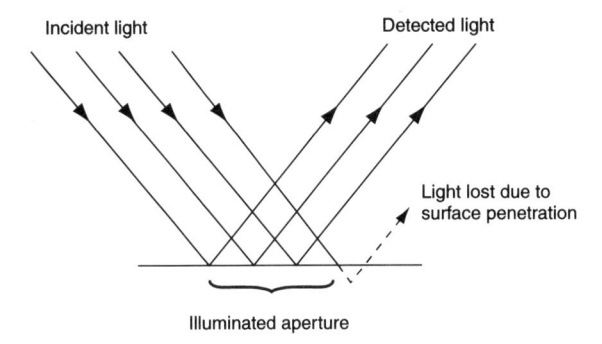

Figure 2.7 Principle of translucency

2.6.4 Normal/diffuse (0/D)

The specimen is illuminated by a beam whose axis is at an angle not exceeding 10° from the normal to the specimen. The reflected flux is collected by means of an integrating sphere. The angle between the axis and any ray of the illuminating beam should not exceed 5°. The integrating sphere may be of any diameter provided the total area of the ports does not exceed 10% of the internal reflecting area of the sphere.

These recommendations reduce to two different geometries: 45/0 and D/0. The Helmholtz theory of reciprocity allows the sample and the light source to be interchanged without affecting the measured reflectance due to the reversibility of optical paths.

Sphere geometry provides for a general-purpose colour-measurement instrument and is preferred for match prediction, as it has several advantages over 45/0. The use of diffuse illumination minimises the effects of surface texture and appearance that arise with uniform illumination (Figure 2.8). The CIE has allowed the viewing beam to be up to 10° away from the normal so that a gloss port can be provided which enables the specular (gloss) component of reflection to be either included or excluded (Figure 2.9). It is for this reason that most D/0 instruments are actually D/8 rather than true D/0. They meet the CIE specification for D/0 as the sample beam is less than 10° from the normal, but their inclusion of a specular port provides an increased measurement

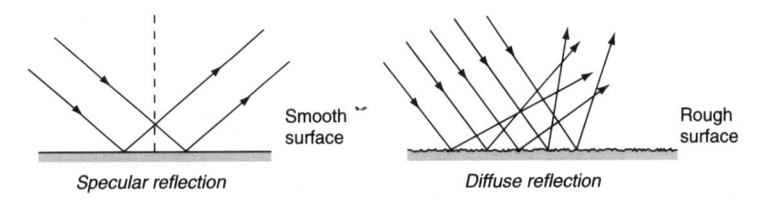

Figure 2.8 Comparison between specular and diffuse reflection

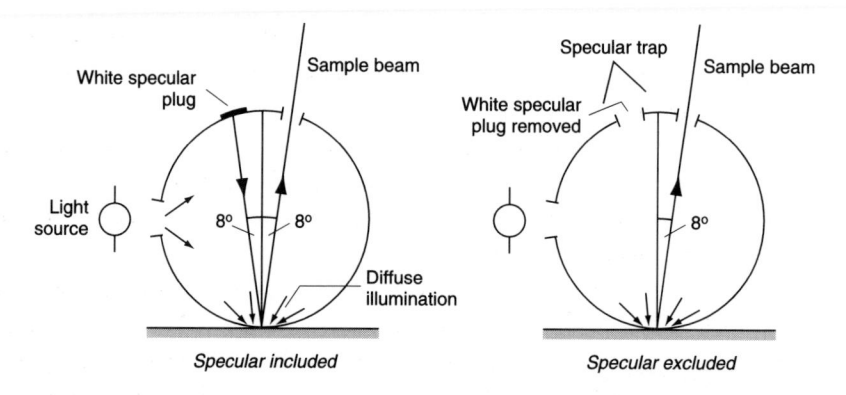

Figure 2.9 Comparison between D/8 specular-included and specular-excluded modes

flexibility that has led to the establishment of the D/8 geometry as the most popular choice for laboratory spectrophotometers.

The 45/0 geometry is always gloss-excluded. This is a better simulation of the way a colourist compares samples in a light booth, and hence it tends to be used in quality-control applications.

2.7 THE DUAL-BEAM SPECTROPHOTOMETER

In order to measure reflectance it is necessary to measure both the incident and reflected light from the sample. Thus in a traditional dual-beam instrument a second beam, the reference beam, is used to measure light that is reflected from the sphere wall. This gives a measurement of the light in the sphere that is incident on the sample. It is possible to measure reflectance with a single-beam instrument, but in this case a sphere correction factor has to be calculated to compensate for coloration of the light in the sphere due to light reflected back from the sample into the sphere, which will then reilluminate the sample.

In a dual-beam instrument the use of a reference beam also increases the instrument measurement stability; any drift of the electronics or light source intensity is cancelled out as both beams will be equally affected. A dual-beam instrument is also less sensitive to changes in sphere efficiency caused by the barium sulphate coating discolouring with age.

2.8 THE SPECTRAL ANALYSER

The spectral analyser is a device that splits light into its spectral components so they can be measured. It is a key part of the instrument, setting most of the performance characteristics. There have been few improvements over the years in the front end of spectrophotometer design, but the analyser and measurement systems have seen many changes and improvements with time.

Early spectrophotometers, such as the G E Hardy and Zeiss RFC3 designs, used mechanical techniques to scan the spectrum over a single light detector, usually a photomultiplier tube. Measurements were then made sequentially, and it could take several minutes to scan from 400 to 700 nm.

Several technologies are available to split light into its spectral components. Prisms, the classic devices for splitting white light into its component colours, were used in many early spectrophotometers. Since the refractive index of the material of a prism depends on the wavelength of the light, the prism deviates light through an angle that is dependent on its wavelength (Figure 2.10). The resultant spectrum is described by a complex mathematical relationship. It can be gradually scanned across the detector by rotating the prism. A slit is placed in front of the detector to define the bandwidth of the incident light, and often a cam arrangement was used to linearise the scan.

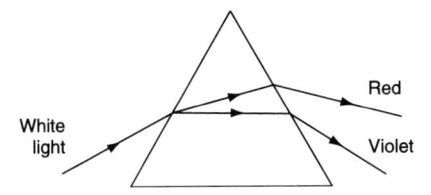

Figure 2.10 Dispersion of white light through a prism

This was the basic principle behind the G E Hardy design of 1928, which was put on the market by General Electric in 1935. This was the first commercially available colour-measurement spectrophotometer, and thanks to its precision and repeatability it remained the mainstay of many laboratories until its replacement in the last few years by instruments such as the Spectraflash 500. The Hardy spectrophotometer used what we would now call reverse optics. To reduce sample heating from the long measurement time, Hardy put the detector at the sphere and illuminated the sample by lowintensity monochromatic light whose wavelength was scanned using a prism: that is, the detector and light source were reversed as compared with a modern instrument.

As an alternative method of producing a spectral measurement, the Zeiss RFC3 spectrophotometer uses a filter wheel with between 16 and 20 narrow-bandwidth interference filters that are sequentially placed in front of the detector. Again, mechanical scanning leads to a lengthy measurement time.

As a variation on this several instruments have been designed that use a rotating circular variable filter (CVD). This is an interference filter in which the transmission bandpass varies linearly with position around a circular disc on which an interference film has been laid down as a wedge. Unless the illumination slit width is varied for each measurement channel, the bandwidth is proportional to centroid wavelength.

Most modern instruments have replaced mechanical scanning by measuring in parallel using an array of silicon diodes, thus making measurements much more quickly – in a matter of a few seconds – and with no moving parts, thus gaining greater reliability. The dispersing element is usually a grating, although sometimes discrete filters are still used.

2.8.1 Diffraction gratings

The diffraction grating as we know it today was invented in 1821 by Joseph von Fraunhofer. A simple plain grating can be made by ruling a large number of grooves with a diamond point into a glass surface. Light shone through this grating will be diffracted through an angle θ that is dependent on its wavelength (Figure 2.11).

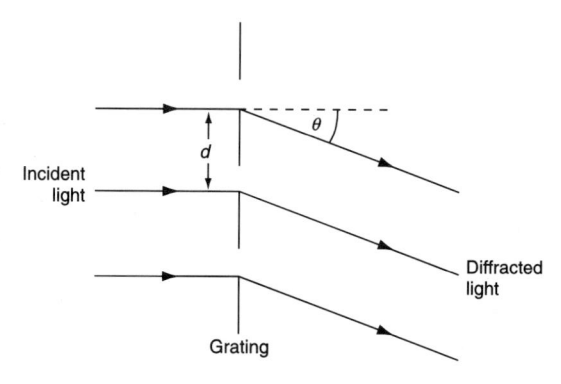

Figure 2.11 Diffraction of light through a grating

As the light passes through the many grooves of the grating, a process of diffraction and interference at each groove causes light of different wavelengths to be diffracted through different angles. Thus light is split into its spectral components (Eqn 2.3):

$$\sin \theta = \frac{n\lambda}{d} \qquad (2.3)$$

where θ = the angle of diffraction
λ = the wavelength of the incident light
d = the groove spacing
n = the integer defining the spectral order.

Eqn 2.3 shows there is a simple relationship between the angle of diffraction and the wavelength.

In a colour-measurement spectrophotometer gratings with groove densities of the order of 300 lines per millimetre are used, producing a visible spectrum of around 10 mm in length.

Modern laser techniques allow the manufacture of concave holographic gratings consisting of an aluminium layer on glass. Here a concave glass blank coated with photoresist is exposed to laser light from two directions simultaneously. These laser beams interfere with each other to produce a groove pattern in the photoresist. This is then developed and coated with aluminium. These gratings combine both the imaging and objective lens systems with the grating in a spectral analyser into a single optical component. Careful control of this process allows parameters such as groove density and profile to be optimised to correct for astigmatism and blaze for peak efficiency at a desired wavelength and to produce a flat field suitable for imaging on to to a diode array. Figure 2.12 shows a comparison between a plain grating and its modern concave counterpart.

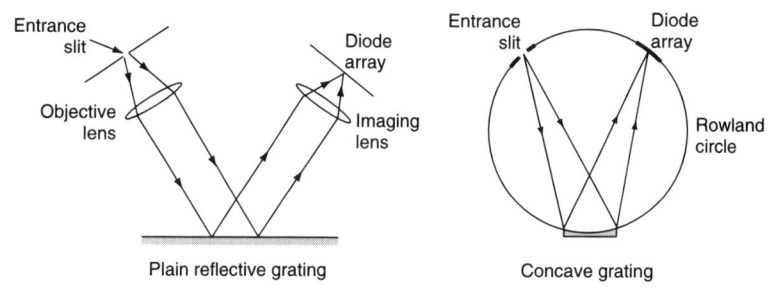

Figure 2.12 Comparison between plain-grating and concave-grating spectral analysers

In the process of replication, developed in the 1950s, gratings are stamped out like records, each having the identical groove pattern of the master. The process has dramatically reduced the cost of gratings, including the concave holographic type.

Gratings are now generally preferred to prisms in today's instruments due to their higher resolving power, greater efficiency and ability to produce a linear spectrum in which the position of light of a particular wavelength can be easily calculated. Also gratings are made by a much more flexible and controllable technology, allowing the instrument designer far greater freedom.

2.9 THE CE3000 SPECTROPHOTOMETER

The Macbeth CE3000 (Figure 2.13) derives from the MS2000, developed in the mid-

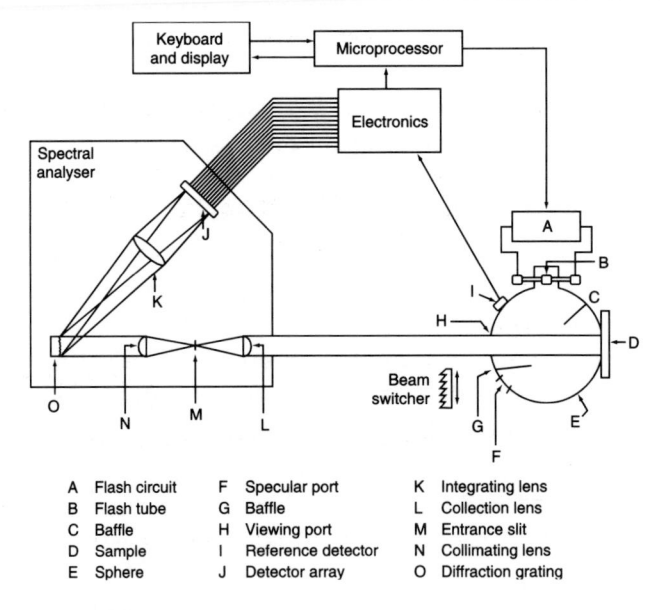

A	Flash circuit	F	Specular port	K	Integrating lens
B	Flash tube	G	Baffle	L	Collection lens
C	Baffle	H	Viewing port	M	Entrance slit
D	Sample	I	Reference detector	N	Collimating lens
E	Sphere	J	Detector array	O	Diffraction grating

Figure 2.13 Macbeth CE3000 spectrophotometer

1970s. It is an interesting instrument, being a pulsed xenon pseudo-dual-beam design, and can be found in many colour laboratories throughout the world. To keep down costs it uses only one analyser based on a flat plain grating with both imaging and objective lens systems. A motor-driven Fresnel-lens beam switcher is used to bend the sample beam to measure the sphere wall. Thus sample and reference-beam measurements can be made sequentially. A simple reference detector is used to calibrate each flash from the light source and compensate for flash-to-flash variations in the lamp.

2.10 THE MICROFLASH SPECTROPHOTOMETER

The Datacolor Microflash portable spectrophotometer is another dual-beam pulsed xenon instrument (Figure 2.14). Designed in the early 1990s, its MC90 analyser uses advanced optics that allow the same concave holographic grating to be used for both the sample and the reference beams. The use of 128-element diode arrays measuring at 3 nm intervals allows software techniques calibrated from the measurement of spectral line sources to be used to generate 31-point 10 nm data. This replaces the mechanical alignment methods used to set the wavelength scale in older instruments.

2.11 FLUORESCENCE

A sample is said to be *fluorescent* when energy absorbed in it at one wavelength is

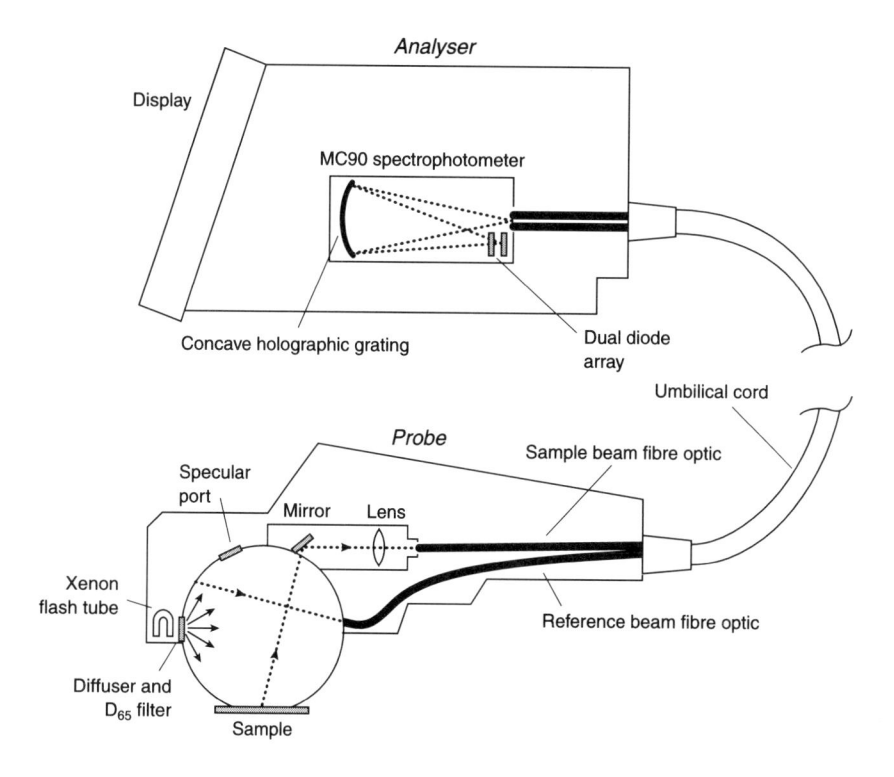

Figure 2.14 Microflash spectrophotometer

re-emitted at another. This process happens at the atomic level and follows the laws of quantum physics. A photon of light is absorbed by an atom promoting its electrons into a higher energy state; the atom then returns to a lower-energy state, emitting a photon of lower quantum energy and hence longer wavelength, as the process is not totally efficient. Thus according to Stokes' law the absorption peak is at a shorter (higher quantum energy) wavelength than the emission peak (lower quantum energy), although in certain so-called 'anti-Stokes' law materials' the absorption is at a longer wavelength than the emission. Typically the Stokes shift (the difference between the absorption and emission peaks) ranges from 70 to 200 nm (Figure 2.15). In nonfluorescent materials absorbed energy is converted to heat.

Where fluorescence is involved it is no longer valid to talk of a reflectance curve. Instead the term *spectral radiance factor* (SRF) is used. The total spectral radiance comprises the sum of the reflected radiance and the luminescent radiance (Eqn 2.4):

$$\beta_{T,\lambda} = \beta_{S,\lambda} + \beta_{L,\lambda} \qquad (2.4)$$

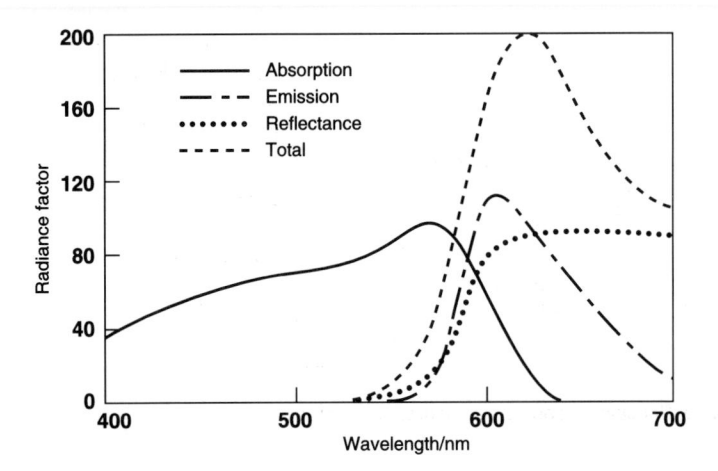

Figure 2.15 Absorption and emission spectra

where $\beta_{T,\lambda}$ = total spectral radiance factor
$\beta_{S,\lambda}$ = reflected radiance factor
$\beta_{L,\lambda}$ = luminescent radiance factor.

One of the largest uses for fluorescent material is in fluorescent brightening agents for the paper, plastics and textile industries. These absorb UV radiation around 350 nm and re-emit around 450 nm in the blue part of the visible spectrum. They increase the apparent whiteness by overcoming the natural yellowing of the raw substrate. Fluorescent dyes and pigments are increasingly used for safety clothing and signs, and in recent years fluorescent dyes for clothing have become fashionable.

The characterisation of fluorescent materials by traditional methods is difficult. In fact instruments of certain designs can give totally erroneous results as they do not allow for the emitted radiance. Any instrument that uses monochromatic illumination and polychromatic detection, such as the Hardy spectrophotometer, will assume that all detected radiation is at the illumination wavelength.

The classical way of measuring the colour of fluorescent materials is to use the double-monochromator method first suggested by Donaldson (Figure 2.16). Here one monochromator is used to measure the light reflected (or emitted) from the sample in the normal way. A second monochromator positioned in the illumination path allows the sample to be scanned with monochromatic light and the reflected (and emitted) spectra measured and recorded for all illumination wavelengths. A matrix of radiance versus illumination spectra can be built, which then allows the total spectral radiance to be calculated for any illuminant.

The recommended instrument geometry for making measurements on fluorescent

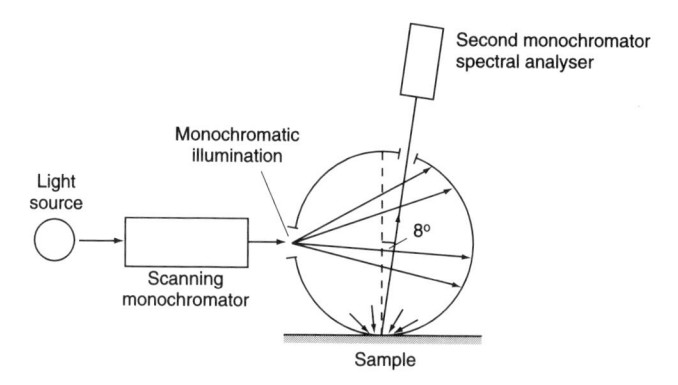

Figure 2.16 Principle of double-monochromator set-up

materials is 45/0. Sphere instruments with D/0 geometry produce erroneous results as the light in the sphere is contaminated and coloured by the radiation emitted from the sample, which can be quite high. Reflectance values of the order of 300% are not uncommon.

Double-monochromator instruments suitable for making colorimetric measurements are currently available only at certain national standardising laboratories. These have usually been designed and built in-house especially for making colorimetric measurements of fluorescent materials. No dual-monochromator instruments are currently commercially available for the measurement of colour. Such instruments are available for chemical analysis applications, but these do not have the necessary accuracy and resolution of the photometric scale for colorimetric work.

It is sometimes necessary to make measurements on fluorescent materials using a sphere instrument because this is the only type of instrument available. It is possible to make comparative measurements and by taking a few precautions the errors caused by self-illumination can be minimised.

(a) Measure using the smallest aperture available; this will reduce the amount of fluorescent radiation in the sphere.
(b) Make specular-excluded measurements; this reduces the sphere efficiency and hence the luminous energy available for self-illumination.
(c) Use an instrument with a light source that correctly simulates the desired illuminant – D_{65}, for example.

The use of filters to remove the energy available at the absorption wavelengths allows independent measurement of total spectral radiance and reflected spectral radiance. It is thus possible to calculate the luminescent radiance. This method works particularly well for measurements on fluorescent brightening agents where the absorption is in the

UV. A 400 nm, or sometimes a 420 nm, cutoff filter will eliminate all the fluorescence and allow a measurement of the reflectance without the luminescence radiance being present. Where coloured fluorescence is concerned, it is possible to use filters up in the visible region to kill the fluorescent component; however, interpretation of the results is more difficult (see section 5.7).

2.12 APPLICATION OF TRANSMISSION SPECTROPHOTOMETRY TO DYES

Transmission measurements, together with application of the Beer–Lambert law (section 1.8.4), have been used for many years in the dyeing industry. The move towards synthetic organic dyes has promoted their use in standardisation procedures during dye manufacture. Standardisation is necessary, as synthetic organic dyes are often contaminated with unreacted intermediaries to a degree that varies from batch to batch and alters the dyeing characteristics of the synthesised dye. Standardisation of dyes then ensures consistency of dyeing between different batches of dye. This allows the user always to use the same recipe to achieve a desired colour instead of adjusting the recipe for each batch of dyes.

Strength differences can be easily adjusted by adding varying amounts of colourless diluent such as salt or dextrin to ensure consistent colouring power. Shade variations are corrected by adding small quantities of other colorants or by blending different batches. Thus commercial dyes are generally not the single pure organic compounds as listed in the *Colour Index* [5].

The standardisation of dyes in solution relies on the use of absorbance measurements defined by the Beer–Lambert law. As the absorbance of mixtures of compounds is additive provided there is no chemical interaction, these techniques can also be used to determine the concentration of individual dyes in a dyebath, which will usually contain a mixture of three dyes.

The use of a flow cell connected to a dyebath allows continuous monitoring of the dyeing process by absorbance-measuring equipment. Measurement at the peak absorbance of the dyes allows changes in dye concentration to be monitored as the process proceeds to dye exhaustion. This allows very sophisticated computer control of dyeing, including optimisation of chemical additions and dyebath temperature profile.

The underlying principles can be understood by reference to Figure 2.17, which shows the absorbance curves at various wavelengths of separate solutions of known concentration of two dyes, a red and a blue, and also for a mixture of the same two dyes in unknown concentration. It is these absorbance values that are to be determined, using Eqns 2.5 and 2.6 below, with the following notation:

APPLICATION OF TRANSMISSION ON SPECTROPHOTOMETRY TO DYES

Red dye
Concentration = c_r
Wavelength of maximum absorption = λ_1
Absorbance at $\lambda_1 = A_r$
Absorbance at $\lambda_2 = a_r$

Blue dye
Concentration = c_b
Wavelength of maximum absorption = λ_2
Absorbance at $\lambda_2 = A_b$
Absorbance at $\lambda_1 = a_b$

For the mixture of red and blue dyes
Measured absorbance at $\lambda_1 = A_1$ Unknown concentration of red dye = x
Measured absorbance at $\lambda_2 = A_2$ Unknown concentration of blue dye = y

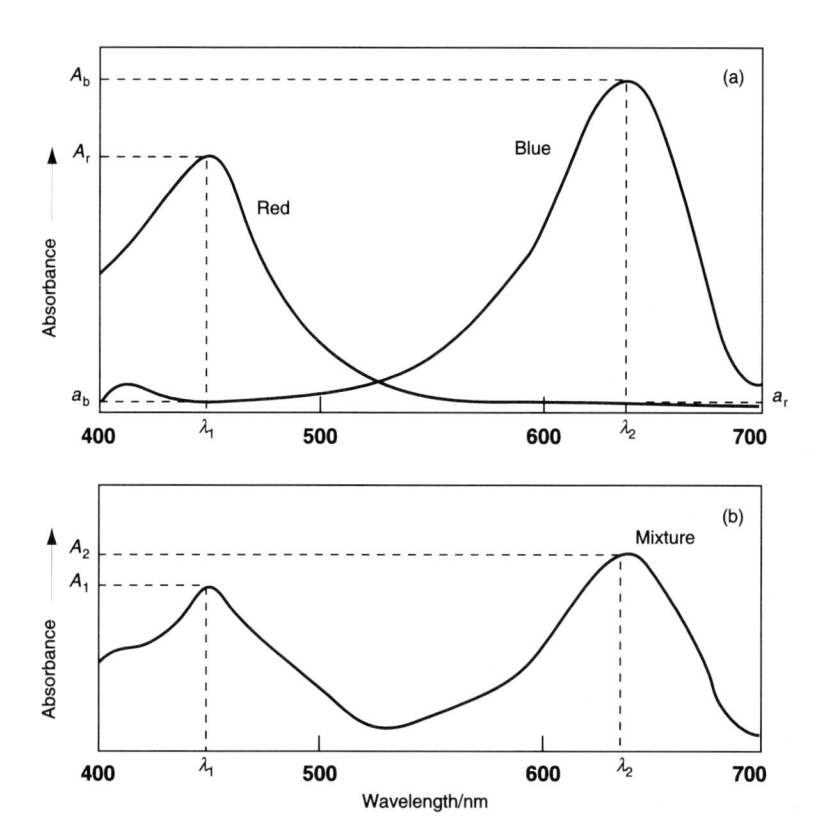

Figure 2.17 Plots of absorbance against wavelength: (a) separate solutions of known concentration of red and blue dyes; (b) a solution of a mixture of the same two dyes in unknown concentration

Then since the absorbances of the two dyes are additive, A_1 is given by Eqn 2.5:

$$A_1 = \frac{x}{c_r} A_r + \frac{y}{c_b} a_b \qquad (2.5)$$

and A_2 by Eqn 2.6:

$$A_2 = \frac{x}{c_r} a_r + \frac{y}{c_b} A_b \qquad (2.6)$$

These are simultaneous equations with only two unknowns, the respective concentrations of the red and blue dye in the mixture, x and y.

Likewise, if a third component were present in the mixture, its concentration could be found by making measurements of the absorbance at λ_3, the wavelength of maximum absorbance for the third dye. For each dye in the mixture a calibration curve of a solution of known concentration of that dye alone must be prepared.

This procedure may seem complex as described, but it can be easily automated and reduced to a simple set of measurement by the use of a modern computer system.

2.12.1 Precautions in the analysis of dye solutions

Certain precautions are needed when preparing samples for the spectrophotometric measurements on dye solutions. The first is to make sure that the dyes are in true solution as single molecules and are not present as dispersions. Disperse and vat dyes, having low aqueous solubility, should be particularly carefully checked by passing a narrow beam of light through the supposed solution in a dark room and viewing at right angles to the beam. If the beam can be seen as a line of scattered light, then the solution contains particles and is not a true solution. This scattering can be caused by undissolved dye particles or by a colloidal dispersion of dyebath additives. The remedy is to find another solvent miscible with water (if that is the principal solvent) which will bring all the materials present into solution. Acetone is frequently used.

If a second miscible solvent has to be used, however, recalibration of the solutions of the component dyes may be necessary as the wavelength and level of peak absorption may change with the new solvent. Concentrated solutions of some very soluble basic dyes can deviate from the Beer–Lambert law due to the formation of aggregates consisting of a very small number of molecules. The remedy here is to dilute the solution sufficiently to cause disaggregation, checking this by ensuring there is a linear relationship between absorbance and concentration in the range used.

The colour of many dyes is pH-sensitive, a property that is used when dyes are employed as indicators in volumetric analysis. The pH of a dye mixture being analysed must therefore be the same as that of the calibrating solutions of single dyes. Factors that can affect the pH include the presence of diluents and dispersing agents. If these are present in the mixtures under analysis, it will usually be necessary to add them to the calibrating solutions. The temperatures of the mixture and the calibrating solutions must also be the same.

2.13 COMPARING RESULTS FROM DIFFERENT DESIGNS OF SPECTROPHOTOMETER

In these days of ISO 9000 and tractability to national standards, it may come as a surprise to learn that no two models of spectrophotometer will produce the same result from measurements on the same sample within normal industrial tolerances. The main reason for this is that the measurement definition laid down by bodies such as the CIE is not specific enough. Colour differences of 2 CIELAB units between instruments are not uncommon, whereas commercial colour tolerances of 0.5–1 CIELAB unit are often called for. These colour differences are due to differences in instrument design in the areas of instrument geometry, port size and position, light source (particularly its SPD), channel centroid and bandwidth. Sample properties such as texture, gloss and fluorescence can exaggerate some of these differences. Typical spectrophotometer specifications are given in Table 2.1. Useful features that can also be incorporated into a spectrophotometer include:

(a) large specular port for full exclusion of the gloss component
(b) good sample viewer to ensure that the correct part of the specimen is under examination
(c) versatile range of sample holders
(d) UV filter to allow both UV-included and UV-excluded measurements
(e) UV calibrator
(f) transmission capability.

Table 2.1 Typical specification of a colour-measurement spectrophotometer

Parameter	Range
Geometry	D/0
Spectral range	360–750 nm
Number of channels	40
Channel bandwidth	10 nm
Wavelength accuracy	0.05 nm
Inter-instrumental agreement	0.5 CIELAB units, max. 0.2 average on BCRA tiles
Photometric range	0–200 %
Photometric resolution	0.01%
Measurement repeatability	0.05 CIELAB units
Measurement aperture sizes	25, 10, 5 mm diameter
Measurement time	5 s

In addition, the instrument should not cause the sample to change temperature.

2.14 CONCLUSION

The basics of colour measurement have not changed since the pioneering days of colour science in the 1930s but the technology, both optical and electronic, has seen rapid advances recently. Modern colour-measuring instruments are orders of magnitude smaller, cheaper and quicker than their predecessors of even a few years ago. This new technology has allowed the development of portable instruments that have moved colour measurement out of the laboratory and on to the manufacturing shop floor.

Whilst claims of improvement in measurement accuracy are not possible until we have tighter standard definitions of what 'accurate' really means (i.e. the basic measurement parameters), several manufacturers now market reference spectrophotometers whose repeatability and inter-instrument agreement is now several times better than the stability of the samples under study. This allows colour specification to move away from physical standards and to use numbers, provided the same model of spectrophotometer is in use – an impossibility with previous-generation instruments. Thus colour measurement is being brought into line with the measurement of other physical properties.

The use of colour-measuring equipment is going to become more widespread as instruments become ever smaller and cheaper, spurred on by the many quality initiatives that are an increasingly important part of manufacturing philosophy today.

REFERENCES

1. D Rich et al. in *Spectrophotometry luminescence and colour; science and compliance* Ed. C Burgess and D G Jones (Amsterdam: Elsevier, 1995)
2. CIE Publication No. 15.2 (1986).
3. *Standard practice for computing the colours of objects by using the CIE system*, ASTM E308-95 (Philadelphia: ASTM, 1995).
4. R Griesser, *Col. Res. Appl.*, **19** (6) (1994) 446.
5. *Colour Index*, 3rd Edn (Bradford: SDC, 1971) and supplements.

FURTHER READING

K McLaren, *The colour science of dyes and pigments* (Bristol: Adam Hilger, 1983).
F Grum and C J Bartleson, *Optical radiation measurements* Vol. 2 (London: Academic Press, 1980).
M C Hutley, *Diffraction gratings* (New York: Academic Press, 1982)
A Berger-Schunn, *Practical colour measurement* (New York: John Wiley, 1994).

CHAPTER 3

Colorimetry and the CIE system

Bryan Rigg

3.1 INTRODUCTION

Colour is extremely important in the modern world. We only have to look around us to see the variety of colours produced on textiles, painted surfaces, paper and plastics. Usually the colour is an important factor in the manufacture of the material and often a vital factor in the commercial success of a product, especially where textiles and paints are concerned. In some cases, such as foods, we use the colour to judge the quality of the material. In others, such as packaging, the colour is important in attracting customers and although the exact colour may not be critical, it must nevertheless be uniform and constant from one article to another. Any variation could be perceived as indicating a lack of care in the preparation or storage of the packaging and might imply a corresponding carelessness with respect to the contents.

In all branches of science and engineering, measurement plays an important part. In commerce materials are usually bought and sold by weight or volume. Without standardised systems for measuring mass, length and time, modern life would be very difficult. It is obvious that a standard system for measuring and specifying colour is equally desirable. Yet there are important differences between colour and, for example, length. The length of the standard metre bar, which until 1960 was used as the standard of length, remained constant as long as conditions such as temperature were adequately controlled. The colour of an object depends on many factors, such as lighting, size of sample, and background and surrounding colours. Much more importantly, colour is a subjective phenomenon and depends on the observer. The measurement of subjective phenomena, such as colour, taste and smell, is obviously more difficult than that of objective phenomena such as mass, length and time.

In any application we need to consider the purpose behind the measurement. Time, for example, can be measured to a fraction of a second but this is not of much use when we want to catch a bus. Basically we need to measure things with sufficient accuracy for our purpose; higher accuracy will be possible, but will cost more and hence be wasteful.

One problem with regard to colour measurement is that the human eye is readily available and particularly sensitive to colour; it can distinguish millions of different colours. Any measurement that is less reliable than the unaided eye will be of limited value. Another consideration is that we are often concerned with differences in colour rather than with colour itself. For many purposes the exact colour is not as important as uniformity of colour. When buying a blue shirt, the exact shade of blue is unlikely to matter, but any appreciable difference between the collar, sleeves and other parts would be unacceptable; moreover, small differences between shirts in the same display would give an impression of carelessness on the part of the manufacturer. Similarly when buying paint a customer will often accept any colour within reasonable range of the desired colour, but would be much less tolerant of even a small difference between the paint from two different cans. With colour, the final objective is to produce something that is pleasing or satisfactory to the eye. If the colour *looks* wrong, it *is* wrong.

When assessing the usefulness of the colour specification system to be described in this chapter, we have to consider how far the system enables us to deal with real problems with respect to colour and how far it fails to deal with the subjective nature of colour. For the present we will simply examine how a system of colour specification can be set up, ignoring for the moment factors such as texture and gloss. We will return to the question of how far such a system fulfils our needs at the end of the chapter.

Almost all modern colour measurement is based on the CIE system of colour specification. The initials come from the French title of the international committee (Commission Internationale de l'Éclairage) that set up the system in 1931. Although additions have been made since, the basic structure and principles are unchanged and the system is widely used. It is essentially an empirical system based on experimental observations rather than on theories of colour vision.

When discussing colour in general, we could be considering coloured lights, coloured solutions or coloured surfaces such as painted surfaces, plastics and textiles. In almost all practical situations we are concerned with coloured surfaces although, as we shall see, the properties of coloured lights are used in the specification of surface colours. The colour of a surface depends on the light source used to illuminate it, the particular observer who views it, and the properties of the surface itself, and of these the nature of the surface is obviously the most important. A piece of white paper will look white under all normal light sources when viewed by any observer with normal colour vision. (This statement is not completely true; for example, small areas of 'white' paper viewed as part of a pattern formed by a variety of colours may look as if it is a different colour. In trying to understand the principles involved in colour measurement it is probably best to ignore such exceptions for the present and adopt a simplified view. In certain practical applications, however, such factors must be considered.) Because our white paper does remain white, there is a tendency to think of colour as a property of the surface only. In considering the measurement of colour, however, the light source and the observer cannot be ignored; we should think about surface, light source *and* observer.

3.2 BASIC PRINCIPLES

It would obviously be difficult to design a system of colour measurement that attempted to describe the colours that we see. We simply have to think how we use words to describe colours. Colour names such as red, yellow, green and blue are reasonably well understood, but names such as rose, salmon and cerise are less well standardised. What might be called rose by one person could be called pink by another. Even for red there is a considerable range of colours which any one person might accept as red. This is not necessarily the same range as for a second person. Experiments with spectrum colours have shown that if one person chooses the wavelength considered closest to a true yellow, a second person might well not agree and claim that the chosen colour is too green or too orange.

Strictly speaking this discussion concerns the naming of colours rather than describing what people actually see. The latter is much harder to pin down, but there is no doubt that different people see colours differently. Even for one person, the appearance of a particular coloured surface can change with changing circumstances, such as the surrounding colour. The aim of the CIE system is to tell us how a colour might be reproduced (by a mixture of three primary light sources) rather than described. The amounts of the three primaries required to match a particular colour provide a numerical specification of that colour. A different colour would require different amounts of the primaries and hence the specification would be different. It turns out that in many applications this is all that is required. As we will see later, some idea of the colour seen can be deduced from CIE colour specifications; furthermore we would never attempt to reproduce a colour by actually mixing the CIE primaries.

Colour is three-dimensional, a property that is apparent in various ways. Colour atlases arrange colours using three scales (hue, value and chroma in the case of the Munsell system, for example). If we apply a range of concentrations of, say, a yellow dye we can produce a range of yellow samples, but there will generally be many different yellows that cannot be matched – for example, those that are slightly redder than our yellows. With mixtures of our yellow with a red dye we will be able to produce a range of reds, oranges and yellows, but there will be many oranges that we cannot match, including colours that are browner or greyer or less saturated than those actually produced. If we use mixtures of three dyes, however, such as our yellow and red dyes together with a blue dye, we can match a wide range of colours: yellows, oranges, reds and blues, and also browns, greys and so forth. In fact, the only colours that we will not be able to match will be very saturated or very pure colours, such as a purer yellow than that produced by our yellow dye. By choosing particularly bright or pure dyes we will reduce to a minimum the range of colours that cannot be matched. (The widest range of colours can be matched if magenta, yellow and cyan are used as the primaries.)

Ignoring the few colours that cannot be produced in this way, we could imagine a system of colour specification in which the concentrations of three specific dyes required on a particular substrate to match a colour could be used to specify that colour. Of course we would need to specify the light source under which the colour was seen, but the three concentrations would give a numerical specification of the colour; the specification would be different for different colours and it would be possible for someone else to reproduce the colour. Unfortunately there would be serious disadvantages to such a system. The relationship between dye concentrations and any measurable properties of a colour are complex, although definable with modern computers (see Chapters 5 and 6). Moreover, dyes or pigments are rarely pure, and their precise colour depends on the method of manufacture. Even repeat batches from the same plant will not match exactly, and again properties of mixtures are not completely predictable.

In contrast, coloured lights are much easier to define and reproduce. Imagine a red light obtained by isolating the wavelength 700 nm from the spectrum. Any laboratory in the world capable of measuring wavelength accurately (an objective physical measurement) could produce the same red colour. A green colour corresponding to 546.1 nm could be produced even more readily, since a mercury lamp emits light at only four wavelengths in the visible region (404.7, 435.8, 546.1 and 577.8 nm). By filtering out the other three, the required green wavelength could be obtained. The wavelengths 404.7 and 435.8 nm could be obtained in a similar manner. Small variations in the operating conditions have no significant effect on the wavelengths emitted by a mercury lamp. Hence three primary light sources could be defined simply as appropriate wavelengths and easily reproduced.

Mixtures of three coloured lights can be produced in various ways, but the simplest is to imagine three spotlights shining on the same area on a white screen (Figure 3.1). The colour produced would be a mixture of the three colours, and a wide range of colours could be produced by varying the amounts of the three primaries. The colours to be matched could include surface colours illuminated by a particular light source.

3.3 ADDITIVE AND SUBTRACTIVE MIXING

Most of us are familiar with the colours produced by mixing paints or coloured solutions. Many readers of this book will be equally familiar with the colours produced on fibres by mixtures of dyes. Obviously the results depend on the exact colours mixed, but roughly:

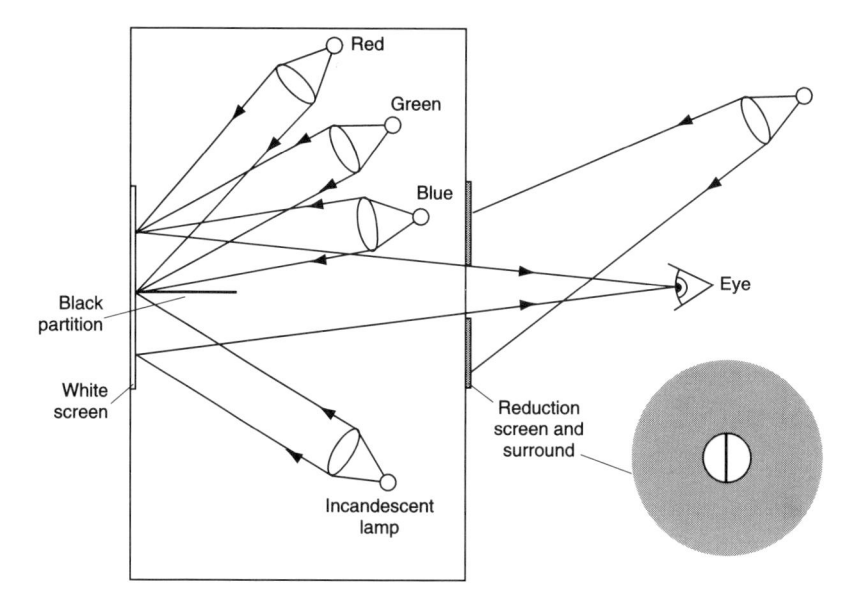

Figure 3.1 Additive colour mixing using three coloured lights, the inset showing the observer's visual field; the left half of the inner circular field is the colour stimulus originating in the incandescent lamp, the right half is the mixture of the red, green and blue stimuli; each lamp is connected in series with a variable resistor to control the amount of each stimulus

> red + blue gives purple
> red + yellow gives orange
> yellow + blue gives green

while red + yellow + blue in the correct proportions gives grey or black (Plate 1).

To people accustomed to mixing dyes or pigments, the colours produced by mixing lights may sometimes be surprising. For example, a blue light mixed with a yellow light might well give white, while red and green lights could be mixed to give a yellow. (Yellow and blue dyes would be expected to give green, while red and green dyes would probably give a dirty brown colour.) Quite obviously, mixing dyes and pigments is fundamentally different from mixing coloured lights. Since the CIE system is based on mixtures of lights, these must be considered further.

Additive mixing occurs when two or more coloured lights are shone at the same time so that we see the two lights together. Consider red and green lights shining on to a white screen using an arrangement similar to that shown in Figure 3.1. The screen will reflect almost all the light incident upon it, and the mixture of red plus green in the same appropriate proportions will reach the observer's eye. If the colour seen (yellow) is surprising this is simply because we are not used to mixing colours in this way. The two colours do not interact with each other at all. If the red and green are single wavelengths, both wavelengths reach our eye and do not interfere with each other in any way. We see the red wavelength plus the green wavelength: hence we call the mixture an additive mixture.

Another simple method of demonstrating additive colour mixing is the Maxwell disc. This is a disc made of sectors of various colours, which is spun at increasing speed. Above a certain speed we see the colours blending together in an additive manner. The colours produced can be varied by altering the relative areas of the differently coloured sectors.

Subtractive mixing is much more common but usually involves more complex processes. In the simplest case, we shine light through two coloured glass filters (Figure 3.2). The light passes through the two filters in succession: only the light transmitted by the first filter (F_1) reaches the second filter (F_2). Each filter 'subtracts' light and the only light seen is that which has succeeded in passing through both. This is completely different from additive mixing, where all the light from the mixture reaches the eye.

Suppose that F_1 is made of red glass and only transmits light of wavelengths longer than 650 nm. The light reaching F_2 will therefore consist only of wavelengths longer than 650 nm and will appear red. Suppose that F_2 is green and transmits only wavelengths between 500 and 550 nm. If white light were to be shone on to this filter the wavelengths between 500 and 550 nm would be transmitted and would look green. In our example the only light reaching F_2 is the red light of wavelengths greater than 650 nm, and therefore no light is transmitted through F_2. Hence no light reaches the eye and the two filters together produce black. In this case F_1 has subtracted all the wavelengths except those of 650 nm or longer and F_2 has subtracted the rest. It is easy to see that the mixture should be considered to be a subtractive mixture. The order in which the filters are placed does not affect the final colour as long as the light seen has passed through both filters.

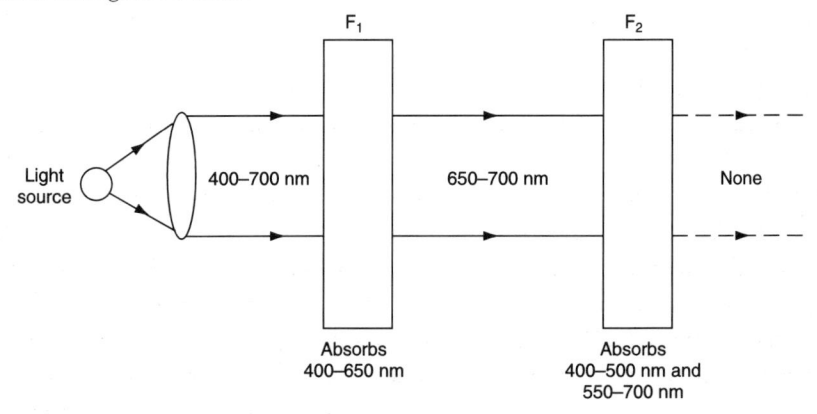

Figure 3.2 Subtractive effect of coloured filters

Not all red glass transmits long wavelengths only. It is possible that some red glass (F_3) may transmit light of all wavelengths. As long as there is a preponderance of red wavelengths, the transmitted light will look red. If F_1 is replaced by F_3 some light of all wavelengths will reach F_2, and the wavelengths between 500 and 550 nm will reach the eye. In this case a subtractive mixture of red plus green will give green. This will be a very dark green because F_3 will only transmit a small proportion of each of the green wavelengths. If we use a new green filter (F_4) which transmits some light at all wavelengths, the result is more difficult to predict. If the fractions of light transmitted by F_3 and F_4 at a particular wavelength are f_3 and f_4 respectively, the fraction transmitted by the subtractive mixture is $f_3 \times f_4$ (ignoring the small amount of light reflected from the surfaces of the filters). For example if, at a particular wavelength, F_3 transmits half the light ($f_3 = 0.5$) and F_4 transmits 10% ($f_4 = 0.1$), then the two filters together transmit only $0.5 \times 0.1 = 0.05$, i.e. one-twentieth of the light. Although we can carry out this calculation at all wavelengths throughout the visible region it is not always easy to deduce the colour of the mixture from the result: even in simple cases like this, the effects of subtractive mixing are often difficult to predict.

With coloured solutions, possible chemical reactions make the situation even more complicated. For example, the addition of (colourless) phenolphthalein to (colourless) sodium hydroxide solution gives a bright pink/magenta colour. Other indicators give equally unpredictable results. Obviously these are extreme examples, but with certain dyes interactions do occur, making the results of mixing not completely predictable.

The most important examples of subtractive mixing are the mixing of paints, and dyeing with a mixture of dyes. The results are often predictable on the basis of everyday experience, but the details of the process are much more complicated than those discussed so far. Consider a red dye and a green dye, applied first separately and then in mixture, to samples of a white substrate. (In this and in all similar examples, if the light source is not specified it should be considered to be a white light consisting of approximately equal amounts of all the wavelengths in the visible region; daylight is one such source.) The red dye on its own would produce a sample which reflected some light at all wavelengths, but less light at certain wavelengths (particularly around 500 nm) than others. Our eyes would see a mixture (additive!) of all the wavelengths, but because most of the reflected light would be red, the colour seen would be red. Similarly the substrate sample dyed with green dye alone would absorb some of the light at all wavelengths, but reflect more light of the green wavelengths than the others, giving a green colour. For the mixture of the two dyes we need to consider each wavelength in turn. For any one wavelength both dyes will absorb some of the light, but the amounts absorbed will be different at different wavelengths. Details of how the amount of light reflected can be calculated are given in section 5.1. At this point it is sufficient to note that the calculations are not simple and are not necessarily accurate – if, for example, the dyes interact in any way.

Qualitatively we would expect that both dyes will subtract from the incident light, particularly at the wavelengths where they absorb strongly (around 500 nm for the red dye and around 400 and 600 nm for the green dye). Hence the mixture is called a subtractive mixture and we see the light that has not been absorbed by either dye; in this case we would see a dirty brown colour. The fact that the results of subtractive mixtures are normally predictable simply comes from our experience; accurate calculations of the results are difficult (even ignoring interactions) and the effects of interactions are not predictable from the colours of the dyes being mixed.

Some examples of the colours obtained by additive and subtractive mixing are illustrated in Plate 1.

3.4 PROPERTIES OF ADDITIVE MIXTURES OF LIGHT

As indicated earlier, coloured lights are easy to define and hence seem to be suitable for use as primaries in a system of colour specification. The properties of additive mixtures of coloured light have been studied for many years [1-5], and those that are particularly relevant to their use in systems of colour specification are considered here.

We can match a wide range of colours using a mixture of, say, red, green and blue primaries. Suppose our primaries are single wavelengths and we use them to match white light consisting of a mixture of all the wavelengths in the visible region using an arrangement such as that shown in Figure 3.1. Although the mixture is physically quite different from the white light, by carefully adjusting the amounts of the primaries we can match the white light: that is, we can produce with the mixture a white that looks identical to the white light. If we change the colour or lightness of the surround, our white colours change in appearance. Over a wide range of conditions, however, the match holds: if the colours change, they both do so by the same amount. This was recognised by Grassman, who stated in 1853: 'Stimuli of the same colour (that is, same hue, same brightness, and same saturation) produce identical effects in mixtures regardless of their spectral composition'. Hence we can deal with colours without considering their spectral composition, at least in many applications. Grassman's law also implies that if colour A matches colour B and colour C matches colour D, then colour A additively mixed with C matches colour B mixed with D. This is vital when we consider that normal colours are additive mixtures of all the wavelengths in the visible spectrum. We need to consider the effect of the additive mixture of all the wavelengths.

It cannot be stressed too strongly that modern colorimetry is based on the properties of additive mixtures of coloured lights, and that these properties have been determined by experiment. The main properties were established well over a century ago. Subsequent work has confirmed that the simple properties described above are indeed valid, but has defined much more closely the range of experimental conditions under which the simple laws hold [6–8]. Theories of colour vision must attempt to explain such laws, and also their exceptions. The fact that colours can be produced using only three primaries implies that there are three types of receptor among the cone cells of the eye, and that variations in the magnitude of the responses from the three types produce the range of colour sensations that we call colour vision. This theory was originally put forward by Young and elaborated by Helmholtz. Modern theories suggest that the visual mechanism is complicated, particularly with respect to the ways in which the receptors are linked to the brain. Modern colorimetry and the CIE system are based on the experimental *facts*, not on any particular theory of colour vision.

So far in this chapter any mathematical treatment of colour matching has been avoided, but some mathematics is essential. Suppose we represent our red, green and blue primary light sources by [R], [G] and [B]. If we use these to match a colour using a mixture of the primaries, we can represent the *amounts* of our primaries by R, G and B respectively. We can then write Eqn 3.1:

$$C[C] \equiv R[R] + G[G] + B[B] \qquad (3.1)$$

which is equivalent to saying that C units of the colour [C] can be matched by R units of the red primary [R] additively mixed with G units of the green primary [G] together with B units of the primary [B]. (In different textbooks different symbols are used in this type of equation. Sometimes the = symbol is used instead of ≡; the latter is preferred here and should be read as 'is matched by'. It is important to distinguish carefully between the primaries themselves, such as [R], and the *amounts* of the primaries used in a match, such as R. The amounts used of each primary, R, G and B, are known as the *tristimulus* values of the colour [C]. These values depend on the colour [C]. If the values are known, they give an indication of the colour. Thus if R and B are high and G is low, the colour can be matched using a lot of the red and blue primaries and only a little of the green primary: thus the colour is some sort of purple. The exact colour obviously depends on the exact nature of our primaries [R], [G] and [B], and if these are very pure the colour is likely to be a saturated purple.

In most respects, equations such as Eqn 3.1 can be treated as ordinary algebraic equations. Thus if we write Eqn 3.2:

$$C_1[C] \equiv R_1[R] + G_1[G] + B_1[B] \qquad (3.2)$$

and Eqn 3.3:

$$C_2[C] \equiv R_2[R] + G_2[G] + B_2[B] \qquad (3.3)$$

then an additive mixture of C_1 units of [C_1] with C_2 units of [C_2] can be matched by $R_1 + R_2$ units of the red primary [R], additively mixed with $G_1 + G_2$ units of the green primary [G], together with $B_1 + B_2$ units of the blue primary [B] (Eqn 3.4):

$$C_1[C] + C_2[C] \equiv (R_1 + R_2)[R] + (G_1 + G_2)[G] + (B_1 + B_2)[B] \qquad (3.4)$$

3.5 A POSSIBLE COLOUR SPECIFICATION SYSTEM

If we were to select and define three particular primaries [R], [B] and [G], then the amounts of these required to match any colour (the tristimulus values R, G and B) could be used to specify the colour. Each different colour would have a different set of tristimulus values, and with practice we could deduce the appearance of the colour from the tristimulus values. Such a system would appear to suffer from several defects, however. These will be considered below, together with descriptions of how potential problems are overcome in the CIE system.

3.5.1 Use of arbitrarily chosen primaries

Different results would be obtained by any two observers using different sets of primaries. Sets of tristimulus values obtained using one set of primaries can, however, be converted to the values that would have been obtained using a second set, provided that the amounts of one set of primaries required to match each primary of the second set of primaries in turn are known. Hence either we could insist that the same set of primaries is always used, or we could allow the use of different sets, but insist that the results are converted to those that would have been obtained using a standard set. In practice this does not matter, as we will see later.

3.5.2 Inadequacy of real primaries

Even if we use pure colours for our set of primaries, there will still be some very pure colours that we cannot match. For example, a very pure cyan (blue-green) might be more saturated than the colours obtained by mixing the blue and green primaries. Adding the third primary [R] would produce even less saturated mixtures. A possible solution in this case would be to add some of the red primary to the pure cyan colour, and then match the resultant colour using the blue and green primaries (Eqn 3.5):

$$C[C] + R[R] \equiv G[G] + B[B] \qquad (3.5)$$

In practice, following this procedure allows all colours to be matched using one set of primaries, the only restriction in the choice of primaries being that it must not be possible to match any one of the primaries using a mixture of the other two. Rearranging Eqn 3.5 gives Eqn 3.6:

$$C[C] \equiv -R[R] + G[G] + B[B] \qquad (3.6)$$

Hence the tristimulus values of C are – R, B and G: that is, one of the tristimulus values is negative. Negative values are undesirable. It would be easy to omit the minus sign or fail to notice it. Careful choice of primaries enables us to reduce the incidence of negative tristimulus values. The best primaries are red, green and blue spectrum colours. Although mixtures of these give the widest possible range of colours, however, there is no set of real primary colours that can be used to match all colours using positive amounts of the primaries. In other words, no set of real primaries exists that will eliminate negative tristimulus values entirely.

Since it is possible to calculate tristimulus values for one set of primaries from those obtained using a second set, there is no need to restrict ourselves to a set of real primary colours. We can use purely imaginary primaries; it is only necessary that these have been defined in terms of the three real primaries being used to actually produce a match. This is not just a hypothetical possibility. Negative tristimulus values would be a nuisance in practice and in the CIE system imaginary primaries are indeed used so as to avoid negative values. It is therefore worth considering this point a little further.

It is difficult to draw (or visualise) three-dimensional diagrams, and colours are commonly represented in two-dimensional plots such as chromaticity diagrams (considered in section 3.12). Any two-dimensional representation of colour must omit or ignore some aspect of colour and should therefore be treated with caution. Two-dimensional plots are normally used to represent the *proportions* of primaries used rather than the amounts. Equal proportions of [R], [G] and [B] could look neutral, but the mixture could be very bright or almost invisible depending on the amounts used. Similarly for surface colours: a very dark grey and a very light grey would require roughly the same proportions of three primary dyes, but would require very different amounts. (Students often confuse proportions and amounts, but the distinction should always be maintained.)

A two-dimensional plot can illustrate the problem under discussion, and its solution (Figure 3.3). Suppose [R], [G] and [B] represent our three primaries, and positions on the diagram represent the proportions of the primaries used to produce the colour corresponding to the position at any point. The proportions of the primaries [R] [G] and [B] can be represented by r, g and b (Eqn 3.7):

$$r = \frac{R}{R+G+B} \qquad g = \frac{G}{R+G+B} \qquad b = \frac{B}{R+G+B} \qquad (3.7)$$

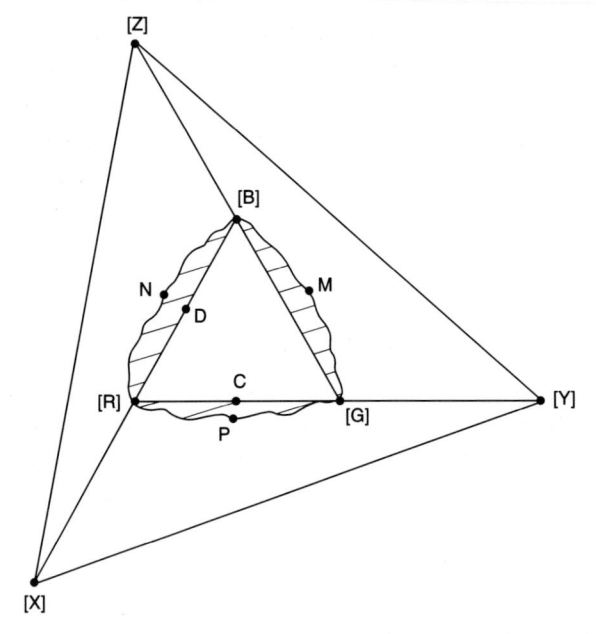

Figure 3.3 Two-dimensional representation of primary colours and additive colour mixing

For example the point C, halfway between [R] and [G], represents the colour formed by mixing equal amounts of [R] and [G]. (Actual amounts are not shown on this plot.) Thus for C we can say that $r = 0.5$, $g = 0.5$, $b = 0$. Similarly for [R] $r = 1$, $b = 0$, $g = 0$. All points within the triangle [R][G][B] can be matched using the appropriate proportions of the primaries. Suppose also that the boundary of real colours (strictly those real colours for which $r + b + g = 1$) is denoted by the shape [R]N[B]M[G]P[R]. Points within the shaded area correspond to real colours, but cannot be matched by positive proportions of the three primaries. The point M, for example, might require $r = -0.2$, $b = g = 0.6$, i.e. equal quantities of [B] and [G] together with a negative amount of [R], the proportions adding up to unity.

Consider the straight line [B]D[R][X], where [R][X] = [B][R]. For all points on the line, $g = 0$. For [B], $b = 1$ and $r = 0$; for D, $b = 0.5$ and $r = 0.5$; for [R], $b = 0$ and $r = 1$, while for [X] $b = -1$, $r = 2$. Thus although [X] is well outside the boundary of real colours, its position can be specified simply and unambiguously using r, b and g. Points [Y] and [Z] can be defined similarly, and by drawing the triangle [X][Y][Z] we can see that all real colours fall within the triangle; all real colours can be matched using positive proportions of three imaginary primaries situated at [X], [Y] and [Z] respectively. Obviously there are many alternative possible positions for [X], [Y] and [Z], all simply specified and allowing all real colours to be matched using positive proportions of the primaries.

If the problem is considered in three dimensions, the corresponding diagram is much more complicated, but the argument is similar. The volume (rather than the area) corresponding to all real colours is somewhat larger than that represented by positive amounts (note, amounts not proportions) of any three real primaries. It is however possible to specify in the three-dimensional space positions for three imaginary primaries such that all real colours can be matched by positive amounts of the three primaries.

3.5.3 Inadequacy of visual observation

From sections 3.5.1 and 3.5.2 we can see that the potential problems associated with the use of different primaries, and with the use of negative amounts of primaries, can be overcome (as indeed they are in the CIE system). We also have to consider how a sample could actually be measured, or how a specification could be arrived at. We seem to be required to produce a visual match, i.e. to use an instrumental arrangement whereby we may adjust the amounts of three suitable primaries (mixed additively) until in our judgement a mixture is obtained that matches the colour to be measured or specified. Such an instrument is called a visual tristimulus colorimeter. The amounts of the primaries required could be noted, and the results converted to the equivalent values for a standard set of primaries. A procedure like this is perfectly possible, the main problem lying in the precision and accuracy achievable. The results will vary from one observer to another because of differences between eyes. Even for one observer repeat measurements will not be very satisfactory. Under the controlled conditions necessary in such an instrumental arrangement (usually one eye, small field of view and low level of illumination) it is impossible to achieve the precision of unaided eyes under normal conditions, for example, when judging whether a colour difference exists between two adjacent panels on a car body under good daylight.

Although metamerism is dealt with in detail in section 3.15 it is necessary to point out here that the matches in the instrumental arrangement being considered are likely to be highly metameric (physically quite different) and this gives rise to many of the problems. The widest range of colours can be matched using primaries each corresponding to a single wavelength. If three such primaries are used to match a colour consisting of approximately equal quantities of all wavelengths in the visible spectrum, the two colours are physically very different even though they look the same to the observer. Not surprisingly it turns out that such a pair of colours is unlikely to match for a second observer. Even for one observer, the differences between the different parts of his eyes are likely to cause problems (Chapter 8). These problems can be minimised by using a small (<2°) field of view, but then the precision of matching is reduced by a factor of about 5, compared with that obtained using a 10° field of view.

Some of these problems can be overcome by using more than three primaries (as in the Donaldson six-filter colorimeter [9]). Using more primaries allows a wide range of colours to be matched even when the primaries are not monochromatic. The degree of metamerism can be greatly reduced and a large field of view can be used. Such instruments are tedious to use, however, while the results are still subjective and obviously can never be better than those obtained using the unaided eye under normal conditions. Although these instruments have played an important role in research into many aspects of colour vision, they have found little application in routine industrial colour measurement.

The CIE realised even before 1931 that this was likely to be so, and decided that the system adopted should allow calculation of tristimulus values from measured reflectance values. In fact the system was set up in 1931 because the required data became available at that time. The following sections describe how the information required to define a *standard observer* was actually obtained and how such information can be used to calculate tristimulus values, and summarises the CIE system as set up in 1931. It must be stressed that the CIE system incorporates the features described earlier, a colour being specified by the amounts of the [X], [Y] and [Z] primaries required, in additive mixture, to match it. The 'standard observer' data is added to the framework already described.

3.6 STANDARD OBSERVER: COLOUR-MATCHING FUNCTIONS

For a surface colour such as that of a textile fabric we can determine how much light is reflected at each wavelength throughout the visible region (Chapter 2). If we knew the tristimulus values for each wavelength, we could calculate the tristimulus values for the sample. The tristimulus values of any one wavelength are the amounts of the three chosen primaries required to match the light of the particular wavelength. Obviously the amounts required depend on the observer, and results for an average (or 'standard' observer) are required. These values were actually determined as follows.

Imagine a visual tristimulus colorimeter similar to the arrangement shown in Figure 3.1, in which one half of the field of view consists of a mixture of the [R], [G] and [B] primaries, while the colour in the other half is a single wavelength. To produce a match experimentally it may well be necessary to add some of [R], [G] or [B] to the wavelength to be matched. This is quite possible experimentally (section 3.5.2), and the resultant tristimulus values for the wavelength will include at least one negative value. Such experiments were carried out by Wright [10] and by Guild [11]. They used somewhat different techniques, and in particular different primaries. Both considered light of many wavelengths throughout the visible spectrum and averaged results from a several observers. The results differed from one observer to another (as expected), but when the results from the experiments were converted to a common set of primaries, the agreement was considered to be satisfactory. The results were expressed as the tristimulus values for an equal-energy spectrum, using primaries [R], [G] and [B], and the results were expressed as the amounts (called distribution coefficients) $\bar{r}$, $\bar{g}$ and $\bar{b}$ required to match one unit of energy of each wavelength throughout the visible region (Figure 3.4). Since [R], [G] and [B] were real primaries (actually 700, 546.1 and 435.8 nm respectively) some of the values were negative, as shown in Figure 3.4. As we will see later, the CIE adopted three unreal primaries [X], [Y] and [Z]; the colour-matching functions in terms of these primaries are denoted by $\bar{x}$, $\bar{y}$ and $\bar{z}$ and are always positive. This ensures that tristimulus values for all real colours are always positive.

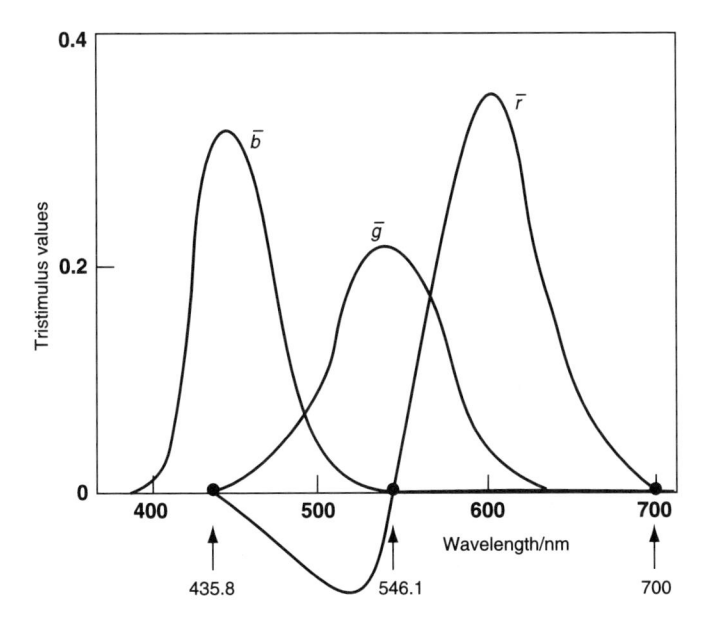

Figure 3.4 Tristimulus values $\bar{r}$, $\bar{g}$ and $\bar{b}$ of spectral stimuli of different wavelengths but constant radiance measured by an average observer with normal colour vision using as primaries the stimuli [R] at 700 nm, [G] at 546.1 nm and [B] at 435.8 nm

3.7 CALCULATION OF TRISTIMULUS VALUES FROM MEASURED REFLECTANCE VALUES

Suppose that we have a sample, such as a painted surface, and that we have measured the fraction of light reflected at each wavelength, R_λ. (Many instruments give readings in terms of the percentage of light reflected, i.e. $R_\lambda \times 100$, but fractions are easier to use in the present discussion.) Provided that the sample is not fluorescent, the R_λ values will be completely independent of the light shone on the sample. For example, a white paint will reflect about 90% of the incident light (i.e. $R_\lambda = 0.9$) at, say, 500 nm whether illuminated with strong daylight or with weak tungsten light. Thus the R_λ values are independent of the actual light source used in the spectrophotometer. The actual amount of light reflected will be different for different light sources, however. Suppose that the sample is now viewed under a light source for which the light emitted at each wavelength is E_λ. Then the amount reflected at each wavelength will be $E_\lambda \times R_\lambda$.

Now if we consider only light of wavelength λ, one unit of energy of λ can be matched by an additive mixture of $\bar{x}_\lambda$ units of [X] together with $\bar{y}_\lambda$ units of [Y] and $\bar{z}_\lambda$ units of [Z] (Eqn 3.8):

$$1[\lambda] \equiv \bar{x}_\lambda[X] + \bar{y}_\lambda[Y] + \bar{z}_\lambda[Z] \tag{3.8}$$

This follows from section 3.6. Hence (Eqn 3.9):

$$E_\lambda R_\lambda[\lambda] \equiv E_\lambda \bar{x}_\lambda R_\lambda[X] + E_\lambda \bar{y}_\lambda R_\lambda[Y] + E_\lambda \bar{z}_\lambda R_\lambda[Z] \tag{3.9}$$

in agreement with the properties of light described earlier. It also follows from the properties of additive mixtures of lights that the light reflected at two wavelengths λ_1 and λ_2, $E_{\lambda 1}R_{\lambda 1}[\lambda_1] + E_{\lambda 2}R_{\lambda 2}[\lambda_2]$, can be matched by $E_{\lambda 1}\bar{x}_{\lambda 1}R_{\lambda 1} + E_{\lambda 2}\bar{x}_{\lambda 2}R_{\lambda 2}$ units of [X] mixed with $E_{\lambda 1}\bar{y}_{\lambda 1}R_{\lambda 1} + E_{\lambda 2}\bar{y}_{\lambda 2}R_{\lambda 2}$ units of [Y] mixed with $E_{\lambda 1}\bar{z}_{\lambda 1}R_{\lambda 1} + E_{\lambda 2}\bar{z}_{\lambda 2}R_{\lambda 2}$ units of [Z].

The total amount of energy reflected over the visible spectrum is the sum of the amounts reflected at each wavelength. This can be represented quite simply mathematically (Eqn 3.10):

$$\sum_{\lambda=380}^{760} E_\lambda R_\lambda \tag{3.10}$$

where the sigma sign (Σ) means that the $E_\lambda R_\lambda$ values for each wavelength through the visible region should be added together, and the limits of $\lambda = 380$ and 760 nm are the boundaries of the visible region. Strictly the spectrum should be divided into infinitesimally small wavelength intervals (dλ) and the total amount of light given by Eqn 3.11:

$$\int_{380}^{760} E_\lambda R_\lambda \, d\lambda \tag{3.11}$$

but in practice the summation form is used. Representing the amounts of [X], [Y] and [Z] in a similar manner, the light reflected from our paint surface can be matched by Eqn 3.12:

$$\sum E_\lambda \bar{x}_\lambda R_\lambda + \sum E_\lambda \bar{y}_\lambda R_\lambda + \sum E_\lambda \bar{z}_\lambda R_\lambda \tag{3.12}$$

CALCULATION OF TRISTIMULUS VALUES FROM MEASURED REFLECTANCE VALUES 97

Since the light reflected from our paint sample can also be matched by X[X] + Y[Y] + Z[Z], it follows (Eqn 3.13):

$$X = \sum E_\lambda \bar{x}_\lambda R_\lambda \quad Y = \sum E_\lambda \bar{y}_\lambda R_\lambda \quad Z = \sum E_\lambda \bar{z}_\lambda R_\lambda \quad (3.13)$$

Hence, since E_λ, $\bar{x}_\lambda$, $\bar{y}_\lambda$, $\bar{z}_\lambda$ and R are known, the X, Y and Z tristimulus values can be calculated.

The calculation can be illustrated by reference to Figure 3.5. Suppose we have measured the reflectance curve of a sample and obtained the results shown in Figure 3.5(a). The R values indicate the *fraction* of light reflected by the sample at each wavelength. At the wavelengths around 500 nm the sample is reflecting a high proportion of the light that is shone on it, no matter how much or how little light that may be. Similarly the fraction reflected at 600–700 nm is low, again irrespective of the amount of light shone on to the surface. To calculate how much light is actually reflected, we need to know how much light is shone on the surface.

Suppose that the surface is illuminated by a source whose energy distribution is shown in Figure 3.5(b), i.e. the source contains relatively less energy at the short-wavelength end of the visible region and relatively much more at the longer wavelengths. The amount of light reflected by the sample at each wavelength will be $E_\lambda R_\lambda$ and this is also plotted against wavelength in Figure 3.5(b).

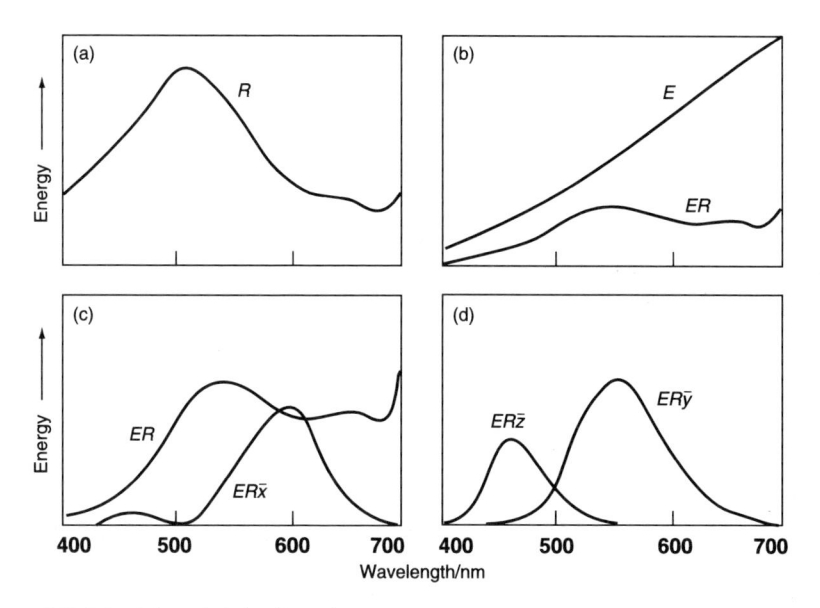

Figure 3.5 Calculation of tristimulus values

We can see that while the curve resembles the R_λ curve, with a maximum at 540 nm and a minimum at 680 nm, the balance between the longer and shorter wavelengths is quite different. The R values are roughly the same at 400 and 620 nm, while the $E_\lambda R_\lambda$ value at 620 nm is almost ten times the corresponding value at 400 nm.

In the calculation of the X tristimulus value we multiply $E_\lambda R_\lambda$ by $\bar{x}_\lambda$. The $E_\lambda R_\lambda$ values are replotted (on an enlarged scale) in Figure 3.5(c), which also shows the corresponding $E_\lambda \bar{x}_\lambda R_\lambda$. The $E_\lambda \bar{x}_\lambda R_\lambda$ curve roughly corresponds to the $\bar{x}_\lambda$ curve shown in Figure 3.6 (page 101). There are two quite distinct parts to the curve with maxima around 460 and 600 nm, but the relative sizes of the two peaks have changed. The curves for $E_\lambda \bar{y}_\lambda R_\lambda$ and $E_\lambda \bar{z}_\lambda R_\lambda$ are shown in Figure 3.5(d). Again the curves roughly resemble the $\bar{y}_\lambda$ and $\bar{z}_\lambda$ curves shown in Figure 3.6 but the relative sizes have changed. The $E_\lambda \bar{z}_\lambda R_\lambda$ peak is considerably lower than that for $E_\lambda \bar{y}_\lambda R_\lambda$. The converse is true for $\bar{z}_\lambda$ compared with $\bar{y}_\lambda$.

The areas under the $E_\lambda \bar{x}_\lambda R_\lambda$, $E_\lambda \bar{y}_\lambda R_\lambda$ and $E_\lambda \bar{z}_\lambda R_\lambda$ curves are proportional to the X, Y and Z tristimulus values respectively. It is obvious that Z is considerably smaller than X or Y. (In fact the E_λ curve corresponds to tungsten light and the approximate tristimulus values are $X = 38$, $Y = 45$ and $Z = 21$.) We can now see why the standard observer is so important, and how the tristimulus values can be obtained without actually producing a visual match for our colour. Provided that we know the energy distribution of the light source under which the specification of our paint sample is required, and also the distribution coefficients $\bar{x}_\lambda$, $\bar{y}_\lambda$ and $\bar{z}_\lambda$ defining the standard observer, we only need to measure R_λ. We do not need to produce a match, nor do we need to use the primary colours [X], [Y] and [Z] or indeed any other set of primaries. The calculation of tristimulus values emphasises the point made earlier that the colour depends on the light source and the observer as well as on the object itself.

3.8 THE 1931 CIE SYSTEM

We have seen that it is possible to calculate tristimulus values of a specified sample, that is, the amounts of three primaries which, if additively mixed, would match the colour of the sample. The CIE had to define standard primaries, standard light sources and a standard observer, together with standard observing and viewing conditions.

3.8.1 Standard primaries

As discussed earlier, the CIE had considerable latitude in selecting three primaries. For the moment, the main property to note is that all real colours can be matched using positive amounts of the chosen primaries [X], [Y] and [Z]. These were defined by Eqns 3.14 to 3.17:

$$C_{\lambda 1} \equiv 0.73467[X] + 0.26533[Y] + 0.00000[Z] \qquad (3.14)$$

$$C_{\lambda 2} \equiv 0.27376[X] + 0.71741[Y] + 0.00883[Z] \qquad (3.15)$$

$$C_{\lambda 3} \equiv 0.16658[X] + 0.00886[Y] + 0.82456[Z] \qquad (3.16)$$

$$S_E \equiv 0.33333[X] + 0.33333[Y] + 0.33333[Z] \qquad (3.17)$$

where λ_1 = 700 nm
  λ_2 = 546.1 nm
  λ_3 = 435.8 nm
  S_E = equal-energy stimulus, which is a stimulus having equal amounts of energy at all wavelengths through the visible spectrum.

The relative sensitivity of the eye to light of different wavelengths had been determined previously, and a particular curve (denoted by V_λ) was adopted as standard by the CIE in 1924. The CIE 1931 standard colorimetric system was made consistent with the 1924 V_λ curve by the choice of primaries such that the $\bar{y}_\lambda$ curve was identical to the V_λ curve.

3.8.2 Standard light sources and standard illuminants

In section 3.7 we saw that, in order to calculate tristimulus values from measured reflectance values, we require the energy distribution of the light source under which the sample is to be matched. If tristimulus values are to be used to specify a colour, some standard set of E_λ values must be used; different sets of E_λ would give different tristimulus values for the same sample.

In practice many different light sources are in use, particularly various phases of daylight and various types of fluorescent tube and tungsten light. If we wish to check that a paint sample is the correct colour, it would probably be satisfactory to check using one form of daylight, tungsten light and a fluorescent tube. Any sample that was satisfactory under all three would almost certainly be satisfactory if seen under any other light source. Hence the CIE needed to specify only a small number of light sources, rather than all possible sources. (The CIE distinguishes between sources and illuminants. A *source* is a physical emitter of light, such as the sun or a lamp, while the term *illuminant* refers to a specified spectral energy distribution. Thus an illuminant can readily be specified, but may not be realisable in practice.) For standardisation purposes the minimum number is desirable. In 1931 fluorescent tubes were unimportant and the CIE specified three standard illuminants as follows.

CIE standard illuminant A

The spectral energy distribution of illuminant A is given in Appendix 1 and plotted in Figure 1.9. Illuminant A represents a black-body radiator at an absolute temperature of 2856 K. Source A can be realised by a gas-filled coiled tungsten-filament lamp operating at a correlated colour temperature of 2856 K. The energy distributions of source A and illuminant A can be very close if a calibrated lamp is used.

CIE standard illuminants B and C

Illuminants B and C correspond to different phases of daylight; illuminant B is intended to represent direct sunlight with a correlated colour temperature of 4874 K and illuminant C average daylight with a correlated colour temperature of 6774 K. The CIE gave details of how sources B and C may be obtained in the laboratory [12]. Neither corresponds very closely to real daylight, however, particularly in the near-UV region. The spectral energy distributions of illuminants B and C are plotted in Figure 1.9. The differences between the illuminants should be compared with the differences between the expected reflectance curves for different colours. Since the amount of light of any one wavelength reaching the eye is proportional to $E_\lambda R_\lambda$, it follows that the amounts of any given wavelength of light reaching the eye from a single surface illuminated by two different sources can be quite different. Thus, considering all wavelengths of the visible spectrum, the tristimulus values for a surface under two different illuminants may well differ widely even though, after allowing time for adaptation, the colours seen under the two sources may look very similar.

3.8.3 Standard observer

The experiments leading to the standard observer were outlined earlier. The results expressed in terms of the CIE [X], [Y] and [Z] primaries are denoted by $\bar{x}_\lambda$, $\bar{y}_\lambda$ and $\bar{z}_\lambda$, the distribution coefficients. These are plotted in Figure 3.6. Note that all values are positive: this results from the choice of primaries, as does the fact that $\bar{z}$ is zero for almost half of the spectrum. The values are also given in Appendix 2.

3.8.4 Standard illumination and viewing conditions

The CIE specified that opaque samples should either be illuminated at 45° from the normal to the specimen surface and viewed at an angle close to the normal, or be illuminated at an angle close to the normal and viewed at an angle 45° to the normal (Figure 2.6).

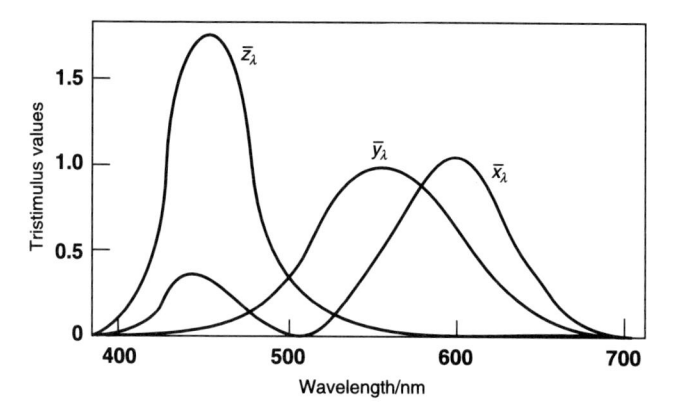

Figure 3.6 Tristimulus values $\bar{x}$, $\bar{y}$ and $\bar{z}$ derived by a linear transformation from the tristimulus values $\bar{r}$, $\bar{g}$ and $\bar{b}$ shown in in Figure 3.4; values $\bar{x}$, $\bar{y}$ and $\bar{z}$ refer to imaginary primaries [X], [Y] and [Z] and define the colour-matching properties of the 1931 standard observer

3.8.5 Units

So far the question of units has been avoided. Using the CIE system as described we could calculate the tristimulus values for a sample using Eqn 3.13. The units used for E_λ and $\bar{x}_\lambda$, $\bar{y}_\lambda$ and $\bar{z}_\lambda$ in Appendices 1 and 2 and Figure 3.6 are arbitrary, however. The values of E_λ for one wavelength relative to another are correct, but the absolute values have not been specified. Similarly the $\bar{x}_\lambda$ values for one wavelength are correct relative to other $\bar{x}_\lambda$ values at other wavelengths, and correct relative to the $\bar{y}_\lambda$ and $\bar{z}_\lambda$ values at the same wavelength; the absolute size of the values is arbitrary. For opaque samples (object colours) the usual practice is to normalise the tristimulus values using Eqn 3.18:

$$X = \frac{\sum E_\lambda \bar{x}_\lambda R_\lambda}{\sum E_\lambda \bar{y}_\lambda} \quad Y = \frac{\sum E_\lambda \bar{y}_\lambda R_\lambda}{\sum E_\lambda \bar{y}_\lambda} \quad Z = \frac{\sum E_\lambda \bar{z}_\lambda R_\lambda}{\sum E_\lambda \bar{y}_\lambda} \quad (3.18)$$

Thus if R_λ is expressed as a percentage, Y runs from zero (for a sample which reflects no light) to 100 (for a sample which diffusely reflects all the light incident on it). The ranges for X and Z depend on the illuminant. For a sample reflecting all the light incident upon it (R = 100% at all wavelengths), the X, Y and Z values for the sample under illuminants A, C and D_{65} (the last named is discussed in section 3.9.1) are given in Appendix 3: the Z tristimulus values in particular vary greatly with the illuminant.

The fact that the tristimulus values of a sample take no account of the intensity of light incident on the sample causes no problems in normal practice. If, for example, we produce a sample with the same tristimulus values as a target (for a particular illuminant), then the two will match for all normal levels of illumination by the chosen illuminant. Indeed the actual appearance of object colours is almost independent of the level of illumination. A piece of white paper looks white in weak sunlight. If the intensity of the sunlight increases, we recognise this, but the paper does not look any lighter. A medium grey is recognised as such whether viewed in daylight on a very dark day or on a very bright day. This is true even though the amount of light reflected by the grey sample on a bright day might well be more than that reflected by the white paper on a dark day. To a very large extent the appearance of object colours seems to be judged relative to the light source. We separate the properties of the sample from those of the source. This applies to changes in the distribution of light throughout the spectrum as well as to changes in the intensity of the source. Our white paper will still look white under tungsten light even though the distribution reaching the eye is vastly different from that when the paper is seen in daylight.

In summary, the CIE have defined a standard observer, standard illuminants, standard illuminating and viewing geometries and a particular set of primaries [X], [Y] and [Z]. The tristimulus values of a particular colour are the amounts of [X], [Y] and [Z] required to match the colour under standard conditions and are referred to as X, Y and Z. In practice, X, Y and Z for coloured materials are calculated from measured reflectance values together with $\bar{x}_\lambda$, $\bar{y}_\lambda$ and $\bar{z}_\lambda$ and E_λ for standard illuminants and observers. For the standard observer an additive mixture of X units of the [X] primary together with Y units of the [Y] primary and Z units of the [Z] primary would look the same as the sample illuminated by the appropriate standard illuminant.

3.9 ADDITIONS TO THE CIE SYSTEM

Since the CIE system of colour specification was adopted in 1931 the basic system has remained unchanged, but with increasing experience some additions have been made.

3.9.1 D illuminants

Illuminants B and C were intended to represent different phases of daylight. Later measurements showed that neither represents any common phase of daylight at all closely, particularly in the near-UV region [13]. This latter aspect is particularly important when considering samples treated with fluorescent brightening agents.

Illuminant D_{65} is based on measurements of the total daylight (sun plus sky) in several countries. Except for times near sunrise and sunset, the relative spectral energy distribution generally corresponds to correlated colour temperatures between 6000 and 7000 K. If we consider illumination by only part of the sky (a portion of blue sky from a north-facing window, for example, or direct evening sunlight from a west-facing window), the correlated colour temperature and energy distribution can be quite different.

Judd et al. showed that the different relative energy distributions could be represented quite closely by a series of curves dependent only on the correlated colour temperature of the particular form of daylight [13]. Taking Judd's work into account, the CIE has defined a series of D illuminants with correlated colour temperatures ranging from 4000 to 25 000 K. In the interests of standardisation, the CIE recommends that D_{65} should be used whenever possible. The energy distribution for D_{65} is given in Appendix 1 and Figure 1.9.

There seems to be no doubt that the D illuminants represent a substantial improvement over illuminants B and C. Illuminant B has rarely been used, and in recent years illuminant C has been almost completely replaced by the D illuminants, D_{65} in particular. The major problem with the D illuminants in that there is no satisfactory way of obtaining, say, D_{65} in the laboratory. Problems occur with metameric pairs. A pair that are a close match according to tristimulus values calculated for D_{65} might well be seen to be a poor match when viewed under a so-called daylight source in a viewing cabinet: the match might well look better if inspected under real daylight, however.

3.9.2 1964 supplementary standard observer (10°)

The original 1931 CIE standard observer was based on experiments using a 2° field of view. This is a much narrower field of view than that normally used for critical colour appraisal. In addition, a few problems were encountered using the 1931 observer, possibly due to the distribution coefficients being too low at short wavelengths [14]. New colour-matching experiments were therefore carried out by Stiles and Burch [15] and by Speranskaya [16]. The experiments were similar to those by Wright and Guild for the 1931 standard observer in that each wavelength in turn was visually matched using an additive mixture of the primary light sources. The main difference was the much wider field of view (10°). As indicated earlier, the use of wide fields with metameric matches causes problems. These were largely overcome by ignoring the centre 2° of the field of view. The two sets of experimental results were combined and used to define the 1964 (10°) CIE supplementary standard observer. Values of $\bar{x}_{10,\lambda}$, $\bar{y}_{10,\lambda}$ and $\bar{z}_{10,\lambda}$ are given in Appendix 4. The subscript 10,λ is used to distinguish the 10° data from the original 1931 2° standard observer data.

It is recommended that the 1964 observer is used whenever a more accurate correlation with visual colour matching of fields greater than 4° is required. There is probably little to choose between the two standard observers for some applications.

3.9.3 Standard illuminating and viewing conditions

The original CIE recommendation was that the sample should be illuminated at 45° to the surface and the light viewed normally, i.e. at right angles to the surface. This mode can be represented by '45/0'. It was assumed that the opposite mode (0/45) would give the same result, but this is not the case if the incident light is polarised [17]. Four possible sets of conditions are now recommended. These are 45/0, 0/45, d/0 and 0/d (Figure 2.6). In the third case the sample is illuminated by diffuse light while in the last case the light reflected at all angles is collected (using an integrating sphere).

3.9.4 Standard of reflectance factor

The CIE recommends that reflectance measurements should be made relative to a perfect diffuser, i.e. a sample that diffusely reflects all the light incident upon it. No such surface exists, but working standards of known spectral reflectance factors are normally used, allowing the correct results to be obtained. (For example, if the working standard reflects 98% of the light of a particular wavelength all values measured relative to the working standard need to be multiplied by 100/98.) In practice, instrument manufacturers supply calibrated white tiles with their instruments. Using these, corrected R values are obtained automatically.

3.10 CALCULATION OF TRISTIMULUS VALUES FROM R VALUES MEASURED AT 20 nm INTERVALS

In Appendices 1, 2 and 4 the values for E_λ and $\bar{x}_\lambda$, $\bar{y}_\lambda$ and $\bar{z}_\lambda$ are given at 5 nm intervals. If R_λ values are also available at 5 nm intervals, tristimulus values can be calculated exactly as described in section 3.7. Many spectrophotometers give reflectance readings at 10 or 20 nm intervals; in this case there is no official CIE advice and different tables of values have been published.

Stearns has described a method which allows appropriate adjustments to the tables of $E_\lambda \bar{x}_\lambda$, etc. values to be calculated for any wavelength intervals [18,19]. The results from using these adjusted tables are consistent with those obtained using 5 nm interval tables and interpolated R values where necessary. Values from other sources are slightly different, and it is difficult to argue that one set is better than another. Even very small differences can cause problems, however (for example, when checking computer programs), and it is highly desirable that everyone uses the same set of values. In the first edition of this book, values calculated using Stearns' method were given. In an attempt to promote uniformity of practice, the Colour Measurement Committee of the Society of Dyers and Colourists has recently recommended that the corresponding values recommended by the American Society for Testing and Materials (ASTM) should be used. These values are given in Appendix 5 for illuminants A, C, D_{65}, some representative fluorescent tubes, and both the 1931 (2°) standard observer and the 1964 (10°) supplementary standard observer. An example of calculating tristimulus values from reflectance values using these tables is given in Appendix 6. Readers requiring more detail regarding the calculation of tristimulus values should refer to ASTM E308-95. The address is given at the beginning of the appendices in this book.

3.11 RELATIONSHIPS BETWEEN TRISTIMULUS VALUES AND COLOUR APPEARANCE

It is difficult to relate the tristimulus values of a sample to the colour appearance in any simple way. One reason is that the colour depends not only on the stimulus, but also on surrounding colours and the state of adaptation of the eye. Even ignoring such factors, the three-dimensional nature of colour makes determining relationships difficult and it is usual to simplify any relationship involving colour by considering only one or two dimensions at a time. It was indicated earlier that in choosing primaries for the CIE system the wide range of possibilities had been used to produce certain desirable features in the final system. A consequence of ensuring that the $\bar{y}_\lambda$ curve corresponded to the V_λ curve was that the Y tristimulus value should roughly represent the lightness of a sample; the higher the Y value, the lighter the sample appears. The scale is far from uniform (as will be discussed in section 3.14), and caution should be exercised when comparing the lightness of quite different colours such as reds and greys; nonetheless, the principle holds in general. Thus if Y = 80 we can be sure that the sample will appear light, while if Y = 3 the sample will look dark.

3.12 CHROMATICITY DIAGRAMS

To represent the other two dimensions of colour it is usual to first define *chromaticity coordinates* (x, y and z) and then plot y against x (Eqn 3.19):

$$x = \frac{X}{X+Y+Z} \qquad y = \frac{Y}{X+Y+Z} \qquad z = \frac{Z}{X+Y+Z} \qquad (3.19)$$

From Eqn 3.19 it follows that $x + y + z = 1$ for all colours; it is therefore only necessary to quote two of the chromaticity coordinates, and these can of course be plotted on a normal two-dimensional graph. It can also be shown that X and Z can easily be calculated from x, y and Y; hence the latter set is an acceptable form of specification, and consideration of Y values and plots of y against x should cover all possible colours. A plot of y against x is called a *chromaticity diagram*. Such a plot is shown in Figure 3.7, in which the spectrum colours are plotted. The line joining the spectrum colours is known as the *spectrum locus*. The x and y values for each wavelength were obtained from the corresponding distribution coefficients (1931 standard observer in this case) (Eqn 3.20):

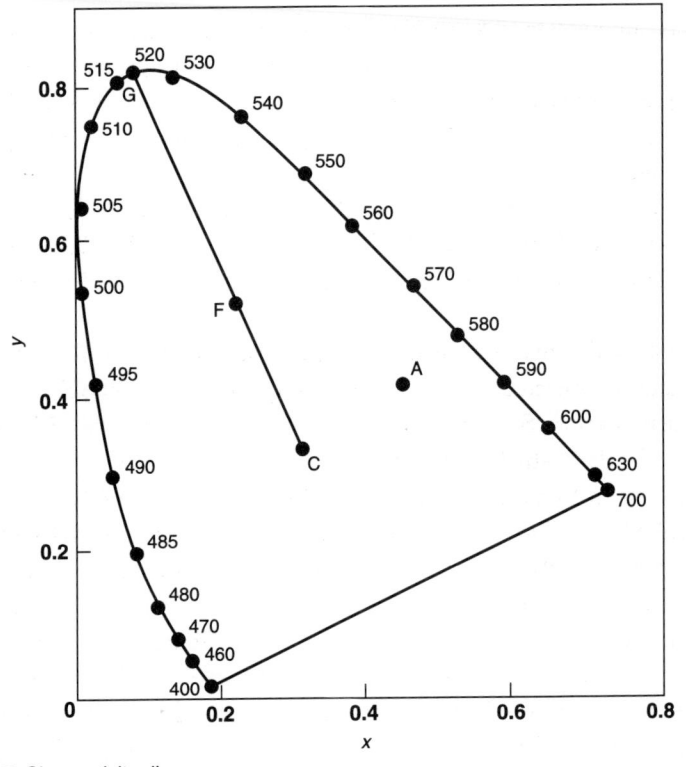

Figure 3.7 Chromaticity diagram

$$x = \frac{\bar{x}_\lambda}{\bar{x}_\lambda + \bar{y}_\lambda + \bar{z}_\lambda} \qquad y = \frac{\bar{y}_\lambda}{\bar{x}_\lambda + \bar{y}_\lambda + \bar{z}_\lambda} \qquad (3.20)$$

Given a knowledge of the colours of the spectrum wavelengths, we can immediately begin to see where pure colours fall on the chromaticity diagram. Wavelengths around 480 nm look blue, wavelengths around 520 nm look green, while wavelengths from 630 nm to the end of the spectrum look red. Colours with x and y values close to the spectrum locus will be very saturated colours with hues close to those of the corresponding spectrum colours. For other colours the problem is more difficult. From the tristimulus values given in Appendix 3 for a sample reflecting all the light incident upon it (and therefore presumably white) we can calculate that, for the sample illuminated by illuminant A, $x = 0.448$ and $y = 0.408$; for the same sample illuminated by illuminant C the values are $x = 0.310$ and $y = 0.316$. These points are marked in Figure 3.7 and denoted by A and C respectively. Thus the same colour appearance (white in this case) can be represented by two quite different points on the chromaticity diagram (or quite different sets of tristimulus values). Consequently. in attempting to predict colour appearance from chromaticity coordinates or tristimulus values, we must be careful to ascertain which illuminant has been used. Strictly speaking, in any application we should also be careful to ascertain which standard observer and which set of observing and viewing conditions are appropriate, but these are not normally as important as the illuminant.

If we now consider colours relative only to illuminant C (surface colours illuminated by illuminant C and coloured lights with the eye adapted to illuminant C), the positions for other colours can be deduced from a simple property of the chromaticity diagram.

If two coloured lights are represented by two points on the chromaticity diagram, then any additive mixture of the two will correspond to a point on the straight line joining the two points. Since the spectrum locus is always concave, it follows that all real colours (each of which must correspond to one or more wavelengths additively mixed) must fall within the area bounded by the spectrum locus and joining the ends. Mixing white light (illuminant C) with monochromatic light of wavelength 520 nm will give points exactly on the line CG in Figure 3.7. Since light of 520 nm looks green, the mixtures will appear various shades of green, from white through pale greens to the saturated green of the spectrum colour. (The points will fall exactly on the line; the colours seen will not necessarily look exactly the same hue. What is seen depends on many factors, but generally mixing white light and a spectrum colour will produce a slight but significant change in hue.) All colours lying on the line CG may be described as colours having a dominant wavelength of 520 nm. Similarly mixing white light and light of wavelength 700 nm (red) will produce a range of pinks and reds. In general, the more the colour resembles the spectrum colour, the closer will the point be to the spectrum locus, while near-neutral colours will correspond to points close to C. For colour F in Figure 3.7 this attribute is defined by the ratio CF : CG, known as the *excitation purity* of colour F. As the excitation purity increases the colour will look less like a neutral colour and more like the corresponding spectrum colour. Samples with excitation purity as low as 0.1 (or 10%) will look distinctly different from neutral. Even very saturated-looking samples, particularly greens, will have excitation purities far from 1 (or 100%).

For the sample used as an example for the calculation of tristimulus values in section 3.7, $X = 38$, $Y = 45$ and $Z = 21$, hence $x = 0.365$ and $y = 0.433$. Remembering that the tristimulus values were calculated for illuminant A, we can see that the dominant wavelength is about 500 nm and hence the sample is a green-blue. If, however, the illuminant was mistakenly taken to be C the dominant wavelength would have been estimated to be about 580 nm and the colour judged to be yellow!

It was stated in section 3.11 that the Y scale is far from uniform. The same applies to the *xy* diagram; equal distances in the diagram do not correspond to equal visual differences. For a fixed difference in x and y the difference seen would be much smaller for a pair of green samples than for pairs of blue or grey samples. This problem will be considered further in section 3.14.

It has been emphasised that colour is three-dimensional. Thus no two-dimensional plot can represent colour completely. In the case of the chromaticity diagram it is simplest to regard the missing factor as the Y tristimulus value. Consider a sample where R = 10% at all wavelengths. If the sample is illuminated by illuminant C the tristimulus values are simply one-tenth of the corresponding values for the sample described in Appendix 3 (where R = 100% at all wavelengths) and the chromaticity coordinates are the same: x = 0.310 and y = 0.316. Both samples are neutral and the difference between the two is indicated by the Y tristimulus values. A neutral sample with a Y value of 100 would be white, while one with a Y value of 10 would be a darkish grey. All other samples with similar chromaticity coordinates would look neutral, but could be white, black or any intermediate shade of grey. (All samples with constant R values will look neutral, but the converse does not hold; many neutral-looking samples have R values that vary considerably with wavelength.) Similarly a colour fairly close to neutral but with a dominant wavelength of 650 nm would look a pale pink if the Y value was very high (the colour was very light), but a reddish grey if the Y value was low (the sample was dark).

In general, any one point on the chromaticity diagram corresponds to a range of colours differing in lightness, and this should always be kept in mind when trying to visualise the colours corresponding to particular chromaticity coordinates. The relationships between x, y and Y values on the one hand and the visual appearance on the other could be developed much further, but it is recommended that, if possible, students should measure their own samples. With modern instruments a student can measure dozens of samples in an hour, and compare the readings obtained with the visual appearance of the samples. This is far better than relying on the vague terms such as grey, red, pink and so forth that have to be used in a textbook. Particular attention should be paid to colours such as browns, fawns and purples. After a little practice it is instructive, for each new sample, to estimate the dominant wavelength, excitation purity, x, y and Y *before* making measurements.

3.13 USEFULNESS AND LIMITATIONS OF THE CIE SYSTEM

In many ways the CIE system of colour specification has been remarkably successful. Almost all important applications of colour measurement use it, and the basic system has survived unchanged for over 60 years. The additions made since 1931 have led to improvements in some respects, but have not changed the basic principles of the system in any way. It is unlikely that any major changes will be made in the foreseeable future.

On the other hand the system has certain limitations. These stem basically from the limited objectives of the system, rather than from a failure to meet those objectives.

The CIE tristimulus values for a sample are related to the colour of the sample, but ignore other important features such as surface texture, gloss and sheen. Thus a gloss paint sample and a matt paint sample might have the same tristimulus values, but obviously will not look the same. Whether the colours of the two samples will look the same depends critically on the geometrical arrangements for illuminating and viewing them. Only if the instrumental geometry of illumination and viewing conditions is similar to that used for visual matching will the colours be seen to be close. An instrument will always average out the light reflected from the area being measured (typically a 2 cm diameter circle). In judging a colour visually some sort of averaging takes place, but the observer is always conscious of any non-uniformity over the area viewed. Thus a matt paint surface, a woven textile surface and a pile fabric will always look different from one another, but their measured tristimulus values could be the same.

Ignoring all features other than colour, the tristimulus values for a sample give only a limited amount of information. The tristimulus values tell us the amounts of three imaginary primaries which if additively mixed will give the same colour as a surface illuminated by a standard source and viewed by a standard observer using one of the standard geometries (assuming that the instrument does correspond to the standard conditions). It follows that the mixture of the CIE primaries would be unlikely to match the surface if it was illuminated by a different source or if the 'match' was viewed by an individual observer or if different illuminating or viewing geometry was used. Obviously a degree of control can be exercised over the source and the geometry. If these are important, we must try to ensure that the instrumental conditions correspond as closely as possible to those to be used when viewing the object visually. The only choice as far as the standard observer is concerned is whether to use the 1931 (2°) observer or the 1964 (10°) observer. Neither is likely to correspond closely to any individual observer, but either may well correspond reasonably closely with the average judgement of real observers, bearing in mind that in many applications the product is mass-produced and will be seen by many different observers.

A full specification of a colour requires X, Y and Z values (or equivalent sets such as x, y and Y or L^*, a^* and b^*) for several different illuminants. The results are still valid only for the standard observer and could be unsatisfactory for a real observer. This should not be a problem in practice since the need is usually for the colours to be acceptable to a large number of potential customers; the standard observer is probably a better guide to the population in general than any one observer. It often happens that one particular individual (the head dyer or senior buyer for a chain store, for instance) inspects fabrics, and problems may arise if that individual's colour vision is appreciably different from that of the standard observer. Problems will be most severe for highly metameric pairs of samples.

Strictly speaking, the tristimulus values tell us nothing about the colour of a sample although, as discussed above, with experience we can make a reasonable estimate of the colour from either X, Y and Z or x, y and Y values. It is then essential that the illuminant used for the measurements is known. Chromaticity coordinates of $x = 0.314$ and $y = 0.331$ correspond to a neutral colour if derived from measurements under illuminant D_{65}, but to a blue colour if derived from measurements under illuminant A.

In many applications the aim is to match a particular target, which might be defined by a set of tristimulus values. If we produce a sample and wish to test whether this matches the target, the sample and target measurements must correspond to exactly the same conditions (illuminant, standard observer, illuminating and viewing geometries, and, in practice, the same instrument). If, for example, the sample really is a good match to the target but the tristimulus values are measured using different standard observers for the sample and target, the resulting tristimulus values and chromaticity coordinates would be appreciably different. Again the sample and target might have different surface structures, such as those of matt paints, gloss paints or pile carpets: the tristimulus values could then be identical but the surfaces would look appreciably different. Whenever possible sample and target should have the same surface structure.

It is usually important that the colours within one batch of fabric, and between repeat batches, should match closely. In these cases the samples will of course be of the same material and the same dyes or pigments will have been used. It would be natural to measure all the samples using the same instrument and (if a spectrophotometer was used) to calculate the tristimulus values for the same standard observer and standard illuminant. Under these conditions, if the tristimulus values for a sample are very close to those for the standard, then the sample will be a close visual match to the standard for any normal observer viewing under a light source roughly equivalent to the standard illuminant used to calculate the tristimulus values. (Exceptions are known: for example, the appearance of metallic paints depends very much on the illuminating and viewing geometries.) If the variation of appearance with, say, viewing angle is different for the sample and the standard, they may not match visually even though the instrumental results (obtained with a different viewing angle) indicate that they should.

3.14 NON-UNIFORMITY OF THE CIE SYSTEM: COLOUR DIFFERENCES

In many ways the chief limitation of the CIE system is its non-uniformity. Equal changes in x, y or Y do not correspond to equal perceived differences. Many attempts have been made to provide a more uniform system. The basic approach has always been to start with the tristimulus values or chromaticity coordinates from the CIE system, and to transform these in some way to give a more uniform system. The end result is a colour-difference formula which, for a pair of samples, gives a number that is intended to be proportional to the difference seen.

Many colour-difference formulae have been proposed, and some are discussed in Chapter 4. The CIE $L^*a^*b^*$ (CIELAB) formula was one of two recommended for general use by the CIE in 1976 [20]. The formula was recommended by the Colour Measurement Committee of the Society of Dyers and Colourists in 1976 [21]. Although it was known to be far from ideal, it was the best available at that time. One feature of the formula that has proved to be most useful has been the associated colour space. This is obtained by using L^*, a^* and b^* as mutually perpendicular axes (Figure 4.5) where L^*, a^* and b^* are defined (except for very low tristimulus values) by Eqn 3.21:

$$L^* = 116(Y/Y_n)^{1/3} - 16$$
$$a^* = 500[(X/X_n)^{1/3} - (Y/Y_n)^{1/3}] \quad (3.21)$$
$$b^* = 200[(Y/Y_n)^{1/3} - (Z/Z_n)^{1/3}]$$

where X_n, Y_n and Z_n are the tristimulus values, for a particular standard illuminant and observer, for a sample reflecting 100% of the light at all wavelengths (Appendix 3). The lightness of the sample is represented by L^* on a scale running from zero for black to 100 for white. The other attributes can be represented on a plot of b^* against a^*. Neutral colours plot close to the origin for any illuminant ($a^* = b^* = 0$).

Colour differences can be quantified by ΔE according to Eqn 3.22:

$$\Delta E = [(\Delta L^*)^2 + (\Delta a^*)^2 + (\Delta b^*)^2]^{1/2} \quad (3.22)$$

and ΔE is intended to be proportional to the difference seen. An idea of the size of ΔE units can be gained by considering that the difference between black and white corresponds to 100 ΔE units, while commercial tolerances are often between 1 and 2 ΔE units.

The other formula recommended by the CIE in 1976 was the CIE $L^*u^*v^*$ formula (except for very low tristimulus values) (Eqn 3.23):

$$L^* = 116(Y/Y_n)^{1/3} - 16$$
$$u^* = 13L^*(u' - u'_n) \quad (3.23)$$
$$v^* = 13L^*(v' - v'_n)$$

where u' and v' are given by Eqn 3.24:

$$u' = \frac{4X}{X + 15Y + 3Z}$$
$$v' = \frac{9Y}{X + 15Y + 3Z} \quad (3.24)$$

and u'_n and v'_n are the u' and v' values calculated using X_n, Y_n and Z_n.

The associated colour space is formed by using L^*, u^* and v^* as mutually perpendicular axes, while colour differences are calculated from Eqn 3.25:

$$\Delta E = [(\Delta L^*)^2 + (\Delta u^*)^2 + (\Delta v^*)^2]^{1/2} \quad (3.25)$$

A plot of u' against v' has the useful properties associated with chromaticity diagrams described earlier, but distances in this plot correspond more closely to perceived differences.

Since 1976, better colour-difference formulae have been developed (Chapter 4). Most take the form of modifications of CIELAB, and the $L^*a^*b^*$ space is still widely used to represent colours. Modern colour-measuring instruments will normally display L^*, a^* and b^* as well as X, Y, Z and x, y. Sometimes just L^*, a^*, and b^* are displayed, but it should be remembered that these have been calculated from the tristimulus values. Again, it should be emphasised that students should measure a wide range of coloured materials and see for themselves the relationships between all the measured coordinates and the perceived colours.

3.15 METAMERISM

Metamerism, mentioned in section 1.9.4, is the phenomenon of two colours that match under one set of conditions, but fail to match under a different set. The two lights must be physically different. With coloured lights, a white light could be a mixture of approximately equal amounts of all wavelengths and could be matched by a mixture of two complementary wavelengths. With surface colours, a sample dyed with a given set of dyes could be matched, under certain conditions, with a sample dyed with different dyes. In each case the match would be physically different, and the match would fail to hold when the conditions changed. For the white light a match for one observer would probably not be accepted as a match by a second observer. For the dyed samples, the match would probably not hold if the light source were to be changed; the match might be satisfactory in daylight, for example, but very poor under fluorescent light. Again, a good match for one observer might be perceived as unsatisfactory by a second.

Suppose we have two objects whose tristimulus values (for a specified standard illuminant and standard observer) are identical. The two objects illuminated by the specified source and viewed by the specified standard observer would look identical, yet their reflectance curves might differ significantly. For example, one object could have constant R values of, say, 30% at all wavelengths (and would look grey under all normal conditions), while the other object could be a piece of nylon dyed to match the first using a mixture of yellow, red and blue dyes. We can generally produce any particular grey colour using such a mixture of dyes, but it is likely that the R values will vary with wavelength. (The lowest R values would probably be found at the wavelengths of maximum absorption for each dye, say 450, 520 and 620 nm respectively.) We can match most coloured objects (under specified conditions) using a mixture of three suitable dyes, but the reflectance curve of the dyed fabric will not be the same as that of the object to be matched. In these cases the dyeing will normally not match the target colour if the conditions are changed, and the two samples are said to be metameric, or to form a metameric pair. In terms of colour measurements, the two samples have the same tristimulus values for a specific combination of illuminant and observer, but the reflectance curves are different. Usually such pairs of reflectance curves cross at least three times. In calculating the tristimulus values the effects of higher R values for the dyeing at some wavelengths must be balanced by the effects of lower R values at other wavelengths. If the illuminant used in the calculation is changed, not surprisingly the tristimulus values will usually change as well. This is also true if the observer is changed.

Four types of metamerism are recognised:
(a) illuminant metamerism
(b) observer metamerism
(c) field size metamerism
(d) geometric metamerism.

Illuminant metamerism is the most important type and occurs when tristimulus values for a metameric pair are calculated using two or more illuminants, or when we look at a metameric pair illuminated successively by two or more sources. A similar problem can arise if a metameric pair match for, say, standard illuminant D_{65}, but fail to match when looked at under a 'daylight' source with a different spectral energy distribution.

Observer metamerism is exhibited when a metameric pair matches for one person, but fails to match when seen by a second person. In this case the colour-matching functions of the two people are different. This is quite normal; it is also normal for the colour-matching functions of any one person to be different from those of the standard observer. Hence it is possible for the tristimulus values for a metameric pair to be identical when calculated for the standard observer, but for the pair to fail to match for any available real observer, even if the correct light source is used. Observer metamerism can give rise to serious problems if the two people are, for example, a supplier and a customer.

In *field size metamerism* a satisfactory match is lost when the field size changes, for example from 2° to 10°. A metameric pair that match when seen at a distance (small field of view) may no longer match when the observer is closer to the samples (large field of view).

Geometric metamerism may be observed when the viewing geometry changes. Metallic paints, for example, may match the target colour for one particular angle of illumination and angle of viewing, but no longer match if either angle is changed.

Metamerism is important in practice because we often do match a sample using a different set of dyes (or pigments). This might be done to enable cheaper dyes to be used, or to produce a match with better light-fastness properties.

Different degrees of metamerism may occur. A metameric pair may match perfectly under one set of conditions, but be a slight mismatch under a different set: such a pair would be said to be slightly metameric. Another pair could match perfectly under one set of conditions, be a slight mismatch under a second set but be a very bad mismatch under a third set. Such a pair would be said to be highly metameric. It is normally desirable to use a new set of dyes with the required cost or fastness properties, but giving a low degree of metamerism. How much metamerism can be tolerated depends on the application. If the two samples will never be seen together, metamerism can be ignored; if, say, the trousers and jacket of a man's suit were to be dyed with different dyes, even a small degree of metamerism would be unacceptable. Whenever we use a different set of dyes to match a sample, some metamerism is inevitable, but this can be minimised by careful choice of the new dyes.

Quantification of the degree of metamerism is difficult. In general, for highly metameric pairs the differences between the reflectance curves are large. Differences in different parts of the visible spectrum are not equally significant; the wavelength regions close to 400 and 700 nm are much less important than those at, say, 550 nm, a wavelength to which the eye is highly sensitive. Hence it is difficult to judge from a pair of reflectance curves just how metameric the corresponding samples will be.

Alternatively, we might consider the colour differences seen over a range of conditions (with different illuminants or different observers): it is difficult, however, to know which illuminants to use. For most purposes it should be adequate to use three illuminants, one corresponding closely to daylight, one to tungsten light and one to a high-efficiency fluorescent tube such as the Philips TL84 lamp, but the illuminants used should be as different as possible from each other. Practical circumstances should also be taken into account. If a customer is known to check matches under a particular source, then that source should be one of those used.

When real samples are available the matches can be checked visually, or the differences for the required illuminants can be calculated from the appropriate tristimulus values using a colour-difference formula. In recipe-prediction programs alternative recipes to match a particular target are often calculated. The likely degree of metamerism can then be predicted before any samples are dyed by calculating the colour differences for the required illuminants.

In all the foregoing discussion it has been assumed that the metameric pair matches perfectly under one particular set of conditions. In practice this is not necessarily so. A pair of samples may well be visibly different under all sets of conditions, and additional complications occur in the estimation of the degree of metamerism. In these cases some sort of correction must be made to the measured coordinates for the second (or third) illuminants to correct for the mismatch under the first illuminant.

Consider, for example, the $L^*a^*b^*$ values given in Table 3.1. Sample 1 is one L^* unit lighter than target 1 under illuminant D_{65}. Hence the fact that it is also lighter (by 1.5 L^* units) than the target under illuminant A is an indication more of the original mismatch than of a significant degree of metamerism. Sample 2 is two a^* units redder than target 2 under illuminant D_{65}. Hence the fact that under illuminant A the a^* value is too low (the sample is greener than the target) is an indication that sample 2 is appreciably metameric with respect to target 2, and that some sort of correction must be made to the measured coordinates for the samples to correct for the mismatch under the first illuminant. In our first example we may correct by subtracting one unit from the L^* values of sample 1 for all illuminants before calculating the differences; in the second example we may do so by subtracting two units from the a^* values for sample 2.

Table 3.1 Metamerism in samples and targets

	Illuminant D_{65}			Illuminant A		
	L^*	a^*	b^*	L^*	a^*	b^*
Target 1	51	10	5	52	8	6
Sample 1	52	10	5	53.5	8	6
Target 2	51	5	10	51	6	11
Sample 2	51	7	10	51	4	11

3.16 COLOUR CONSTANCY AND CHROMATIC ADAPTATION

Colour constancy and metamerism are closely linked and sometimes confused. The distinction between them is clear: metamerism concerns a pair of samples or colours, while colour constancy is a property of a single sample. In section 3.15 metamerism was discussed in terms of the possible differences in colour between pairs of samples which matched under one particular set of conditions.

Most natural objects appear to be more or less the same colour when illuminated by different light sources. We talk of white paper, white chalk and so forth, implying that the paper and chalk always look white, and indeed they do under all normal conditions. This property of colour constancy is such a normal part of everyday experience that it comes as a surprise to realise that it only arises by a complicated process.

As indicated earlier, the percentage of light reflected at any wavelength by a nonfluorescent object is constant, irrespective of the amount or type of light incident upon it. Consider a piece of white chalk. Let us suppose that the chalk reflects 95% of the light at all wavelengths. If the chalk is illuminated by daylight corresponding to illuminant D_{65} the amount of light reflected is $0.95E_\lambda$ where E_λ is the relative energy of illuminant D_{65} at a wavelength λ. From Figure 1.9 we can see that E_λ is relatively high at the wavelengths in the middle of the visible region and rather lower at the ends of the visible region. The energy distribution of the light reflected by the chalk follows the same pattern, and it seems somewhat surprising that the response produced by this should be 'white'.

Now consider the same piece of chalk illuminated by tungsten light, corresponding to illuminant A. The relative energy of illuminant A across the visible region is also shown in Figure 1.9, the amount of energy emitted at the longer wavelengths being far greater than that emitted at the shorter wavelengths. Since the percentage of light reflected is still 95% at all wavelengths, a far greater amount of light will be reflected at the longer wavelengths than at the shorter wavelengths. The distribution of the light reflected is obviously completely different from that of the light reflected when the chalk was illuminated by daylight. Although the light reaching the eye is different in the two conditions, we normally see the chalk as being white under both illuminants.

Obviously some sort of compensation process is taking place. The eye and/or the brain in some way compensate for the change of illumination. If our eyes are adapted to daylight and we go into a darkened room, we see little at first. Our eyes gradually adapt to the low level of illumination over a period of about half an hour and our vision improves greatly. A similar process occurs if our eyes are first adapted to daylight, and we then enter a room lit by tungsten light. Again our eyes gradually adapt to the new light. This process is known as *chromatic adaptation*. The situation is more complicated if we change light sources quickly. If for a while we look at a sample of white paper in a viewing cabinet using a daylight source and then the source is changed to tungsten light, the paper looks more or less white after only a second or two. This adaptation is far quicker than normal chromatic adaptation. Since our eyes adapt slowly for up to half an hour there will be further changes in the perceived colour, but these are normally relatively slight; our white paper continues to look white. The simplest explanation seems to be that in some way we compare the light reflected from a surface with that emitted by the light source. If the energy distributions are similar we judge the object to be neutral in colour. (This obviously cannot apply to experiments with coloured lights in which there is no light source in the normal sense.)

We know from experience that while most objects remain more or less colour-constant under normal light sources, some objects do change colour appreciably. This can be very annoying if, for example, we have bought an article because we liked the colour as it appeared in the shop, but find it less attractive when we see it under tungsten light at home. The question arises: how can we deduce from colour measurement whether an object will be colour-constant? and if not, how great will the change of colour be? As with metamerism, the degree of nonconstancy can vary. Other things being equal, we would like to select sets of dyes which give a reasonably colour-constant sample.

We have already seen that the R values for all nonfluorescent samples are independent of the light source, and hence give no direct guide to colour constancy. The *amount* of light reflected from all objects changes greatly as the light source changes, even for colour-constant objects. Similarly the tristimulus values and chromaticity coordinates for the same object under different sources are quite different and again give no direct guide to colour constancy (Appendix 3: note that the Z tristimulus value changes by a factor of 3 when the illuminant changes from A to D for a sample which would look white under both sources).

The $L^*a^*b^*$ values change much less, but the relationships between changes in these coordinates and changes in colour appearance are not known with any certainty. Even if the $L^*a^*b^*$ coordinates of an object under two different light sources are identical, it cannot be concluded that the object is colour-constant. Similarly, significant differences between the $L^*a^*b^*$ coordinates do not necessarily imply that the object is *not* colour-constant. It is necessary to know the relationship between the tristimulus values (or chromaticity coordinates or $L^*a^*b^*$ values) for colours which look exactly the same to an observer adapted to different illuminants. Such colours are called *corresponding colours*. Given the tristimulus values for an object under one light source, we could calculate the tristimulus values for a corresponding colour under a second source. Any difference between these and the measured values for the object under the second source would be a measure of the degree of colour change or colour nonconstancy. (The difference could be expressed as a ΔE value calculated using a colour-difference formula.)

Considerable experimental work related to chromatic adaptation and colour constancy has been carried out [22–26]. A brief discussion of two of the techniques used will enable the difficulties involved in such studies to be appreciated. Working with coloured objects, such as painted papers, some sort of memory method must be used. An observer, after a suitable adaptation period, describes the colour seen in terms of Munsell coordinates, for example. The experiment is repeated with the observer adapted to a second light source and the difference between the Munsell coordinates given for the two sources is a measure of the colour nonconstancy of the sample. The accuracy of the results depends on how well the observers can memorise the Munsell system.

The demands on the observer's memory are much less if only a short period of time is allowed for adaptation. If an observer views a sample in a viewing cabinet and is allowed to switch backwards and forwards between two light sources, any change in colour should be evident and the change could be quantified – by reference to a grey scale, for example. Under these conditions, however, there is no doubt that the observer's eyes are not properly adapted to either of the light sources and it is doubtful how far such results would apply to a normal viewing situation where an observer is properly adapted to one light source. If we buy an article which looks the desired colour under tungsten light in the home but which changes markedly when taken into a kitchen lit by a fluorescent tube, we might not be too worried if the original colour apparently returned after a few minutes, but would be much more concerned if the difference in colour remained.

In experiments with visual colorimeters the field of view can be arranged so that one eye, adapted to one light source, sees a test colour, and at the same time the other eye, while adapted to a second light source, sees a second test colour. The observer adjusts the second test colour until it appears to match the first test colour. The two test colours are now 'corresponding colours' and their coordinates can be recorded.

The results of such experiments appear to be significantly different depending on whether the colours involved are surface colours or coloured lights in colorimeters. Results from experiments with surface colours are likely to correspond more closely with what is required than those from experiments using visual colorimeters; the conditions of the former experiments are much closer to everyday viewing conditions. Unfortunately when proper time is allowed for adaptation to the light source the results will be less precise.

Several different chromatic adaptation formulae have been proposed [22,24,27,28]. Each uses tristimulus values for one illuminant to calculate the corresponding colour for a second illuminant. The results from different formulae can be quite different. Prediction errors are typically of the order of 0.02 in x and y. This corresponds roughly to ten ΔE units (1976 CIELAB formula), which is larger than the colour shifts normally observed for surface colours. It seems that more accurate formulae are required before

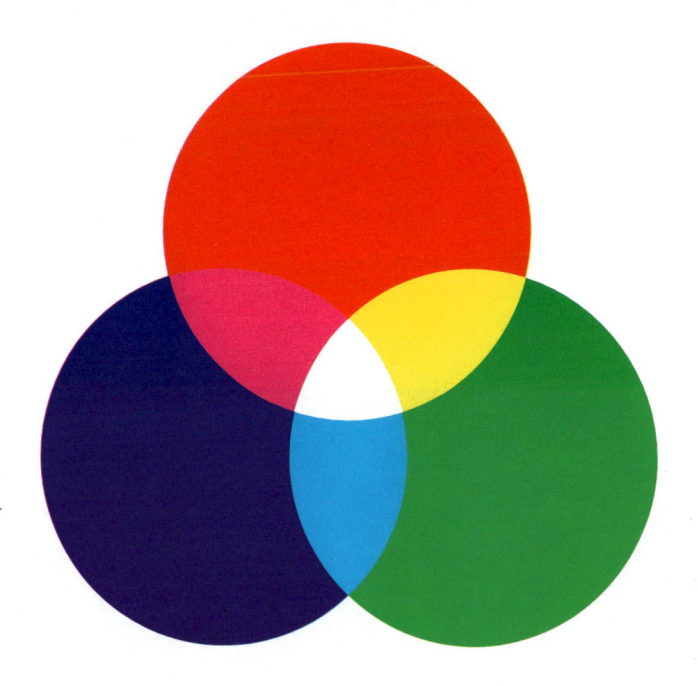

Additive

Subtractive

Plate 1 Additive and subtractive colour mixing

Plate 2 The Munsell Color Tree
(photo courtesy of Macbeth Division of Kollmorgen Instruments Corpn, New Windsor, NY, USA)

prediction of the degree of colour constancy of surface colours can be useful, and recent work suggests that a satisfactory chromatic adaptation formula may indeed now be available [29].

3.17 COLOUR RENDERING OF LIGHT SOURCES

Most normal samples look more or less the same colour under the various phases of daylight and under tungsten light, while under the light from a fluorescent tube some of these samples no longer appear to be their 'true' colour. Under these circumstances we tend to blame the light source rather than the sample. We say that the fluorescent lamp has poor 'colour-rendering' properties.

The fact that some samples do not appear to be their true colours under some fluorescent tubes is related to the spectral composition of the light from the tubes. It is quite possible for the light from a fluorescent tube to look very similar to daylight, even though the spectral composition may be quite different. Whereas for daylight there is very roughly the same energy at each wavelength, for fluorescent tubes the energy tends to be concentrated at four wavelengths (corresponding to the mercury emission lines at 405, 436, 546 and 578 nm). Thus the spectral composition of the light reflected from a sample lit by a fluorescent lamp may well be quite different from that of the light reflected from the sample in daylight. Not surprisingly, the colour of the sample might well appear to be different under the two sources.

A wide range of fluorescent lamps is available today. Some are intended to look like daylight and some like tungsten light. For any particular appearance, however, most of the light can be concentrated at the mercury emission wavelengths (generally giving a more efficient lamp in that less electricity is used); alternatively, the proportion of the light spread over the whole spectrum can be much higher. Such a lamp would be less efficient in the use of electricity, but its colour-rendering properties would be better. There is an official CIE method for specifying the colour-rendering properties of light sources relative to a reference illuminant [30].

The reference illuminant should be nearly the same chromaticity as the lamp to be tested. The reference illuminant for light sources with correlated colour temperatures below 5000 K should be a Planckian radiator (section 1.3); for higher correlated colour temperatures the appropriate D illuminant should be used. The method specifies 14 test colour samples. In all cases the difference between the perceived colour illuminated by the light source and that of the sample illuminated by the reference illuminant is calculated.

The CIE method aims at estimating the degree to which the light source imparts to the test samples their 'true' colours. For some purposes this is not satisfactory. For example, a light source to be used in a butcher's shop should obviously make the meat look fresh; the colour of other materials is much less important. More generally we are often concerned with the attractiveness of colours, rather than whether they are rendered accurately. Many samples look more attractive under a source which makes them look brighter and more colourful.

These remarks should not be taken as a criticism of the CIE methods. Clearly no one method can be satisfactory for all possible applications. Rather they should be taken as a warning that when dealing with any aspect of colour, we should carefully consider exactly what we require. The techniques available for the measurement of colour and colour differences should be used to help us solve problems, and not allowed to become an end in their own right.

REFERENCES

1. H Grassman, *Poggendorf's Ann.*, **89** (1853) 69.
2. H Grassman, *Phil. Mag.*, **7** (4) (1853) 254.
3. F Blottiau, *Rev. d'Opt.*, **26** (1947) 193.
4. P W Trezona, *Proc. Phys. Soc.*, **B66** (1953) 548.
5. P W Trezona, *Proc. Phys. Soc.*, **B67** (1954) 513.
6. G S Brindley, *J. Physiol.*, **122** (1953) 332.
7. P W Trezona, *Color metrics* (Soesterberg, Holland: AIC, 1972) 36.
8. W S Stiles and G Wyszecki, *Acta Chromatica*, **2** (1973) 155.
9. R Donaldson, *Proc. Phys. Soc. London*, **59** (1947) 554.
10. W D Wright, *Trans. Opt. Soc. London*, **30** (1928/29) 141.
11. J Guild, *Phil. Trans. Roy. Soc.*, **A230** (1931) 149.
12. R Davis and K S Gibson, Miscellaneous Publication 114 (Washington, DC: National Bureau of Standards, 1931).
13. D B Judd, D L McAdam and G Wyszecki, *J. Opt. Soc. Amer.*, **54** (1964) 1031.
14. D B Judd, *J. Opt. Soc. Amer.*, **39** (1949) 945.
15. W S Stiles and J M Burch, *Optica Acta*, **6** (1959) 1.
16. N I Speranskaya, *Opt. Spectr.* (USSR), English translation, 7 (1959) 424.
17. S Johnson *et al.*, *J.S.D.C.*, **79** (1963) 731.
18. W H Foster Jr *et al.*, *Col. Eng.*, **8** (1970) 35.
19. E I Stearns, *Clemson University Rev. Ind. Manage. Text. Sci.*, **14** (1975) 79.
20. CIE Publication 15, Supp. 2 (E1.3.1) (TC1.3) (Paris: Bureau Central de la CIE, 1978).
21. K McLaren and B Rigg, *J.S.D.C.*, **92** (1976) 337.
22. H Helson, D B Judd and M H Warren, *Illuminating Eng.*, **47** (1952) 221.
23. D L McAdam, *J. Opt. Soc. Amer.*, **46** (1956) 500.
24. R W Burnham, R M Evans and S M Newall, *J. Opt. Soc. Amer.*, **47** (1957) 35.
25. C J Bartleson, *Col. Res. Appl.*, **4** (1979) 119.
26. M R Pointer, *Col. Res. Appl.*, **7** (1982) 113.
27. C J Bartleson, *Col. Res. Appl.*, **4** (1979) 143.
28. Y Nayatani, K Takayama and H Sobagaki, *Col. Res. Appl.*, 9 (1984) 106.
29. M R Luo, M C Lo and W G Kuo, *Col. Res. Appl.*, in press.
30. CIE Publication 13.2 (TC3.2) (Paris: Bureau Central de la CIE, 1974).

CHAPTER 4

Colour-order systems, colour spaces, colour difference and colour scales

K J Smith

4.1 COLOUR-ORDER SYSTEMS AND COLOUR SPACES

4.1.1 Introduction

Any discussion of colour requires reference to a logical framework that allows the interrelationships of colours to be unambiguously expressed. The logical framework is usually provided by a *colour-order system*. This allows the communication of colour precepts over distance and through time, even where physical specimens do not exist or have changed in colour with age.

A technical committee of the International Organisation for Standardisation, ISO/TC187 (Colour Notations), has defined a colour-order system as a set of principles for the ordering and denotation of colours, usually according to defined scales [1]. For our purposes, we may amplify this and say that a colour-order system is a set of principles that defines:

(a) an arrangement of colours according to attributes such that the more similar their attributes, the closer are the colours located in the arrangement, and
(b) a method of denoting the locations in the arrangement, and hence of the colours at these locations.

The purpose of a colour-order system determines the number of attributes that must be considered, each attribute defining one dimension of the system. For example, a one-dimensional system may be adequate in the design of lighting systems, where it is sometimes sufficient to consider only the single attribute of CIE luminance factor (Y), which is a function of the total reflectance of each surface within the volume to be lit. Colour is three-dimensional, however, and for a complete colour specification a colour-order system such as that given by CIE x, y and Y is necessary (Chapter 3): x, y and Y are attributes of a colour and each is used to define a dimension of the system. The dimensions are arranged by means of three mutually perpendicular axes. The three attributes are *fundamental* to the system because they define it, and they are *orthogonal* (that is, each may be varied without having to change any other).

Other attributes may be derived from these three but they cannot be orthogonal to them and are therefore not part of the definition of the system. Thus the chromaticity coordinate z may be derived from, but is not orthogonal to, x and y because $z = 1 - x - y$. Similarly, the attributes of dominant, or complementary, wavelengths (λ_d, λ_c) and excitation purity (p_e) may be derived from, but are not fundamental to, a system defined by x and y. We may however define a colour-order system by means of the orthogonal attributes λ_d (or λ_c), p_e and Y, with x and y then being derived attributes. Whether the system be defined by means of x, y and Y, or λ_d (or λ_c), p_e and Y, the relationships between colours in it are the same. Each defines the same *colour space*, that is, the geometric representation of colours in three dimensions [2]. There is a large overlap between this definition of a colour space and that given earlier of a colour-order system: this is unsurprising, since any three-dimensional colour-order system necessarily defines a colour space and any colour space allows colours to be ordered. The distinction often made is that a colour-order system is primarily defined by a set of material colour standards, whereas a colour space is essentially a conceptual arrangement.

Over the years, more than 400 colour-order systems have been compiled [3]. The first to be recorded was devised by Aristotle about 350 BC. It was vaguely three-dimensional and white was placed opposite black; red, however, was placed between black and white, red being the colour of the sky between the states of night and day. Leonardo da Vinci (1452–1519) is said to have painted sequences in which closely related colours were placed near each other. Newton (1642–1727), whose discovery of the nature of white light may be regarded as having begun the science of colour physics, arranged all the hues in a circle, with complementary hues opposite and white at its centre. These arrangements were two-dimensional, however, and could not therefore include all colours.

One three-dimensional arrangement, with many variations, is used in all the successful systems developed this century. The discovery of this arrangement is attributed to Forcius, in 1611, and its inherence is demonstrated by its repeated rediscovery. One rediscovery was made by Munsell, and resulted in the colour-order system bearing his name. Munsell apparently first described his system in a lecture given in 1905 in which he listed its advantages, including:

(a) colour names based on natural objects (which often vary in colour) are replaced by a definite notation;
(b) each colour is named by its notation and can be recorded and transmitted by it, enabling contracts for coloration to be closely specified;
(c) the system can be expanded to accommodate new colours.

This colour-order system will be discussed to the exclusion of others because it became that most widely used and remains the standard against which more recent systems are judged. (For descriptions of other colour-order systems, see Derefeldt [4], Hunt [5] and Judd and Wyszecki [6].)

4.1.2 Munsell colour-order system: concept

A model of the Munsell colour-order system is shown in Plate 2. The three fundamental orthogonal attributes defining the system are called Munsell value V, Munsell chroma C and Munsell hue H. Each is scaled with the aim of perceptual uniformity, so that equal changes in any one of the attributes represent the same perceived difference in colour. Unlike the CIE xyY system, it uses cylindrical coordinates; for readers meeting this coordinate system for the first time, a brief explanation follows. (In the following three paragraphs, x and y are general variables, not CIE x and y.)

We are all familiar with two-dimensional plotting of y against x, with the axes of x and y perpendicular to each other (Figure 4.1). In this so-called rectangular coordinate system, the point A = [2,2] (meaning x = y = 2) is located, relative to the origin (O = [0,0]), by measuring two units along the x-axis to the point X_A, and then from X_A two units parallel to the y-axis. The location of A may alternatively be specified using polar coordinates, i.e. in terms of a distance and an angle. Taking the positive part of the x-axis as a starting line, we may specify the location of A unambiguously in terms of the distance OA and the angle in degrees between the x-axis and OA, conventionally measured anticlockwise. By Pythagoras's theorem, the distance OA = $(2^2 + 2^2)^{1/2} = 8^{1/2}$ units and the angle is 45°, so in a polar coordinate system we write A = $(8^{1/2}, 45)$. In the Munsell colour-order system, we write the polar coordinates of any point (B in Figure 4.1) as (C,H), where C is the distance OB and H is the angle OB makes with the positive part of the x-axis. C and H are derived from the rectangular coordinates x and y by Eqn 4.1:

$$C = (x^2 + y^2)^{1/2}$$
$$H = \arctan(y/x) \tag{4.1}$$

In a polar coordinate system, a point having one or both of its rectangular coordinates negative is specified using values of H > 90. Thus the point with rectangular coordinates [−2, −2] becomes $(8^{1/2}, 225)$ in polar form.

Colour, however, is three-dimensional. In a rectangular coordinate system the third dimension is introduced by adding a third axis, perpendicular to those of x and y and passing through O. If exactly the same thing is done in a polar coordinate system, it becomes a cylindrical coordinate system in which, by convention, the principal axis

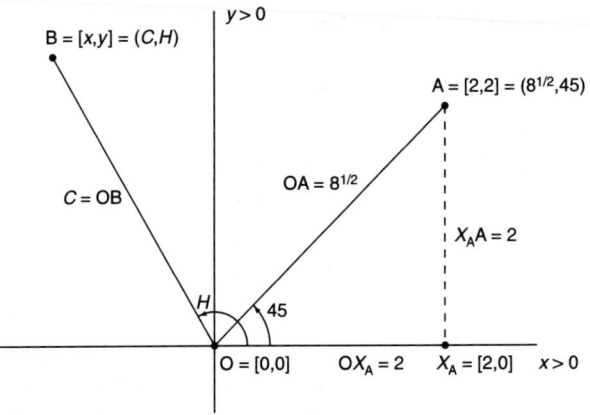

124 COLOUR-ORDER SYSTEMS, COLOUR SPACES, COLOUR DIFFERENCE AND COLOUR SCALES

Figure 4.1 Rectangular and polar coordinate systems, showing the points $A = [2,2] = (8^{1/2},45)$ and $B = [x,y] = (C,H)$, where $C = OB$ and the angle H is measured anticlockwise from the reference line OX_A

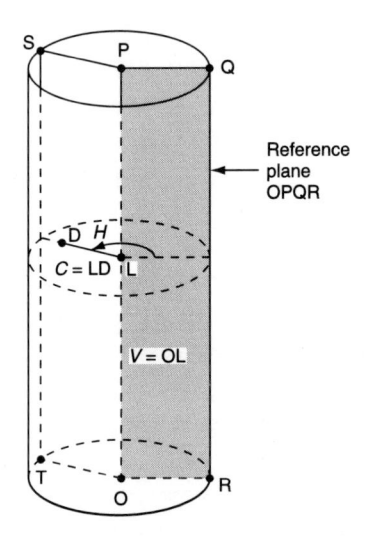

Figure 4.2 Cylindrical coordinate system, showing the point $D = (V,C,H)$, where $V = OL$ (L being the point of intersection of the principal axis OP of the cylinder and the plane containing D which is perpendicular to it), $C = LD$ and H is the angle LD makes with the reference plane OPQR

(OP) of the cylinder is oriented vertically (Figure 4.2). The general point is now $D = (V,C,H)$, where V is the distance OL (L being the point where the principal axis intersects the horizontal plane containing the point D), C is the distance LD and H is measured anticlockwise from the reference plane OPQR, bounded by the principal axis. (Cylindrical coordinates, though perhaps unfamiliar, have the advantage of making the structure of colour space much easier to work with; it is worth persevering with the concept.)

In the Munsell system the vertical axis of the cylinder is the V-axis. Its lower end (V = 0) represents the perfect black, a term often used to indicate a uniform reflectance of 0%, and its upper end (V = 10) an approximation to the white of the *perfect reflecting diffuser*. The CIE defines the latter as the ideal isotropic diffuser (that is, radiation reflected from it is equal in intensity in all directions in the hemisphere in which it occurs) with a uniform reflectance of 100% [2]. The intermediate points on the V-axis represent the infinite number of achromatic colours (that is, colours that resemble only black and white [1]) corresponding, *inter alia*, to uniform percentage reflectances of R ($0 < R < 100$), which are perceived as blacks (if $R \approx 0$), whites (if $R \approx 100$) and greys. All colours with a given V, whether achromatic or chromatic (*chromatic* being the opposite of achromatic, that is, colours possessing hue, even if only slightly [2]), fall on the horizontal plane that contains the given V. The V-coordinate of a coloured surface is determined by its *lightness*, which is a function of the total reflectance of the surface, weighted according to the response of the human visual system to stimuli of different wavelengths. The lightness of a colour is a measure of how it would appear, for example, in a black and white photographic print, provided all the processes leading to the print exactly emulated the human visual process. If two colours, say an orange and a grey, appear identical in such a print they have the same lightness, and hence the same V.

Originally, V was defined in terms of the equivalent of CIE Y ($0 \leq Y \leq 100$) as $V = Y^{1/2}$ [7], so that $0 \leq V \leq 10$. This is an application of a relationship first recognised in 1834 by Weber, and known as Weber's law. It describes many percepts other than lightness, including such common ones as length and weight, and states that identical changes in the ratios of stimuli (not in their absolute values) are equally perceptible. Thus, according to Weber's law, the change accompanying an increase in weight from 1 kg to 2 kg is as perceptible as that resulting from an increase from 2 kg to 4 kg (not 3 kg). For any stimulus S the percept of which obeys Weber's law, Eqn 4.2 applies:

$$\frac{\Delta S}{S} = k \qquad (4.2)$$

where ΔS is a change in the stimulus and k is a constant.

In 1937, the perceptual spacing of the Munsell system was improved by redefining V by an implicit function known as the Judd polynomial (Eqn 4.3) [8]:

$$Y = 1.2219V - 0.23111V^2 + 0.23951V^3 - 0.021009V^4 + 0.0008404V^5 \qquad (4.3)$$

At that time Y was defined with the luminance factor of magnesium oxide, Y(MgO), set to 100. Since, relative to the perfect reflecting diffuser, Y(MgO) is in fact about 97.5, V = 10 corresponds to CIE $Y \approx 102.6$. The Judd polynomial cannot be solved analytically for V but it can be solved, to any required precision, by iterative methods.

We shall next discuss the angular dimension (H), which is determined by the hue of the colour. Forcius and Newton drew hue circles. The more general term is *hue circuit*, which implies a loop but one which is not necessarily circular. The outline of the CIE chromaticity diagram (Figure 3.7) is a noncircular loop. It consists of an open curve (the spectrum locus) closed by the straight-line locus of the nonspectral purples, and contains all the hues possible.

Before defining hue, we must first define the term *elementary colour*, which is a colour that can be defined only by reference to itself [1]. There are six such colours: white, black, red, yellow, green and blue. We may thus identify two elementary achromatic colours (white and black) and four elementary chromatic colours (red, yellow, green and blue). *Hue* is then defined as the attribute of a chromatic colour according to which it appears to be similar to one of the elementary chromatic colours or to a combination of two of them [2]. The hues of most chromatic colours appear to be combinations of two elementary chromatic colours which are always adjacent in the hue circuit. Opposite pairings (red and green, and yellow and blue) do not occur. Thus orange colours possess simultaneously the attributes of redness and yellowness. In all red colours redness predominates, but most have also some yellowness or some blueness; for example, redness predominates in scarlet colours but there is some yellowness too. Some reds appear to possess only the attribute of redness, however, and are called *unitary* (or *unique*) reds. Similarly, unitary yellows, greens and blues are each defined as the hue of a colour which resembles only one of the four elementary chromatic colours [1].

In the CIE xy diagram, the colours in the combination of the spectrum locus and that of the nonspectral purples are arranged in such a way that the less the distance between any two points on the loci, the more their hues resemble each other. In this diagram, however, equal steps do not generally correspond to visually equal changes in hue, whether made along the loci or in angle swept relative to the achromatic point.

The Munsell system has a qualitatively similar hue circuit. It surrounds the achromatic axis but it is a circle in which equal steps do correspond to visually equal differences in hue. In it, the five so-called Munsell principal hues – red (R), yellow (Y), green (G), blue (B) and purple (P) – are equally spaced around the circle and arranged clockwise in the order given when viewed from 'above' (Figure 4.3). Lying halfway between each pair of adjacent principal hues is one of the five intermediate hues (YR, GY, BG, PB and RP). Together, the principal and intermediate hues constitute the ten major hues of the system. Each is subdivided into ten equal parts, so that the whole

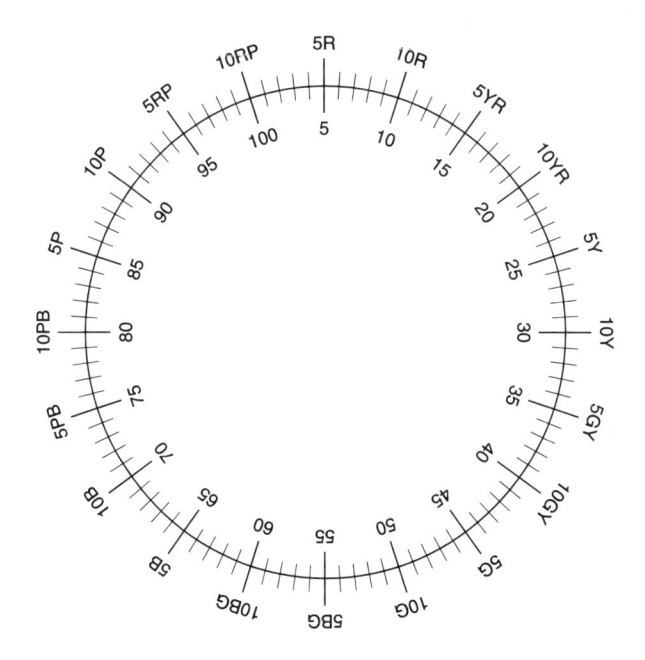

Figure 4.3 Organisation and alternative denotation of the Munsell hue (*H*) circuit, the notation outside the circle being the more common

circle is divided into a total of 100 equal angular intervals. The attribute of Munsell hue (*H*) may thus be specified by means of an angular scale of $0 < H \leq 100$ with 100 (equivalent to 0) representing a hue midway between RP and R, R itself being at $H = 5$, YR at $H = 15$, and so on. More usually, however, a system is used in which one of the major hues is given preceded by a number *n* ($0 < n \leq 10$). If $n = 5$, the major hue itself is indicated (5R, for example, denotes the major hue red). A designation of $n > 5$ implies a hue clockwise from the given major hue (and $n < 5$, anticlockwise); thus, for example, 7R denotes a red shade that is yellower than the major red hue 5R.

To define the significance of the dimension C of the Munsell system, we return to the chromaticity diagram. The colours of the spectrum locus and the nonspectral purples resemble one of the elementary chromatic colours (or more usually only two adjacent ones). Most colours of a given lightness, however, also resemble that achromatic colour which has the same lightness. An orange colour, say, may possess only the attributes of redness and yellowness in a given ratio, whereas a brown which possesses these two attributes in the same ratio also has the attribute of greyness. This leads us to two further definitions:

(a) a *full chromatic colour* is a colour that resembles only the elementary chromatic colours, and (so) does not at all resemble grey

(b) *chroma* is the attribute of a colour that expresses the degree by which it differs from an achromatic colour of the same lightness [1].

The Munsell chroma (C) of a colour dictates the distance from the achromatic axis at which it is placed in the system. It is a measure of the extent by which the colour differs from the achromatic colour of the same V. The orange colour mentioned above has a higher C than the brown and hence lies further from the achromatic axis, all truly achromatic colours having C = 0.

For nonfluorescent colours the theoretical upper limit to C is determined by the so-called *optimal colour stimuli*. These are the colours of imaginary objects which reflect perfectly in some parts of the spectrum and absorb completely in the remainder, so that everywhere either R = 100 or R = 0. They are the (nonfluorescent) colours of highest possible p_e for given Y and λ_d (or λ_c). Maximum theoretical C is different for different H, and for given H is different for different V, generally being higher at intermediate V.

Until now, we have denoted the general point in the Munsell system by V, C and H, but in practice the position of a chromatic colour is denoted by HV/C, so that our orange colour (with, say, H = 2.5YR, V = 6 and C = 12) would be denoted 2.5YR 6/12. Neutral (achromatic) colours are denoted by NV/, so that an achromatic colour with V = 6 would be designated N6/.

4.1.3 Munsell colour-order system: realisation

Much of the importance of the Munsell system is due to its physical exemplification by a collection of coloured specimens or *colour atlas*, defined as the arrangement of coloured specimens according to a colour-order system [1]. The specimens, or 'chips', systematically sample the colour space defined by the conceptual system; indeed, some systems are defined by a collection of chips.

The atlas of the Munsell system is the *Munsell book of color*, first published in 1915. The current edition is available with either about 1500 glossy or about 1300 matt chips. It systematically samples Munsell colour space by means of a collection of neutral colours (C = 0) and 40 charts, each of constant H (H = 2.5R to 10RP in steps of 2.5) and representing a vertical part-plane bounded on one side by the V-axis, such as those labelled OPQR and OPST in Figure 4.2. The principles of this layout are shown in Plate 2. On each chart chips are displayed at intervals of V = 1 and C = 2, arranged with those in each row of constant V and those in each column of constant C (Figure 4.4). The inequality of the intervals of V and C arises because although Munsell designed the three fundamental attributes of his system with the aim of perceptual uniformity, he deliberately chose scales such that V = 2C = 3H at C = 5. The scaling of H relative to that of V and C needs to be qualified (at C = 5) because

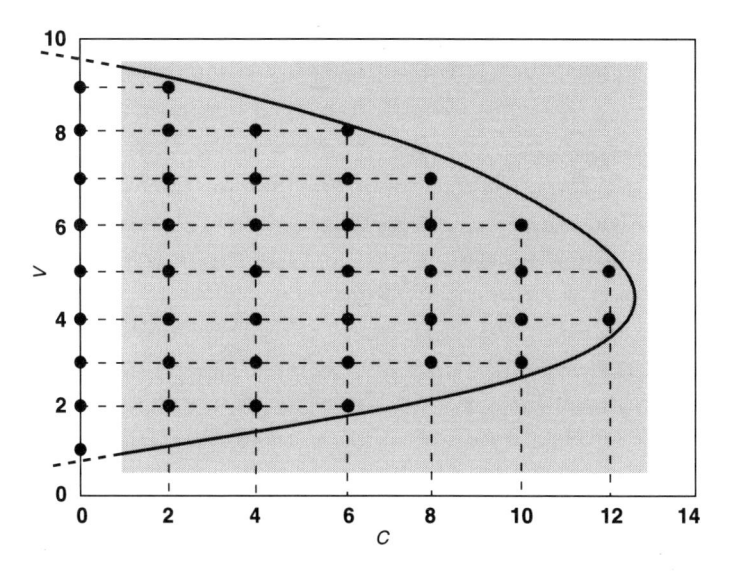

Figure 4.4 Schematic of *Munsell book of color* constant-hue (*H*) chart, shown as grey rectangle, in relation to Munsell constant-hue plane: the limit of the gamut attainable (at this *H*) using suitably stable colorant combinations (curve) restricts the chips that may be displayed to those at the intersections marked ●

Munsell colour space is specified by a cylindrical coordinate system. Although in any such system there is a constant angle between any two planes, each of constant hue, the distance between them is zero at the principal axis and increases with increasing distance from it, whereas in a perceptually uniform system it is equality of distance (not of angle) that corresponds to visually equal differences in colour.

In each hue chart, the chip of highest C for each V has a lower C than that of the optimal colour stimulus for this H and V. The C of real surface colours is necessarily lower because they are produced using real colorants, which neither absorb nor reflect perfectly at all visible wavelengths. Additionally, in any colour atlas the maximum C illustrated is restricted by the need for the colour of the chips to be maintained throughout a production run, to be reproduced between runs, and to be stable to the various agencies to which they are exposed during use.

If their visually perceived spacings are not to vary, the chips in each region of colour space also need to show similar changes in colour with changes in the quality of the illumination under which they are viewed. Many colour atlases have prescribed viewing conditions, usually including illumination resembling D_{65}, but the difficulty of producing a source that exactly resembles D_{65} leaves scope for lamps from different manufacturers to resemble it in different ways. If each of the chips in an atlas could be made using only the relevant selection of colorants from a gamut consisting of, say, one of each of the six elementary colours, it might be possible to select colorants that allowed the spacings of the chips to be maintained under a limited range of qualities of illumination. Except near the hues of the four elementary chromatic colours, however, such a gamut would restrict the maximum C that could be illustrated. In order to produce chips of C high enough to satisfy users of the atlas, the chromatic colorants used to produce the majority of chips must be supplemented by ones of high C at intermediate hues. It is frequently difficult to find suitably stable colorant combinations that are not metameric with respect to each other.

The chips of highest C on each Munsell constant-H chart form a curved boundary, such as that in Figure 4.4, which is different for each chart. The boundaries of each pair of adjacent charts are similar, however, so that all 40 boundaries form a smooth, but irregular, three-dimensional locus called a *colour solid*: that is, a three-dimensional representation of that part of colour space which can be achieved by means of coloured objects [1]. This is illustrated by the model in Plate 2.

4.1.4 Visual assessments of colour

For the visual determination of the specification of an unknown specimen, a colour atlas is viewed alongside the specimen under prescribed conditions. An approximate specification of the specimen is given by the denotation of the chip judged closest to it in colour. The human visual system can distinguish several million colours, however, whereas most colour atlases contain at best a few thousand chips. Even the most comprehensive systematic sampling of colour space yet produced, the *ICI colour atlas*, illustrates only about 27 000 colours [9]. For a more precise specification, interpolation between chips is therefore almost always necessary.

Whilst the human visual system is an excellent null detector (being very capable of assessing whether or not colours match), it is not good at estimating the relative magnitude of differences in colour. If several assessors interpolate to arrive at a specification their reports usually differ, often significantly, and even the same assessor assessing on different occasions usually gives a variety of specifications. In an experiment, the *ICI colour atlas* coordinates of six specimens were assessed, to the nearest chip, by five experienced assessors under identical conditions on each of six occasions; not one assessor chose the same three coordinates for any specimen on all occasions, and one assessor did not choose the same three coordinates, for three of the specimens, on any occasion.

Lack of correlation between interpolations is one of the principal problems of visual methods of specifying colours. Nevertheless colour atlases continue to be used because it is necessary in some applications, and desirable in others, to have available a collection of chips systematically sampling colour space.

4.1.5 CIE xyY colour space

Many of the problems associated with the use of colour atlases could be minimised by the application of instrumental colour measurement, were it possible to interconvert the result of measurement of a colour expressed in terms of CIE x, y and Y, for example, and its coordinates in the colour-order system which the atlas samples. Relatively simple and reliable formulae for the interconversion are generally unavailable, however. In the Munsell system, whilst the Judd polynomial (Eqn 4.3) provides an easy means of interconverting CIE Y and Munsell V, simple formulae have not yet been found for interconverting chromaticity coordinates and Munsell H and C.

The reason for this is that CIE xyY colour space is not perceptually uniform, nor was it designed to be. The non-uniformity of Y as a measure of lightness is implied by the nonlinearity of the Judd polynomial (Eqn 4.3) but is more simply exemplified by the fact that a grey of $Y \approx 20$ (not 50) is perceived to be equally different from the perfect black ($Y = 0$) and the perfect white ($Y = 100$). Furthermore, if Munsell colour space is accepted as a good approximation of visual uniformity, and if the chromaticity diagram were uniform, then for any given V measurements of chips of constant H should yield straight lines of constant dominant wavelength, and measurements of chips of constant C should give circles centred at the achromatic point. They do not. Loci of constant H curve significantly and those of constant C are very distorted ellipses.

Since the introduction by the CIE in 1931 of the xyY system, many investigators have searched for simple formulae which transform its coordinates, or the tristimulus values (X, Y and Z) from which they are derived, into a more perceptually uniform system. None of the resulting colour spaces has, other than incidentally, been exemplified by a colour atlas, and none approaches the benchmark uniformity of the Munsell system. Of the many formulae developed, only some of those which were at some time in widespread use, or were significant in the development of CIE recommendations, are discussed here.

4.1.6 Judd triangular and MacAdam rectangular UCS diagrams

Early effort concentrated on linear transformations of CIE x and y, resulting in so-called uniform chromaticity scale (UCS) diagrams. These maintain some of the properties of the xy-diagram including that of additivity, important in industries such as television. One transformation proposed by Judd in 1935 considerably improved the uniformity of the chromaticity diagram, but was awkward to use because it involved plotting triangular coordinates.

In 1937, MacAdam, recognising both the benefit of the end results of the triangular system and the difficulty of deducing them, devised a two-coordinate system which yielded an almost identical, but rectangular, UCS diagram. This was a direct precursor of CIELUV (section 4.1.13), one of the two colour spaces currently recommended by the CIE. Its two coordinates are u (abscissa) and v (ordinate), defined by Eqn 4.4:

$$u = \frac{4x}{P} = \frac{4X}{Q}$$
$$v = \frac{6y}{P} = \frac{6Y}{Q}$$
(4.4)

where $P = 3 - 2x + 12y$ and $Q = X + 15Y + 3Z$, x and y being CIE chromaticity coordinates, and X, Y and Z the tristimulus values [10].

4.1.7 Hunter L$\alpha\beta$ and Scofield Lab colour spaces

The opponent theory of colour vision developed by Hering led to two independent developments in 1942. In Hering's theory the six elementary colours are arranged in three opposing pairs, white and black, red and green, and yellow and blue [11]. Any colour may then be specified by quantifying its attributes of whiteness or blackness (its lightness), redness or greenness, and yellowness or blueness. If the resulting parameters are arranged by means of three mutually perpendicular axes, with the perfect black at [0,0,0], the result is an *opponent colour space* (so called because it is defined by opponent attributes) which is qualitatively similar to CIE xyY and Munsell spaces (Figure 4.5).

At the time Hunter was attempting to develop a direct-reading colorimeter, and seeking a method suitable for the analogue conversion of photoelectric signals to the coordinates of a colour space. He proposed a formula in which lightness is quantified by the Munsell V function of the time ($Y^{1/2}$) and both redness–greenness and yellowness–blueness are nonlinear transforms of variables which he called α and β respectively, these being themselves transforms of CIE x and y. In terms of tristimulus values, Hunter modelled redness minus greenness as a function of $X - Y$ and yellowness minus blueness as a function of $Y - Z$. He found it necessary to scale the latter by a factor of 0.4 relative to the former, in order to achieve reasonable uniformity in his colour space [12].

This formula was modified in the following year by Scofield, who was the first to adopt the variables widely used today: L for lightness, and a and b for the opponent chromatic attributes, where $L = 10Y^{1/2}$ (which, for $0 \leq Y \leq 100$, establishes a scale of $0 \leq L \leq 100$ for lightness) and a and b are functions of α and β respectively, still nonlinear and maintaining the factor of 0.4 used by Hunter [13].

4.1.8 Adams chromatic value colour space

The other development of 1942 was a result of an earlier suggestion by Adams, who linked Hering's opponent colour vision theory and the Young–Helmholtz trichromatic

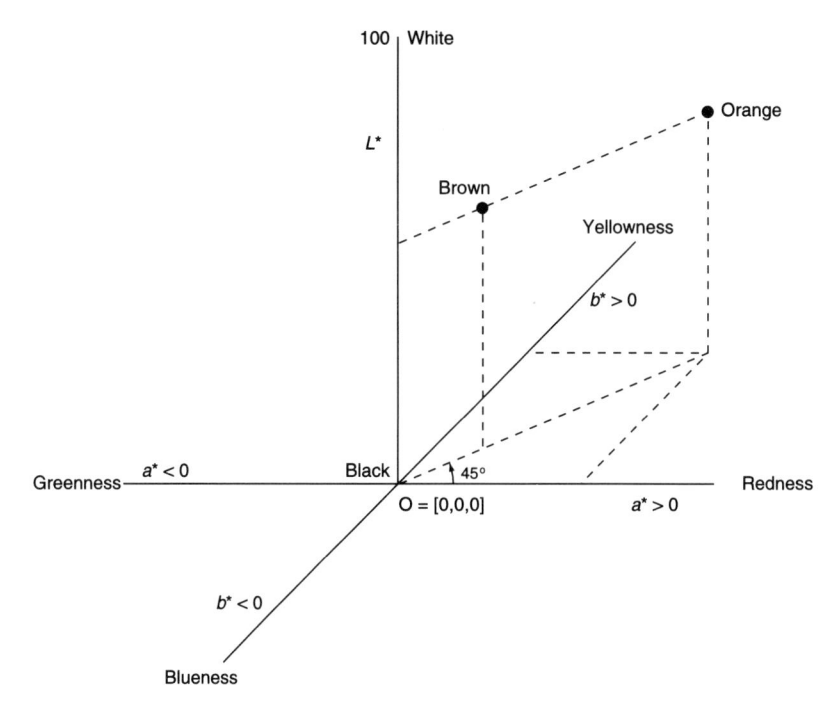

Figure 4.5 Rectangular axes of general opponent colour space (verbal axis labels) and CIELAB colour space ($L^*a^*b^*$ labels), also showing approximate relative locations of orange and brown colours of equal lightness

theory. The last-named postulates three types of receptor in the retina, one sensitive to each of red, green or blue stimuli. Adams suggested that the red–green opponent response was a function of the difference between the stimulations of the red- and green-sensitive receptors, and the yellow–blue response a function of the difference between those of the green- and blue-sensitive receptors. He modelled these differences as functions of $X - Y$ and $Z - Y$ respectively, the order of the latter function being reversed relative to Hunter's because Adams wished to emulate the configuration of Munsell space (which has its hue circuit reversed relative to that of the CIE chromaticity diagram, which Hunter had emulated).

Adams's formula incorporated two important advances. First, each of the tristimulus values of a colour under a given illuminant and observer was normalised by dividing it by the corresponding tristimulus value of the perfect reflecting diffuser, conventionally denoted by a subscript n. Employing the resulting ratios (X/X_n, for example) causes all achromatic colours to fall at the same point in a UCS diagram, irrespective of the combination of illuminant and observer; they otherwise fall at different points for different illuminant/observer combinations, as exemplified in Figure 3.7. Secondly, in order to devise a more uniform colour space, Adams applied the Munsell V function of the time ($Y^{1/2}$) to the X and Z tristimulus values, thus defining the three functions as shown in Eqn 4.5:

$$V_x = (X/X_n)^{1/2}$$
$$V_y = V = Y^{1/2} \quad \text{(by definition } Y_n = 100 \text{ for } 0 \le Y \le 100)$$
$$V_z = (Z/Z_n)^{1/2}$$
(4.5)

Adams modelled lightness as V_y, the opponent chromatic variable of redness minus greenness as $V_x - V_y$ and that of blueness minus yellowness as $V_z - V_y$. To the last, he applied the same scaling factor (0.4) employed by Hunter (for the same reason) to define what became known as Adams chromatic value colour space. We have mentioned above (section 4.1.5) that in the CIE xy diagram colours of different Munsell H, but constant V and C, plot as distorted ellipses. Adams tested his formula by plotting colours of constant V and C, and found the resulting loci to be much closer to circular. An additional indication of the non-uniformity of xyY colour space is that for constant C the size of the ellipses decreases with increasing V. Adams plotted the more nearly circular loci at various levels of V and found that, for constant C, their radii were nearly constant [14].

4.1.9 Hunter Lab colour space

In 1958 Hunter brought together some of these ideas in a new formula that became widely used throughout the next two decades, especially in the USA. It was originally given in a form suitable for use under CIE standard illuminant C and the CIE 1931 (2°) standard observer, but it is given in the general form which allows its application under other combinations in Eqn 4.6:

$$L_H = 100 Y_R^{1/2}$$
$$a_H = 175(X_R - Y_R)\left(\frac{0.0102 X_n}{Y_R}\right)^{1/2}$$
$$b_H = 0.4 \times 175(Y_R - Z_R)\left(\frac{0.00847 Z_n}{Y_R}\right)^{1/2}$$
(4.6)

in which L_H is a measure of lightness ($0 \le L_H \le 100$), a_H and b_H are measures respectively of redness minus greenness and yellowness minus blueness (the subscript H indicating that these are Hunter coordinates) and X_R, Y_R and Z_R are the respective tristimulus ratios X/X_n, Y/Y_n and Z/Z_n [15]. The factors 0.0102 and 0.00847 were included because, for illuminant C and the 2° observer (referred to as C/2 hereafter), $X(MgO) \approx 0.0102^{-1}$ and $Z(MgO) \approx 0.00847^{-1}$.

4.1.10 Adams–Nickerson (ANLAB) colour space

In 1944, Nickerson and Stultz proposed replacing the Weber derivations of V_x, V_y and V_z of the Adams chromatic value space (Eqn 4.5) by the Judd polynomial obtained by substituting, in turn in Eqn 4.3, X_R, Y_R and Z_R for Y and V_x, V_y and V_z for V; they also suggested scaling the resulting lightness coordinate by a factor of 0.23 to make unit steps in lightness more consistent with those of the opponent chromatic scales [16]. In 1952, Glasser and Troy proposed reversing the term for blueness minus yellowness so that the order of the hues in their circuit conformed to that in CIE xyY and Hunter $L\alpha\beta$ colour spaces [17]. The three opponent scales were then denoted L, A and B, in line with the nomenclature used by Scofield. The resulting formula was widely used, largely due to publicity given it by Nickerson, and as a result became known as the ANLAB (Adams–Nickerson *LAB)* formula, given in Eqn 4.7:

$$L_{AN} = 0.23 S V_y$$
$$A_{AN} = S(V_x - V_y) \qquad (4.7)$$
$$B_{AN} = 0.4 S(V_y - V_z)$$

where V_x, V_y and V_z are derived from measurements of X, Y and Z relative to MgO.

The factor S, applied equally to all three attributes, allows the whole colour space to be scaled according to its applications. Originally, Nickerson and Stultz suggested $S = 42$ to give correlation between unit distance in the resulting colour space and the (USA) National Bureau of Standards unit of colour difference. This unit had been defined by Judd in 1939 as of such size that smaller colour differences could be ignored in textile colour matching to average commercial tolerances. Later, however, Nickerson proposed $S = 40$, which became the factor most widely used, although other factors have been applied. Results from the ANLAB formula are therefore ambiguous unless S is clearly stated, the usual notation being (for $S = 40$) ANLAB(40) or one of its abbreviated forms, AN(40) and AN40.

4.1.11 Early cube-root colour spaces

At this time several investigators came to consider the Judd polynomial (Eqn 4.3) unnecessarily complicated, and suggested cube-root approximations to it. For the attribute of lightness (L), these are of the form $L = k_1 Y^{1/3} - k_2$, where k_1 and k_2 are positive constants; the variations among the constants chosen by different investigators probably arose because their data were derived under different viewing conditions. The most widely used was probably that of Glasser *et al.*, commonly known as the Glasser cube-root formula (Eqn 4.8):

$$\begin{aligned} L_G &= 25.29Y^{1/3} - 18.38 \\ a_G &= 106\,[(1.02X)^{1/3} - Y^{1/3}] \\ b_G &= K[Y^{1/3} - (0.847Z)^{1/3}] \end{aligned} \quad (4.8)$$

where K = 42.34 ($\approx 106 \times 0.4$), and L_G = 100 for Y(MgO) = 102.56 [18].

4.1.12 CIE 1960 UCS diagram and CIE 1964 (U*V*W*) colour space

In 1960 the CIE, concerned by the many formulae then in use, made an attempt at standardisation. From several candidates of similar uniformity, MacAdam's 1937 rectangular *uv* UCS diagram was chosen because of the simplicity of its transformation from the CIE *xy* diagram. It then became known as the CIE 1960 UCS diagram. In 1964 it was extended by the CIE to three dimensions, by including a cube-root formula for the attribute of lightness, in this formula denoted W^*, represented by a third axis perpendicular to those of U^* and V^* (which are nonlinear transforms of the CIE 1960 UCS variables u and v). The resulting system was called the CIE 1964 ($U^*V^*W^*$) uniform colour space (Eqn 4.9):

$$\begin{aligned} W^* &= 25Y^{1/3} - 17 \\ U^* &= 13W^*(u - u_n) \\ V^* &= 13W^*(v - v_n) \end{aligned} \quad (4.9)$$

where $1 \leq Y \leq 100$ is measured relative to the perfect reflecting diffuser, and u_n and v_n are respectively the values of u and v for it, being included so that, for all achromatic colours, $U^* = V^* = 0$ [19].

These attempts at standardisation were unsuccessful. Over the next few years, several new formulae were devised, and the relative merits of various formulae were extensively investigated. Although there was not a consensus, the evidence favoured ANLAB and its cube-root derivatives.

4.1.13 CIE 1976 UCS diagram, CIELUV and CIELAB colour spaces

In 1973 the CIE agreed to MacAdam's suggestion that a cube-root version of ANLAB be adopted. The resulting colour space (CIELAB) cannot, however, have a UCS diagram associated with it, because the derivation of its opponent chromatic variables involves nonlinear transforms of CIE tristimulus values. In view of the need for a UCS diagram in several applications, and the failure of its earlier attempts at standardisation, the CIE also agreed to adopt a modification of the 1960 UCS diagram based on the work of Eastwood, and to extend it to three dimensions by incorporating the same cube-root lightness function as in the cube-root version of ANLAB. These proposals became CIE recommendations in 1976 and are still in force for the definition of approximately uniform colour spaces.

Eastwood had shown that the spacing of the Munsell chips in CIE 1964 ($U^*V^*W^*$) space could be made more uniform by increasing v by half. The first recommendation, defined in Eqn 4.10, is therefore the rectangular CIE 1976 UCS diagram, with abscissa $u' = u$ and ordinate $v' = 1.5v$, u and v being the coordinates of the CIE 1960 UCS diagram [19]:

$$u' = u = \frac{4x}{P} = \frac{4X}{Q}$$
$$v' = 1.5v = \frac{9y}{P} = \frac{9Y}{Q} \tag{4.10}$$

where P and Q are as defined in Eqn 4.4.

The second recommendation is for a so-called approximately uniform colour space, created from this UCS diagram by the addition of an axis of lightness (L^*) perpendicular to u^* and v^* (which are derived from u' and v' respectively), known as the CIE 1976 ($L^*u^*v^*$) colour space and officially in abbreviated form as CIELUV (Eqn 4.11):

$$\begin{aligned} L^* &= 116(Y_R)^{1/3} - 16 \quad \text{if } Y_R > 0.008\,856 \\ L^* &= 903.3 Y_R \quad \text{if } Y_R \leq 0.008\,856 \\ u^* &= 13 L^* (u' - u'_n) \\ v^* &= 13 L^* (v' - v'_n) \end{aligned} \tag{4.11}$$

where $Y_R = Y/Y_n$ ($0 \leq Y \leq 100$) and Y_n is the tristimulus value of 'a specified white object colour' (usually the perfect reflecting diffuser, so that $Y_n = 100$), and u'_n and v'_n are the values of u' and v' respectively, for it [19].

The third recommendation is the approximately uniform CIE 1976 ($L^*a^*b^*$) colour space, officially abbreviated to CIELAB. Its three mutually perpendicular opponent-colour axes are L^*, a^* and b^*, where L^* is defined exactly as in Eqn 4.11, including the exception for $Y_R \leq 0.008\,856$. The exception condition was proposed by Pauli (and is usually referred to as the Pauli extension), to overcome a problem of most cube-root approximations: that, for very small values of Y_R, $L^* < 0$. The opponent chromatic scales are defined by Eqn 4.12:

$$a^* = 500[f(X_R) - f(Y_R)]$$
$$b^* = 0.4 \times 500[f(Y_R) - f(Z_R)] \quad (4.12)$$

where, writing T_R for each of the tristimulus ratios $X_R = X/X_n$, $Y_R = Y/Y_n$ and $Z_R = Z/Z_n$ (X_n, Y_n and Z_n being the tristimulus values of the specified white object colour), the relevant T_R are given by Eqn 4.13 [19]:

$$f(T_R) = (T_R)^{1/3} \quad \text{if } T_R > 0.008856$$
$$f(T_R) = 7.787 T_R + \frac{16}{116} \quad \text{if } T_R \leq 0.008856 \quad (4.13)$$

The asterisk superscript on each of the explicit variables in the CIELUV and CIELAB formulae differentiates these parameters from the similar ones used in others. The common metric lightness scale is such that $0 \leq Y \leq 100$ leads to $0 \leq L^* \leq 100$ (although the two are not linearly related). The perfect reflecting diffuser has $L^* = 100$ and, like all other achromatic colours, it has $u^* = v^* = a^* = b^* = 0$, the perfect black having $L^* = 0$. In CIELAB space, colours with $a^* > 0$ possess the attribute of redness, those with $a^* < 0$ greenness, those with $b^* > 0$ yellowness, and those with $b^* < 0$ blueness (Figure 4.5). Thus, an orange colour, possessing simultaneously the attributes of redness and yellowness, has a^* and $b^* > 0$.

Since the CIELAB formula is now so important and because the application of the exception conditions is sometimes misunderstood, two example calculations follow, in which each output is given to one decimal place. The first is representative of the majority of calculations and is simple. Suppose, for colour E, $X_R = 0.86$, $Y_R = 0.85$ and $Z_R = 0.80$. Each of these clearly exceeds 0.008 856, so that none of the exceptions applies and, using the subscript E to denote the relevant CIELAB values of the colour, we have (Eqn 4.14):

$$L^*_E = 116 \times 0.85^{1/3} - 16 = 93.9$$
$$a^*_E = 500(0.86^{1/3} - 0.85^{1/3}) = 1.9 \quad (4.14)$$
$$b^*_E = 0.4 \times 500(0.85^{1/3} - 0.80^{1/3}) = 3.8$$

Thus colour E is of relatively high lightness and possesses a small amount of the attribute of yellowness, together with an even smaller amount of redness; it is in fact similar to the colour of bleached cotton.

The second is more complicated. Suppose, for colour F, $X_R = 0.0086$, $Y_R = 0.0085$ and $Z_R = 0.0090$. Before calculating its lightness, we notice that $Y_R = 0.0085 < 0.008\,856$, so we use the exception (Eqn 4.15):

$$L^*_F = 903.3 \times 0.0085 = 7.7 \qquad (4.15)$$

For a^*_F we realise that $X_R, Y_R < 0.008\,856$. In the equation for a^*_F, we apply the exception to both X_R and Y_R, and (Eqn 4.16):

$$a^*_F = 500\left[\left(7.787 \times 0.0086 + \frac{16}{116}\right) - \left(7.787 \times 0.0085 + \frac{16}{116}\right)\right] = 0.4 \quad (4.16)$$

For b^*_F, $Y_R = 0.0085 < 0.008\,856$ but $Z_R = 0.0090 > 0.008\,856$. We therefore apply the exception only to the term in Y_R (Eqn 4.17):

$$b^*_F = 0.4\left\{500\left[\left(7.787 \times 0.0085 + \frac{16}{116}\right) - 0.0090^{1/3}\right]\right\} = -0.8 \quad (4.17)$$

Colour F, being of very low lightness and almost achromatic, is sensibly black. In textile dyeing, it is likely that its intense black shade could be matched only on raised-pile fabrics such as velvet.

So far, we have discussed the two colour spaces (CIELUV and CIELAB) in terms of three mutually perpendicular axes, but it is much easier to visualise the structure of colour space if cylindrical coordinates are used. Even before 1976 some workers had been using cylindrical systems derived from the perpendicular coordinates of existing colour spaces, and in that year the CIE defined approximate metric correlates of chroma (C^*_{uv}) and hue (h_{uv}) in CIELUV space as given in Eqn 4.18 [19]:

$$\begin{aligned}C^*_{uv} &= (u^{*2} + v^{*2})^{1/2} \\ h_{uv} &= \arctan(v^*/u^*)\end{aligned} \qquad (4.18)$$

In CIELAB colour space, the two correlates (C^*_{ab} and h_{ab}) are analogously defined by substituting a^* for u^* and b^* for v^* in Eqn 4.18, the subscripts (uv and ab) being used to distinguish one pair of correlates from the other.

The correlates of hue (h_{uv} and h_{ab}, which we shall denote collectively as h), neither of which is superscripted, are expressed in degrees. They are scaled such that $0 \le h < 360$; $h = 0$ indicates a colour on the positive part of the u^* or a^* axes, h being measured anticlockwise therefrom, so that the positive part of the v^* or b^* axes is at 90, and so on. None of the four unitary hues, however, usually falls exactly on an axis, unitary reds, yellows, greens and blues having, under C/2, $h_{ab} \approx 25, 85, 165$ and 260 respectively.

The correlates of chroma (C^*_{uv} and C^*_{ab}, collectively C^*) are such that $C^* = 0$ indicates a truly achromatic colour, and $C^* > 0$ for all other colours. In the CIELAB system, the orange and brown colours of equal lightness we met whilst discussing Munsell colour space (section 4.1.2) both have $0 < h_{ab} < 90$, and may both have $h_{ab} = 45$. If this were so, they would be differentiated only by the parameter C^*_{ab} (Figure 4.5). Typically, for the brown $C^*_{ab} \approx 10$, and for the orange $C^*_{ab} \approx 70$, some fluorescent orange colours having $C^*_{ab} > 100$.

In the CIELUV system, there remains one more approximate correlate to define, namely that of CIE 1976 *uv* saturation (s_{uv}) defined in Eqn 4.19 [19]:

$$s_{uv} = 13[(u' - u'_n)^2 + (v' - v'_n)^2]^{1/2}$$
$$= \frac{C^*_{uv}}{L^*} \qquad (4.19)$$

Saturation is related to chroma. The ISO/TC187 definition of chroma was given in section 4.1.2, but it is here more useful to use the CIE definitions of both saturation and chroma. Each of these, however, involves the terms *chromaticness* (or colourfulness) and *brightness*, which we must first define. The CIE defines these as the attributes of a visual sensation according to which, for chromaticness, the perceived colour of an area appears to be more (or less) chromatic and, for brightness, an area appears to emit more (or less) light. The CIE then defines saturation and chroma as the chromaticness of an area judged as a proportion, for saturation, of its (own) brightness, and, for chroma, of the brightness of a similarly illuminated area that appears white. Eqn 4.19 is approximately consistent with these definitions because the CIE defines lightness (of which L^* is an approximate correlate) as the brightness of an area judged relative to the brightness of a similarly illuminated area that appears white [2].

4.1.14 Residual non-uniformity of CIELUV and CIELAB colour spaces

Although the spacing of the Munsell colours in both CIELUV and CIELAB spaces is far more uniform than in CIE *xyY* space, it is certainly not perfect. Chips of constant Munsell C, which plot as distorted ellipses of size varying inversely with V in the *xy* diagram, plot as distorted circles of similar radii in CIELAB space, and the loci of chips of constant H, which in the *xy* diagram curve significantly, do so – though more gently – in CIELAB. The search continues for the perceptually uniform colour space, even though several workers have concluded that such a space is an impossibility.

4.2 COLOUR-DIFFERENCE EVALUATION

4.2.1 Introduction

The determination of the difference between the colours of two specimens is important in many applications, and especially so in those industries, such as textile dyeing, in which the colour of one specimen (the batch) is to be altered so that it imitates that of the other (the standard). This is usually an iterative process. An estimate, based on either the experience of a colourist or the output of computer match prediction (Chapters 5 and 6), is made of the amounts of each of (usually) three colorants needed to bring about the required colour change. These amounts are then applied to the batch, and the resulting colour is compared with that of the standard. If it does not match the estimate is adjusted, taking account of the extent and nature of the mismatch. The revised amounts of the colorants are then applied, the result evaluated, and so on, until the colour of the batch does match that of the standard.

The time taken to perform this process, and hence its cost, clearly increases with the number of iterations required. The use of the term 'match' in industry is something of a misnomer, however. Although customers are pleased if the colour of the batch is exactly that of the standard, they are usually prepared to tolerate some colour difference in order to reduce the average number of iterations the supplier performs and hence the prices they are asked to pay. The task of the supplier is to produce a batch of a colour that differs from its standard to an extent sufficiently small that it is within a nonzero tolerance of it. The size of the tolerance is usually prescribed by the customer, who is the ultimate arbiter of whether a batch is within tolerance (a pass) or outside it (a fail).

The penalties of wrong pass/fail decisions by the supplier are often severe, especially when they occur beyond the stage of trial coloration in the laboratory. There are two categories of wrong decisions by the supplier: batches may be wrongly passed or wrongly failed. If a batch is wrongly passed, faulty material will be delivered to the customer, who may reject it, demand a discount or, if the process allows it, return the goods for correction. In any case, delivery is delayed. If the contract implies deliveries be of the correct shade, and the error is discovered only after the customer has added value to the batch, a large claim on the supplier may result. A batch wrongly failed may be unnecessarily reprocessed or sold at a reduced price; inevitably, delivery is again delayed. Whatever the nature of the wrong decision, the supplier's profits will be reduced and – often of greater consequence – his relationship with the customer will be damaged.

4.2.2 Reliability of visual colour-difference assessments

As we have seen (section 4.1.4), the human visual system is excellent at assessing whether two specimens match. It is, however, nowhere near as good at quantifying the magnitude of a mismatch and at assessing whether the colour of the batch is within tolerance. If the supplier and customer assess its colour difference from standard visually, they are likely to disagree.

Over the years several sets of data have been published, quantifying both the repeatability and reproducibility of visual assessments of colour differences. (For replicate visual assessments, *repeatability* is a measure of the extent to which a single assessor reports identical results, and *reproducibility* is the corresponding measure for more than one assessor.) On the basis that a decision that differs from that of the majority of the group under study is a wrong decision, these studies reported individual wrong decisions ranging from about 13 to 24%. Furthermore, most of these studies were made on pairs of specimens coloured using the same colorants and which were therefore nonmetameric mismatches. In industrial situations, however, the batch is often metameric with respect to the standard, increasing the probability of disagreements. Judgements will therefore be more subject to observer metamerism and, even if the viewing conditions for the assessments by supplier and customer are nominally the same, they are unlikely to be exactly identical. So illuminant metamerism will add to the confusion.

4.2.3 Reliability of instrumental colour-difference evaluation

The results from instrumental methods are much less variable than those from visual assessments. A study conducted in 1965 by a subcommittee of the American Society for Testing and Materials (ASTM) took the form of an analysis of the results of measurements made by 16 operators using five different models of instrument on five ceramic tiles at each of seven colour centres, the pairs of tiles at each centre exhibiting colour differences typical of those encountered in industrial pass/fail decisions. The standard deviation of colour-difference measurements was found to be comparable to the smallest differences detectable visually [20], a somewhat surprising result given the outcome of later similar studies. Nevertheless, since then both the repeatability and reproducibility of instrumental colour-difference measurements (defined analogously to their visual counterparts, section 4.2.2) have much improved; those of visual assessments can be assumed to be largely unchanged.

It is however neither the repeatability nor the reproducibility of instrumental results that presents the real problem in industrial colour-difference determination: rather, it is the difficulty of devising a formula transforming the measurements into end-results that accord sufficiently with the majority visual assessment to be of practical value. As we shall see, until recently the formulae available tended to give more wrong decisions than the average visual assessor.

4.2.4 Colour-difference evaluation: fundamentals

The most elementary of all colour spaces is constructed by arranging the CIE tristimulus values by means of three mutually perpendicular axes. If the X, Y and Z values of a batch (B) are equal to those of its standard (S), the two are by definition a perfect match under the geometry used to measure them and the combination of illuminant and observer used to derive the tristimulus values. They will appear to be a perfect match to an assessor whose visual responses precisely emulate those of the relevant standard colorimetric observer, provided he views them under exactly the same conditions of geometry and illumination used to measure them and to transform the measurements. Having identical tristimulus values, B and S necessarily occupy the same point in XYZ colour space.

Suppose, however, that B and S are very close to each other but do not precisely coincide. Because of the gradation of the attributes from which XYZ space is constructed they must cause very similar, but not identical, colour stimuli. They mismatch to a small extent. As B and S are removed farther and farther from each other, so they cause progressively less similar visual stimuli and they mismatch increasingly. The distance between B and S in XYZ colour space is thus a measure of the difference in colour between the two. Since XYZ space is Euclidean, the distance BS may be calculated by applying Pythagoras's theorem in three dimensions (Eqn 4.20):

$$BS = [(X_B - X_S)^2 + (Y_B - Y_S)^2 + (Z_B - Z_S)^2]^{1/2} \qquad (4.20)$$

where X_B, Y_B and Z_B are the tristimulus values of B, and X_S, Y_S and Z_S those of S. Colour difference is conventionally denoted by ΔE_σ, where the Greek character Δ (delta) stands for difference, E is the initial letter of the German word *Empfindung* (meaning sensation) and σ represents subscript character(s) used, whenever there may be doubt, to clarify the derivation of ΔE_σ. Since the symbol Δ is used to represent differences in general, we may extend its use to differences in the component attributes of B and S, and Eqn 4.20 may be rewritten as Eqn 4.21:

$$\Delta E_{XYZ} = [(\Delta X)^2 + (\Delta Y)^2 + (\Delta Z)^2]^{1/2} \qquad (4.21)$$

In the evaluation of colour differences in industry, however, XYZ colour space suffers from two disadvantages, one of which is so serious that ΔE_{XYZ} is hardly ever used, at least at the output stage of colorimetric calculations.

The first disadvantage is that the significance of the component differences (ΔX, ΔY and ΔZ) cannot readily be related to the nature of the components of colour difference perceived by assessors. Often a knowledge of the nature of the components is of almost as much consequence as that of the overall colour difference itself. For example, a colourist trying to correct a failed batch needs to be able to determine a modified formulation of colorants, and this he cannot do unless the component differences are meaningful to him. However, even the most elementary transform (that to CIE *xyY* space) and the calculation of ΔE_{xyY} and its components partially obviates the problem. Further transformation allows the calculation of the parameters $\Delta \lambda_d$ (or $\Delta \lambda_c$), Δp_e and ΔY, which for relatively small colour differences are qualitative correlates of hue, chroma and lightness differences respectively.

The more significant disadvantage of XYZ colour space is that it is not uniform with respect to the magnitude of visual assessments of colour differences. This manifests itself in two ways. Firstly, if two batches are visually assessed equal in total colour difference from the same standard but differ significantly in the nature of their differences from it, evaluation of ΔE_{XYZ} for each of the batches relative to the standard usually gives unequal numerical results. Secondly, if visual assessment indicates equality of the magnitude of the colour difference between a pair of specimens at one colour centre and that between a pair at another centre that is not close to the first, measurement usually yields values of ΔE_{XYZ} which are not numerically the same for the two pairs. This non-uniformity of XYZ colour space is shared by xyY space, and therefore by λ_d (or λ_c) $p_e Y$ space.

Experimental studies indicate that where batches are visually assessed as having equal total colour differences from a given standard but removed from it in different directions, they always fall in xyY space on an ellipsoid centred about the standard. This explains the first manifestation. The reason for the second is that the ellipsoids that correspond to visually equal colour differences from standards at distinct colour centres vary greatly in size, eccentricity and orientation. The evidence presented by these investigators is exemplified, in two dimensions, in Figure 4.6, which is based on the work of MacAdam [21]. The lengths of the longest and shortest semi-axes of any of the ellipsoids represent total colour differences perceived equal; the ratio of the two lengths, estimated by MacAdam to be about 30 : 1, thus provides a measure of the non-uniformity of xyY space. Later, McLaren estimated the same ratio for XYZ space.

Lack of uniformity in a colour-difference formula poses serious problems in the determination of colorimetric pass/fail limits in colour matching. When the permissible extent of mismatch to a standard is neither specified by the customer nor able to be estimated even approximately by reference to historical data for a similar colour, the colourist must determine the relevant tolerance volume. This can be achieved by correlating instrumental results and visual decisions for a series of batches. Graphical methods can provide a scatter diagram (Figure 4.7) from which, given enough batches, the tolerance ellipse for this colour may be determined. The task is usually complicated by the need to consider all three dimensions of colour differences and, of course, by any wrong decisions that are made. Since the size, eccentricity and orientation of the tolerance ellipsoid are known at best only very approximately in advance, and since those batches which are clear passes or clear fails do not contribute significantly to the determination of its boundary, a large number of batches is usually

COLOUR-DIFFERENCE EVALUATION 145

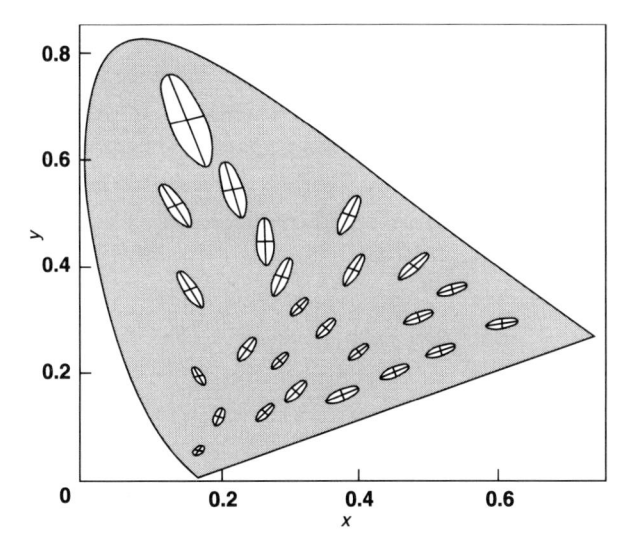

Figure 4.6 CIE 1931 chromaticity diagram, showing MacAdam's 1942 data for standard deviation of colour matching: the distances between the boundaries of all the ellipses (shown enlarged 10×) and their respective centres represent colour differences perceived equal in magnitude [21]

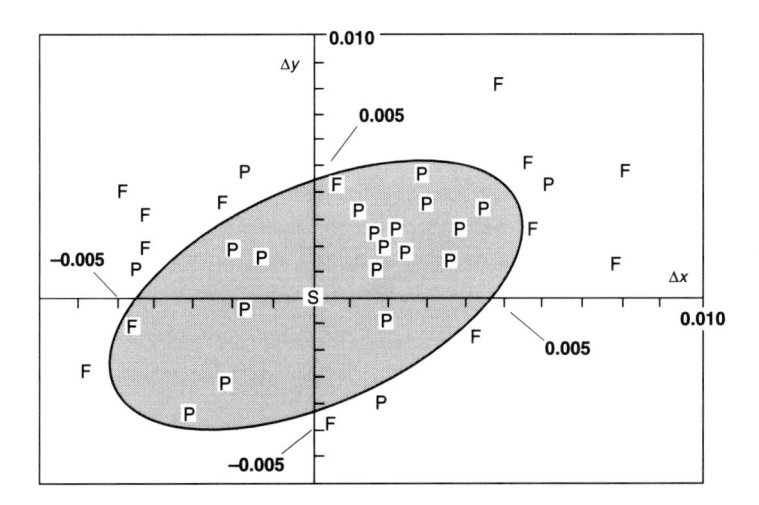

Figure 4.7 Determination of tolerance ellipse, in (part of) CIE chromaticity diagram, from scatter of lots visually passed (P) and failed (F) relative to standard (S): fewer than half the 35 lots contribute significantly to the definition of the boundary and there are six wrong decisions (17%)

needed to define it with reasonable confidence. So lengthy is the process that production may well be completed before enough batches have been coloured to allow its conclusion. And even when the number of batches does allow the process to be completed, it must be started afresh if the next shade to be matched is significantly different.

4.2.5 Single-number shade passing

Since the introduction of the CIE system in 1931, much effort has been devoted to the development of a colour-difference formula in which all tolerances (for a given end-use) are numerically equal, regardless of the colour of the standard and the direction of the colour difference of the batch from it. In any Euclidean colour space σ, the total colour difference ΔE_σ between B and S is calculated, according to the general form of Eqn 4.21, as the Pythagorean sum of the differences (ΔA_i) between batch and standard in each of any three orthogonal attributes (A_i, $i = 1, 2, 3$) of the space (Eqn 4.22):

$$\Delta E_\sigma = \left[\sum_{i=1}^{3} (\Delta A_i)^2 \right]^{1/2} \quad (4.22)$$

where $\Delta A_i = A_{i,B} - A_{i,S}$. In a completely uniform Euclidean colour space, pass/fail boundaries at all colour centres would be spherical shells of uniform radius. A specific form of Eqn 4.22 based in such a space would make it theoretically possible for the colour measurement of just one batch, visually assessed as a borderline pass/fail at any colour centre and irrespective of the direction of its difference from standard, to establish the relevant colorimetric tolerance in every direction at all colour centres. This goal is often referred to as *single-number shade passing* (SNSP).

Many of the colour-difference formulae developed since 1931 were associated with the evolution of the more uniform Euclidean colour spaces discussed in sections 4.1.6 to 4.1.13. Indeed, their evolution was driven primarily by the desire for increasingly uniform colour-difference formulae, the evolution of the colour spaces usually being a by-product. Of these, only the CIELAB colour-difference formula remains in widespread use in the surface-coloration industries. For the rest, it is sufficient to note that one or more colour-difference formulae may be derived by substituting orthogonal variables of the space for $A_{i,B}$ and $A_{i,S}$ in Eqn 4.22.

4.2.6 Development of CIELAB and CIELUV colour-difference formulae

In 1967 the Colorimetry Committee of the CIE, concerned by the problems caused by the variety of formulae then being used in colour-difference evaluation, commenced a programme of work that was eventually to lead to largely unified practice, at least for a few years. Four formulae were selected for initial investigation, and over the next few years their performance was independently investigated by several workers; it became apparent, however, that none was as reliable as the average assessor.

Meanwhile the need for standardisation was made more pressing year by year by the increasing use of instrumental colour-difference evaluation. So urgent became the need that by 1973 it was clear that considerably the most important contribution that the Committee could make was to reach agreement on a formula, almost regardless of which. Agreement was reached in favour of a new cube-root formula, proposed by MacAdam. The choice was influenced both by the tendency of the comparative studies to indicate the Adams chromatic value lineage to be consistently near the top of reliability rankings, and by the simplicity of cube-root formulae relative to the Judd polynomials (Eqn 4.3) used in ANLAB. The result was the publication, in 1976, of two CIE recommendations, CIELAB and CIELUV, for approximately uniform colour spaces and colour-difference calculations.

4.2.7 Calculation of CIELAB and CIELUV colour difference

Formulae have already been given for the calculation of the coordinates of the point representing a given colour, in both rectangular and cylindrical form, in each of CIELAB and CIELUV spaces (Eqns 4.11–4.13). The colour difference between a batch (B) and its standard (S) is defined, in each space, as the Euclidean distance between the points (B and S) representing their colours in the relevant space. The formulae for the calculation of colour difference and its components in the two spaces are identical in all but the nomenclature of their variables. We shall therefore detail only those pertaining to the calculation of colour difference in CIELAB; the corresponding formulae for CIELUV colour difference may be obtained by substituting, in Eqns 4.23 to 4.30 as appropriate, ΔE^*_{uv} for ΔE^*_{ab}, Δu^* for Δa^*, Δv^* for Δb^*, ΔC^*_{uv} for ΔC^*_{ab}, Δh_{uv} for Δh_{ab} and ΔH^*_{uv} for ΔH^*_{ab}, ΔL^* being common [19].

If L^*_B, a^*_B and b^*_B are the CIELAB rectangular coordinates of a batch, and L^*_S, a^*_S and b^*_S those of its standard, substituting $\Delta L^* = L^*_B - L^*_S$, $\Delta a^* = a^*_B - a^*_S$ and $\Delta b^* = b^*_B - b^*_S$ in Eqn 4.22 gives Eqn 4.23:

$$\Delta E^*_{ab} = [(\Delta L^*)^2 + (\Delta a^*)^2 + (\Delta b^*)^2]^{1/2} \qquad (4.23)$$

where ΔE^*_{ab} is the CIELAB colour difference between B and S. Here L^*, Δa^* and Δb^*, and hence ΔE^*_{ab}, are in commensurate units.

The usefulness of cylindrical variables has already been emphasised (section 4.1.2). If, however, we make the substitution using the corresponding cylindrical component differences, we immediately encounter a problem. The cylindrical variable ΔL^* is identical to that of the rectangular form and ΔC^*_{ab} shown in Eqn 4.24:

$$\Delta C^*_{ab} = C^*_{ab,B} - C^*_{ab,S} \qquad (4.24)$$

is by the definition of Eqn 4.18, in the same units. However, the hue angle difference Δh_{ab} (Eqn 4.25):

$$\Delta h_{ab} = h_{ab,B} - h_{ab,S} \qquad (4.25)$$

is in degrees, and so is incommensurate with the other two variables: the substitution is mathematically invalid.

The definition of CIELAB colour difference includes two methods of overcoming the problem. The first uses radian measure to obtain a close approximation to a hue (not hue angle) difference ΔH^*_{ab} in units commensurate with those of the other variables (Eqn 4.26):

$$\Delta H^*_{ab} = C^*_{ab} \, \Delta h_{ab} \frac{\pi}{180} \qquad (4.26)$$

but its use is restricted to 'small colour differences away from the achromatic axis'. It is rarely used because of both this restriction and the general applicability of the second method.

Suppose there does exist a variable ΔH^*_{ab} representing, in units commensurate with the other variables of CIELAB colour difference, the hue difference between batch and standard, and that it is orthogonal to both ΔL^* and ΔC^*_{ab}. Then ΔE^*_{ab} must be the Pythagorean sum of these three component differences (Eqn 4.27):

$$\Delta E^*_{ab} = [(\Delta L^*)^2 + (\Delta C^*_{ab})^2 + (\Delta H^*_{ab})^2]^{1/2} \qquad (4.27)$$

We know the value of each of the first three variables in Eqn 4.27 (ΔE^*_{ab} from the output of Eqn 4.23, ΔL^* as one of the inputs to Eqn 4.23, and ΔC^*_{ab} from Eqn 4.24, each without knowledge of ΔH^*_{ab}). By rearranging Eqn 4.27 we can define ΔH^*_{ab} (Eqn 4.28):

$$\Delta H^*_{ab} = [(\Delta E^*_{ab})^2 - (\Delta L^*)^2 - (\Delta C^*_{ab})^2]^{1/2} \qquad (4.28)$$

Unfortunately, this method also has its problems. The other four components of CIELAB difference are defined as differences and are thus signed so that, for example, $\Delta L^* > 0$ if $L^*_B > L^*_S$ but $\Delta L^* < 0$ if $L^*_B < L^*_S$, while ΔH^*_{ab} is defined (by Eqn 4.28) as a square root, the sign of which is indeterminate. The CIE states that 'ΔH^*_{ab} is to be regarded as positive if indicating an increase in h_{ab} and negative if indicating a decrease'. This may be interpreted as implying that the sign of ΔH^*_{ab} is that of Δh_{ab}, so that $\Delta H^*_{ab} > 0$ if the batch is anticlockwise from its standard, and $\Delta H^*_{ab} < 0$ if clockwise. Thus, for example, for batch B1a and standard S1 (Figure 4.8), where $h_{ab,B1a} = 30$ and for $h_{ab,S1} = 10$, $\Delta h_{ab} = 30 - 10 = 20$ (greater than zero), so that the sign of ΔH^*_{ab} is positive. Since B1a is anticlockwise relative to S1, this is reasonable. Now consider batch B1b ($h_{ab,B1b} = 350$) and the same standard. Here $\Delta h_{ab} = 350 - 10 = 340$,

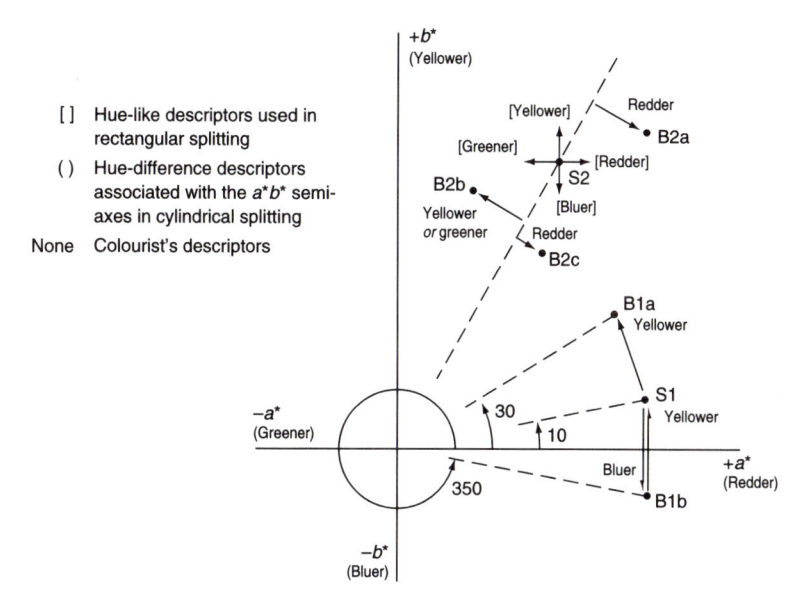

Figure 4.8 CIELAB a^*b^* diagram, showing locations of standards (S1, S2), batches (B1a, B1b, B2a, B2b, B2c) and qualitative hue differences between them

which is again greater than zero so that ΔH^*_{ab} is positive. The hue vector from S1 to B1b must, however, clearly be considered clockwise, so that Δh_{ab} and ΔH^*_{ab} should be negative. This problem arises whenever $\Delta h_{ab} > 180$. The definition therefore presents problems, but for many years it offered the only way of calculating ΔH_{ab}.

Another method was based on the work of Huntsman. Equating the right-hand sides of Eqns 4.23 and 4.27, followed by manipulation, yields Eqn 4.29 [22]:

$$\Delta H^*_{ab} = 2(C^*_{ab,B} \, C^*_{ab,S})^{1/2} \sin\left(\frac{\Delta h_{ab}}{2}\right) \quad (4.29)$$

Although Eqn 4.29 provides a simpler method of calculating ΔH^*_{ab}, it still suffers from the disadvantage that it outputs the wrong sign of ΔH^*_{ab} when $\Delta h_{ab} > 180$. The correct sign may, however, be determined without knowledge of the value of Δh_{ab}, by testing the relative sizes of the two directed areas $a^*_B b^*_S$ and $a^*_S b^*_B$; denoting the correct sign of ΔH^*_{ab} by s [23] gives Eqn 4.30:

$$\Delta H^*_{ab} = s[2(C^*_{ab,B} \, C^*_{ab,S} - a^*_B a^*_S - b^*_B b^*_S)]^{1/2}$$
$$s = 1 \quad \text{if } a^*_B b^*_S < a^*_S b^*_B \quad (4.30)$$
$$s = -1 \quad \text{if } a^*_B b^*_S > a^*_S b^*_B$$

This method is being increasingly employed.

In recommending two formulae (CIELAB and CIELUV) rather than one, the CIE admitted its failure. Yet two formulae were inevitable, CIELAB being chosen because it was among the formulae recognised as more reliable for surface colours and CIELUV because of its associated UCS diagram. Studies of the performance of CIELUV for surface colours had not then been made, even though such studies might have enabled it – had it achieved at least parity with CIELAB – to be the only recommendation, and the need for standardisation was urgent. Most of the subsequent independent investigations showed insignificant differences in the reliability of the two for the evaluation of surface colour differences, and also that both formulae make more wrong decisions than does the average visual assessor. It is unsurprising that the search continued for formulae of better performance. Indeed, the CIE recognised, in making its dual recommendations in 1976, the probability of the development of better formulae by making its recommendations 'pending the development of an improved coordinate system' [19].

4.2.8 Early optimised colour-difference formulae

While the CIE 1976 recommendations were evolving, several colour-difference formula were being developed outside the CIE. All were derived by applying to each of the component differences of an existing formula weights that maximised the correlation of the overall colour differences given by the modified formula with those from one or more sets of visual data. The details of most of these optimised formulae need not detain us because they have found little application in industry. Meanwhile, however, two optimisations, the CMC($l:c$) and the Marks and Spencer plc (M&S) formulae, were being developed on broadly similar lines. Each has been, and still is, widely used, especially in the textile coloration industries.

4.2.9 M&S colour-difference formulae

Around 1975, Marks & Spencer was considering launching the sale of ready-made separates for men's suitings, an operation requiring close-tolerance matchings across dye lots. At the time, the chain was adopting Philips TL84 fluorescent lighting as standard in its stores, thereby causing dyers great difficulties in obtaining matches under that illuminant that were sufficiently nonmetameric with respect to daylight and tungsten. It was decided that the risks were unacceptable without the objectivity of reliable instrumental evaluation, but none of the colour-difference formulae then available was considered adequate for the purpose. Thus it was that, in 1976, the first version of the M&S formula was developed, jointly by Marks & Spencer and Instrumental Color Systems (ICS, now Datacolor International). It has since been steadily modified in the light of experience of its application. Earlier versions were largely limited to in-house use by M&S, the first version made more widely available being M&S80 (the digits indicate the approximate year of release). Subsequent major revisions have been M&S82, M&S83, M&S83A and M&S89, all versions being based in AN(40) colour space.

The formulae cannot be discussed in detail because, for commercial reasons, none has been published. In essence, the designers accepted the visual non-uniformity of ANLAB space and mathematically modified a small subvolume of it in the vicinity of any given colour centre (standard); this produced a new local space, the three attributes of which are sufficiently uniform, both within and between themselves, for practical purposes. Each subvolume is limited in extent by the global non-uniformity of ANLAB space from centroid to centroid, and is therefore usually referred to as a microspace, but within each microspace the good approximation to uniformity stretches far enough from its centre to cope with the small-to-moderate colour differences encountered in industry. Suitable scaling between centres allows the goal of SNSP although, while successive M&S formulae were designed with that aim and used as such for many years, M&S has recently reverted to separate tolerances for each of the components of colour difference, considered in addition to an overall limit, to produce a tolerance volume which is a truncated sphere, albeit one of size independent of the colour at its centre. Testing of an early version (probably M&S80) in strictly SNSP yielded fewer wrong decisions than an average visual assessor and a significant improvement on the number given by CIELAB.

4.2.10 CMC(l:c) colour-difference formula

The CMC($l:c$) formula was derived from the JPC79 colour-difference formula, developed within J & P Coats (now part of Coats Viyella plc). Work on JPC79 had started before that on the M&S formulae, and preliminary testing demonstrated an impressive performance: it made slightly fewer wrong decisions than even the most reliable of the Coats assessors.

Since its formation in 1963, the Colour Measurement Committee of the Society of Dyers and Colourists had championed the cause of standardisation in instrumental colour-difference evaluation. It recommended AN(42) in 1970, changed its allegiance to AN(40) in 1972 following an international lead, recommended CIELAB in 1976, and now saw in the performance of JPC79 an opportunity to make a significantly improved recommendation. The original formula had been developed in AN(50) colour space. In order to make it more generally acceptable, the committee suggested its basis be converted to CIELAB. Although this proved impracticable, it was rebased in AN(43.909) space, 43.909 being arguably a factor very closely approximating ANLAB to CIELAB. The original formula had been scaled so that its unit colour difference corresponded to the mean visual pass/fail limit of the J & P Coats in-house assessors for a product that did not require the highest-quality matches. Again at the suggestion of the committee, it was rescaled so that its unit colour difference represented the mean visual pass/fail limit of assessors whose tolerance was believed to represent the highest quality of pass/fail matching in the textile industry.

The revised JPC79 formula is given in Eqn 4.31:

$$\Delta E_{JPC79} = \left[\left(\frac{\Delta L}{S_L} \right)^2 + \left(\frac{\Delta C}{S_C} \right)^2 + \left(\frac{\Delta H}{S_H} \right)^2 \right]^{1/2} \quad (4.31)$$

where ΔL, ΔC and ΔH are respectively the AN(43.909) lightness, chroma and hue differences between batch and standard (calculated analogously to their CIELAB counterparts ΔL^*, ΔC^*_{ab} and ΔH^*_{ab}), and:

$$S_L = \frac{0.08195 L_S}{1 + 0.01765 L_S}$$

$$S_C = 0.638 + \frac{0.0638 C_S}{1 + 0.0131 C_S}$$

and $S_H = S_C T$
where $T = 1$ if $C_S < 0.638$
but $T = k_1 + |k_2 \cos(H_S + k_3)|$ if $C_S \geq 0.638$
where $k_1 = 0.36, k_2 = 0.4, k_3 = 35$ if $H_S \leq 164$ or $H_S \geq 345$
but $k_1 = 0.56, k_2 = 0.2, k_3 = 168$ if $164 < H_S < 345$

L_S, C_S and H_S being respectively the AN(43.909) lightness, chroma and hue angle (in degrees) of the standard.

Use of the JPC79 formula demonstrated two anomalies, however. One was a failure in the evaluation of the lightness differences between pairs of specimens of low lightness. The other arose with pairs of nearly neutral specimens. The discontinuity induced by the definition of T (in Eqn 4.31) sometimes gave rise to significant differences in ΔE_{JPC79} for a pair of specimens just within the limit of $C_S < 0.638$, and another pair, of visually similar magnitude and direction of colour difference, but just outside it [24]. The first anomaly was resolved by setting S_L constant for specimens of low lightness, and the second by replacing the discontinuous function with a continuous one. Four other changes were also introduced in order to increase the domain of applicability of the formula and the likelihood of its wider acceptance.

Firstly, in order to permit its application to acceptability (pass/fail) decisions in different industries and to perceptibility evaluations (which are independent of industry), it was modified to allow relative tolerances to lightness, chroma and hue differences to be independently varied. Colourists differ systematically according to the industry in which they work by having different tolerances (in pass/fail work) to the three cylindrical variables of colour. (It is emphasised that we are here discussing relative tolerances to the components of colour difference, not to be confused with the absolute pass/fail tolerances of acceptability decisions.) Relative tolerances allow the magnitude of the cylindrical components of a colour difference to be adjusted according to the ratio in which a visual pass/fail assessor in a given industry is prepared to tolerate equally perceptible differences in each component. Relative tolerances of less than unity were to be avoided if possible and, since tolerance to hue differences is not usually wider than that to the other components, the relative tolerance to differences in hue (h) was set to unity and the variables of tolerances to lightness and chroma differences (relative to that to hue differences) were expressed as l and c respectively. Since $h = 1$ always, the relative tolerances to all three components are completely specified by giving the values only of l and c.

In perceptibility judgements (typical of which are determinations of fastness-test results) equally perceptible differences in each of the components are of the same importance, whereas in the visual acceptability judgements used to derive JPC79 the assessors were about twice as tolerant to lightness differences as to differences in chroma or hue. Thus the second change made was to halve the value of JPC79 S_L (Eqn 4.31), so that the modified formula quantifies perceptibility when $l = c\ (= h) = 1$, the equivalent of the textile acceptability evaluations implicit in the original being obtained by setting, in the modified formula, $l = 2$, still with $c\ (= h) = 1$.

Thirdly, by the time the committee was ready to publish the revised formula, CIELAB had largely ousted ANLAB in industrial practice and it was therefore rebased in CIELAB space by the simple and completely justifiable assumption that the differences between CIELAB and AN(43.909) are negligible in the context of relatively small colour differences.

Finally, the committee felt that the revised formula needed to be distinguished from its origins, especially since the commercial connotation of the original designation (JPC) might make it unpalatable to some organisations. Accordingly, the revised formula was published in 1984 under the name of CMC($l:c$), CMC being the abbreviation commonly used for the Colour Measurement Committee (Eqn 4.32) [24]:

$$\Delta E_{\text{CMC}(l:c)} = \left[\left(\frac{\Delta L^*}{lS_L} \right)^2 + \left(\frac{\Delta C^*_{ab}}{cS_C} \right)^2 + \left(\frac{\Delta H^*_{ab}}{S_H} \right)^2 \right]^{1/2} \qquad (4.32)$$

where ΔL^*, ΔC^*_{ab} and ΔH^*_{ab} are respectively the CIELAB lightness, chroma and hue differences between batch and standard, l and c are the tolerances applied respectively to differences in lightness and chroma relative to that to hue differences (the numerical values used in a given situation being substituted for the characters l and c, for example CMC(2 : 1), whenever there be possible ambiguity), and:

$$S_L = \frac{0.040975 L^*_S}{1 + 0.01765 L^*_S} \quad \text{if} \quad L^*_S \geq 16$$

$$\text{but } S_L = 0.511 \quad \text{if} \quad L^*_S < 16$$

$$S_C = 0.638 + \frac{0.0638 C^*_{ab,S}}{1 + 0.0131 C^*_{ab,S}}$$

$$\text{and } S_H = S_C(TF + 1 - F)$$

$$\text{where } F = \left[\frac{(C^*_{ab,S})^4}{(C^*_{ab,S})^4 + 1900} \right]^{1/2}$$

$$\text{and } T = k_1 + \left| k_2 \cos(h_{ab,S} + k_3) \right|$$

where the k_i (i = 1, 2, 3) are as defined in Eqn 4.31, and L^*_S, $C^*_{ab,S}$ and $h_{ab,S}$ are respectively the CIELAB lightness, chroma and hue angle (in degrees) of the standard.

The mathematics of the CMC(l : c) formula deserve examination, because they well illustrate the general principles of optimised formulae, currently so important in industrial applications. In CIELAB space, Eqns 4.27 and 4.28 define the shell containing all shades equally acceptable as matches to (or perceived as equally different from) a standard at a given colour centre. Arising from the non-uniformity of CIELAB space, the magnitudes of each of ΔL^*, ΔC^*_{ab} and ΔH^*_{ab} are not usually equal, and ΔE^*_{ab} is therefore a variable which is assumed in CMC(l : c) and most other optimised formulae to define an ellipsoidal shell with its three axes orientated in the directions of the component differences. The non-uniformity of CIELAB space further dictates that an equally acceptable (or perceptible) ΔE^*_{ab}, at another colour centre, is unlikely to define a similar shell. For a formula to allow SNSP, however, we require the overall colour difference to be a constant, so that its locus describes a spherical shell of equal radius at all colour centres. The ellipsoid in CIELAB space may be converted into a sphere by dividing each of its attribute differences (ΔL^*, ΔC^*_{ab} and ΔH^*_{ab}), in turn, by the length of the semi-axis of the ellipsoid in the direction of the relevant attribute difference (S_L for lightness, S_C for chroma and S_H for hue). The inclusion in Eqns 4.27 and 4.28 of the relative tolerances (l and c) yields the first line of the CMC(l : c) formula (Eqn 4.32). This line therefore converts a usually ellipsoidal tolerance volume in CIELAB space into a spherical one in a CMC(l : c) microspace.

The principal difficulty in the design of optimised colour-difference formulae, however, is to derive mathematics allowing the generation of the systematic variation in the relative magnitudes of attribute differences judged equally acceptable (or equally perceptible) at different centres. These mathematics occupy the whole of the remainder of the formula. Their effect is demonstrated in Figure 4.9.

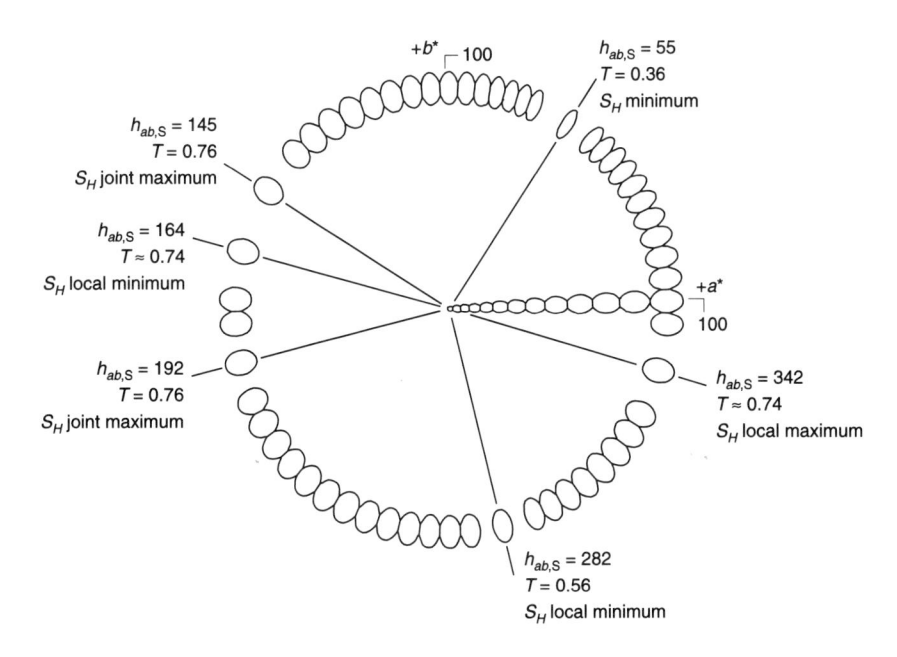

Figure 4.9 Cross-section through their centroids of some ellipsoids of $\Delta E_{CMC(l:1)} = 2.0$ in any CIELAB L^*-plane, showing variations in S_C with C^*_{ab} and in S_H with C^*_{ab} and h^*_{ab} [25]

On testing, the CMC(2 : 1) formula (Eqn 4.32 with $l = 2$ and $c = 1$) yielded marginally fewer wrong decisions than its precursor JPC79 [24]. CMC($l : c$) is probably the best generally available and widely tested colour-difference formula devised to date, certainly for textile applications. It is now used extensively in textile coloration industries world-wide, and to a lesser but significant extent in other industries. In 1988 it formed the basis of (extant) British Standard 6923. In 1989 it was adopted as a test method by the American Association of Textile Chemists and Colorists (AATCC Test Method 173-1989, modified 1992) and by ISO (ISO 105, Textiles – Test for Colour Fastness, Part J03). It has not been adopted by the CIE, however, although a formula derived by consideration of it has recently been approved as a recommendation (section 4.2.11).

4.2.11 CIE 1994 ($\Delta L^* \Delta C^*_{ab} \Delta H^*_{ab}$) colour-difference formula

In 1989 the CIE agreed to establish a technical committee with a remit 'to study existing metrics used in industry to evaluate colour differences between object colours in daylight illumination and to develop a recommendation on this subject'. A tentative two-part proposal was made in 1992. The first part details a new colour-difference formula. The second outlines methodology for incorporating future improvements, either minor or major, the former resulting in evolution of the new formula in the light of fresh evidence, and the latter in changes of formula arising from radical modelling concepts. This part of the proposal therefore foresees an ongoing cycle of establishing, improving and replacing of a model. The proposal was published as a CIE technical report in 1995.

Most colour-difference formulae are based on the assumption that the necessary parameters are those of colour centre, and magnitude and direction of colour difference. Many other influences affect visual assessments, however, including the nature (other than colour) of the specimens and the conditions under which they are viewed. These factors have now been fully recognised for the first time, and parameters have been included in the new formula allowing their influence to be taken into account. Although it is admitted that 'it is not currently possible to account for all these effects', a set of basis conditions has been specified under which the new formula can be expected to perform well, using default values of the parameters. Determination of their values under other conditions is expected to be part of the strategy of continual formula improvement. The basis conditions are those usually pertaining in industrial colour-difference assessments:

(a) that the specimens are homogeneous in colour
(b) that the colour difference between them is such that $\Delta E^*_{ab} \leq 5$
(c) that they are placed in direct edge contact
(d) that each subtends an angle greater than 4° to the assessor
(e) that the assessor's colour vision is normal, and
(f) that they are illuminated at 1000 lux and viewed in object mode, against a uniform neutral grey background of $L^* = 50$, under illumination simulating D_{65}.

The basis of the new formula is the CIELAB colour-difference formulae, the committee regarding as important both the wide acceptance of these formulae, and their incorporation of the perceptual correlates of lightness, chroma and hue differences. The formula replaces those previously recommended for the calculation of small-to-moderate colour differences between coloured materials, but it is a formula for colour-difference evaluation only; it cannot replace CIELAB and CIELUV as recommendations for approximately uniform colour spaces. Its full title is the CIE 1994 ($\Delta L^* \Delta C^*_{ab} \Delta H^*_{ab}$) colour-difference model, with official abbreviation CIE94 and colour-difference symbol ΔE^*_{94}. It includes a new term (ΔV), which is the visually perceived magnitude of a colour difference (Eqn 4.33):

$$\Delta V = k_E^{-1} \Delta E^*_{94} \qquad (4.33)$$

where k_E is an overall visual sensitivity factor, set to unity, making $\Delta V = \Delta E^*_{94}$, under the conditions usually applying in industrial assessments. ΔE^*_{94} is defined by Eqn 4.34:

$$\Delta E^*_{94} = \left[\left(\frac{\Delta L^*}{k_L S_L} \right)^2 + \left(\frac{\Delta C^*_{ab}}{k_C S_C} \right)^2 + \left(\frac{\Delta H^*_{ab}}{k_H S_H} \right)^2 \right]^{1/2} \qquad (4.34)$$

where ΔL^*, ΔC^*_{ab} and ΔH^*_{ab} are the CIELAB lightness, chroma and hue differences between the batch and its standard.

The variables k_L, k_C and k_H, akin to the l, c and h of the CMC(l : c) formula, are here called parametric factors, to avoid any possible confusion with acceptability tolerances. Under basis conditions they are set to $k_L = k_C = k_H = 1$. Other values allow adjustment to be made independently to each colour-difference component to account for deviations from the basis conditions. Thus, for example, a reduction in lightness sensitivity is well established when assessments of textile pairs are made, and it is reasonable to expect better correlation of CIE94 results with those from visual assessments of textile specimens when $k_L = 2$ (and $k_C = k_H = 1$).

The lengths of the ellipsoid semi-axes (S_L, S_C and S_H), termed weighting functions, allow adjustment of their respective components according to the location of the standard in CIELAB colour space, but they are defined differently from their CMC(l : c) counterparts by Eqn 4.35:

$$\begin{aligned} S_L &= 1 \\ S_C &= 1 + 0.045 C^*_{ab,X} \\ S_H &= 1 + 0.015 C^*_{ab,X} \end{aligned} \qquad (4.35)$$

where $C^*_{ab,X} = C^*_{ab,S}$ when the standard of a pair of specimens may be clearly distinguished from the batch, as is usually the case in industrial colour-difference evaluation. The asymmetry of optimised formulae usually causes the colour difference between a pair of specimens (A and B), when calculated with A as standard, to be different from that obtained when B is taken as standard. The difference is anomalous when neither specimen can logically be designated standard, and $C^*_{ab,X}$ may then be defined as the geometric mean of the CIELAB chromas of the pair (Eqn 4.36):

$$C^*_{ab,X} = (C^*_{ab,A} \, C^*_{ab,B})^{1/2} \qquad (4.36)$$

The definitions of S_L, S_C and S_H in the CMC($l:c$) (Eqn 4.32) and the CIE94 (Eqn 4.34) formulae are very different. The CIE94 formula is intended to be generally applicable, and these functions therefore model only effects verified by several independent visual data sets. Three sets were used to test the validity of the functions defining the semi-axes of the CMC($l:c$) formula. In these tests the dependence of S_L on L^*_S was inconsistent, both between the new sets of data and with the model in CMC($l:c$); moreover, no clear indication of the dependence of S_H on $h_{ab,S}$, also modelled in the CMC($l:c$) formula, could be established. The dependence of both S_C and S_H on $C^*_{ab,S}$, again part of the CMC($l:c$) formula, was strong, however [26]. Accordingly, CIE94 was adjusted so that S_L is independent of L^*_S, with S_C and S_H as simple increasing functions of $C^*_{ab,S}$ only.

Subsequent independent studies of textile specimens have however confirmed the dependence of S_L on L^*_S, and it may be that, for textiles, this will be among the first of the envisaged continual empirical improvements made to CIE94.

4.2.12 Descriptors of colour-difference components

It is often difficult to determine the import of the components of a colour difference if they are given only in numerical form. Today most proprietary colour-measurement systems are capable of graphical output, illustrating both the magnitude and direction of a colour difference, but this too can be difficult to understand. Verbal descriptors accompanying the numerical output of component differences can make interpretation much easier. Unfortunately, the potential for confusion is high. In the visual assessment of colour difference, colourists in most industries use one of several cylindrical component systems, none of which is wholly consistent with the cylindrical partitioning of colour difference used in colorimetry. The principal commonality between the variables used by colourists and those of colorimetry is in describing differences in hue. Where commonality does exist, a designer needs to make the descriptors output by his software as consistent as possible with the terms colourists use. Where it does not, the system user should not have to grapple with the use of terminology not wholly compatible with his own.

These considerations do not affect descriptors of lightness differences because colourists rarely employ a variable called lightness, except to imply the visual correlate of metric lightness [27]. Whether the component be derived from rectangular or cylindrical splitting, the batch is described as lighter if $\Delta L^* > 0$ and darker if $\Delta L^* < 0$.

Where rectangular splitting is used, the nature of the residue of the colour difference is usually indicated by comparative adjectives of two of the four hue attributes associated with the chromatic semi-axes of opponent colour space (Figure 4.5). Thus the batch is described as redder if $\Delta a^* > 0$, greener if $\Delta a^* < 0$, yellower if $\Delta b^* > 0$ and bluer if $\Delta b^* < 0$. Ignoring any difference in their lightness, batch B2a (Figure 4.8) is output as redder and yellower than standard S2, and batch B2b as greener and bluer. Some users object to the latter pair of descriptors because they feel it perverse to describe a batch as greener and bluer than a standard that is in the first quadrant of colour space ($0 < h_{ab,S} < 90$), and so possesses the combined attributes of redness and yellowness. (This objection may be overcome by reporting B2b as less red and less yellow.) Colourists usually object to *any* of the foregoing pairs of descriptors, however. Before giving their reasons, we must explain their method of describing hue differences.

When assessing colour difference the colourist, usually subconsciously, first assigns the standard to one of a limited number of notional colour categories. His set may consist of the higher-chroma colour categories shown as sectors in Figure 4.10, supplemented by lower-chroma categories such as brown and olive, and also by white, grey and black. Together his categories must include all colours, so that each is necessarily broad. He allocates to yellow, for example, any higher-chroma colour in which he regards the attribute of yellowness as predominating to such an extent that he cannot allocate it to an adjacent category. The colour would be allocated to the

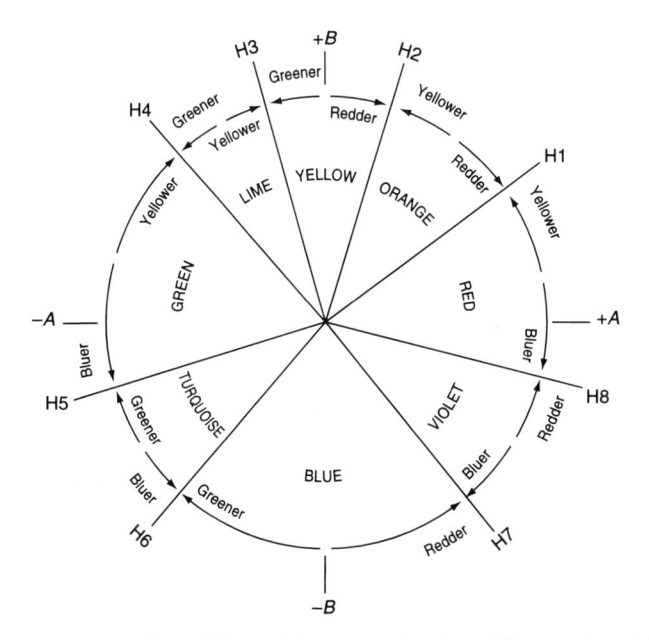

Figure 4.10 Cooper's method of determining appropriate hue-difference descriptors, showing in the ANLAB *AB* diagram his eight colour categories (sectors), their boundaries (at hue angles H1 to H8) and the terms used by colourists to describe hue differences from standards assigned to each category

relevant adjacent category (orange) only when there is also so much redness present that he does not consider it sensibly yellow. In describing all hue differences a colourist uses the comparatives of only the four elementary chromatic colours (redder, yellower, greener and bluer). To describe hue differences from a single standard he permits only two of them, the allowed pair always being those corresponding to the elementary chromatic colours closest in the hue circuit to the category to which he has allocated the standard (never to the same category). Furthermore, to indicate the nature of the hue difference of a single batch from its standard, he uses only one of the allowed pair of terms. Thus, if he categorises its standard orange, he would term a batch either yellower or redder; if as yellow, he would term it either greener or redder.

Different colourists, and the same colourist on different occasions, will categorise some shades differently. The borderline between adjacent categories is necessarily subjective; nevertheless, in describing the hue difference of a single batch from a sensibly chromatic standard, the colourist always uses a single term. This is why a colourist will object to the pairs of hue-difference descriptors output by rectangular-splitting software. For batch B2a (Figure 4.8) the output of redder *and* yellower is confusing because it must be either yellower *or* redder than S2. If he allocates S2 yellow, part of the output (yellower) is meaningless, because he would never term a batch yellower than a sensibly yellow standard. Worse, however, is that he can be misled by part of the output (greener or less red) for batch B2c because he would term it redder than S2. An acceptable alternative system of nomenclature has yet to be found, and the simultaneous use of two of the colourist's hue-difference terms, or ones dangerously like them, must be regarded as an inevitable concomitant of rectangular splitting. This is one of the reasons for the decline in its use. The other is that the structure of colour space, and therefore the nature of colour differences, is so much easier to understand when split cylindrically.

In cylindrical splitting, the residue of colour difference is partitioned into differences in chroma (ΔC^*_{ab}) and hue (ΔH^*_{ab}). While colourists rarely use the variable of chroma, except to imply the visual correlate of metric chroma, acceptable single-word comparatives derived from the noun 'chroma' have so far defied invention and this has led to a variety of descriptors being used. 'Stronger' ($\Delta C^*_{ab} > 0$) and 'weaker' ($\Delta C^*_{ab} < 0$), once widely used, are declining because they can be applied to describe differences in variables other than chroma (section 4.4.2). 'Chromatically stronger' and 'chromatically weaker' overcome the problem but are ugly; moreover, the format of colour-difference output often has to be distorted to accommodate the long fields they require. The descriptors in commonest use are 'higher chroma' and 'lower chroma' (for $\Delta C^*_{ab} > 0$ and $\Delta C^*_{ab} < 0$ respectively). Thus, for example, B2a is output as being of higher chroma than standard S2.

Cylindrical splitting allows the nature of a difference in hue to be indicated by outputting a single hue-difference descriptor, but problems arise in ensuring that it is the term the colourist would employ were he visually assessing the pair. The descriptor output is determined by both the hue angle of the standard ($h_{ab,S}$) and the direction of rotation through lesser angular displacement from standard to batch, indicated by the correct sign of ΔH^*_{ab}.

Many implementations of cylindrical splitting output the hue-difference descriptor associated with the first of the rectangular semi-axes to be crossed, either during rotation from standard to batch or (where a semi-axis is not then crossed) by continuing rotation beyond the batch. Using this method both B2a and B2c are output as redder than S2 and B2b yellower. If the colourist categorises S2 orange, these descriptors all accord with the terms he would use. If he allocates it yellow, however, only the first two outputs are correct. That for batch B2b is yellower and so not immediately meaningful to him, although in this case he may be able to deduce its import. Much more thought is needed when standard and batch straddle a semi-axis: for the higher-chroma pair B1b and S1 (Figure 4.8) the same hue-difference descriptor (redder) is output regardless of which is considered standard. The colourist is then faced with outputs that break two of the tenets of his system:

(a) the term 'redder' should not be used to describe a batch relative to a sensibly red standard
(b) whatever term may be used to describe the hue difference between a batch and its standard, the same term is never used if they are reversed.

Several attempts have been made to overcome these problems. Around 1970, Cooper analysed the hue-difference terms used in a large number of assessments made by many colourists. He then evaluated the mean hue angles at which there was a change in the pairs of terms used, and thereby determined the mean hue angles of the boundaries between the members of the set of colour sector categories shown in Figure 4.10. Each of the hue angles (H1 to H8) being known, the method allows the hue-difference descriptor output for a batch relative to any standard to be chosen from the pair most likely to be used by a colourist. The appropriate descriptor is the comparative of the adjective describing the next elementary chromatic colour sector encountered, after leaving that containing the standard, by rotation in the direction of lesser angular displacement from standard to batch, extended if necessary.

This method disposes of the problem of two specimens close to a semi-axis. For standard S1, batch B1b is output as bluer, while for standard and batch reversed the batch is described as yellower, both according with the terms used by a colourist. It does not avoid the problem caused by different colourists categorising the same standard as orange or yellow. In his analysis Cooper found considerable scatter in the boundary hue angles, both between and within assessors. This was confirmed by a later study by HATRA, which also indicated a significantly different set of mean boundary angles. The largest discrepancy between the boundary angles suggested by the two studies was 20° (for the turquoise-blue boundary), and for the HATRA data alone the widest spread of between-assessor means for any boundary was 34° (for blue-violet). Clearly, it is impossible for the method always to generate the descriptor the colourist would use.

In 1978 Taylor put forward a method in which rotation is continued until two semi-axes are crossed and the descriptors associated with each are output, the second in parentheses. For the comparison of any batch and a sensibly chromatic standard, one of the descriptors always accords with the term the colourist would use; the other term is meaningless to him and can be ignored [28]. For B2b relative to S2, the output is 'yellower (greener)'. If the colourist categorises S2 orange, the first tallies with his term but the second is an epithet he would never choose; if he thinks S2 yellow, the first is meaningless but the second agrees with his term. The method has two disadvantages: the length of output field required and the potential for those unfamiliar with the system to misinterpret the output. The advantage of the method more than compensates, however; its output is never wrong.

4.3 COLOURISTS' COMPONENTS OF COLOUR AND COLOUR DIFFERENCE

4.3.1 Introduction

With the increasing use of instrumental colour measurement, cylindrical systems involving the variables of lightness, chroma and hue are now widely used in the coloration and related industries. The systems are well understood and their use is relatively easily learned. Moreover, their learning requires no knowledge of colorant nature and behaviour. They are thus extremely useful for communication between workers in these industries.

Colourists, however, are primarily concerned with coloration processes. They therefore tend to think of the colour of the specimens with which they work, and the colour differences between them, in terms of the nature and behaviour of the colorants used to colour them. Brown colours, for example, are often matched using a combination of yellow and red colorants with one either sensibly achromatic or of hue complementary to that of brown. If a batch so produced is judged to be yellower than standard, the dyer knows that its hue difference can be corrected by lowering the relative proportion of yellow colorant in its formulation.

The nature of the colorants used in an industry largely dictates which of the several systems its colourists use to describe the other components of colour difference. The textile dyer, for instance, is usually dealing with a mixture of three dyes, often all sensibly chromatic; on the other hand, an opaque paint is commonly coloured by four pigments, one white, one black and two chromatic. Unfortunately the relationships between the variables (other than hue) of any of the systems employed by colourists and those of colorimetry are complex. It is for those used by the dyer that the relationship has been most extensively investigated and these terms are probably the best defined. Those used in other industries often have different meanings for different colourists.

4.3.2 Dyer's components of colour and colour difference

The first study of the relationship between the dyer's variables and those of colorimetry was made in 1943. Like several later investigations, it was based in the CIE xyY system. It is more useful, however, to relate the dyer's variables to those of a more uniform and cylindrically partitioned colour space. Reports of the first such studies were based in ANLAB space [29], but their findings are here discussed using CIELAB, the two spaces being qualitatively similar.

In order to investigate the relationship, we must first examine the *build* (or *build-up*) of dyes on a textile substrate. By this is understood the effect on colour of applying increasing concentrations of a single dye or a compatible mixture of dyes to a given substrate, all other conditions remaining the same. The builds of a selection of dyes on a typical textile substrate are shown, schematically as solid lines, in the $L^*C^*_{ab}$ and a^*b^* diagrams in Figures 4.11 and 4.12, which together define the build loci of the dyes in the three dimensions of CIELAB colour space. Most undyed textiles are sensibly white and, although not strictly achromatic (b^* usually being slightly positive), we may conveniently consider the starting point of each locus to lie at $L^* \approx 90$ and $C^*_{ab} = 0$. The other end of each locus represents the practical upper limit of concentration of the dye on the given substrate, the constraint being that further addition of dye to the bath does not cause significant change in the colour of the resulting dyeing. Applying increasing concentrations of dye causes:

(a) a decrease in L^* of the dyeing, although that for the dye labelled Yellow 1 is very small except at higher concentrations, typifying the behaviour of greenish-yellow dyes
(b) an initial increase in C^*_{ab} followed by a decrease (except for truly achromatic dyes)
(c) usually a change in h_{ab} (again with the exception of truly achromatic dyes); the rate of change at first increases with concentration and later decreases, the direction of change always being through the lesser angular displacement towards $20 \leq h_{ab} \leq 60$ (the value of h_{ab} being almost constant for a given substrate).

Increasing concentration is often used to define *depth* (or depth of shade), arguably the dyer's most important perceived variable. The SDC's Terms and Definitions

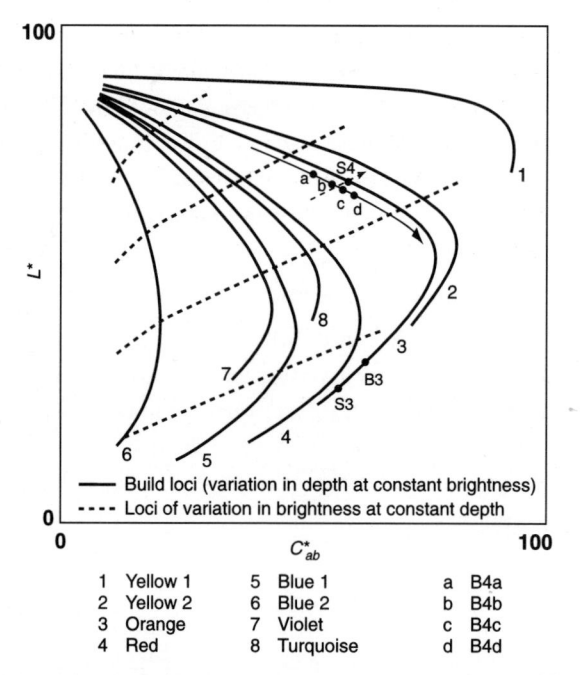

Figure 4.11 Build loci of typical dyes on textile substrate assumed to be at $L^* \approx 90$ and $C^*_{ab} = 0$, in the CIELAB $L^*C^*_{ab}$ diagram [29]

1	Yellow 1	5	Blue 1	a	B4a
2	Yellow 2	6	Blue 2	b	B4b
3	Orange	7	Violet	c	B4c
4	Red	8	Turquoise	d	B4d

— Build loci (variation in depth at constant brightness)
---- Loci of variation in brightness at constant depth

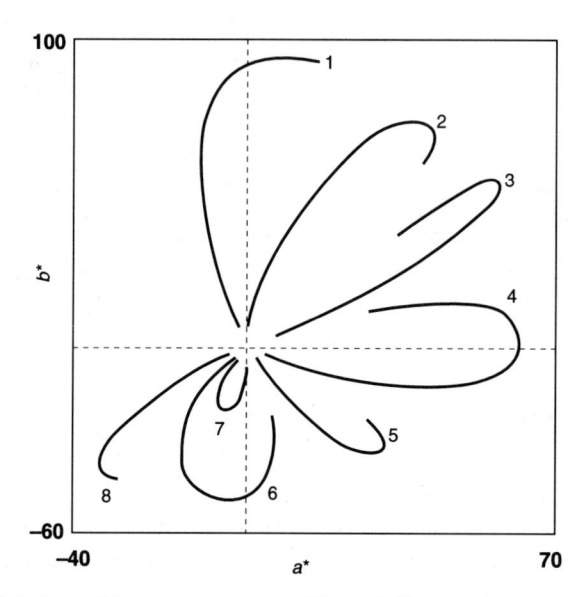

Figure 4.12 Build loci, of the same dyes as in Figure 4.11 and on the same substrate, in the CIELAB a^*b^* diagram [29]

Committee defines depth as 'that colour quality, an increase in which is associated with an increase in the quantity of colorant present all other conditions remaining the same' [30]. This definition should not be taken too literally, however. Firstly, although a change in h_{ab} almost inevitably accompanies an increase in concentration, it is not seen by the dyer as part of a change in depth, the change in hue being a hue difference *per se* and hence assigned to one of the other variables. Secondly, a similar caveat distinguishes a change in depth from one in the remaining variable brightness/dullness, discussed below. Thirdly, although the definition might be taken as implying that the concept of relative depth is applicable to the comparison of dyeings in which all but the concentrations of dye(s) remain constant, it is not so restricted, nor can it be. The concept necessarily transcends all conditions, extending to the evaluation of specimens where one or all of the dyes, their concentrations, the substrate, the application method and even the viewing conditions are different. For example, in colour matching the dyer must assess the equality, in all three of the component variables, of a batch on its substrate and a standard, frequently on a different substrate, produced using unknown colorants of unknown concentrations applied by an unknown method. And increasingly the colourist is asked to judge the colour constancy of the batch alone, under a range of light sources.

The dyer's remaining variable is that of *brightness*, defined by the Terms and Definitions Committee as the converse of dullness, which is defined in turn as that colour quality an increase in which is comparable to the effect of the addition of neutral grey colorant [30]. Care is needed in the interpretation of this definition too. The addition of a quantity of any nonfluorescent dye, including an achromatic one, to any nonfluorescent dyeing necessarily increases its depth. Thus, although the addition of a relatively small quantity of an achromatic dye to a bright dyeing does increase its dullness, it also – however slightly – increases its depth. The ratio of the increase in depth to that in dullness increases as the brightness of the original dyeing decreases until, in the ultimate, the addition of any quantity of a strictly achromatic dye to a strictly achromatic dyeing can cause only an increase in its depth. The loci in colour space of increasing dullness are represented in Figure 4.11 as dashed lines, which show that an increase in the dyer's perceived variable of dullness represents, for all chromatic dyeings, a decrease in both L^* and C^*_{ab} [29]. Thus differences in either depth or brightness usually involve differences in both L^* and C^*_{ab}.

Dyers, however, usually use single-word comparatives to describe differences in depth or brightness. A survey showed that most prefer the terms *fuller* and *thinner* to indicate higher and lower depth respectively, and almost as many favoured *brighter* and *flatter* for higher and lower brightness; other terms such as *deeper*, *paler* and *duller* were deprecated by most respondents.

We can now explore the second reason for not taking too literally the definition of depth given above. Loci of increasing dullness and of increasing depth are both loci of decreasing L^* and C^*_{ab} beyond the depth at which a dye yields dyeings of maximum C^*_{ab}, and they become increasingly similar as depth increases. Thus the heavier the depth of a dyeing, the more the dyer's two concepts become confused. A dyer investigating the build of a single dye, labelled 'Orange', would consider the batch represented by the point B3 (Figure 4.11) to be thinner than the standard at S3, because he knows S3 requires a much higher concentration of dye. If he were shade matching, he would be more likely to consider B3 brighter, because the most economical method of adjusting B3 to approximate the colour of S3 is not to increase the concentration of the dye(s) in B3 but to add a small amount of an achromatic dye, or one of complementary hue.

Even below maximum C^*_{ab}, however, there is a complex relationship between the dyer's variables of depth and brightness and the colorimetric ones of lightness and chroma. Many users of the latter make the assumption that lighter equates with thinner, darker with fuller, higher-chroma with brighter and lower-chroma with flatter. But these assumptions are quantitatively wrong for the comparison of most batches and a given standard, and even qualitatively wrong for about half of all batches. Consider the standard S4 and the batches B4a to B4d in Figure 4.11. The solid line through S4 and that through the batches both represent loci of increasing depth (in the direction of the arrow), and therefore of equal brightness. The dashed line through S4 is a locus of increasing brightness (again, in the direction arrowed) and of equal depth. Batches B4a and B4b are thinner than S4, and batches B4c and B4d are fuller; moreover, all four batches are flatter. For B4a, $\Delta L^* > 0$ and $\Delta C^*_{ab} < 0$, and for B4c, $\Delta L^* < 0$ and $\Delta C^*_{ab} < 0$. It follows that the assumptions are qualitatively correct for these two batches. For B4a, however, $|\Delta L^*| \ll |\Delta C^*_{ab}|$, whereas its difference in depth (from S4) is clearly greater than that in brightness; for B4c, $|\Delta L^*| \approx |\Delta C^*_{ab}|$, but its difference in depth is much less than that in brightness. The assumptions are thus quantitatively wrong. For batches B4b and B4d, one of the assumptions is in each case qualitatively incorrect. For B4b $\Delta L^* < 0$ but it is thinner, and for B4d, $\Delta C^*_{ab} > 0$ but it is flatter.

Translation of colorimetric component differences, other than in hue, to the dyer's terms is in general clearly impossible by inspection of the former alone. Reliable conversion was enabled by the development around 1970, by ICI, of a mathematical model, then called HBS (hue, brightness and strength, strength being one of the terms the dyer recognises in context as synonymous with depth). In 1979, ICI installed a proprietary (ICS) colour-measurement system and agreed to allow ICS limited distribution of the software to third parties. The model was included in the version of the M&S colour-difference formula (M&S80) released the following year, and it has remained part of all subsequent M&S formulae. At that time the Society's Colour

Measurement Committee also examined the HBS model; it suggested that, in order to avoid the confusion caused by 'strength' being used with various meanings in several contexts (section 4.4.2), the term be replaced by 'depth'. Accordingly ICI renamed the model DBH (depth, brightness and hue), the changed order of the variables reflecting the primary importance to the dyer of depth. Details of the model remain unpublished.

4.3.3 Paint colourist's components of colour and colour difference

The relationship between the paint colourist's colour variables and those of colorimetry is, like that of the dyer's, complex and is further complicated by three factors alien to the dyer. Firstly, the paint colourist works with two distinct pigment concentration systems, namely variable loading and fixed loading. The former is similar to the dyer's concentration system in that the ratio of the total quantity of pigments, relative to that of medium, may vary. In the latter the ratio of the amount of all pigments, taken together, is kept constant relative to that of medium, colour being altered by changing the ratios of the pigments relative to each other. Secondly, whilst the build loci of all organic and some inorganic chromatic pigments are similar to those of organic dyes (Figure 4.11, solid lines), those of most inorganic chromatic pigments are (as exemplified by the line WC in Figure 4.13), much more nearly straight lines,

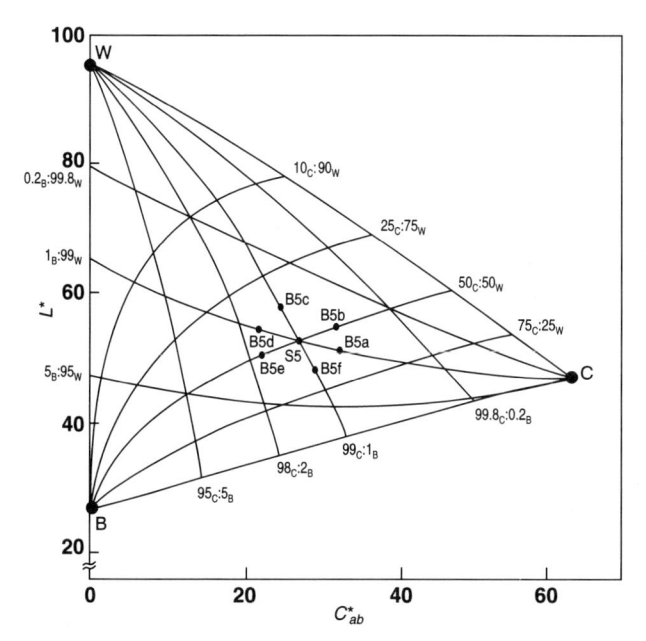

Figure 4.13 The paint colourist's variables of colour difference, shown in the CIELAB $L^*C^*_{ab}$ diagram, for a fixed total loading of an inorganic chromatic pigment (C) and achromatic black (B) and white (W) pigments

C^*_{ab} always increasing with increase in pigment concentration. The cause of the differences in build-loci shape is ascribed to wavelength-dependent differences in the refractive indices of the pigments.

The third and most significant complication results from the greater number of colorants used by the paint colourist. Both the paint colourist and the dyer usually work with two chromatic colorants plus a black (or, more usually in the case of the dyer, a colorant of hue approximately complementary to the colour to be matched), but the paint colourist employs a white pigment too. Both use the same variable of hue, which is common to their systems and to that of colorimetry, to indicate differences in colour caused by differences in the relative amounts of their two chromatic colorants, so that in the $L^*C^*_{ab}$ plane the dyer's system reduces to two colorants and the paint colourist's to three. In addition to the variable of hue, therefore, the dyer needs only two degrees of freedom (expressed as depth and brightness, section 4.3.2), while the paint colourist requires three, one for the description of colour differences caused by changes in the amounts of each of chromatic, black and white pigments. The terms most commonly used by paint colourists to describe colour differences caused by changing the (relative) concentrations of pigments in a formulation are [16]:

(a) *stronger* (and *weaker*) for more (or less) chromatic pigment
(b) *dirtier* (and *cleaner*) for more (or less) black
(c) *whiter* (and *deeper*) for more (or less) white.

The relationship between the paint colourist's variables, for a fixed-loading system, and the colorimetric ones L^* and C^*_{ab} is shown in Figure 4.13, in which the points C, B and W represent respectively 100% (nominal) loadings of inorganic chromatic pigment and perfectly achromatic black and white pigments in a clear medium. Each line originating from C (or B or W) is a locus of decreasing concentration of chromatic (or black or white) pigment and increasing, and compensating, concentration of a constant ratio of black to white (or chromatic to white, or chromatic to black) pigments, the ratio being shown at the end of each line most distant from C (or B or W, all respectively). The pigment colourist would describe, relative to the standard S5, batches B5a as stronger and B5d as weaker, B5e as dirtier and B5b as cleaner, and B5c as whiter and B5f as deeper. Changes in any one of these variables usually involve differences in both L^* and C^*_{ab}. In particular, whilst a difference described by the pigment colourist as stronger or weaker (to imply a higher or lower chromatic pigment content) almost always has both $\Delta L^* \neq 0$ and $\Delta C^*_{ab} \neq 0$, it is only the inequality in C^*_{ab} that is ever described by the same terms used in the colorimetric context.

It can be shown that the variations in colour that the paint colourist describes as whiter and deeper correspond to the dyer's thinner and fuller. The paint colourist's other two variables, taken together, are then equivalent to the dyer's brightness/dullness. For relatively small changes in the black content of a formulation of high C^*_{ab}, however, the paint colourist's descriptors of 'dirtier' and 'cleaner' are, for practical purposes, equivalent to the dyer's 'flatter' and 'brighter'. Although the addition of a small quantity of black to a dyeing (of high C^*_{ab}) must cause some increase in depth, the increase is so small that the dyer cannot possibly detect it in the presence of the overwhelmingly larger increase in dullness the addition causes.

4.4 EVALUATION OF DEPTH AND RELATIVE DEPTH

4.4.1 Introduction

The value of being able to make reliable quantitative assessments of depth differences during colour matching is obvious but the requirement exists in several other processes too. Arguably the three most important of these ultimately concern the appraisal of various relative properties of colorants *per se* rather than, as in colour matching, the results of their application. The three processes are:

(a) the evaluation of the relative strengths of colorants, enabling their standardisation
(b) the determination of their depth-dependent properties, such as relative fastness
(c) the estimation of their relative cost-effectiveness in coloration.

It is useful to distinguish between what we shall call similar and dissimilar colorants and colours. *Similar* colorants contain the same single coloured species or mixture of species in the same, or very nearly the same, proportions. The application of similar colorants produces similar colours, which may or may not (nearly) match. If they do, their reflectance curves are (almost) identical. If they do not, their curves can be made (almost) identical by altering the depth of either colour. Conversely, *dissimilar* colorants contain different coloured species, or the same several species in different proportions. The application of dissimilar colorants produces dissimilar colours, which again may or may not match: whether they do or not, their reflectance curves differ, and cannot be made identical by altering the depth of either colour.

4.4.2 Colorant strength and standardisation

It would be unacceptable if each batch of a given colorant performed differently in coloration processes. Colorant manufacturers therefore aim to make each batch such that its performance is, within commercially acceptable limits, equal to that of a standard. The performance of an unstandardised batch may differ from that of its standard in many ways, such as ease of solution or dispersion, but only the relationship of the colours it imparts in a given application concerns us. One factor considered by the standardising colourist is the relative depth of colorants, from which the strength of the batch relative to that of its standard is deduced.

For present purposes, we shall define the *strength* of a colorant as a measure of its capacity, when applied at a given concentration in a given coloration process, to influence the depth of the resulting colour. For coloured dyes or pigments applied to a sensibly white substrate or in conjunction with a white pigment, the effect is to increase depth; for white pigments applied in admixture with a coloured pigment, however, the depth is decreased. The relationship between strength F, concentration c and depth D, evaluated such that $D > 0$ for the application of any c (> 0) of a colorant, and $D < 0$ for a white, is Eqn 4.37:

$$F = \frac{kD}{c} \tag{4.37}$$

where, to make $F > 0$, $k > 0$ for coloured and $k < 0$ for white colorants, and $|k|$ depends on the required scaling of F given those of c and D.

The resulting value of F may not be valid for all c and may be true only for the given c. To understand the reason for this, we must examine the relationship between c and D, illustrated by the schematic build curve labelled α in Figure 4.14. Usually, over some

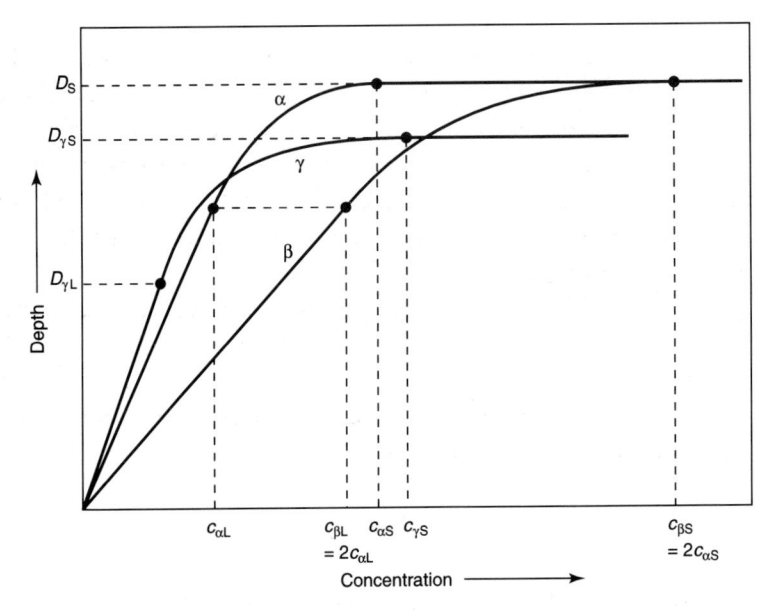

Figure 4.14 Build curves (some measure of depth D against concentration c); curves α and β are for similar colorants

range of c, $0 \leq c \leq c_{\alpha L}$ where $c_{\alpha L}$ is a finite concentration, most industrial colorants exhibit linear build – that is, D is directly proportional to c, all other relevant conditions being maintained. Often, however, for $c > c_{\alpha L}$ the build is nonlinear, the rate of increase in D continuously diminishing with increase in c until, at $c = c_{\alpha S}$, it becomes zero and the colorant is said to be saturated with respect to depth attainable. The three segments of the build curve (linear, nonlinear and saturated) are thus delimited by two concentrations: $c_{\alpha L}$, the limit of linearity of build, and $c_{\alpha S}$, the saturation limit of the colorant. Therefore F is constant for all $c \leq c_{\alpha L}$ but thereafter continuously decreases.

In standardisation, however, it is the strength of a batch B of colorant relative to that of its standard S that is of interest. Relative strength is usually assessed at equal depth. One reason is that visual quantification of depth difference is unreliable except at or near equality (see also section 4.4.5). A second is that, to enable adjustment of his existing colour-matching formulations based on S, the user of B needs to know that concentration c_B of B required to produce depth equal to that yielded by a given concentration c_S of S. We may deduce from Eqn 4.37 that the strength F_B of B relative to that of S (which is usually assigned a nominal value of 100%) is inversely proportional to the ratio of c_B to c_S (Eqn 4.38):

$$\%F_B = \frac{c_S}{c_B} \times 100 \qquad (4.38)$$

provided assessment is made at equal D. Since B and S are similar colorants, the build curve of B (β in Figure 4.14) is to a good approximation simply a scaling of that of S (α) in the direction of c. Approximately the same value of F_B is therefore obtained at all $D < D_S$ (the common saturation depth of B and S). Most colorants are, however, standardised at D on the linear segment of their build curves because the rate of change in D with change in c is there greatest, so that the effect on the resulting value of F_B of a given error in assessment of D is less than at higher D.

For equal D ($< D_S$) of B and S represented by the curves β and α, $c_B = 2c_S$, and F_B = 50%. The user of B can now appropriately adjust formulations based on S. The required adjustment is an inverse one, doubled concentrations being needed when using the half-strength B. There is, however, another expression of relative strength (F'_B) which may be applied directly by the user (Eqn 4.39):

$$F'_B = \frac{c_B}{c_S} \times 100 \qquad (4.39)$$

without a percentage sign. Thus, for our example B and S, F'_B = 200.

Confusion is rarely caused by having two methods of reporting relative strength.

The principal use of the F_B method is to distinguish versions of similar colorants deliberately made at different strengths and sold under otherwise identical designations; paradoxically, the percentage sign is more often omitted than not. Thus, for example, if S be sold as a 100-strength brand, B may be sold without further adjustment of strength under the designation '50'. The F'_B method is hardly ever employed in this way but is used in most other contexts. The context therefore usually identifies the method. Moreover, different conventions are used in reporting the two methods. (Although colourists recognise several synonymous terms qualifying differences in strength, by far the most common practice in respect of both dyes and pigments is to describe B as *stronger* if $c_B < c_S$ and *weaker* if $c_B > c_S$, c_B and c_S yielding equal D.[1])

A report on the results of the F_B method usually includes the word 'strength', omits the comparatives, and should include a percentage symbol. The format for the F'_B method is much more rigid. A typical oral report for our example is 'B is weaker 200 to 100 against S', the general written form being 'B $ F'_B:100 vs S', where $ = Str (stronger) when $F'_B < 100$, $ = Eq (equal) when $F'_B = 100$, and $ = W (weaker) when $F'_B > 100$; for example, A Str 80:100 vs S.

Whatever the strength relationship between B and S, the colours they yield may differ in attributes, such as the dyer's brightness and hue, as well as in depth. Colourists refer to these other differences collectively as differences in *shade*. Two types of shade difference are of interest. The first is that between the colours as they stand, regardless of depth difference (ΔD). The second is the shade difference that would exist were the colour produced by B adjusted so that $\Delta D = 0$. This is often called the *residual colour difference* (ΔE_R). The two are the same only if, without adjustment, $\Delta D = 0$, or if B builds without change of shade. The latter is an unusual condition, most colorants building redder (Figure 4.12). The colourist can usually make a fair estimate of ΔE_R, but its colorimetric evaluation depends on the ability both to evaluate ΔD, several methods for which are discussed in sections 4.4.6 to 4.4.8, and to predict the colour of

---

1 We have now met four usages of strength and its associated comparatives, stronger and weaker: the colorimetrist's, to describe higher and lower chroma (section 4.2.12); the dyer's, to indicate depth differences, although he more often uses other terms (section 4.3.2); the pigment colourist's, to qualify differences in the chromatic-pigment content of a formulation (section 4.3.3); and now the colorant manufacturer's and user's, to describe differences in the colouring power of colorants. The last three are deeply ingrained in colourists' jargon and distinction between them is usually obvious from the context. Change is therefore neither likely nor desirable. The colorimetric usage often overlaps the contexts of the other three, however, and there is obvious potential for confusion. Writers of software for proprietary colour-measurement systems have the opportunity to avoid this confusion by using terms, such as those involving 'chroma' listed in section 4.2.12. Where there is any possibility of ambiguity, users of existing software must ensure that their intended meaning is clearly conveyed.

B when $\Delta D = 0$, for which there are far fewer methods. One of them, DBH splitting, was discussed in section 4.3.2. We shall meet another in section 4.4.7.

4.4.3 Standard depths for fastness testing of colorants

In determining the colorants to be used in matching, the colourist's choice is influenced both by the relative performance of colorants substitutable in his coloration process (for example, the sensitivity of the yield of dyes to variations in liquor ratio) and by their relative ability to meet the criteria required of its end-product, principally those of fastness. Colorant manufacturers' literature usually indicates the in-coloration and fastness properties of each colorant, but since fastness (and many other properties) depend on depth, such information would be of doubtful value unless it pertained to sensibly the same depth for all colorants.

Fortunately, colorant manufacturers publish property data at an arbitrary fixed depth, known as 1/1 standard depth (SD), which most colourists regard as being of approximately medium depth. From these data, given that many colorant properties vary similarly and only slowly with depth, the colourist can usually make an estimate of relative properties at other depths that is adequate for practical purposes. Fastness to light, however, often changes rapidly with depth. In addition to being given at 1/1 SD, therefore, light-fastness data are usually presented at some of a limited number of simple ratios of 1/1 SD. Those in more frequent use are 1/200, 1/25, 1/12, 1/9, 1/6, 1/3 and 2/1, and are so designated. In addition, data for those blue and achromatic colorants used mainly in the production of navy and black colours are often given at appropriate applied depths, greater than 2/1 SD, designated navy blue/light (N/L), navy blue/dark (N/Dk), black/light (B/L) and black/dark (B/Dk).

The concept of standard depth was born of collaboration between the major German and Swiss dye manufacturers, who by 1930 had agreed ten physical colours of approximately equal depth, plus higher depth navies and blacks, and by 1960 the 1/25, 1/12, 1/6, 1/3, 1/1 and 2/1 SD series commonly used for textile dyes had been established very much in their present form.

The current internationally recognised physical references for these ratios are defined by DIN 54 000, published in 1969. Each series consists of between twelve and eighteen dyeings on wool piece (gaberdine). About two-thirds of them are nonfluorescent high-chroma colours illustrating, between them, most of the hue sectors shown in Figure 4.10, the remainder being olives, browns and greys. The references for N/L, N/Dk, B/L and B/Dk are also defined by DIN 54 000. Mention is made of all these ratios in ISO 105, Part A01: *General Principles of Fastness Testing*, which stipulates that dye manufacturers shall publish fastness data obtained by testing 1/1 SD specimens wherever possible, supplemented by data for one or more of the other ratios as required; exceptions are made only for navy and black dyes, which should if possible be tested at both the appropriate (N or B) depths (L and Dk).

The visual method for determining any ratio of 1/1 SD for an unknown colorant is comparatively simple. The usual procedure is iterative and consists of assessing the equality of the depth of a test specimen, prepared by applying the unknown colorant, and that of the member of the relevant reference series most closely matching its colour, placed adjacently and viewed under illumination approximating daylight. Interpolation between the references most closely straddling the colour of the test is used when none of the references alone is judged a close enough match. The method is the essence of AATCC Evaluation Procedure 4, Standard Depth Scales for Depth Determination, which, however, omits reference to the N and B depths.

Some of these ratios of SD, notably 1/25, 1/3 and 1/1, are also frequently used in the pigment industries, often supplemented by the ratios 1/200 and 1/9, which are virtually exclusive to pigment applications.

4.4.4 Relative cost of colorants

Depth assessment is also important in determining the relative cost of colorants in a coloration process, perhaps where a user has established one manufacturer's colorant (E) in an application and is offered an alternative (A) by another. Provided A can fulfil the required performance criteria, the user's decision to switch or not is usually influenced by the relative cost considerations.

Relative cost depends on relative price and relative strength. The concept of relative strength, explained for similar colorants in section 4.4.2, may be extended to dissimilar ones. Visually, relative strength is most reliably evaluated by assessment of equal-depth colours; so too is relative cost. If A (unit price P_A) and E (unit price P_E) are each applied alone – that is, in so-called straight shades or self shades – at concentrations c_A and c_E respectively that yield equal depth, the cost of A relative to that of E, assigned the value of 100%, is R, given by Eqn 4.40:

$$\%R = \frac{c_A P_A}{c_E P_E} \times 100 \qquad (4.40)$$

where $R > 100\%$ indicates A to be more costly than E, and $R < 100\%$ *vice versa*.

The method works well if A and E are similar colorants but less well if they are dissimilar, because two problems often then occur. Even so, it is widely used to provide a cost comparison that, although crude if A and E are dissimilar, is usually adequate to indicate whether it is worthwhile to apply a more refined but more time-consuming method.

The first problem is that there may be a significant difference in the builds of A and

E, and thus in the strength relationship between them. If, in Figure 4.14, the curve labelled γ represents the build of A and α that of E, the strength relationship varies from approximately A Str 80:100 vs E ($0 < D \leq D_{\gamma L}$), thereafter rapidly decreasing to W 200:100 ($D = D_{\gamma S}$), and does not exist if $D > D_{\gamma S}$. A proportionate range of R obtains regardless of the ratio of P_A to P_E. The second problem arises because colorants are not usually used alone but in admixture with others, and A and E, being dissimilar colorants, are almost certainly different in shade. Thus a formulation using A is likely to require concentrations of its other, common components, different from those in a matching formulation containing E, because A partly replaces, or needs to be replaced by, the common components. Any cost advantage of A in straight shades may thereby be increased or decreased, or even reversed.

A comparison of overall colorant costs (R') eliminating both problems, for any given colour, requires the single-component products (cP) in Eqn 4.40 to be replaced by the sums of those of all components (Eqn 4.41):

$$\%R' = \frac{\sum_{A=1}^{a} c_A P_A}{\sum_{E=1}^{e} c_E P_E} \times 100 \qquad (4.41)$$

where there are a and e colorants in the matching formulations. The resulting value of R' obtained from other pairs of matchings using A and E will almost certainly be different, however. If, therefore, E is currently being used to match a range of colours, a valid overall cost comparison can be obtained only by determining separate values of R' for a representative selection and calculating an average weighted by the amount of each in typical production. The calculation is often too time-consuming to be practicable unless computer prediction can be used.

4.4.5 Visual assessment of relative depth

Little substantive evidence exists regarding the repeatability and reproducibility of visual assessments of relative depth. Nevertheless consensus among colourists is at least as unlikely in depth assessments as in any other type of visual colour-difference judgement.

Visual colour matching and colorant standardising are iterative processes in which the aim is an overall colour difference, and hence a difference in any component, that is approximately zero. A mistake made during iteration must therefore become evident before the final iteration. Whether or not it can be corrected, additional cost is incurred. The higher reliability of the visual system in null decisions is the main reason why standardising iterations, especially the last, are made at equal depth. Typically, the colours resulting from the application of five concentrations of the batch, at ratios of 90, 95, 100, 105 and 110% of the concentration that the previous iteration indicated necessary to equal the depth given by a prescribed concentration of the standard, may be assessed against the colour given by the prescribed concentration. Provided the depth of the centroid batch colour is within about 10% of that forecast, the colourist can then judge which of the real batch colours, or of the notional ones halfway in depth between them, is closest in depth to the standard colour. His judgement is therefore effectively made at 2.5% relative concentration intervals.

It is pointless to prepare a more closely spaced set. A 2.5% difference in depth, or one perceived equivalent in any other colour attribute, is the least that can be detected by a trained colorist with near certainty. This difference is known as a *trace*. Given the likelihood of error in the assessment of relative depth, the possible error in visual standardisation of strength is about 2.5%, even ignoring all other errors.

A trace difference is usually much less than that demanded in colour matching using the batch. Most matchings involve more than one colorant, however, and the error in overall colour resulting from the application of a mixture can exceed the sum of those resulting from separate applications of its components. Using match-prediction data, Smith modelled opposing 2.5% strength differences for a yellow and a blue reactive dye, both building linearly beyond the concentrations chosen, the weaker yellow and stronger blue yielding equal depth. The values of ΔE_{MS89} were 0.12 for the resulting pair of yellow colours and 0.13 for the blues. Using half the concentration of the stronger yellow together with half that of the weaker blue and, separately, the other pair of half-concentrations, two green colours were then synthesised, with $\Delta E_{MS89} = \Delta H_{MS89} = 0.64$.

In attempting to determine the concentration of a colorant necessary to yield either a given ratio of 1/1 SD or a depth match to the colour from a given concentration of another colorant, the aim is also equality of depth; but it must often be assessed in the presence of a difference in shade. The shade given by the colorant being investigated is unlikely to be close to that of any of the SD reference specimens, and the shades from two colorants being evaluated for relative cost may be very different. For example, the 'yellow' component of a combination used to produce drab colours may fall anywhere in the area bounded roughly by yellowish-orange, greenish-yellow, yellowish-olive and brown. The visual system is easily confused when judging the partition of colour differences. Even experienced observers can mistake redness for depth, for example, and most colorants build redder (Figure 4.12). Colourists are often unwilling to qualify the direction of a minor component, let alone to quantify it, in the presence of a much larger difference in another. The average limit for voluntary qualification of a minor component seems to be that it must be at least one-fifth the magnitude of the largest.

Whenever the end-point is a significant difference in shade, visual estimation of equality of depth is therefore extremely difficult.

Reliable estimates of depth differences are both valuable in various applications and difficult to achieve by visual means, and colorimetric methods have continually been sought that can augment or replace them. Instrumental methods are much more repeatable and more reproducible. But it has always been difficult to find a colorimetric method that gives results in sufficiently good agreement with those of visual assessments.

The three main types of colorimetric method are discussed in the sections that follow.

4.4.6 Single-wavelength methods

The absorption of light by solutions of colorants is routinely measured by manufacturers during standardisation, and also by users to check their deliveries. Solution methods are quick to perform, and also reliable. Kuehni, for example, estimated the mean standard deviations of relative strength evaluations of single preparations, each performed once by absorbance and reflectance measurements, to be respectively 0.5% and 3.5% [31]. The chief limitations of solution methods are that the colorants must be soluble and, for the results to be meaningful, the relative performance of the batch (B) and standard (S) in solution must adequately imitate that in the relevant coloration process. Hydrolysis of reactive dyes is probably the most frequent cause of lack of correlation. Most dyes, however, contain coloured species in addition to their principal component; these may differ in quality and quantity between B and S, and their performance in dyeing is not always in proportion to their contribution to colour in solution.

The solution technique most often employed is to compare the absorbances of B and S at their wavelengths of maximum absorption. The method ignores performance at other wavelengths, and so is limited to comparisons of similar colorants. It can be extended to include all visible wavelengths (λ, nm) by, for example, evaluating depth in solution as the sum of absorbances (A_λ) at each of a limited number of wavelengths, usually equally spaced (Eqn 4.42):

$$D_S = k \sum_{\lambda=400}^{700} A_\lambda \qquad (4.42)$$

where D stands for colorimetrically evaluated depth, subscripts being used to differentiate various D values, and k is a constant that enables scaling of D_S as desired. Eqn 4.42 is rarely used for dissimilar colorants, however, because of lack of correlation between their relative performances in solution and coloration.

Where the correlation is inadequate, reflectance-based techniques are employed. The simplest possible such method, devised independently by Beresford and Taylor, defines depth (D_{BT}) by Eqn 4.43:

$$D_{BT} = k R_{min} \qquad (4.43)$$

where R_{min} is minimum reflectance in the visible waveband [32,33]. Results from this method are not generally in good agreement with those of visual assessments of the depth differences between dissimilar colours, however. It is useful to quantify the extent of disagreement in a way that enables comparison of the efficiencies of this and other colorimetric formulae. We shall assume that the eighteen 1/1 SD reference specimens defined by DIN 54 000 are of equal visual depth, and therefore that a perfect formula would return equal D for all. We can then express the extent of disagreement (E) of a formula as the half-range (S) of D it returns, for all references, as a percentage of the midpoint $\hat{D}$ of the range. Thus for $\hat{D}_{BT}$ = 3.1 and S_{BT} = 1.0 [34], E_{BT} = ±100 (1.0/3.1) = ±32% – a high level of disagreement. Even in the evaluation of relative depth of similar colours, D_{BT} does not provide an easy means of quantifying the inequality of depth of B and S because depth is not linearly related to reflectance.

There is however a function of reflectance (R_λ, $0 < R_\lambda \leq 1$) that provides a relationship linear with concentration, and thus with depth, over approximately the same range of concentrations that visual assessors regard the build of a given colorant as linear. It is the Kubelka–Munk function $(K/S)_\lambda$, fully discussed in Chapters 5 and 6 (Eqn 4.44):

$$(K/S)_\lambda = \frac{(1-R_\lambda)^2}{2R_\lambda} \qquad (4.44)$$

which in this chapter may be regarded as a single variable.

Several methods of evaluating D are based on the $(K/S)_\lambda$ function. In each of them, if it is D of a coloured specimen that is required, then the relevant $(K/S)_\lambda$ is that given by Eqn 4.44 with $R_\lambda = R_{M,\lambda}$, where $R_{M,\lambda}$ is the measured reflectance of the specimen. If, however, the substrate has significant depth and the requirement is D imparted to it by the application of a colorant, $(K/S)_{c,\lambda}$, then Eqn 4.45 holds:

$$(K/S)_{c,\lambda} = \frac{(1-R_{M,\lambda})^2}{2R_{M,\lambda}} - \frac{(1-R_{S,\lambda})^2}{2R_{S,\lambda}} \qquad (4.45)$$

where $R_{S,\lambda}$ is the measured reflectance of the uncoloured substrate. If it is the depth imparted to a substrate by a colorant alone that is wanted, $(K/S)_{c,\lambda}$ (Eqn 4.45) is appropriate if $R_{S,\lambda}$ is the measured reflectance of a so-called blank-treated substrate, that is, a substrate that has been subjected to the coloration process in the absence of colorant.

The simplest application of the $(K/S)_\lambda$ function is to assume depth $(D_{K/S})$ proportional to $(K/S)_{\lambda(Rmin)}$ (Eqn 4.46):

$$D_{K/S} = k(K/S)_{\lambda(Rmin)} \qquad (4.46)$$

The method is used in various standards for the comparison of B and S similar colorants.

Use of Eqn 4.46 is restricted to the comparison of similar colours. For comparisons of dissimilar colours, it is again necessary to use a method that takes into account the whole of the visible waveband. Such methods are of two types, namely those that incorporate summation of $(K/S)_\lambda$ at all relevant λ (section 4.4.7), and those that involve the conversion of all R_λ to some orthogonal tricoordinate system of complete colour specification (section 4.4.8).

4.4.7 $(K/S)_\lambda$ summation methods

In the simplest and most widely used of the $(K/S)_\lambda$ summation methods, depth $(D_{\Sigma K/S})$ is evaluated as the sum of $(K/S)_\lambda$ (Eqn 4.47):

$$D_{\Sigma(K/S)} = k \sum_{\lambda=400}^{700} (K/S)_\lambda \qquad (4.47)$$

for an equally spaced selection of wavelengths in the visible waveband.

The visual perception of depth, like that of any other colour quality, is affected by the light by which it is viewed and the visual system. In 1973, Garland suggested a modification of Eqn 4.47, now known as the 'Integ' method. In this method each $(K/S)_\lambda$ is weighted by the product of illuminant energy (E_λ) and the sum of the CIE colour-matching functions $(\bar{x}_\lambda + \bar{y}_\lambda + \bar{z}_\lambda)$ at the relevant λ [35], depth D_I then being given by Eqn 4.48:

$$D_I = k \sum_{\lambda=400}^{700} (K/S)_\lambda \, E_\lambda (\bar{x}_\lambda + \bar{y}_\lambda + \bar{z}_\lambda) \qquad (4.48)$$

It is worth exemplifying some of the major benefits of the Integ technique. It is associated with the Colour Map, a CIELAB a^*b^* diagram in which, by definition, all colours are shown at equal D_I ($D_I = 20$, say) (Figure 4.15) [36]. The spots are the positions of individual colorants, gY and rY indicating respectively greenish- and reddish-yellows, and R and B red and blue. Imagine the combination rY, R, B employed by a user to produce a wide variety of colours. A manufacturer wishes him to buy a certain colorant gY instead of rY, which he currently buys elsewhere. Since D_I is a measure of depth, it may be used to quantify the build curves of gY and rY, and hence to quantify the strength and cost relationship between them at any D_I. Since D_I is a function of $(K/S)_\lambda$, and $(K/S)_\lambda$ is an additive function (Chapters 5 and 6), it follows that any two colorants (P and Q) separately yielding $D_{I,P} < 20$ and $D_{I,Q} = 20 - D_{I,P}$, produce if applied together total $D_I = 20$. Binary loci at $D_I = 20$ may then be drawn on the Colour Map by joining the locations of relevant pairs of individual colorants by (invariably gently curving) lines passing through the locations of a sufficient number of binary combinations of P and Q. These locations may be determined by reading from the build curve of each colorant those concentrations necessary to give $D_{I,P}$ and $D_{I,Q} = 20 - D_{I,P}$; they are applied together in the relevant coloration process and the resulting colour is measured. A much quicker method is to use computer synthesis from match-prediction data for each colorant.

The guiding principle in the interpretation of a Colour Map at depth D_I is that any combination of three colorants is capable of yielding any colour at depth D_I, provided its location falls within the area bounded by their three binary loci, but incapable of producing any colour that falls outside it. Figure 4.15 shows that, in combination with the existing R and B, gY is capable of giving all the colours available with rY, except for a small deficiency in reddish-yellow and orange colours. This is more than offset,

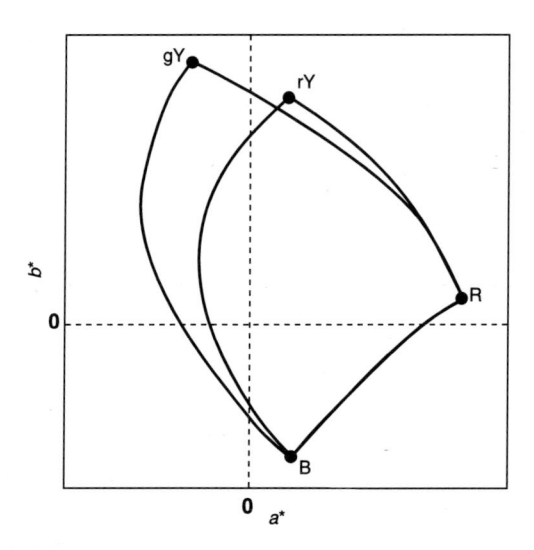

Figure 4.15 Colour map (at fixed D_I) illustrating relative shade gamuts attainable with greenish-yellow (gY) and reddish-yellow (rY) colorants used in combination with common red (R) and blue (B) colorants

however, by the much larger range of greenish- and mid-yellows, greens and olives that only gY can produce. The combined techniques of Integ and the Colour Map allow all aspects of the strength, cost and colour-generating relationships between colorants to be investigated and demonstrated graphically and convincingly.

The Integ technique has not been incorporated in any national or international standard, however, although a variant has been adopted by six major European dye manufacturers for determinating the relative strengths of colorants. It includes a method of estimating residual colour difference ($\Delta E^*_{ab,R}$) (section 4.4.2). Scaling of the $(K/S)_\lambda$ summation of B is first applied so that B yields D equal to that of S, the resulting $(K/S)_\lambda$ values of B and S are then converted to reflectances (using Eqn 5.2) from which $\Delta E^*_{ab,R}$ is calculated [37]. Although $\Delta E^*_{ab,R}$ is partitioned into the components ΔL^*, ΔC^*_{ab} and ΔH^*_{ab}, rather than the dyer's variables of brightness ΔB_D and hue ΔH_D, it is reasonable to assume that ΔB_D is given by Eqn 4.49:

$$\begin{aligned}\Delta B_D &= \left[(\Delta E^*_{ab})^2 - (\Delta H^*_{ab})^2\right]^{1/2} \\ &= \left[(\Delta L^*)^2 + (\Delta C^*_{ab})^2\right]^{1/2}\end{aligned} \quad (4.49)$$

since depth difference is by the definition of normalisation zero and, necessarily, $\Delta H_D = \Delta H^*_{ab}$. The scope of the method is limited to comparisons in which, before normalisation, $\Delta E^*_{ab} \leq 2.0$ and B and S are similar colorants. The most cogent reason for the latter restriction affects all $(K/S)_\lambda$ summation techniques. In nearly every case in which dissimilar colours match under some combination of illuminant and observer, and are therefore of equal depth under that combination, all $(K/S)_\lambda$ summation methods indicate some difference in D, often a large one. Conversely, if a $(K/S)_\lambda$ summation method reports equal D for colours of unlike shade, the colours may be very unequal in perceived depth. For the Integ formula (Eqn 4.48) the extent of disagreement $E_I \approx \pm 38\%$.

4.4.8 Methods based on colour-order systems and colour spaces

Methods based on orthogonal tricoordinate colour-specification systems have the advantage that they cannot yield unequal D when applied to B and S the system deems to match under a given set of conditions. The first was suggested by Godlove in 1954, but achieved little use in colorimetry. It was followed in 1957 by a development by Rabe and Koch of a colour-difference formula based in the widely used DIN colour-order system. The geometrical arrangement of the DIN system is conical. Its three orthogonal variables, in the order in which they are conventionally specified, are:
(a) DIN-Farbton (hue, T), all colours of a given T having the same CIE λ_d (or λ_c) under specified conditions, originally C/2

(b) DIN-Sattigungsstufe (saturation, S), a function of the distance of the specimen from the achromatic point in the relevant CIE chromaticity diagram, and
(c) DIN-Dunkelstufe (darkness, d), defined by Eqn 4.50:

$$d = 10 - 6.1723 \log_{10}(40.7 Y/Y_o + 1) \qquad (4.50)$$

where Y is the CIE Y of the specimen and Y_o that of the optimal colour stimulus (section 4.1.2) of the same T [38].

In this formula, depth (D_{RK}) is defined by Eqn 4.51 [39]:

$$D_{RK} = \frac{S(10 - 1.2d)}{9} + 1.06d \qquad (4.51)$$

and, alone among the multiple-wavelength methods discussed, is independent of hue. Its development was based on evaluation of the eighteen 1/1 SD references (most of which are of higher DIN-S), for which it returns an extent of disagreement $E_{RK} \approx \pm 4\%$, a very low figure [34]. It is historically important because it was used, in part, to define ratios of SD other than 1/1. Otherwise, its application has been limited, partly because of its poor performance for visually equal-depth colours of lower DIN-S than the majority of those in the 1/1 SD series [40].

The next significant formula, proposed by Gall and Riedel in 1965, defines depth D_{GR} according to Eqn 4.52:

$$D_{GR} = Y^{1/2}(10 - sa_\phi) \qquad (4.52)$$

where Y is CIE Y ($0 \leq Y \leq 100$), s is given by Eqn 4.53:

$$s = 10\,[(x - x_n)^2 + (y - y_n)^2]^{1/2} \qquad (4.53)$$

and a_ϕ is an empirically determined value that depends on illuminant, observer, D_{GR} and ϕ, which in turn is defined by Eqn 4.54:

$$\phi = \arctan \frac{y - y_n}{x - x_n} \qquad (4.54)$$

The terms x and y are the CIE chromaticity coordinates of the specimen under given illuminant and observer, and x_n and y_n are those of the achromatic point [40].

The formula was developed using data from visual assessments of lacquered specimens. It is the basis of several DIN standards for pigment colour performance, and is incorporated in a computer program (FIAF, Farbstarke durch Iterativen Angleich der Farbetiefen) that is routinely used by several European manufacturers in pigment standardisation. Estimates of the efficiency of the technique vary, but its more general use is limited by the absence of published tables for a_ϕ, other than for a few D_{GR} values.

The final method we shall discuss resulted from a request by ISO/TC38/SC1 in 1976 for a colorimetric method of determining 1/1 SD, with the proviso that it should yield $E_{ISO} \leq \pm 10\%$. A formula was proposed by Christ in 1985. It defines a 1/1 SD surface in CIELAB colour space by means of eight subsurfaces, each of which occupies a unique sector in the a^*b^* diagram, and all meeting at $a^* = b^* = 0$, each subsurface being specified by empirically determined coefficients. In order to determine if B is of 1/1 SD, its CIELAB cylindrical coordinates (L^*_B, $C^*_{ab,B}$ and $h_{ab,B}$) are found, and L^*_B compared with the L^* (L^*_{SD}) of the point with coordinates $C^*_{ab,B}$ and $h_{ab,B}$ on the relevant subsurface, where L^*_{SD} is defined by Eqn 4.55:

$$L^*_{SD} = 20.4 + C^*_{ab,B} P + 6 \exp(X)$$
$$X = -\frac{C^*_{ab,P} P}{6}$$
$$P = \sum_{m=0}^{3} K_{n,m}(h_{ab,B} - h_n)^m \qquad (4.55)$$

and $h_n \leq h_{ab,B} < h_{n+1}$ and the values of $K_{n,m}$ are shown for each limiting h_n in Table 4.1. Then, if $\Delta L^* = L^*_B - L^*_{SD} \approx 0$, B is of 1/1 SD for practical purposes and, if $\Delta L^* \geq 0$ (or $\Delta L^* \leq 0$), B is of lesser (or greater) than 1/1 SD [41].

Published results indicate $E_{ISO} \approx \pm 20\%$ [41,42]. In the absence of a better proposal, the method was adopted in 1995 as ISO 105, Part A06: Instrumental Determination of 1/1 Standard Depth, with $|\Delta L^*| \leq 0.5$ for specimens to be sensibly of 1/1 SD. It is an

Table 4.1 Limiting angles (h_n) and coefficients ($K_{n,m}$) of the polynomial P in Eqn 4.55 (from ISO 105, Part A06)

n	h_n	$K_{n,0}$	$K_{n,1}$	$K_{n,2}$	$K_{n,3}$
1	0	3.27×10^{-1}	1.73×10^{-3}	-2.13×10^{-5}	8.68×10^{-7}
2	52	4.81×10^{-1}	6.57×10^{-3}	1.14×10^{-4}	-3.50×10^{-6}
3	79	6.73×10^{-1}	5.08×10^{-3}	-1.69×10^{-4}	1.07×10^{-6}
4	135	6.14×10^{-1}	-3.83×10^{-3}	1.02×10^{-5}	6.40×10^{-8}
5	203	4.21×10^{-1}	-1.56×10^{-3}	2.32×10^{-5}	-4.82×10^{-7}
6	267	2.90×10^{-1}	-4.51×10^{-3}	-6.93×10^{-5}	2.43×10^{-6}
7	302	1.51×10^{-1}	-4.40×10^{-4}	1.85×10^{-4}	-2.30×10^{-6}
8	340	2.76×10^{-1}	3.69×10^{-3}	-7.66×10^{-5}	9.22×10^{-7}
9	360	3.27×10^{-1}	1.73×10^{-3}	-2.13×10^{-5}	8.68×10^{-7}

alternative to, and not a replacement for, the visual method implied in ISO 105, Part A01. It is limited to 1/1 SD. Attempts to initiate work leading to parallel methods for other ratios failed on grounds of resource. Although it is not part of the specifications in A06, however, several organisations are routinely producing specimens purporting to be at ratios lower than 1/1 SD by using A06 to determine the concentration of a colorant necessary to give 1/1 SD and multiplying by the relevant fraction. The practice is justifiable only when it is known that the colorant builds linearly, or very nearly so, to at least 1/1 SD.

The method has another disadvantage in that it defines L^*_{SD}, at any given C^*_{ab} and h_{ab}, absolutely. Thus if agreement between the instruments used to determine L^*_{SD} and L^*_B is less than perfect, the method can give erroneous evaluations. For example, using a spectrophotometer in good order, Smith found E_{ISO} to be, not ±20%, but −29, +11%. A similar problem can occur with any formula that claims to evaluate depth absolutely.

All the $(K/S)_\lambda$ summation methods and tricoordinate methods discussed yield qualitatively similarly shaped equal-depth surfaces when mapped in CIELAB colour space, exemplified by the $D_I = 20$ ($D_{65}/10$) surface shown in Figure 4.16 [36]. The surfaces are steeply concave when viewed from $L^* = 100$ and $a^* = b^* = 0$, with achromatic and high C^*_{ab} blue to violet colours at the lowest levels of L^*, and high C^*_{ab} yellows at the highest.

None of the methods discussed in this section gives results that always correlate well

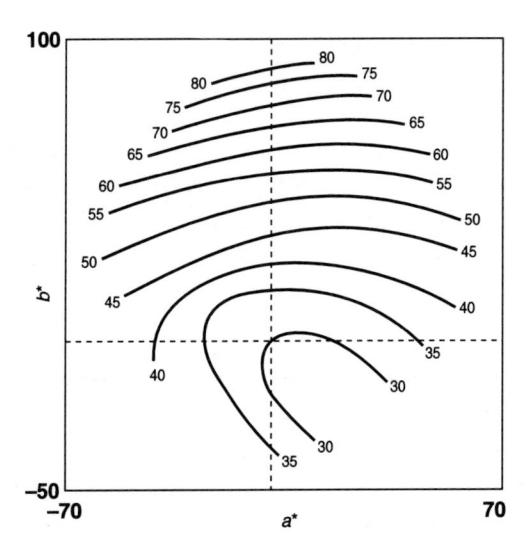

Figure 4.16 CIELAB a^*b^* diagram showing L^* contours of equal-depth surface defined by $D_I = 20$ ($D_{65}/10$) (after Derbyshire and Marshall [36])

with mean visual assessments. Apart from D_{GR}, which has other problems, none returns adequately uniform values for all the colours in the 1/1 SD series. Methods based on $(K/S)_\lambda$ summation are easy to apply, additive and, applied to a concentration series of a given colorant, yield D usually in good agreement with visual estimates of relative depth; but they are unable to handle fluorescent colours (because the $(K/S)_\lambda$ function fails if, for example, $R_\lambda > 1$ in Eqn 4.44), and they tend to report different D for matching colours. Tricoordinate methods cannot suffer from this disadvantage, but all those discussed are limited in applicability by the nonlinearity of their output values of relative D with visual depth differences, and hence their non-additivity; moreover, there are often difficulties in deriving the required input coordinates from colour-measurement data, and limits on the number of depths they can evaluate.

4.5 EVALUATION OF FASTNESS-TEST RESULTS

4.5.1 Introduction

Even a perfect colour match may be rendered worthless if the product changes colour significantly during further processing or in use by its purchaser, and equally so if colour removed from it causes a significant change in the colour of other articles. The latter phenomenon is of great importance in the textile industry, where most relevant agencies have a chance to cause both a change in the colour of a specimen and staining of adjacent material. Numerous agencies can cause these changes. Colour-fastness tests that may be applied in the laboratory to simulate the effects of many of these agencies have been progressively agreed locally, nationally and increasingly internationally.

The method currently used for most visual assessments of fastness-test results relies on comparing the contrast between tested and untested specimens with those of the relevant one of two internationally agreed physical scales, each consisting of pairs of grey chips having graded contrasts. One of these scales is defined by ISO 105: Textiles – Tests for Colour Fastness, Part A02: Grey Scale for Assessing Change in Colour, and the other by Part A03: Grey Scale for Assessing Staining.

4.5.2 Grey scales for assessing change in colour and staining

Each of the two grey scales consists of five pairs of grey chips arranged as shown schematically in Figure 4.17. The left-hand member of each pair is identical to the right-hand member of the pair denoted 5, that pair therefore showing zero contrast. The contrast between members of the pairs increases the farther to the right the pair, so that the pair on the far right, denoted 1, has the greatest contrast. In the Part A02 scale the right-hand chip of each pair is progressively lighter the lower the grade the

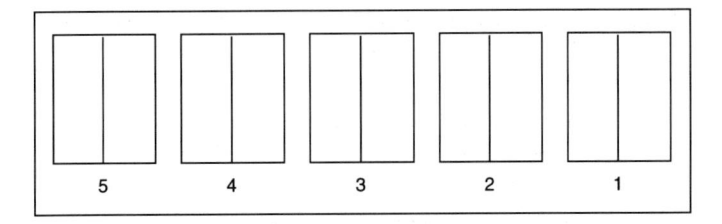

Figure 4.17 Five-step grey scale for visual assessment of change in colour (ISO 105:Part A02), or staining (ISO 105:Part A03)

pair represents, while in the Part A03 scale it is darker. Grade 5 of the Part A03 scale is sensibly white.

In the assessment the appropriate scale is placed as near as possible to the tested and untested specimens and in the same plane, which is illuminated at an intensity of at least 600 lux and at about 45° by a source representing north sky light (in the northern hemisphere). Viewing is made approximately normally to the plane with the colour of the surrounding field approximating Munsell N5. The contrast between the tested and untested specimens then most closely resembles either the real contrast of one of the five pairs in the grey scale, in which case the rating is recorded as the denotation of that pair (for example, 3), or the imaginary contrast midway between those of two adjacent pairs, when the rating accorded is those of both pairs separated by a hyphen (for example, 3–4). Grade 5 is given only when there is imperceptible contrast. There is increasing use of nine-step scales, in which the imaginary contrasts of the five-step scales are replaced by corresponding real pairs of chips; the number of possible ratings remains nine.

Grade 5 of each scale has always been specified in terms of CIE Y, the current values being $Y = 12 \pm 1$ for the A02 scale and $Y \geq 85$ for the A03 scale. The contrasts of the pairs of each scale are defined by the CIELAB formula (section 4.2.7), as shown in Table 4.2, all measurements being made with a d/0 spectrophotometer, specular component included, and transformed under $D_{65}/10$.

4.5.3 Development of the grey scales

The first proposal for a single grey scale for assessing differences in depth of colour, regardless of the colour itself, was made by Cunliffe in 1947. It resulted in the SDC Grey Scale No. 2, employed in assessment of results of migration tests, and was designed to be used by the contrast method. Application of other scales being suggested at that time relied on finding members which matched in depth each of the untested and tested specimens, and therefore contained many more members. The

EVALUATION OF FASTNESS-TEST RESULTS

Table 4.2 Colour-difference specification of grey scales used in the assessment of fastness-test results

Grade	Scale for change in colour (ISO 105:A02) ΔE_{ab}	Tolerance	Scale for staining (ISO 105:A03) ΔE_{ab}	Tolerance
5	0	+0.2	0	+0.2
4–5	0.8	±0.2	2.2	±0.3
4	1.7	±0.3	4.3	±0.3
3–4	2.5	±0.35	6.0	±0.4
3	3.4	±0.4	8.5	±0.5
2–3	4.8	±0.5	12.0	±0.7
2	6.8	±0.6	16.9	±1.0
1–2	9.6	±0.7	24.0	±1.5
1	13.6	±1.0	34.1	±2.0

spacing of the steps of all these scales was an arithmetic progression of colour differences; thus, for a five-step scale, we have (Eqn 4.56):

$$\Delta E_{5,R} = (5 - R)\,\Delta E_{5,4} \tag{4.56}$$

where $\Delta E_{5,R}$ is the colour difference of the pair defining a rating of R, that defining a rating of 5 being zero.

It was McLaren who raised the possibility of using scales based on a geometric progression in which each successive pair shows twice the perceived difference of the previous pair (Eqn 4.57):

$$\Delta E_{5,R} = 2^{4-R}\,\Delta E_{5,4} \tag{4.57}$$

except that $\Delta E_{5,5}$ is, again, zero by definition [43]. The advantage of a geometric scale is that it allows the commercially important higher grades to be more closely spaced in a scale with the same maximum contrast. Substituting $R = 1$ in Eqns 4.56 and 4.57 and equating their right-hand sides gives Eqn 4.58:

$$\Delta E_{5,4(A)} = 2\Delta E_{5,4(G)} \tag{4.58}$$

where the subscripts A and G indicate the arithmetic and geometric versions of the scale respectively. Grade 3 on the geometric scale is therefore exactly equivalent in contrast to grade 4 on the arithmetic one. McLaren also demonstrated statistically that the contrast method is valid not only for assessing differences of depth, but for any type of difference [44].

The principle of a single five-step geometric grey scale for the assessment of change in colour was adopted by ISO/TC38/SC1 in 1951, and for the assessment of staining in 1952, and these scales have since achieved almost universal use.

4.5.4 Reliability of assessments of fastness-test results

Even at the time of the adoption of the scales, it was realised that visual assessment could introduce uncertainties but it was then considered that methods based on instrumental measurement would be even more unreliable. In the succeeding years, colour measurement became both more widely accessible and more reliable. In 1976, the Society's Fastness Test Coordinating Committee established its Instrumental Assessment of Colour Fastness Subcommittee (IASC), with a remit to investigate instrumental methods relevant to all aspects of the evaluation of fastness test results. IASC at first concentrated on developing a method for evaluating staining, regarding this as most likely to yield a satisfactory outcome because it is limited to a relatively small volume of colour space around 'white'.

In the IASC experiments, nearly 350 pairs of stained and unstained specimens were independently assessed visually using the Part A03 scale by an average of about 35 experienced assessors from nine laboratories, giving a total of over 12 000 assessments. In most cases, there was better agreement between the individual assessments made within each laboratory than between the means from different laboratories. This is not surprising, partly because newcomers to a laboratory are likely to be trained by existing personnel. In addition, assessors within a laboratory are likely to compare ratings, whether deliberately or not, during routine work and to adjust future assessments to reduce disagreements. Both reasons contribute to an equalisation of the so-called 'severity' of assessors within a laboratory, even though their collective severity may not then agree with that of groups of assessors elsewhere. Most significant, however, was the variation among the individual assessments: about 16% of the individual visual assessments were one or more steps from their means. The greatest variation was found for assessments between about 3 and 4, which is particularly disturbing as this region brackets a disproportionate number of commercial fastness-test pass/fail limits. This tendency was later confirmed by a study which also showed that the variability of visual assessment generally exceeds that caused by the performance of the test itself [45].

The colour differences of the 350 pairs of unstained and stained specimens were evaluated on eleven instruments. A statistical investigation showed that the best correlation between the mean visual assessments and the mean instrumental results was obtained using Eqn 4.59:

$$SSR = 7.05 - 1.43 \ln(4.4 + \Delta E^*_{ab}) \qquad (4.59)$$

where SSR is the staining scale rating derived from instrumental measurement of the CIELAB colour difference ΔE^*_{ab} between the stained and unstained specimens. The relative reliability of the instrumental method was demonstrated by comparing the standard deviations σ_i of the individual instrumental ratings (from Eqn 4.59) from the mean visual assessment of each pair, assumed true, with those (σ_v) of the individual visual assessments from the same means. The instrumental method is then more reliable than the visual if $\sigma_i/\sigma_v < 1$, a result obtained for about 90% of the pairs. For about 30% of the pairs $\sigma_i/\sigma_v < 0.5$, indicating the instrumental method to be more than twice as reliable.

4.5.5 Colorimetric evaluation of staining test results

A method of instrumental assessment based on Eqn 4.59 was submitted by the UK to the meeting of ISO/TC38/SC1 in 1981, but at its meeting in 1984 objections were raised by the USA and Switzerland.

These objections led to work in which the same pairs of stained and unstained specimens were visually assessed by several observers working independently and instrumentally measured in several laboratories in various countries. Analysis of the visual assessments confirmed the tendency found by the IASC for the severity of individual laboratories to differ significantly, but identified even more significant differences between countries. These results were discussed at the meeting of ISO/TC38/SC1 in 1987 when, in addition to the original formula proposed by the UK, two others were debated. One of these was accepted and is now incorporated in ISO 105, Part A04: Method for the Instrumental Assessment of Degree of Staining of Adjacent Fabrics (Eqns 4.60 and 4.61):

$$SSR_c = 6.1 - 1.45 \ln(\Delta E_{GS}) \quad \text{if } SSR_c \leq 4 \qquad (4.60)$$

$$SSR_c = 5.0 - 0.23 \Delta E_{GS} \quad \text{if } SSR_c > 4 \qquad (4.61)$$

where ΔE_{GS} is defined by Eqn 4.62:

$$\Delta E_{GS} = \Delta E^*_{ab} - 0.4[(\Delta E^*_{ab})^2 - (\Delta L^*)^2]^{1/2} \qquad (4.62)$$

According to A04, the stained and unstained specimens are measured on an instrument with d/0, 0/45° or 45°/0 geometry and with light resembling illuminant C or D_{65}, ΔE^*_{ab} and ΔL^* being calculated using C/2 or $D_{65}/10$. The value of SSR_c is calculated to two decimal places but reported (as SSR) to the nearest half-step, in line with the practice for visual assessment; thus, for example, SSR = 4–5 if 4.25 < SSR_c < 4.74.

4.5.6 Colorimetric evaluation of change in colour

Finding an acceptable instrumental method of determining ratings for change in colour was expected to be rather more difficult. It would have to be capable of quantifying a change, not just from white, but from whatever was the colour of the specimen under test. It would also have to be able to quantify a change in any direction from that colour. The application of many fastness tests to colours produced using mixtures of colorants often causes wholly or mainly a reduction in depth. Other changes can predominate, however; for example, one of the colorants may be more sensitive than the others to the test agency, and some tests can cause an increase in depth.

The first proposal for an instrumental method was made in 1984, when the UK suggested the use of the CMC($l:c$) colour-difference formula (Eqn 4.32). The visual assessment of the Part A02 test is a judgement of relative perceptibility, and the UK argued that with $l = c = 1$ this formula had been shown to be more reliable than most others for quantifying such judgements.

In 1987 a British experiment and a much larger one in Switzerland were reported, each demonstrating the method to be more reliable than visual assessment. The work in Switzerland led to the development by the Schweizerische Echtheits-Kommission (SEK, Swiss Fastness Commission) of an alternative proposal, the SEK1 formula, based on the cylindrical coordinate version of the CIELAB colour-difference formula (Eqn 4.27) with ΔC^*_{ab} modified. Evidence was presented indicating the higher reliability of this formula, summarised by comparing the arithmetic means of the standard deviations (σ) expressed in Grey-Scale Rating (GSR) units, of the individual visual measurements ($\sigma = 0.57$), the corresponding instrumental results converted using CMC(1:1) ($\sigma = 0.44$) and SEK1 ($\sigma = 0.34$), each from the means of visual assessments. Concern was expressed that the SEK1 formula had been tested employing only the data used to derive it, a practice that often leads to an overestimate of the reliability of a formula. In view of the large differences in the severity of assessors from different countries found during the development of the staining method, there was also concern that most of the visual assessors participating in the experimental work during the development of the SEK1 formula had been Swiss. An international experiment was therefore agreed.

The results were reported in 1989, and showed that significant differences in severity did indeed exist between countries. Analysis of the results led SEK to develop a more reliable formula, SEK2, generally similar to SEK1 but with the term involving CIELAB hue difference ΔH^*_{ab} now also modified (Eqns 4.63 to 4.65):

$$\Delta E_F = [(\Delta L^*)^2 + (\Delta C_F)^2 + (\Delta H_F)^2]^{1/2} \tag{4.63}$$

$$\Delta C_F = \frac{\Delta C_K}{1 + (20C_M/1000)^2}$$

$$\Delta C_K = \Delta C^*_{ab} - D$$

$$C_M = \frac{C^*_{ab,t} + C^*_{ab,u}}{2}$$

$$D = \frac{\Delta C^*_{ab} \, C_M \exp(-X)}{100}$$

$$X = \left(\frac{h_M - 280}{30}\right)^2 \quad \text{if } |h_M - 280| \le 180 \tag{4.64}$$

$$X = \left(\frac{360 - |h_M - 280|}{30}\right)^2 \quad \text{if } |h_M - 280| > 180$$

$$h_M = \frac{h_{ab,t} + h_{ab,u}}{2} + 180Y \text{ where } Y = 0 \text{ if } |h_{ab,t} - h_{ab,u}| \le 180$$

but, if $|h_{ab,t} - h_{ab,u}| > 180$, $Y = 1$ if $h_{ab,t} + h_{ab,u} < 360$ and $Y = -1$ if $h_{ab,t} + h_{ab,u} \ge 360$

$$\Delta H_F = \frac{\Delta H_K}{1 + (10C_M/1000)^2} \tag{4.65}$$

$$\Delta H_K = \Delta H^*_{ab} - D$$

in which h_M is a term that depends on the relative CIELAB hue angles of the tested and untested specimens (t and u refer to tested and untested specimens respectively). The resulting ΔE_F is then converted to GSR using Table 4.3 or by means of Eqns 4.66 and 4.67:

$$GSR = 5.0 - \frac{\ln(\Delta E_F/0.85)}{\ln 2} \quad \text{if } \Delta E_F > 3.4 \tag{4.66}$$

$$GSR = 5.0 - \frac{\Delta E_F}{1.7} \quad \text{if } \Delta E_F \le 3.4 \tag{4.67}$$

Table 4.3 ISO 105:A05 conversion of ΔE_F (from SEK2 formulae, Eqn 4.63) to equivalent grey-scale rating (GSR)

ΔE_F	GSR
$\Delta E_F < 0.40$	5
$0.40 \leq \Delta E_F < 1.25$	4–5
$1.25 \leq \Delta E_F < 2.10$	4
$2.10 \leq \Delta E_F < 2.95$	3–4
$2.95 \leq \Delta E_F < 4.10$	3
$4.10 \leq \Delta E_F < 5.80$	2–3
$5.80 \leq \Delta E_F < 8.20$	2
$8.20 \leq \Delta E_F < 11.60$	1–2
$11.60 \leq \Delta E_F$	1

which give virtually identical results when the resulting GSR is rounded to the nearest half-step.

This proposal was adopted and is now incorporated in ISO 105, Part A05: Method for the Instrumental Assessment of the Change in Colour of a Test Specimen. According to A05, the specimens are measured on an instrument consistent with the definitions in CIE Publication 15.2 [19] and the relevant CIELAB parameters are then preferably calculated under $D_{65}/10$, with $D_{65}/2$, $C/10$ and $C/2$ also allowed.

Four points regarding the instrumental methods ISO 105, Parts A04 and A05, are worth emphasising. Firstly, they are permitted alternatives to the corresponding visual methods, A03 and A02 respectively, not replacements for them, although it has been suggested that ISO 105, Part A01: General Principles of Testing be amended to include an instruction to use the relevant instrumental method in the event of visual assessments being disputed.

Secondly, in those cases where the change in colour of a specimen is not wholly or mainly a reduction in depth, the visual method (A02) allows the numerical rating to be qualified by appending descriptor(s) to indicate the nature of the change. A report of 4R, for example, implies that the overall contrast between the tested and untested specimens is judged that of grade 4 but that the tested specimen is wholly or mainly redder (R), without significant loss in depth. Qualification is not incorporated in A05, but can be important because, where a choice exists between two specimens rated quantitatively the same, the one selected is more often that which suffers only a loss in depth.

Thirdly, the formulae of the instrumental methods give continuous intermediate rating results which, according to A04 and A05, are later converted to half-step rating reports analogous to those resulting from visual assessments. Some users of the methods, however, report results to one decimal place, or even more. This practice is to be deprecated because it ignores the imprecision of the fastness tests themselves and that caused by the measurement of unlevel specimens. The latter is usually more of a problem with stains, though it may be reduced (where the size of specimens allows) by averaging measurements from different areas of the specimen.

Lastly, when fluorescent species are present on the specimens being compared both A04 and A05 may yield results that do not agree with visual assessments because of differences between the light source in the instrument and that used for visual assessment. The difference is usually largest in the UV region, so that disagreement is more probable when the fluorescent species is a fluorescent whitening agent (FWA). Whether the FWA is present on the untested speimen or comes from elsewhere, it is likely to be distributed between the specimen, the test liquor (where applicable) and the adjacent during the test, and anomalous instrumentally derived ratings may be obtained as a result.

4.5.7 Visual assessment of light-fastness test results

Attention has recently turned to the development of an instrumental method for the assessment of light-fastness test results. In most such methods of test the specimen is exposed alongside one of two eight-step sets of standards, one developed in Europe and the other in the USA. Each is produced by colouring wool with blue dyes. Light fastness is often affected by humidity, and wool was chosen as the substrate for the standards in the 1920s because, of the substrates then available, dyeings on wool were least affected.

Each member of the set of standards developed in Europe (BWSE) is dyed with a different dye. The standards are denoted 1 to 8, with each standard denoted N ($1 < N \leq 8$) being, at least in theory, twice as resistant to fading as that denoted $N - 1$. Standard 1 is therefore very fugitive, a fade being detectable after exposure to bright sunshine for only a few hours, and standard 8 very fast, a fade not being detectable until exposure has been continued for several weeks. The set of blue wool standards developed in the USA (BWSA) is produced by blending different ratios of fibres dyed with either CI Mordant Blue 1, which is fugitive, or CI Solubilised Vat Blue 8, which is very fast and is in fact the single dye used to produce BWSE 8. These standards are denoted L2 to L9; again, each higher-numbered standard is twice as fast as the preceding one. The two sets of standards are not interchangeable.

Numerous methods are used to determine light fastness. One frequently used procedure is ISO 105, Part B02: Colour Fastness to Artificial Light: Xenon Arc Fading Lamp Test, Method 1. The specimen and the standards are arranged as shown in Figure 4.18, placed in the fading lamp apparatus with an opaque cover (C1) over their central

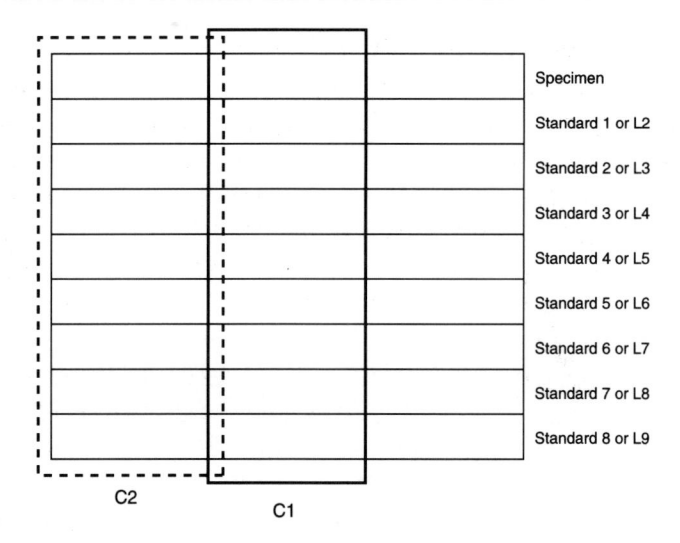

Figure 4.18 Method of covering specimen and standards (BWSE 1–8 or BWSA L2–L9) for light-fastness testing according to ISO 105:Part B02

thirds, and irradiated uniformly at specified relative humidity and black-panel temperature, UV and IR irradiation being controlled by specified filtration. When the contrast between the exposed and unexposed parts of the specimen is judged equal to that of grade 4 of the grey scale for assessing change in colour (ISO 105, Part A02), an additional opaque cover (C2) is placed over the left-hand thirds of the specimen and standards. Irradiation is then continued until the contrast between the still exposed and unexposed portions of the specimen is perceived equal to grade 3, when the test ends. It is thus necessary to monitor the fade of the specimen continuously. The specimen, as well as some of the standards, may be photochromic (that is, their colours may change on exposure to light, reverting to their original colours when exposure ceases); this could lead to an erroneous assessment of light fastness, which is an irreversible change. The specimen and standards are therefore kept for twenty-four hours in the dark, after which two assessments of the light fastness of the specimen are made, under the viewing conditions summarised in section 4.5.2, by identifying the standards that show the same contrast between unexposed and successively the left-hand and right-hand exposed parts as that between the same areas of the specimen. Half-step assessments, such as (BWSE) 4–5, are allowed. The single, final assessment of the light fastness of the specimen is reported as the arithmetic mean of these two records, expressed to the nearest half-step.

4.5.8 Colorimetric evaluation of light-fastness test results

The first published instrumental method for assessing the results of light-fastness tests appeared in 1992. The method allows, as an alternative to frequent visual inspection of the specimen during testing, less frequent instrumental measurements of both specimen and standards, all of which are converted to equivalent ΔE_F by means of the formula in ISO 105, Part A05 (Eqns 4.63 to 4.65). The GSR are output to, say, two decimal places using the continuous function (Eqns 4.66 and 4.67) and then plotted against time of exposure, as exemplified in Figure 4.19. In line with the practice for visual assessment, the light-fastness rating of the specimen is then determined as the arithmetic average of the locations on the plot of its contrasts equivalent to GSR = 4 and GSR = 3, relative to the corresponding contrasts of the standards. For the specimen (X) shown, the individual ratings at these contrasts are, respectively, 4–5 and 5–6, leading to an overall rating of 5.

An experiment was reported in which over 200 specimens, scattered throughout colour space, were each separately exposed to daylight and to xenon and carbon arc light, and then assessed visually by a panel of observers and instrumentally according to the above method. Comparison of the mean assessments, each rounded to the nearest half-step, showed perfect agreement for no fewer than 92% of the exposures, the largest disparity being one step, which occurred in only 1% of the specimens [46]. Further work is being carried out to validate and develop the method.

4.6 EVALUATION OF WHITENESS AND YELLOWNESS

4.6.1 Introduction

Colours that are sensibly white have both high lightness and low chroma, and therefore occupy a very small subvolume of colour space. The ISCC-NBS Method of Designating Colors, for example, allows the name 'white' to be used alone, or in conjunction with any of the five principal Munsell hues to give derivatives such as 'reddish-white', only for colours of Munsell value ≥ 8.5. Additionally, throughout most of the Munsell hue circuit, the name white, including its derivatives, is limited to colours of Munsell chroma ≤ 1.2 or 1.5 [47] (Figure 4.20). Colours that may be designated white, including its derivatives, therefore occupy only about 3% of the total volume of the Munsell colour solid.

In virtually all the coloration industries, however, the commercial importance of white shades is out of all proportion to the small volume of colour space they occupy. The assessment of whiteness is of obvious importance if the final colour of the product is sensibly white. In the paper trade, and when natural fibres are used in textiles, the production of sensibly white shades almost invariably requires the substrate to be

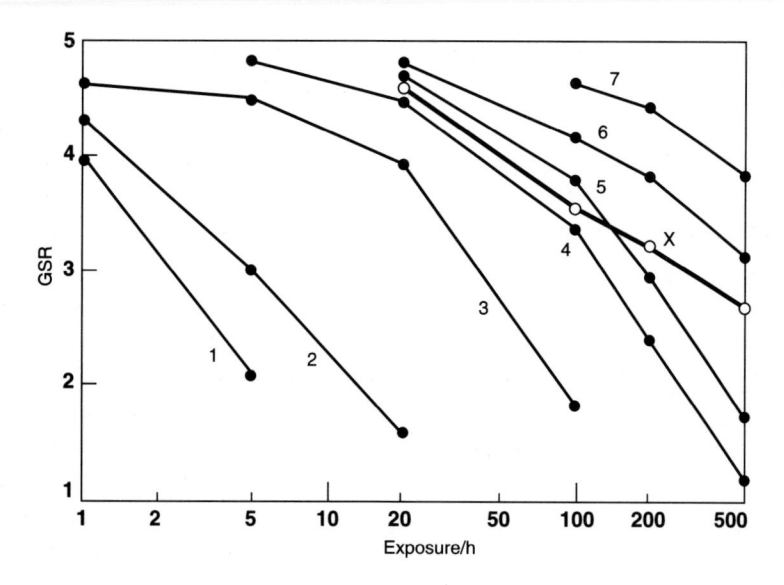

Figure 4.19 Instrumental determination of light fastness rating (after Kubler and Ulshofer [46])

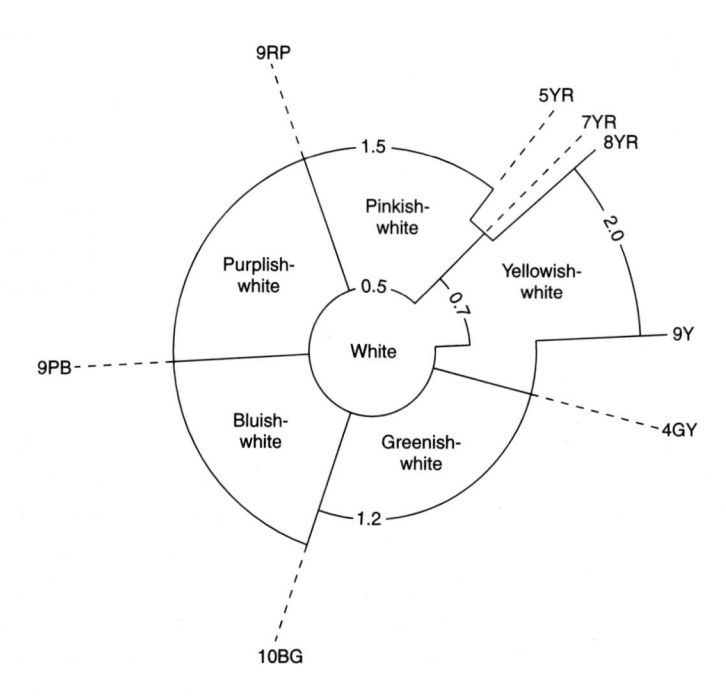

Figure 4.20 Munsell *CH* diagram showing limits of colours called white (all of Munsell $V \geq 8.5$) [47]

bleached, and the process is frequently controlled by assessing whiteness. Whiteness can be equally important when the final colour of the product is not white. In inks, paints and plastics, the quality of the white pigment used determines how clean a tint or pastel shade appears, and in the paper and textile industries the attainable brightness of pale shades is limited by the colour of the substrate. Cotton, for example, still constitutes more than half the global consumption of textile fibres, and about half of it is bleached.

Considerable efforts have therefore been devoted to the development of methods for the assessment of whiteness. Although white shades clearly occupy a three-dimensional subvolume of colour space, whiteness has generally been assessed visually using a one-dimensional scale of standards, or instrumentally by means of a formula giving a single-number output. The colour of a given white product is often influenced by the quantity in it of a specific impurity. Its colour therefore usually falls on or near a line which, although it may be curved in all three dimensions of colour space, may be subdivided to allow an unambiguous single-number report of whiteness. Unfortunately, the loci of different products do not coincide. In addition, although yellowness is always considered the antithesis of whiteness, an increase alone in blueness (which is complementary to yellowness) is not necessarily the preferred deviation towards a 'better' white. Some assessors prefer more violet or more cyan tints of white: these assessors are said to have respectively red or green hue preference, and those who prefer deviation towards blue alone neutral hue preference. An assessor with red hue preference will rate a reddish-white more highly than will one with neutral hue preference, who in turn will rate it more highly than will an assessor with green hue preference. Many ways of assessing whiteness have therefore been developed, differing in the direction of their hue preferences, in their relative weightings of the influences of each of the three dimensions of colour and – because they were developed largely independently – in their scalings.

4.6.2 Perception of whiteness

If two specimens that are sensibly white differ only in lightness, the lighter one appears whiter. If they differ only in blueness, the bluer one appears whiter. If one is lighter but less blue, it may or may not appear the whiter, depending on the balance of the two. The reflectance curves in Figure 4.21 exemplify those of cotton in its grey (unbleached) state (G) and after bleaching (B). Whilst bleaching increases reflectance across the whole of the visible waveband, it does so especially at the 'blue' end of the spectrum: hence it significantly increases both lightness and blueness. There is therefore a large increase in the whiteness of the cotton, although it still appears to possess the attribute of yellowness because it is still absorbing more strongly in the blue region than elsewhere.

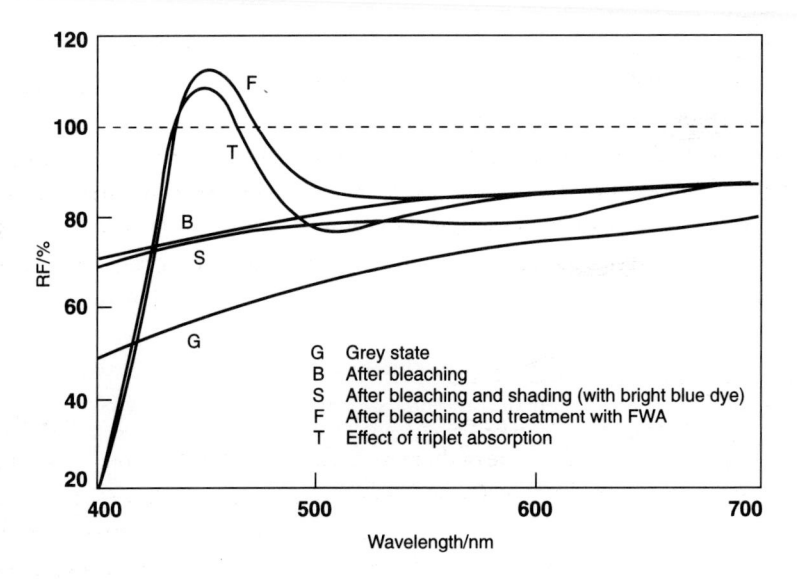

198 COLOUR-ORDER SYSTEMS, COLOUR SPACES, COLOUR DIFFERENCE AND COLOUR SCALES

Figure 4.21 RF curves of cotton in grey state and after various treatments

G Grey state
B After bleaching
S After bleaching and shading (with bright blue dye)
F After bleaching and treatment with FWA
T Effect of triplet absorption

Its whiteness can be further increased in two ways. The first is to shade it with a small quantity of a bright blue dye, which (as shown by the resulting reflectance curve S in Figure 4.21) absorbs selectively in regions of the spectrum other than blue. If the dye is nonfluorescent, this decreases the lightness of the shaded product and, in absolute terms, leaves its blueness virtually unchanged. Its relative blueness is increased, however, and the positive effect of the increase in apparent blueness on perceived whiteness outweighs the negative one of the decrease in lightness, so that the shaded product appears whiter.

Alternatively, whiteness may be increased by treating the product with an FWA. These are colourless substances that strongly absorb energy in the near UV (from about 320 to 400 nm) and re-emit it by fluorescence as visible violet or blue light. As exemplified by the curve F (Figure 4.21), they thus cause a large increase in the blueness of the product (and a much smaller increase in lightness), and thereby a significant increase in perceived whiteness.

The light from a specimen treated with FWA is no longer entirely reflected, so that strictly speaking we cannot describe curve F as a reflectance curve. It is correctly termed a radiance factor (RF) curve, where RF ($\beta_{e,\lambda}$ at wavelength λ) is defined by Eqn 4.68 [2]:

$$\beta_{e,\lambda} = \beta_{S,\lambda} + \beta_{L,\lambda} \tag{4.68}$$

where $\beta_{S,\lambda}$ is the energy truly reflected and $\beta_{L,\lambda}$ that contributed by fluorescent emission, each expressed as a fraction of the light incident at λ. The subscript S stands for substrate, because in many applications it is the true reflectance of the substrate that wholly or largely determines $\beta_{S,\lambda}$, and L is for luminescence, the collective term for fluorescence and related phenomena.

Since the FWA causes an increase in whiteness by absorbing in the UV, the smaller the ratio of UV to visible energy falling on the specimen, the less the apparent increase in whiteness will be. Visual assessments of FWA-containing specimens are therefore usually made either in actual daylight or under a lamp that closely emulates daylight in both the visible and the UV.

4.6.3 Visual assessment of whiteness

Many scales for the visual assessment of whiteness have been developed. Although it is no longer generally available, one of those most commonly used in the past will serve as an example. This was the Ciba-Geigy Plastic White Scale, which consists of twelve plastic plaques ranging in whiteness, expressed in terms of Ciba whiteness units, from −20 to +210 in unequal steps. On this scale, the perfect reflecting diffuser has a value of 100, as with most other methods of evaluating whiteness. A value of 70 is typical of that of cotton which has been bleached but not FWA-treated. The least difference in whiteness perceptible to an experienced assessor is five units. The first four plaques (−20 to +50) contain a yellow pigment (iron(II) oxide) and are not fluorescent. The rest contain increasing concentrations of the FWA Uvitex SFC (Ciba). They constitute a series of approximately neutral whites, each successive step being of higher excitation purity (p_e) but all having dominant wavelength (λ_d, $D_{65}/10$) about 470 nm. The plaques are fast to light and are washable. One side of each is glossy and the other matt, the side used being determined by the surface characteristics of the specimen being assessed.

Visual assessment is made by viewing the specimen and the appropriate side of the plaques against a light background under, according to the original method, north sky light (in the northern hemisphere) behind glass 'or, better still, in the open'. The whiteness of the specimen, reported in terms of Ciba whiteness units, is that of the plaque most closely resembling it. Interpolation in multiples of five units is allowed if the whiteness of the specimen is judged to fall between that of two adjacent scale steps.

The method also allows reporting the tint of whites judged unequal to that of the scale, by suffixing whiteness with a letter indicating the direction of the difference (R for redder or G for greener) followed by a single digit from 1 to 5 indicating its magnitude, with 1 corresponding to a trace (section 4.4.5) [48].

4.6.4 Development of colorimetric methods of evaluating whiteness

The first instrumental method for the assessment of whiteness consisted of plotting the position of the specimen relative to rectangular axes according to its CIE Y value (ordinate) and colorimetric purity, p_c (abscissa), the latter being related to p_e by Eqn 4.69:

$$p_c = \frac{p_e y_b}{y_s} \qquad (4.69)$$

where y_b and y_s are respectively the CIE y chromaticities of the specimen and of the point where the line of its λ_d intersects the boundary of the chromaticity diagram. The background to the plot showed contours of equal whiteness deduced from the mean visual assessments of a panel of assessors of each of 36 specimens of cotton fabric. The specimens varied significantly only in Y and p_c; all were sensibly white and of $574 \leq \lambda_d \leq 576$ nm [49].

By 1966 well over a hundred whiteness formulae had been suggested, most of them having been developed empirically by correlating instrumental measurements with visual assessments. A selection of those that have been, or are, in fairly widespread use is given in Table 4.4. Each of these formulae yields a whiteness (W) of 100 for the perfect reflecting diffuser. That, however, is nearly all they have in common, as exemplified by the results of their application under C/2 to four of the RF curves in Figure 4.21. Clearly, this is inconvenient for communication.

4.6.5 CIE whiteness formula

No formula can adequately evaluate the whitenesses of more than a very limited collection of specimens unless it takes account of all three dimensions of colour. Of the seven formulae in Table 4.4, however, only two (Eqns 4.70 and 4.74) do so. In 1972 Ganz proposed a new generic formula, the Ganz linear whiteness formula (Eqn 4.77):

$$W = DY + Px + Qy + C \qquad (4.77)$$

which does involve the three dimensions (CIE xyY) of the colour of the specimen [57]. The coefficients D, P, Q and C may be varied to adjust the hue preference and scaling of the formula, as well as allowing any combination of illuminant and observer to be used.

At about the same time the CIE, recognising the disadvantages of the diverse methods then in use, became interested in Eqn 4.77 as an acceptable way of unifying practice. International studies on behalf of the CIE indicated that the coefficient D should be set to unity. Ganz suggested three pairs of values of the coefficients P and Q (Table 4.5), the choice being determined by the hue preference desired of the resulting formula. If the values of P and Q for red or green hue preference are substituted in Eqn 4.77, the formula then imitates the hue preference of respectively the Stensby (Eqn

Table 4.4 A selection of whiteness formulae (W = whiteness)

				W of RF curve (Figure 4.21)				
Eqn	Formula,	W =	Ref.	G	B	S	F
4.70	Berger[a]	$1.060Z + 0.333Y - 1.277X$	50	25	64	75	132
4.71	Blue reflectance[b,c]	B	51[d],52	58	78	77	100
4.72	Harrison[e]	$100 + R_{430} - R_{670}$	50,53	75	89	89	102
4.73	Hunter[f]	$L_H - 3b_H$	54	54	79	86	126
4.74	Stensby[g]	$L_H + 3a_H - 3b_H$	55	54	78	85	140
4.75	Stephansen[e]	$2R_{430} - R_{670}$	50	30	65	65	91
4.76	Taube[h,i]	$4B - 3G$	51[d],52,54,56	23	62	74	145

(a) Where XYZ are the CIE (C/2) tristimulus values of the specimen.
(b) Where B is the percentage reflectance, relative to that of the perfect reflecting diffuser, of the specimen when illuminated with light of spectral power distribution equal to that of CIE illuminant C weighted by the CIE 1931 $\bar{z}_\lambda$ colour–matching function.
(c) More commonly written as $W = Z_\%$, where $Z_\% = Z/1.18103$, Z being defined as in note (a), and in this form its output is often called %Z brightness.
(d) AATCC Test Method 110–1979 has been superseded by AATCC Test Method 110–1989 (reaffirmed 1994), which no longer includes this formula.
(e) Where R_{430} and R_{670} are the percentage reflectances of the specimen at 430 and 670 nm respectively.
(f) Where L_H and b_H are respectively the Hunter lightness and yellowness coordinates of the specimen calculated under C/2 (Eqn 4.6).
(g) Where L_H and b_H are defined as in note (f) and a_H is the Hunter redness coordinate of the specimen, similarly calculated.
(h) Where B is defined as in note (b), and G is defined analogously by substituting the CIE 1931 $\bar{y}_\lambda$ colour–matching function.
(i) May also be calculated by the equivalent formula $W = L_H(L_H - 5.7b_H)/100$, where L_H and b_H are as defined in note (f).

Table 4.5 Coefficients P and Q for red, neutral and green hue preference versions of the Ganz linear whiteness formulae (Eqn 4.77) [58]

Hue preference	P	Q
Red	800	–3000
Neutral	–800	–1700
Green	–1700	–900

4.74) or Berger (Eqn 4.70) formulae, which are among the most extreme in hue preference.

In the interests of uniformity of practice, however, in 1982 the CIE adopted Eqn 4.77 with the values of P and Q corresponding only to neutral hue preference. The CIE whiteness formula is usually written in the form of Eqn 4.78:

$$W = Y + 800(x_n - x) + 1700(y_n - y) \qquad (4.78)$$

where the coefficient C, explicit in Eqn 4.77, is implied by setting (Eqn 4.79):

$$C = Px_n + Qy_n \qquad (4.79)$$

where x_n and y_n are the chromaticity coordinates of the achromatic point of the chosen observer (2° or 10°), always under D_{65} because the assessment of fluorescent whites under any other illuminant is usually irrelevant [19]. The higher the value of W, the higher is the whiteness of the specimen. For the perfect reflecting diffuser, $W = 100$. A specimen containing FWA can have $W \gg 100$. The difference just perceptible to an experienced visual assessor equates to about 3 CIE whiteness units, comparable to about 5 Ciba whiteness scale units (section 4.6.3).

4.6.6 CIE tint formula

In the meantime, Ganz and Griesser had suggested a generic formula for the instrumental determination of the tint (T_w) of sensibly white specimens (Eqn 4.80):

$$T_w = mx + ny + k \qquad (4.80)$$

where x and y are the CIE chromaticity coordinates of the specimen. The coefficients m, n and k may be varied to allow the formula to imitate different scales for the visual assessment of tint [59], such as the Ciba method. Experiment indicated that, in the chromaticity diagram, lines of equal tint are almost parallel to that of $\lambda_d = 466$ nm, along which white specimens are considered of approximately neutral (bluish) hue, allowing suitable values of the coefficients to be determined. Also in 1982, the CIE adopted Eqn 4.81:

$$T_w = a(x_n - x) - 650(y_n - y) \qquad (4.81)$$

where $a = 1000$ for $D_{65}/2$ and 900 for $D_{65}/10$ [19]. This, again, is a specific version of the generic formula (Eqn 4.80) rewritten, this time implying the coefficient k. For neutral whites, including the perfect reflecting diffuser, $T_w = 0$. If $T_w > 0$ the specimen is a greenish-white; if $T_w < 0$ it is reddish.

In 1987 the CIE whiteness (Eqn 4.78) and tint (Eqn 4.81) formulae were incorporated in ISO 105, Part J02: Method for the Instrumental Assessment of Whiteness. Two years later the whiteness formula (alone) was written into AATCC Test Method 110-1989 (reaffirmed 1994): Whiteness of Textiles, replacing the Blue Reflectance and Taube formulae (Eqns 4.71 and 4.76) of earlier editions.

4.6.7 Limitations of CIE whiteness formula

The CIE, ISO and AATCC methods all contain broadly similar caveats. The spectral power distribution of the illumination of the specimen in the measuring instrument must approximate that of D_{65} and, for results to correlate with visual assessment of specimens containing FWA, it must do so not just in the visible waveband but in the UV too. Specimens must be sensibly white, such that $40 < W < (5Y - 280)$ and $-3 < T_w < 3$. (For example, application of Eqn 4.78 to measurements of 1/1 Standard Depth blue colours yields the erroneous result $W \approx 300$.) The transformation of measurements to W and T_w is preferred under $D_{65}/10$, but $D_{65}/2$ is also allowed. The method yields reports of W and T_w which are relative, not absolute, implying that it can be used only to compare specimens. Even so, it is limited to the comparison of specimens that do not differ much in colour or fluorescence, and they must be measured on the same instrument at nearly the same time.

The reason for these last two restrictions is that, although sources have been developed which closely simulate D_{65}, none has proved suitable for use in commercial instruments. Available instruments, even instruments of the same model, therefore differ in the extent to which they illuminate the specimen with energy approximating D_{65}, especially in the relative amounts of visible and UV irradiation; over a period of time, differences occur even with a single instrument because of ageing of its light source. The whiteness of a specimen containing FWA is a function of the distribution of its RF (Eqn 4.68) which, at any visible wavelength at which fluorescence occurs, is the sum of the true reflectance ($\beta_{S,\lambda}$) of the specimen and the emittance ($\beta_{L,\lambda}$) of the energy from the UV which excites the FWA. The measurement of $\beta_{S,\lambda}$ remains the same whatever the distribution of light striking the specimen (because reflectance is the ratio of reflected to incident light, both at the given wavelength). If, however, the ratio of UV to visible energy in the illumination decreases, the excitation of the FWA is reduced and it fluoresces less at the given visible wavelength, hence reducing $\beta_{L,\lambda}$. Thus the RF of the specimen at that wavelength decreases, and a lower value of measured whiteness results.

Gartner and Griesser measured the whiteness of a stable FWA-containing specimen on three models of spectrophotometer over a period of just over a year. The maximum difference in whiteness they found, between instruments at any time, converts to about

25 Ciba whiteness scale units. The results from all three instruments drifted lower during the period by the visual equivalent of between two and six traces [60].

4.6.8 Evaluation of absolute whiteness

These observations led Gartner and Griesser to develop a method of maintaining the balance of visible to UV illumination of the specimen. It uses a filter that is virtually transparent in the visible waveband but opaque in the UV, which is placed partly into the path of the energy from the source to the specimen. The amount by which the filter is interposed is adjustable, so that the balance of visible to UV energy striking the specimen may be altered. It is adjusted, whenever necessary, to give consistent readings of the whiteness of a stable FWA-containing reference. Provided specimens of unknown whitenesses absorb UV and emit visible energy at wavelengths similar to those at which the reference absorbs and emits, their instrumentally evaluated whitenesses may then be compared, even if they are measured at very different times [60].

Once the balance of visible to UV energy in an instrument has been adjusted, the coefficients of the Ganz linear whiteness formula (Eqn 4.77) may be adjusted to compensate for any remaining differences between it and a reference instrument. This is achieved by measuring the whitenesses of a series of reference specimens first on the reference instrument and then on the other, and performing linear regression on the resulting values of whiteness. Provided the balance of specimen illumination in each is correctly maintained thereafter, any number of instruments – even if they are of different models – can be made to give closely similar values for specimens, so that for practical purposes each instrument may be regarded as yielding absolute whiteness values [61].

The combination of the methods of adjustable filtration and modifying the coefficients of the Ganz formula thus removes the restriction that specimens to be compared must be measured on the same instrument at nearly the same time. Although neither method has been accepted as a standard, the hardware and software required for their combination has been widely used in industry for several years, generally with good results. Much of the combined methodology is currently being considered for inclusion in an international standard for the paper industry.

The method of balancing the UV and visible energy demands that the source in the instrument emit more UV energy than required. This has largely restricted its application to instruments with xenon lamps. Adjustable filtration of the visible light in instruments with a UV-deficient source is theoretically possible; in many instruments using tungsten or tungsten–halogen lamps, however, the amount of filtration required would reduce the visible energy to unacceptably low levels.

4.6.9 Triplet absorption

Instruments incorporating pulsed-xenon lamps can yield whiteness results that do not accord with visual assessment. This is due to an effect called triplet absorption, which is caused by the high intensity of the light pulse. The problem does not arise in instruments with xenon lamps which continuously illuminate the specimen because the intensity of illumination is then comparable to that used in visual assessment.

A mechanism for triplet absorption has been proposed by Thommen. Under bombardment by photons, a molecule of FWA absorbs UV energy, which promotes it from the singlet ground state (S_0 in Figure 4.22) to an excited state, usually the first excited singlet (S_1). It then usually reverts to its ground state by one of two processes. The first is vibrational relaxation (V), in which its absorbed energy is lost thermally. The second is fluorescence (F) – the reverse of the process of absorption – but this takes place only after some of the energy of the molecule has been lost by vibrational relaxation, so that emission occurs at a higher, and therefore visible, wavelength. All these processes take place extremely quickly.

Sometimes, however, the molecule drops initially to the first triplet state (T_1) and only then to its ground state, either by vibrational relaxation (V') as before or by phosphorescence (P), which again causes emission at visible wavelengths but which is a much slower process. Even so, under the levels of illumination in an instrument with a continuous xenon lamp or during visual assessment, it is unlikely that a molecule in its first triplet state will be struck by a further photon. Thus the instrument produces an RF curve exemplified by that labelled F in Figure 4.21, which yields a whiteness value in agreement with visual assessment.

The pulse of light that illuminates the specimen in a pulsed-xenon instrument, however brief by human standards, is long compared with these processes, and it is very intense. There is therefore a significantly increased probability of a molecule in the first triplet state absorbing the energy of another photon (A') and thereby being promoted to a higher triplet state (such as T_2 in Figure 4.22). In some species of FWA this triplet state has significant absorption in the visible waveband (hence the term triplet absorption) and the instrument therefore produces an RF curve (T in Figure 4.21) which leads to a whiteness value that does not accord with visual assessment [62]. Most manufacturers of pulsed-xenon instruments are aware of this problem and their more recent designs use pulses of lower intensity to mitigate it.

4.6.10 Evaluation of yellowness

The most common colour change of sensibly white products is an increase in yellowness. Among the causes of yellowing are the physical and chemical processes the product undergoes during subsequent manufacture, chemicals released from packaging

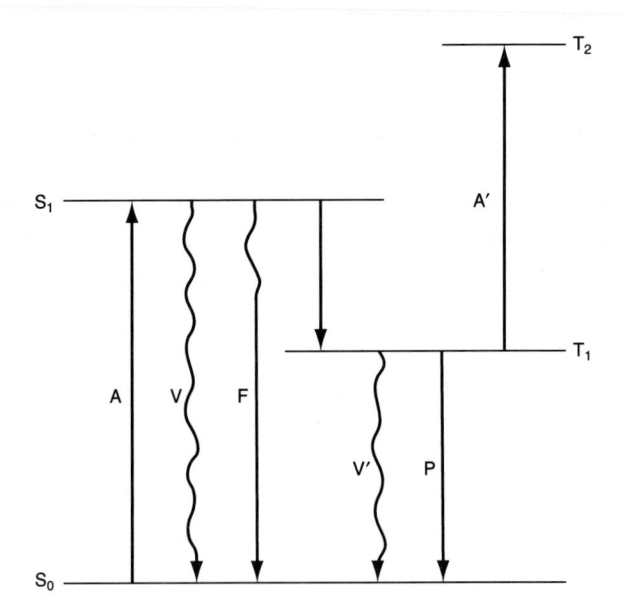

Figure 4.22 Mechanisms of FWA luminescence under normal levels of illumination (involving S_0, S_1 and T_1 only) and triplet absorption under intense illumination in short-duration pulsed-xenon instrument (also involving T_2), according to Thommen [62]

materials and atmospheric contaminants during storage, and light and soiling during use. Paradoxically, they also include the physical and chemical processes applied by the user whilst attempting to restore the initial whiteness of the product. The user processes are applied because yellowness in whites is associated with age, contamination and product degradation, and is therefore perceived as undesirable. Various formulae have been developed to allow the instrumental quantification of yellowness, although not as many as for whiteness.

The perception of yellowness is caused by higher absorption in the blue region than elsewhere in the visible waveband. Most whiteness formulae give weight to the lower absorption of a specimen in this region by including a term in either $cZ - X$ or $cZ - Y$, where c is a formula-dependent constant and X, Y and Z are the tristimulus values of the specimen. It is not surprising therefore that yellowness formulae usually include a term in either $X - cZ$ or $Y - cZ$. This is exemplified by the two formulae that are probably most commonly used.

The first is ASTM D1925-70: Standard Test Method for the Yellowness of Plastics, which, despite its title, is often used to assess the yellowness index (Y_i) of other types of specimen (Eqn 4.82):

$$Y_i = \frac{1.28X - 1.06Z}{Y} \times 100 \tag{4.82}$$

where X, Y and Z are the C/2 CIE tristimulus values of the specimen.

The second is ASTM E313-73, Standard Test Method for Indexes of Whiteness and Yellowness of Near-White, Opaque Materials, which includes a method for the determination of yellowness of near-white opaque specimens without limitation of type using Eqn 4.83:

$$\begin{aligned}Y_i &= \left(1 - \frac{B}{G}\right) \times 100 \\ &= \frac{Y - 0.847Z}{Y} \times 100\end{aligned} \tag{4.83}$$

where B and G are defined as in Table 4.4, and Y and Z are the C/2 tristimulus values of the specimen [52].

Although the results given by these two formulae are necessarily not in general equivalent, their outputs are qualitatively similar. In both, the higher the value of Y_i the greater the yellowness of the specimen, both give $Y_i \approx 0$ for the perfect reflecting diffuser, and in both $Y_i < 0$ indicates blueness.

REFERENCES

1. Document WG1/N8, ISO/TC187 (colour notations), WG1 (terms and definitions) (1989).
2. *International lighting vocabulary*, 4th Edn, CIE Publication No. 17.4, Vienna (1987).
3. F W Billmeyer Jr, *AIC annotated bibliography on color-order systems* (Beltsville, MD: Mimeoform Services, 1985).
4. G Derefeldt, 'Vision and visual dysfunction' in *The perception of colour*, Ed. P Gouras (CRC Press: Boca Raton, Florida, 1991).
5. R G W Hunt, *Measuring colour*, 2nd Edn (Ellis Horwood: Chichester, 1991).
6. D B Judd and G Wyszecki, *Color in business, science and industry*, 3rd Edn (John Wiley: New York, 1975).
7. G Priest, K S Gibson and H J McNicholas, Technical paper no. T167, National Bureau of Standards, Washington DC (1920).
8. S M Newhall, D Nickerson and D B Judd, *J. Opt. Soc. Amer.*, **33** (1943) 385.
9. K McLaren, *J. Oil. Col. Chem. Ass.*, **45** (1971) 879.
10. D L MacAdam, *J. Opt. Soc. Amer.*, **27** (1937) 294.
11. E Hering, *Outlines of a theory of light sense* (trans. L M Hurvich and D Jameson) (Cambridge, MA: Harvard University Press, 1964).
12. R S Hunter, Circular no. C429 (National Bureau of Standards: Washington, DC, 1942).
13. F Scofield, Scientific Circular no. 664 (National Paint, Varnish, Lacquer Association, USA, 1943).
14. E Q Adams, *J. Opt. Soc. Amer.*, **32** (1942) 168.

15. R S Hunter, *J. Opt. Soc. Amer.*, **48** (1958) 597.
16. D Nickerson and K F Stultz, *J. Opt. Soc. Amer.*, **34** (1944) 550.
17. L G Glasser and D J Troy, *J. Opt. Soc. Amer.*, **42** (1952), 652.
18. L G Glasser et al., *J. Opt. Soc. Amer.*, **48** (1958) 736.
19. *Colorimetry*, 2nd Edn (Vienna: CIE, 1986).
20. A M Illing and I Balkanin, *Amer. Ceram. Soc. Bull.*, **44** (1965) 956.
21. D L MacAdam, *J. Opt. Soc. Amer.*, **32** (1942) 247.
22. R Seve, *Col. Res. Appl.*, **16** (1991) 217.
23. M Stokes and M H Brill, *Col. Res. Appl.*, **17** (1992) 410.
24. F J J Clarke, R McDonald and B Rigg, *J.S.D.C.*, **100** (1984) 128, 281.
25. R McDonald, *Text. Chem. Col.*, **20** (6) (1988) 31.
26. Report of CIE TC1-29 'Industrial colour-difference evaluation', Full Draft No. 4, Vienna (1994).
27. Report on colour terminology, Committee of Phys. Soc. (1948).
28. K McLaren and P F Taylor, *Col. Res. Appl.*, **6** (1981) 75.
29. K McLaren, *J. Col. App.*, **1** (4) (1972) 12.
30. *J.S.D.C.*, **101** (1985) 367.
31. R G Kuehni, *Text. Chem. Col.*, **16** (1984) 41.
32. J Beresford, *J. Oil Col. Chem. Assocn*, **53** (1970) 800.
33. M E Taylor, *Text. Chem. Col.*, **2** (1970) 149.
34. R G Kuehni, *Text. Chem. Col.*, **10** (1978) 75.
35. E E Garland, *Text. Chem. Col.*, **5** (1973) 227.
36. A N Derbyshire and W J Marshall, *J.S.D.C.*, **96** (1980) 166.
37. W Baumann et al., *J.S.D.C.*, **103** (1987) 100.
38. DIN 6164 (1980).
39. P Rabe and O Koch, *Melliand Textilber.*, **38** (1957) 173.
40. L Gall and G Riedel, *Die Farbe*, **14** (1965) 342.
41. H A Christ, *Textilveredlung*, **20** (1985) 241.
42. W Griesser, B Meyer and J van Diest, *Textilveredlung*, **22** (1987) 104.
43. K McLaren, *J.S.D.C.*, **68** (1952) 203.
44. K McLaren, *J.S.D.C.*, **68** (1952) 205.
45. S M Jaeckel, *J.S.D.C.*, **96** (1980) 540.
46. W Kubler and H Ulshofer, *Textilveredlung*, **26** (1991) 95.
47. National Bureau of Standards Circular 553 (Washington, DC: US Dept of Commerce, 1955).
48. Ciba-Geigy Publication 09083 (1973).
49. D L MacAdam, *J. Opt. Soc. Amer.*, **24** (1934) 188.
50. A Berger, *Die Farbe*, **8** (1959) 187.
51. AATCC Test Method 110-1979 (Research Triangle Park, NC: AATCC, 1989).
52. ASTM Test Method E313-73 (1994).
53. V G W Harrison, PATRA Reports, London, **2** (1938) 1; **3** (1939) 1.
54. R S Hunter, *J. Opt. Soc. Amer.*, **50** (1960) 44.
55. P S Stensby, *Soap Cosmetics Chem. Specialties*, **43** (7) (1967) 80.
56. K Taube, *Study of home laundering methods* (Beltsville, MD, Institute of Home Economics).
57. E Ganz, *J. Col. Appl.*, **1** (4/5) (1972) 33.
58. E Ganz, *Appl. Optics*, **18** (1979) 1073.
59. E Ganz and R Griesser, *Appl. Optics*, **20** (1981) 1395.
60. F Gartner and R Griesser, *Die Farbe*, **24** (1975) 199.
61. E Ganz, *Appl. Optics*, **15** (1976) 2039.
62. F Thommen, Article No.309, Turntable, No.3 (Basel: Ciba-Geigy, 1988).

CHAPTER 5

Recipe prediction for textiles

Roderick McDonald

5.1 INTRODUCTION

The commercial application of computer methods to dye recipe formulation in textiles was first disclosed by Alderson, Atherton and Derbyshire in 1961 [1], although the first adequate treatment of the underlying mathematics had been published as early as 1944 by Park and Stearns [2]. The reason for the delay in applying the mathematics was simply that computers were not generally available to industry until around 1960. In the early 1960s the availability of reliable computers at reasonable prices enabled the larger textile- and dye-manufacturing companies to investigate the possibilities of computer colorant formulation, and by 1963 several alternative approaches to the problem had been published [3].

There is still a multiplicity of approaches to the colorant formulation problem. The reason is that computer colorant formulation is not an exact science. Although the mathematics of the subject has been very thoroughly investigated at a theoretical level, in practice real dyes and pigments do not respond ideally in the way suggested by the theory, and therefore several alternative methods for dealing with the deviations from ideality have been devised. Nevertheless certain computer colorant formulation systems do give a very high success rate in first-time matches to target.

The theory and concepts that have led to the development of functions of reflectance linearly related to colorant concentration have been dealt with in Chapter 1. The most useful of these functions is the Kubelka–Munk function (Eqn 5.1):

$$R = \frac{1 - R_g(a - b\coth bSx)}{a - R_g + b\coth bSx} \quad (5.1)$$

where
$a = 1 + (K/S)$
$b = (a^2 - 1)^{1/2}$
R = reflectance of the film of thickness x, expressed in fractional form
K and S = absorption and scattering coefficients of the film
R_g = reflectance of the background over which the film lies.

In practical colorant formulation we normally deal with opaque films or materials, or else take steps to ensure that during measurement the sample thickness is such that increasing the thickness would have no effect on the reflectance values obtained. If we let x tend to infinity in Eqn 5.1 we obtain Eqn 5.2:

$$\begin{aligned} R &= 1 + (K/S) - [(K/S)^2 + 2(K/S)]^{1/2} \\ &= 1 + (K/S) - \{(K/S)[(K/S) + 2]\}^{1/2} \end{aligned} \quad (5.2)$$

and its inverse (Eqn 5.3):

$$K/S = \frac{(1 - R_\infty)^2}{2R_\infty} \quad (5.3)$$

where R_∞ = reflectance (stated in fractional form) of light of a given wavelength by a sample of infinite thickness.

One other valuable feature of Kubelka–Munk theory is that the absorption and scattering coefficients of a coloured material can be built up from the individual absorption and scattering coefficients of the individual pigments or dyes (Eqn 5.4):

$$K/S = \frac{c_1 K_1 + c_2 K_2 + c_3 K_3 + \ldots + K_s}{c_1 S_1 + c_2 S_2 + c_3 S_3 + \ldots + S_s} \quad (5.4)$$

where c = concentrations of colorants
K = absorption coefficients
S = scattering coefficients
subscript s identifies the substrate
subscripts 1, 2, 3... identify the individual colorants.

In the classical Kubelka–Munk analysis it is assumed that the colorant has both absorbing and scattering power, and this is the case with, for example, pigment particles embedded in a paint layer. Textile dyes, however, can be considered to be dissolved in the fibre and therefore have no scattering power of their own. Any scattering of light can be regarded as being only from the fibres. Therefore in the denominator in Eqn 5.4, S_1, S_2, S_3... can be set to zero to give Eqn 5.5:

$$\begin{aligned} K/S &= \frac{c_1 K_1 + c_2 K_2 + c_3 K_3 + \ldots + K_s}{S_s} \\ &= c_1(K_1/S_s) + c_2(K_2/S_s) + c_3(K_3/S_s) + \ldots + K_s/S_s \end{aligned} \quad (5.5)$$

The ratio K/S_s is now a parameter on its own, even though it is a ratio, and amounts to only a single constant instead of the two we had before. This makes it possible to characterise a dye with only one substrate-specific parameter rather than two.

In textile recipe formulations the various K/S_s values, which are specific to each dye on the given substrate are commonly known as *absorption coefficients*.

To simplify the mathematics in this chapter, the following terms are used:
(a) $R = R_\infty$
(b) $f(R) = K/S$ of dyed material $= (1 - R)^2/2R$
(c) $f(R_s) = K/S$ of substrate $= K_s/S_s = (1 - R_s)^2/2R_s$
(d) $a_1 = K/S$ of dye 1 on a given substrate $= K_1/S_s$, etc.

Therefore for a given wavelength λ, Eqn 5.5 can be written in the form of Eqn 5.6:

$$f(R_\lambda) = a_{\lambda,1}c_1 + a_{\lambda,2}c_2 + a_{\lambda,3}c_3 + \ldots + f(R_{s,\lambda}) \tag{5.6}$$

Usually only 16 wavelengths at 20 nm intervals from 400 to 600 nm are used for recipe prediction work, experience having established that there is no gain in accuracy by using 10 nm intervals, which at the same time calls for a considerable increase in computational time and in data storage requirements.

Let us now consider the application of Eqn 5.6 to single-dye reflectance curves for CI Acid Black 60 dyed on nylon (Figure 5.1). If we take the reflectance values at a wavelength of 600 nm and plot these against dye concentration, we obtain a markedly

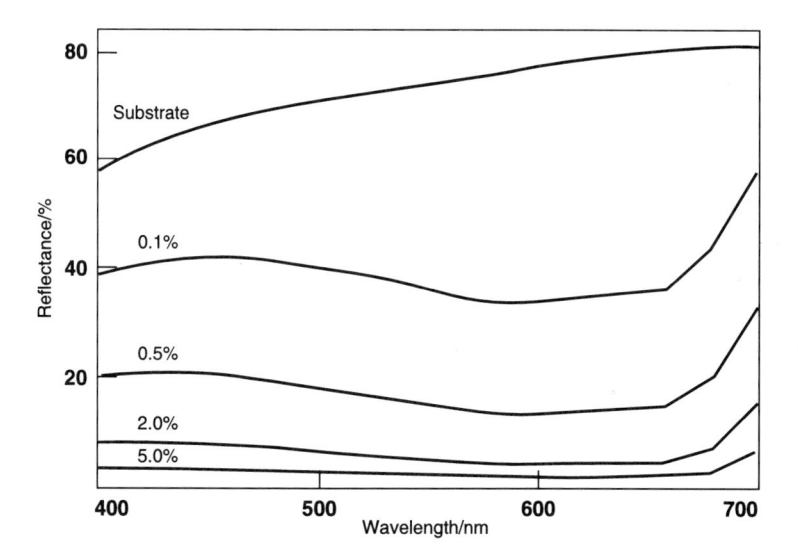

Figure 5.1 Reflectance curves for different concentrations of CI Acid Black 60 on a nylon substrate

nonlinear relationship (Figure 5.2). As the concentration is increased the reflectance decreases very rapidly at first, and at higher concentrations it asymptotically approaches a limiting value.

If, however, we convert the reflectance values to their f(R) equivalents using Eqn 5.3, we obtain a near-linear relationship (Figure 5.3): as the dye concentration tends towards zero, the f(R) values decrease towards a positive minimum value which is the value of the uncoloured substrate $f(R_s)$. In this context the uncoloured substrate value

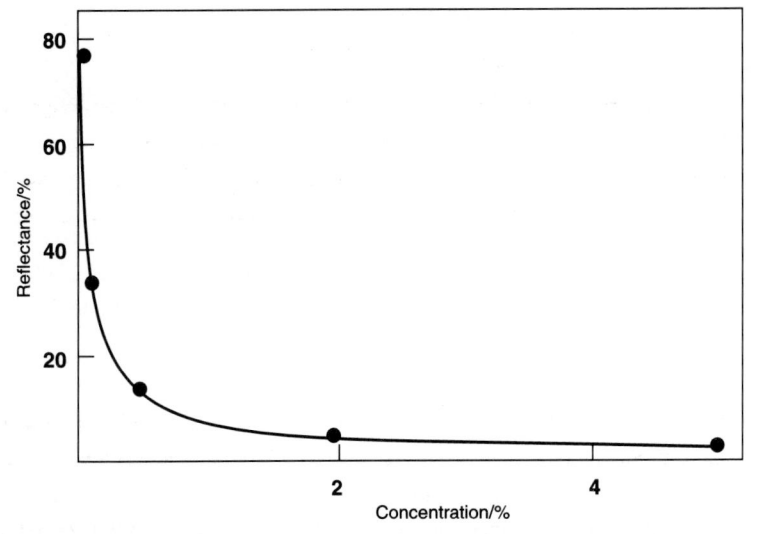

Figure 5.2 Reflectance of different concentrations of CI Acid Black 60 on a nylon substrate at a wavelength of 600 nm

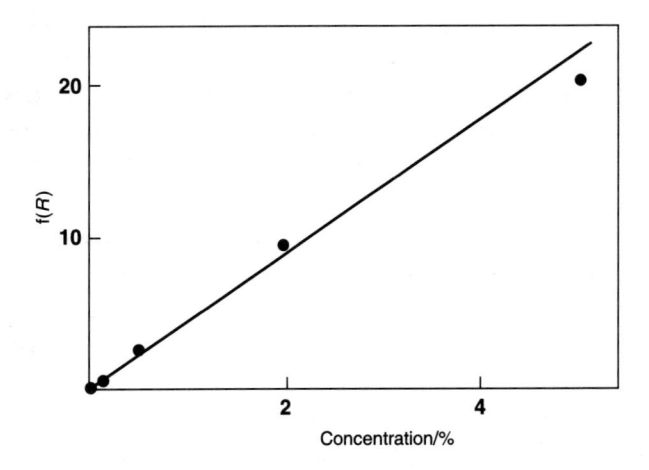

Figure 5.3 Data from Figure 5.2 plotted in terms of f(R) values

is that of a sample of substrate which has been given a full dyeing treatment in a blank dyebath containing all dyeing auxiliary agents except dye; this is necessary because components of the dyeing process other than the dye can alter the colour of the substrate.

The relationship between f(R) and dye concentration at wavelength λ is given by Eqn 5.7:

$$f(R_\lambda) = f(R_{s,\lambda}) + a_\lambda c \qquad (5.7)$$

which is simply a contraction of Eqn 5.6. Clearly, the gradient of the line in Figure 5.3 is in fact the absorption coefficient a_λ.

Given any dye concentration dye c, it is therefore possible to calculate the f(R) values and consequently, using Eqn 5.2, the reflectance value of the dyeing at any wavelength. Conversely, given a reflectance value for a dyeing made with that dye, it is possible to convert this to the f(R) value and calculate the dye concentration from the inverse of Eqn 5.7 (Eqn 5.8):

$$c = \frac{f(R_\lambda) - f(R_{s,\lambda})}{a_\lambda} \qquad (5.8)$$

The proportionality between a and c is an approximation. Where a perfect linear relationship exists between a and c, only one dyeing would in theory be required in order to obtain the required absorption coefficient a. However, deviations from linearity occur much more frequently with reflectance functions such as the Kubelka–Munk function than with Beer's law for solutions. The linearity of the Kubelka–Munk function frequently only applies at low dye concentrations, and deviations are common at high dye concentrations and low reflectance values. These stem from the limited substantivity of some dyes for certain substrates and from saturation effects in dyeing, as well as from optical effects such as the limiting lowest reflectance of a substrate, which is greater than zero even when dyed to the darkest black. Corrections can be introduced to compensate for these deviations from linearity and these are described in section 5.6.

For these reasons, it is usual to dye a range of concentrations of the dye as shown in Figure 5.1 and derive the absorption coefficients from these rather than rely on one dyeing. The 'average' absorption coefficient, which is the gradient of the line fitted through the series of f(R) values plotted against c, can be calculated by least-squares regression techniques.

Absorption coefficients calculated by this method are shown in Figure 5.4 as a function of wavelength. Unlike a single-dye reflectance curve, this absorption curve does not refer to one dyeing only but characterises the spectral absorption properties of the

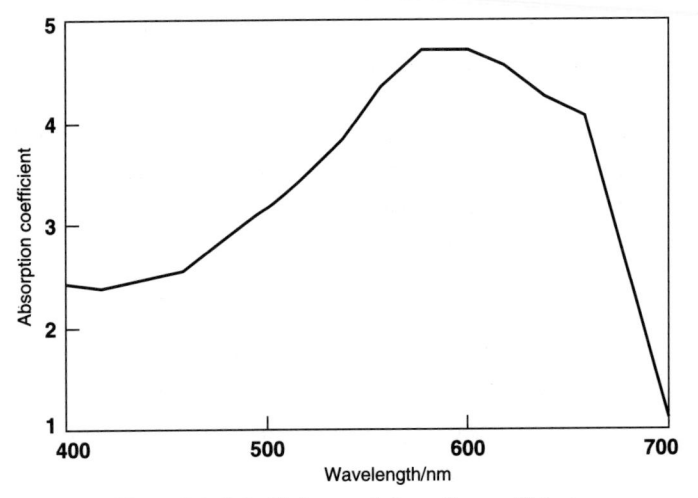

Figure 5.4 Data from Figure 5.1 plotted in terms of absorption coefficients

dye at all concentrations. The colorant can thus be characterised by a single absorption coefficient curve, rather than requiring a series of reflectance curves. In textile dyeing, however, because of the differences in dye uptake resulting from variable substrate absorptivity and different dyeing methods, the absorption coefficients derived as above are specific only to the particular dyeing conditions and fibre type and are not a universal specification of dye absorption.

Although the reflectance curve of any given mixture of dyes can be easily calculated from Eqn 5.6 followed by Eqn 5.2 if the absorption coefficients of the dyes are known, the converse problem of determining the dye concentration to match a given target colour is less straightforward. The difficulties arise when colorimetric or tristimulus matching is necessary rather than spectrophotometric curve matching, as is the case in most practical colour-matching situations: this is because exact reproduction of the target spectral curve is not possible, because the target sample has to be matched with a set of dyes which are different from that used to colour the target.

5.2 SPECTROPHOTOMETRIC CURVE MATCHING

Let us first consider the strictly nonmetameric match in which the target has been dyed on the same substrate as that to be used for the prediction, with the same dyes and dyeing auxiliaries under identical dyeing conditions. We should then be able to predict dye concentrations such that the reflectance curves of target and prediction can be superimposed to give a complete spectral match. If we take a very simple case of only two known dyes in the target shade, we can select two wavelengths at which the absorbing powers of the dyes are distinctly different and write out the corresponding prediction equations at the two wavelengths (Eqn 5.9):

$$f(R_{\lambda_1}) = f(R_{s,\lambda_1}) + a_{\lambda_1,1}c_1 + a_{\lambda_1,2}c_2$$
$$f(R_{\lambda_2}) = f(R_{s,\lambda_2}) + a_{\lambda_2,1}c_1 + a_{\lambda_2,2}c_2 \tag{5.9}$$

We have a set of simultaneous equations in which everything is known except c_1 and c_2, and which can be solved by the normal methods of substitution.

In the case of a nonmetameric match involving three dyes, we can select a set of three equations at three wavelengths where the dyes have distinctly different light-absorbing powers (Eqn 5.10):

$$f(R_{\lambda_1}) = f(R_{s,\lambda_1}) + a_{\lambda_1,1}c_1 + a_{\lambda_1,2}c_2 + a_{\lambda_1,3}c_3$$
$$f(R_{\lambda_2}) = f(R_{s,\lambda_2}) + a_{\lambda_2,1}c_1 + a_{\lambda_2,2}c_2 + a_{\lambda_2,3}c_3 \tag{5.10}$$
$$f(R_{\lambda_3}) = f(R_{s,\lambda_3}) + a_{\lambda_3,1}c_1 + a_{\lambda_3,2}c_2 + a_{\lambda_3,3}c_3$$

or, in matrix notation (Eqn 5.11):

$$G = \begin{bmatrix} g_{\lambda_1} \\ g_{\lambda_2} \\ g_{\lambda_3} \end{bmatrix} = \begin{bmatrix} f(R_{\lambda_1}) - f(R_{s,\lambda_1}) \\ f(R_{\lambda_2}) - f(R_{s,\lambda_2}) \\ f(R_{\lambda_3}) - f(R_{s,\lambda_3}) \end{bmatrix} \quad A = \begin{bmatrix} a_{\lambda_1,1} + a_{\lambda_2,1} + a_{\lambda_3,1} \\ a_{\lambda_1,2} + a_{\lambda_2,2} + a_{\lambda_3,2} \\ a_{\lambda_1,3} + a_{\lambda_2,3} + a_{\lambda_3,3} \end{bmatrix} \quad C = \begin{bmatrix} c_1 \\ c_2 \\ c_3 \end{bmatrix} \tag{5.11}$$

Eqn 5.11 can be written as Eqn 5.12:

$$G = AC$$
$$C = A^{-1}G \tag{5.12}$$

If we let Q = the inverse of matrix A, we have Eqns 5.13 and 5.14:

$$Q = A^{-1} = \begin{bmatrix} q_{\lambda_1,1} + q_{\lambda_2,1} + q_{\lambda_3,1} \\ q_{\lambda_1,2} + q_{\lambda_2,2} + q_{\lambda_3,2} \\ q_{\lambda_1,3} + q_{\lambda_2,3} + q_{\lambda_3,3} \end{bmatrix} \tag{5.13}$$

$$c_1 = q_{\lambda_1,1}g_{\lambda_1} + q_{\lambda_2,1}g_{\lambda_2} + q_{\lambda_3,1}g_{\lambda_3}$$
$$c_2 = q_{\lambda_1,2}g_{\lambda_1} + q_{\lambda_2,2}g_{\lambda_2} + q_{\lambda_3,2}g_{\lambda_3} \tag{5.14}$$
$$c_3 = q_{\lambda_1,3}g_{\lambda_1} + q_{\lambda_2,3}g_{\lambda_2} + q_{\lambda_3,3}g_{\lambda_3}$$

The above technique is very simple, needing only a hand-held calculator to compute the recipe once the inverse matrix Q has been calculated, and many attempts have been made to adapt it to the general matching problem. Its applicability is strictly in the field of nonmetameric matching, however, and the conditions necessary for this are found only rarely in practice, although the method has been used to calculate corrections to batches which are a little off shade.

In practice, the target colour has usually been dyed with different colorants from those to be used for predicting the match. In this case dye concentrations could easily be calculated so that the spectral curve of prediction and target would match exactly at the three chosen wavelengths, but they would probably diverge considerably at other wavelengths and the visual match would be poor. Nevertheless there is still some merit in trying to obtain as near as possible coincidence of the reflectance of the prediction with that of the target, in order to minimise metamerism.

Consider a target and an attempted match prepared with three dyes different from those used for the target (Figure 5.5): the two curves diverge at certain wavelengths and cross over several times. Now, if the curves cross over at least three times we shall get metamerism, which will increase with increasing divergence between the curves. The object of the spectral match approach is to alter the dye concentrations so that the curves are as nearly as possible coincident, i.e. that the sum of the differences between them are at a minimum.

This approach was used very successfully in the Davidson and Hemmendinger COMIC analogue computer in the early 1960s [4]. In this equipment the differences in Kubelka–Munk values at 16 wavelengths were displayed on an oscilloscope screen, and the operator could alter dye-concentration potentiometers on the front of the computer until the displayed differences were at zero at all wavelengths across the spectrum.

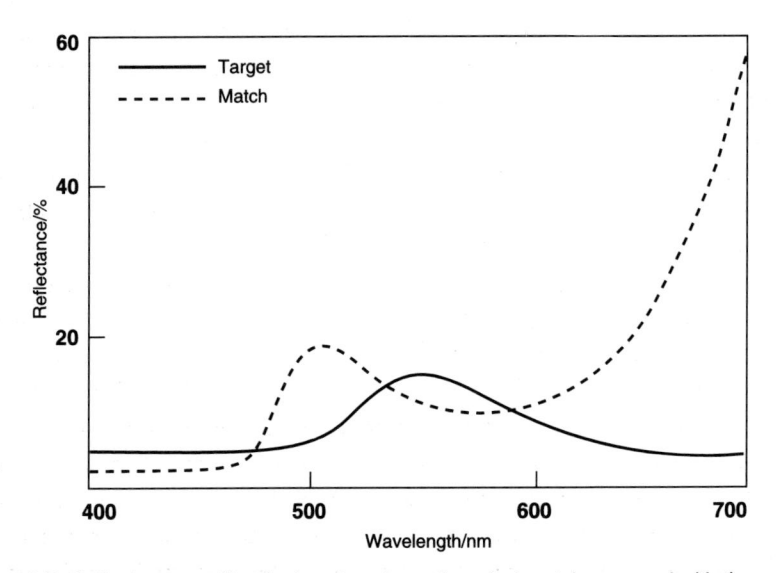

Figure 5.5 Reflectance profile of a target and an attempted match prepared with three dyes different from those used for the target

The operator could plug in boxes containing preset potentiometers simulating the absorption coefficients of alternative dyes, and in this manner could select the combination which gave the best match according to this criterion. Similar equipment was developed by Redifon in the UK [5], but in this equipment the oscilloscope display was in reflectance R or $\Delta R/R$.

Unfortunately, the eye varies in its response to light of different wavelengths across the spectrum, as exemplified by the varying red, green and blue tristimulus responses in the CIE standard observer curves. Divergences between spectral curves at some wavelengths are thus more important than those at other wavelengths in determining any degree of mismatch and metamerism.

5.3 COLORIMETRIC MATCHING

Where metamerism between target and prediction exists, a visual match is obtained under the particular illuminant only when the XYZ values of the prediction equal the XYZ values of the target, and not when arbitrarily chosen parts of the spectral curves coincide; the differences between the spectral curves must be weighted by the visual response curves of the observer and the energy distribution of the illuminant.

In the COMIC and Redifon computers an additional tristimulus difference computer (TDC) was used in conjunction with the dye concentration potentiometers to make the final fine adjustment to minimise ΔX, ΔY and ΔZ between prediction and target.

Since metamerism occurs in most commercial matching situations the colorimetric approach to colorant formulation to minimise ΔX, ΔY and ΔZ has now been universally adopted. This is an iterative approach to prediction and requires a computer to carry out the relatively simple but time-consuming calculations. A flow chart summarising the basis of most prediction systems is shown in Figure 5.6.

At stage 1, the reflectance curve or XYZ values of the target are fed into the computer, together with the names of the selected dyes; this combination may be selected either by the operator or by the computer from predetermined rules for combining suitable dyes. At stage 2 a recipe is generated by the computer using either an arbitrary starting point or a suitable algorithm, while at stage 3 the $f(R)$ values for each wavelength are calculated from the recipe and absorption coefficients using Eqn 5.6. From these the reflectance curve is calculated using Eqn 5.2 and then the XYZ values (stages 4 and 5). At stage 6 the colour difference between the colour of the predicted recipe and that of the target is calculated. If this colour difference is within the predetermined tolerance, the recipe is printed; if not, a correction matrix is generated at stage 7 and used to alter the recipe to reduce the colour difference. This iteration loop is repeated until the colour difference is reduced to an acceptable level, when the final recipe is printed.

218 RECIPE PREDICTION FOR TEXTILES

```
1. Percentage reflectance or XYZ of target (X_t, Y_t and Z_t)
   E_λ x̄_λ ȳ_λ z̄_λ illuminant and tristimulus constants
   Absorption coefficients for dyes 1, 2 and 3 (a_1, a_2 and a_3)
   Percentage reflectance of substrate

2. Starting recipe, e.g.
   c_1 = 2%
   c_2 = 2%
   c_3 = 2%

3. Evaluate at 16 wavelengths:
   f(R_λ) = f(R_{s,λ}) + a_{λ,1}c_1 + a_{λ,2}c_2 + a_{λ,3}c_3

4. Convert f(R) to percentage reflectance at 16 wavelengths

5. Convert percentage reflectance to predicted XYZ (X_p, Y_p and Z_p)

6. Calculate colour difference between prediction and target:
   ΔX = X_t − X_p
   ΔY = Y_t − Y_p
   ΔZ = Z_t − Z_p

   Are ΔX, ΔY and ΔZ < 0.01?  — No → 7. Correct recipe (loops to 3)
   Yes ↓
8. Print recipe
```

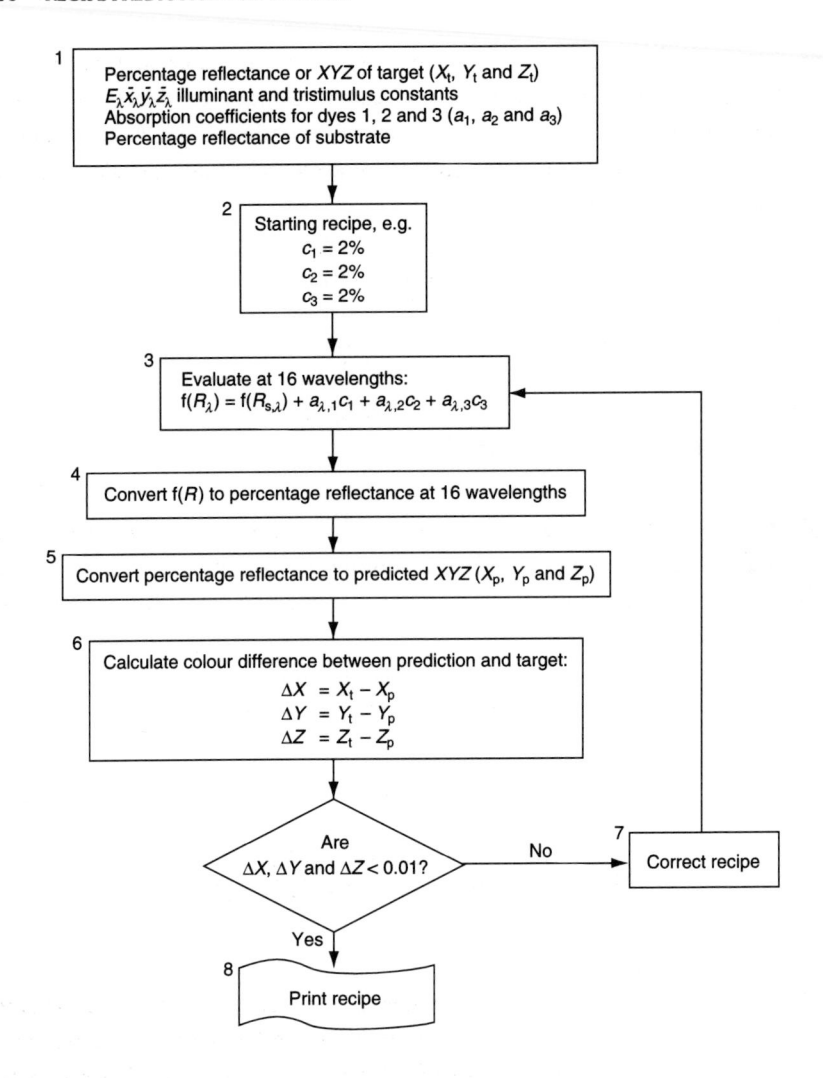

Figure 5.6 Flow chart showing the basis of most computer match-prediction systems

The most intriguing section of the flow chart in Figure 5.6 is the method used at stage 7 to enable the computer to correct the initial approximate recipe on to target. This is accomplished as follows: we first assume that we begin with a starting recipe with dye concentrations c_1, c_2 and c_3, that tristimulus values have been calculated for this recipe (via Eqns 5.2 and 5.6) as being X_1, Y_1 and Z_1 and that we are aiming at a target in colour space with tristimulus values X_t, Y_t and Z_t. Considering the starting recipe: if we were to change one of the dye concentrations by a small amount Δc_1, the resultant recipe would give a change in tristimulus values to X_2, Y_2 and Z_2 as calculated using Eqns 5.6 and 5.2. If we divide the change in X by the change in dye concentration of dye 1, we should obtain a gradient $\Delta X/\Delta c_1$ which could be assumed to be linear over a small distance in colour space and which could be used to predict the change in X resulting from any small change in concentration Δc_1. Similar expressions could be derived for Y and Z (Eqn 5.15):

$$\Delta X = \left(\frac{\partial X}{\partial c_1}\right) \Delta c_1$$

$$\Delta Y = \left(\frac{\partial Y}{\partial c_1}\right) \Delta c_1 \quad (5.15)$$

$$\Delta Z = \left(\frac{\partial Z}{\partial c_1}\right) \Delta c_1$$

If we change the concentrations of all three dyes by small amounts Δc_1, Δc_2 and Δc_3, the consequent change in X, Y and Z is the resultant of the changes in colour coordinates caused by each of the dyes trying to move the colour in different directions across colour space (Eqn 5.16):

$$\Delta X = \left(\frac{\partial X}{\partial c_1}\right) \Delta c_1 + \left(\frac{\partial X}{\partial c_2}\right) \Delta c_2 + \left(\frac{\partial X}{\partial c_3}\right) \Delta c_3$$

$$\Delta Y = \left(\frac{\partial Y}{\partial c_1}\right) \Delta c_1 + \left(\frac{\partial Y}{\partial c_2}\right) \Delta c_2 + \left(\frac{\partial Y}{\partial c_3}\right) \Delta c_3 \quad (5.16)$$

$$\Delta Z = \left(\frac{\partial Z}{\partial c_1}\right) \Delta c_1 + \left(\frac{\partial Z}{\partial c_2}\right) \Delta c_2 + \left(\frac{\partial Z}{\partial c_3}\right) \Delta c_3$$

which Gall called the *influence matrix* [6].

Although the terms in brackets in this matrix could be numerically approximated by assuming, say, 1% changes in each colorant concentration in turn and determining the effect on X, Y and Z as shown above for dye 1, it is better to calculate the matrix terms by differentiation. For two colours a very small distance apart in colour space, having coordinates $X_1 Y_1 Z_1$ and $X_2 Y_2 Z_2$, Eqn 5.17 can be written:

$$X_1 = \sum E_\lambda \bar{x}_\lambda R_{1,\lambda} \quad \text{and} \quad X_2 = \sum E_\lambda \bar{x}_\lambda R_{2,\lambda} \quad (5.17)$$

with similar equations for Y_1 and Y_2, and Z_1 and Z_2. Therefore Eqn 5.18 is valid:

$$\frac{\partial X}{\partial c_i} = \sum E_\lambda \bar{x}_\lambda \left(\frac{\partial R_\lambda}{\partial c_i}\right) \quad (5.18)$$

Eqn 5.19 is then derived by substitution:

$$\begin{aligned}\frac{\partial X}{\partial c_i} &= \sum E_\lambda \bar{x}_\lambda \left(\frac{dR_\lambda}{d[f(R_\lambda)]}\right)\left(\frac{\partial[f(R_\lambda)]}{\partial c_i}\right) \\ \frac{\partial Y}{\partial c_i} &= \sum E_\lambda \bar{y}_\lambda \left(\frac{dR_\lambda}{d[f(R_\lambda)]}\right)\left(\frac{\partial[f(R_\lambda)]}{\partial c_i}\right) \\ \frac{\partial Z}{\partial c_i} &= \sum E_\lambda \bar{z}_\lambda \left(\frac{dR_\lambda}{d[f(R_\lambda)]}\right)\left(\frac{\partial[f(R_\lambda)]}{\partial c_i}\right)\end{aligned} \quad (5.19)$$

To obtain the bracketed terms in Eqn 5.19 at a given wavelength λ, we first differentiate Eqn 5.6 with respect to c_1, c_2 and c_3 to obtain Eqn 5.20:

$$\begin{aligned}\frac{\partial[f(R_\lambda)]}{\partial c_1} &= a_{\lambda,1} \\ \frac{\partial[f(R_\lambda)]}{\partial c_2} &= a_{\lambda,2} \\ \frac{\partial[f(R_\lambda)]}{\partial c_3} &= a_{\lambda,3}\end{aligned} \quad (5.20)$$

i.e. $\partial[f(R_\lambda)]/\partial c_i$ is the absorption coefficient for dye i at wavelength λ. Now we also know that the Kubelka–Munk equation at λ is Eqn 5.21:

$$f(R_\lambda) = \frac{(1-R_\lambda)^2}{2R_\lambda} \quad (5.21)$$

Therefore we can write Eqn 5.22:

$$\frac{d[f(R_\lambda)]}{dR_\lambda} = \frac{R_\lambda^2 - 1}{2R_\lambda^2} \quad \text{and} \quad \frac{dR_\lambda}{d[f(R_\lambda)]} = \frac{2R_\lambda^2}{R_\lambda^2 - 1} \quad (5.22)$$

Since the relationship between $f(R)$ and R is nonlinear, the value of $d[f(R)]/dR$ varies according to the reflectance magnitude and therefore applies only over a very small reflectance range; it is usually calculated from the reflectance curve of the target. If only XYZ values of the target are available, it must be calculated from the reflectance curve corresponding to the concentrations of the current recipe at the end of each iteration loop. By similar calculations, the partial differentials of dye 2 and dye 3 can be obtained.

The influence matrix (Eqn 5.16), which shows how the changes in XYZ are influenced by changes in dye concentration, is not of direct interest in the correction subroutine, but it does provide an essential link in the calculation of the more useful *correction matrix*. This is obtained by application of standard matrix algebra to invert the influence matrix (Eqn 5.23):

$$\Delta c_1 = \left(\frac{\partial c_1}{\partial X}\right)\Delta X + \left(\frac{\partial c_1}{\partial Y}\right)\Delta Y + \left(\frac{\partial c_1}{\partial Z}\right)\Delta Z$$

$$\Delta c_2 = \left(\frac{\partial c_2}{\partial X}\right)\Delta X + \left(\frac{\partial c_2}{\partial Y}\right)\Delta Y + \left(\frac{\partial c_2}{\partial Z}\right)\Delta Z \quad (5.23)$$

$$\Delta c_3 = \left(\frac{\partial c_3}{\partial X}\right)\Delta X + \left(\frac{\partial c_3}{\partial Y}\right)\Delta Y + \left(\frac{\partial c_3}{\partial Z}\right)\Delta Z$$

The new corrected recipe is then:

new c_1 = original c_1 + Δc_1
new c_2 = original c_2 + Δc_2
new c_3 = original c_3 + Δc_3.

At the end of each loop in the iteration procedure the correction matrix is calculated as just described and, if the ΔX, ΔY and ΔZ differences from standard are not within tolerance, they are inserted into the correction matrix to compute the changes in dye concentration required to bring the recipe nearer to target.

If the initial approximate recipe is some distance away from target, then using $\partial c_1/\partial X$ and similar values may not result in the corrected recipe matching the target XYZ values. Several iterations with repeated recalculation of these matrix terms may be necessary before convergence to target is attained, because the linear correction coefficients are approximations, applying over relatively small distances in colour space.

Figure 5.7 illustrates the nonlinear nature of the relationship between X and dye concentration for a given dye. When concentration c_t is applied in Eqn 5.6 it will give the target tristimulus value X_t. Consider a single-dye recipe with dye concentration c_1, which is to be corrected so that the dye concentration will give the target tristimulus value X_t. From Figure 5.7 it can be seen that the target concentration required would be c_t. Application of the correction matrix factor $\partial c_1/\partial X$ calculated at concentration c_1 gives the correction effected along the tangent at c_1: clearly, the recipe will be 'overcorrected' to give too great a concentration change to c'_1. Conversely, a correction from concentration c_2 along its tangent will 'undercorrect' to give too small a concentration change to c'_2. This is why several computer iterations, with recalculation of the correction matrix at each loop, are required to converge on to the target dye concentrations. Although both these corrections are 'wrong' they are closer to target than the original recipe attempt and so a correction matrix calculated about either of the new dye concentrations will be more accurate, leading to a smaller error in the next loop of the convergence procedure.

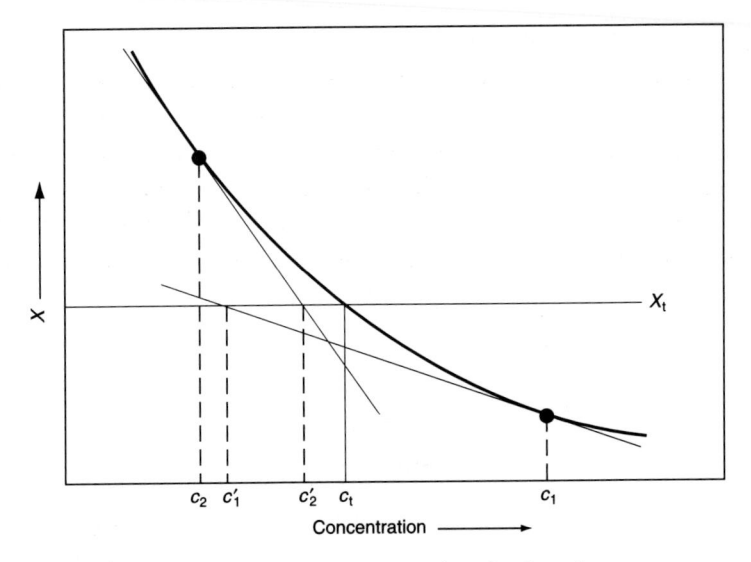

Figure 5.7 Relationship between X and the concentration of a given dye

Alternatively, we might consider a starting recipe with a dye concentration higher than c_1. In this case a correction along the tangent about this dye concentration would lead to an indicated dye concentration even lower than c'_1, which could be further away from the target c_t than the original dye concentration, and could even indicate a negative dye concentration. Thus a poor starting recipe can sometimes lead to a failure to converge on to the target. Attempts to use the recipe correction matrix to correct for large differences in colour between sample and standard may similarly be ineffective.

Gall commented that an improvement in the correction calculation could be achieved if the correction was not made linearly by the customary tangent method [6]. He suggested that consideration could be given to the direction of the required correction and the convexity of the tristimulus value plane in relation to the concentration of the colorants. This is possible by introducing additional quadratic correction coefficients. Instead of the square correction matrix with its nine coefficients described above, however, we would now need correction matrices with 27 coefficients, since mixed coefficients ($\partial^2 c_1/\partial X \partial Y$, for example) have to be taken into account.

If the target reflectance curve is known, then $dR_\lambda/d[f(R_\lambda)]$ can be calculated from the target reflectance and will become more accurate as convergence proceeds. If the target is only available as XYZ values, then the $dR_\lambda/d[f(R_\lambda)]$ must be calculated from the current recipe reflectance curve at the end of each loop.

Although this treatment has been described for three-dye mixtures, it can be extended to combinations with up to six dyes (see section 5.5.1).

5.4 THE STARTING RECIPE

The number of iterations required to attain the desired recipe such that the match criterion is met depends on the effectiveness of the starting recipe. In the early days of computer formulation a starting recipe was often arbitrarily chosen (2% of each of the dyes involved, for instance). In 1966, however, Allen described an extremely successful algorithm for determining the starting recipe, and this has been widely adopted [7].

Allen reasoned that to obtain a tristimulus match we must solve three nonlinear simultaneous equations in three unknowns, the dye concentrations c_1, c_2 and c_3 (Eqn 5.24):

$$f_1(c_1c_2c_3) = X$$
$$f_2(c_1c_2c_3) = Y \quad (5.24)$$
$$f_3(c_1c_2c_3) = Z$$

where f_1, f_2 and f_3 represent certain nonlinear functions of the dye concentrations. The object was to determine a method of linearising these equations to obtain an approximate solution.

The matrices and vectors are first defined (Eqn 5.25):

$$V = \begin{vmatrix} X \\ Y \\ Z \end{vmatrix} \quad T = \begin{vmatrix} x_{400} & \cdots & x_{700} \\ y_{400} & \cdots & y_{700} \\ z_{400} & \cdots & z_{700} \end{vmatrix}$$

$$E = \begin{vmatrix} E_{400} & 0 & \cdots & 0 \\ 0 & E_{420} & \cdots & 0 \\ \cdots & \cdots & \cdots & \cdots \\ 0 & 0 & \cdots & E_{700} \end{vmatrix} \quad R = \begin{vmatrix} R_{t,400} \\ \cdots \\ \cdots \\ R_{t,700} \end{vmatrix} \quad P = \begin{vmatrix} R_{p,400} \\ \cdots \\ \cdots \\ R_{p,700} \end{vmatrix} \quad (5.25)$$

where x, y and z = CIE standard observer visual matching curves for the specified light source
R_t = reflectance of the target to be matched
R_p = reflectance of the computer-predicted recipe

and subscripts 400... represent wavelength in nm.

For a perfect match in the given illuminant, Eqn 5.26 holds:

$$TEP = TER \quad (5.26)$$

and therefore (Eqn 5.27):

$$TE(R - P) = 0 \quad (5.27)$$

Allen argued that if the degree of metamerism between target and prediction is not too great, then the difference between the reflectance of the target and the prediction will be fairly small; we may then assume as an approximation a constant rate of change of f(R) with reflectance over the small difference between the two. We may therefore write Eqn 5.28 with a fair degree of accuracy:

$$\begin{aligned} R_t - R_p &= \Delta R_\lambda \\ &= \left(\frac{dR_\lambda}{d[f(R_\lambda)]} \right) \Delta f(R_\lambda) \\ &= \left(\frac{dR_\lambda}{d[f(R_\lambda)]} \right) (f(R_{t,\lambda}) - f(R_{p,\lambda})) \end{aligned} \quad (5.28)$$

We can also define the matrices given in Eqn 5.29 and 5.30:

$$F = \begin{vmatrix} f(R_{t,400}) \\ \ldots \\ \ldots \\ f(R_{t,700}) \end{vmatrix} \quad G = \begin{vmatrix} f(R_{p,400}) \\ \ldots \\ \ldots \\ f(R_{p,700}) \end{vmatrix} \quad (5.29)$$

$$D = \begin{vmatrix} d_{400} & 0 & \ldots & 0 \\ 0 & d_{420} & \ldots & 0 \\ \ldots & \ldots & \ldots & \ldots \\ 0 & 0 & \ldots & d_{700} \end{vmatrix} \quad (5.30)$$

where $d_\lambda = dR_\lambda/d[f(R_{s,\lambda})]$. We can now write $R_t - R_p$ for all wavelengths as Eqn 5.31:

$$R - P = D(F - G) \quad (5.31)$$

Substituting Eqn 5.31 in Eqn 5.27, we obtain Eqn 5.32:

$$TED(F - G) = 0 \quad (5.32)$$

Transposing, we have Eqn 5.33:

$$TEDF = TEDG \quad (5.33)$$

A further three matrices are now defined as in Eqn 5.34:

$$S = \begin{vmatrix} f(R_{s,400}) \\ \ldots \\ \ldots \\ f(R_{s,700}) \end{vmatrix} \quad C = \begin{vmatrix} c_1 \\ c_2 \\ c_3 \end{vmatrix} \quad A = \begin{vmatrix} a_{400,1} & a_{400,2} & a_{400,3} \\ a_{420,1} & a_{420,2} & a_{420,3} \\ \ldots & \ldots & \ldots \\ a_{700,1} & a_{700,2} & a_{700,3} \end{vmatrix} \quad (5.34)$$

where $a_{400,1}$ = absorption coefficient at wavelength 400 nm of dye 1
c_1 = concentration of dye 1
$f(R_{s,\lambda})$ = Kubelka–Munk function of the undyed substrate.

Referring to Eqn 5.7, we can then write Eqn 5.35:

$$G = S + AC \tag{5.35}$$

Substituting Eqn 5.35 into Eqn 5.33 gives Eqn 5.36:

$$TEDF = TED(S + AC) \tag{5.36}$$

Transposing gives Eqn 5.37:

$$TEDAC = TED(F - S) \tag{5.37}$$

Eqn 5.37 now represents three linear equations in three unknowns (the dye concentrations), which can be solved by the usual calculation of the inverse matrix. Rearranging to obtain C results in Eqn 5.38:

$$C = (TEDA)^{-1} \cdot TED(F - S) \tag{5.38}$$

The $\partial R/\partial[f(R)]$ values used in matrix D on both sides of Eqn 5.37 are obtained from the reflectance curve of the target. Thus the matrix of nine coefficients $TEDA$ cannot be previously established for the three dyes; unfortunately we need to know the reflectance curve of the target to set up this matrix. This is not normally a problem in recipe prediction work where the target is measured on a spectrophotometer; occasionally, however, the target colour specification is available only in the form of colour coordinates. In this case the above approximate starting recipe technique cannot be used and resort must be made to some other algorithm, such as starting with an arbitrary 2% of each dye, or with a dye concentration as some function of depth of shade.

If a poor start recipe is given to the computer then it will take more iterations to converge to the target (see section 5.3). The importance of obtaining a good starting recipe and the effectiveness of the approach described by Allen is illustrated in Table 5.1 for 122 predictions carried out using basic dyes on acrylic yarn. In this exercise the computer iteration was stopped when the difference between target and prediction was less than 0.02 units in X, Y and Z. It can be seen that the number of iterations required for convergence was considerably lower when the Allen start was used. When arbitrary starting recipes were used the number of iterations increased significantly and the computer failed to produce a recipe for 19 of the target colours, and for these a second attempt using a better starting recipe would be required. In eleven cases, where metamerism was probably small, the Allen start gave a recipe within the match criterion of 0.02 units in X, Y and Z and no further iteration was required.

Table 5.1 Basic dyes on acrylic yarn

No. of loops	Allen start frequency	Arbitrary start frequency
0	11	0
1	93	0
2	7	9
3	6	24
4	3	40
5	1	28
6	1	2
Total	122	103

The reason why the arbitrary start technique failed to predict certain recipes was that the start was so far away in colour space from the target that application of the correction matrix derived from the starting recipe over such a large distance in colour space led to negative dye concentrations, which stopped the iteration process.

5.5 REDUCING COST AND METAMERISM BY COMPUTER FORMULATION

The speed of digital computing is such that the complete procedure of initial rough formulation and iteration to minimise ΔX, ΔY and ΔZ to the required degree of accuracy requires less than a second. The time-limiting factors are subsidiary operations such as extracting the relevant absorption coefficients from the disc files and printing the recipe. Computer recipe formulation therefore obviously provides a means of testing alternative dye combinations to find the optimum mixture that will give minimum cost and/or minimum metamerism relative to the target. The prediction theory described has dealt only with three-dye combinations, but it is sometimes advantageous to use combinations of four, or more rarely five or six, dyes to minimise metamerism effects.

The number of recipes to be predicted rises very rapidly as the number of possible dyes is increased. The possible combinations can be calculated from Eqn 5.39:

$$^nC_r = \frac{n!}{r!(n-r)!} \qquad (5.39)$$

where n = number of dyes in the permitted list
 r = number of dyes allowed per recipe.

Table 5.2 Recipes possible for three- and four-dye mixtures

| No. of dyes | No. of combinations possible ||
	Three dyes	Four dyes
3	1	0
4	4	1
6	20	15
10	120	210
20	1140	4875
30	4060	27405

Table 5.2 shows the number of recipes possible for three- and for four-dye mixtures. Obviously not all possible combinations will yield meaningful recipes. For example, there would be little point in trying to predict a recipe for a red target using three blue dyes. Usually some selection from the list of possible dyes is made before computation begins. In computer programming for minimum cost/minimum metamerism facilities, it is usual to enable the computer to predict matches under illuminant D_{65}, to discard all recipes with one or more negative dye concentrations and to list out only, say, the six cheapest or six least metameric recipes according to some numerical index of metamerism under the alternative illuminants of interest.

Normally three-dye mixtures are used. Occasionally, however, none of the three-dye mixtures available provides a match with a sufficiently low degree of metamerism. In these cases it may be worthwhile to consider combinations of four dyes or more.

5.5.1 Four-dye colorimetric matching

In four-dye prediction, the rough match and correction iteration mathematics are expanded to deal with four unknowns, the dye concentrations. We can use the extra dye to minimise residual colour difference in one of the tristimulus values in illuminant A.

Allen has investigated the problem of matching for minimum metamerism [8]. Figure 5.8 shows a plot in illuminant A of all possible matches for a certain target colour. The recipes were predicted to match the target under illuminant D_{65}, but under illuminant A there is evidently a considerable variation in the colour difference from target of the predicted recipes. Figure 5.8 shows only the plot of the tristimulus values X and Z, and there is clearly much more difference among the matches along the X direction than along the Z direction. Figure 5.9 shows a plot of the Y and Z values of the same samples on an expanded scale, and demonstrates that the Y and Z tristimulus values

228   RECIPE PREDICTION FOR TEXTILES

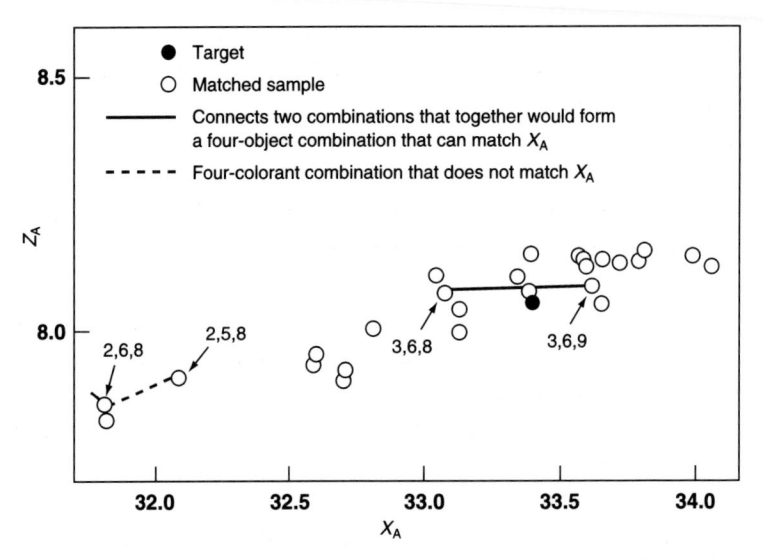

Figure 5.8 Matches of a given sample under illuminant D_{65} plotted for illuminant A colour space, with tristimulus values X and Z; matches were calculated by computer from all possible combinations of nine colorants

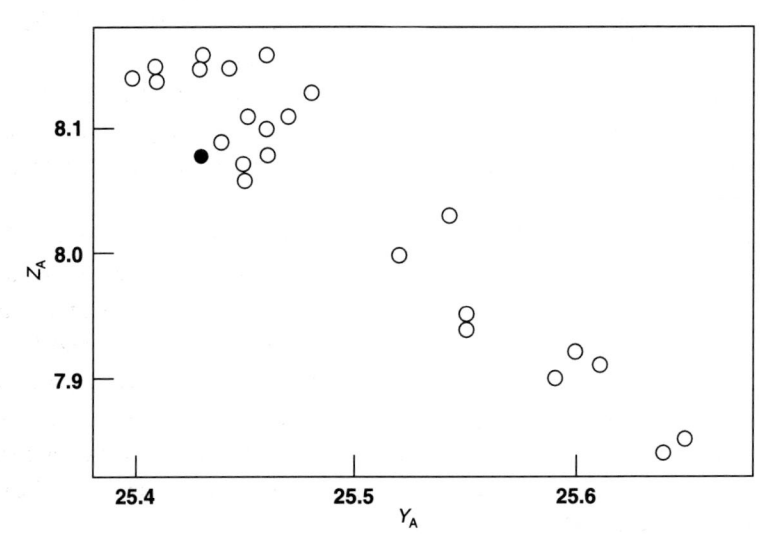

Figure 5.9 Similar type of plot as Figure 5.8, but showing tristimulus values Y and Z under illuminant A

under illuminant A show about the same degree of variation, which is significantly smaller than the variation in X value. It follows that there is a greater possibility of reducing the metamerism shown in daylight and tungsten light with a four-dye

combination by using the X tristimulus value under illuminant A than by using the Y or Z tristimulus value. The prediction should therefore be made to minimise:
- ΔX, ΔY and ΔZ (illuminant D_{65})
- ΔX (illuminant A).

This approach to prediction can be extended to matching with up to six dyes by minimising ΔX, ΔY and ΔZ under illuminants D and A.

As can be seen from Table 5.2, if there are ten dyes in the selection, then for four-dye mixtures the computer would have to calculate 210 possible combinations. The considerable amount of computation can be reduced by the following procedure described by Allen [8] and illustrated in Figure 5.8.

It can be assumed that every four-dye combination that matches the target under illuminant D_{65} consists of two three-dye combinations each of which matches the target under illuminant D_{65}, these three-dye combinations differing by only one dye and having the other two dyes in common. In order to be able to match the X tristimulus value under illuminant A, these two three-dye combinations must lie on either side of the X (illuminant A) value. Allen therefore programmed the computer to search through all possible three-dye matches and select pairs of these that differ by one dye and that lie on either side of the X (illuminant A) value. These pairs provide the four-dye combinations which are then used to calculate matches.

Holdaway has suggested that, since most metameric matches can be adequately satisfied using only three carefully selected dyes, it is reasonable to expect that the addition of only one or two further dyes would produce a satisfactory improvement for most practical applications [9,10].

Bearing in mind the variability of dye absorption coefficients, a comparison of the spectral reflectance characteristics of the target and the predicted match is primarily of theoretical significance, being useful only to give an indication of the potential to produce a good nonmetameric match after adjustment to minimise ΔX, ΔY and ΔZ. For example, a predicted match in which the reflectance curve oscillated several times above and below that of the target would be considered suspect, whereas if the general shape of the reflectance curves of target and match were similar it would be reasonable to expect that the similar characteristics would be retained after adjustment of the trial dyeings to minimise ΔX, ΔY and ΔZ.

The procedure used by Holdaway was to assemble combinations of three dyes from which three-dye recipes were produced and stored up to a specified number. If the operator requested a maximum number of four dyes per recipe, the three-dye recipes, taken in order of increasing metameric index, were then combined with a fourth dye initially at 0% concentration. Convergence was continued to produce a match under illuminant D_{65}. The recipe was accepted and stored only if the metameric index was reduced compared with the starting approximation. If the operator specified a maximum of five dyes per recipe, all three- and four-dye recipes were again stored in order of increasing metameric index and the four-dye recipes only were combined with a fifth dye at 0% concentration to give a starting approximation for a five-dye recipe. The finally adjusted recipe would be accepted and stored only if there was a reduction in the metameric index compared with the starting approximation.

Table 5.3 shows the reduction in colour difference under illuminant A of dyeings computed to match under illuminant D obtained by increasing the number of dyes.

It is considerably more difficult to compute a recipe which will match the target under different illuminants since the introduction in the late 1970s of high-efficacy lamps such as TL84 into retail stores. The spectral distributions of these lamps are completely different from those of daylight or tungsten lamps. The difficulties of producing a match under the two illuminants D_{65} and A are substantially increased if a match is also required under a high-efficacy lamp.

Table 5.3 Colour difference in ANLAB(40) units under illuminant A

Shade no.	Three dyes	Four dyes	Five dyes
1	1.02	0.52	
2	1.15	0.52	
3	1.15	0.18	0.06
4	1.44	0.26	
5	1.48	0.49	
6	1.57	0.63	
7	0.41	0.26	
8	0.27	0.23	0.19
9	2.63	0.43	0.12
10	2.63	0.63	
11	2.68	0.18	0.06
12	2.68	0.43	
13	2.75	0.49	
14	2.75	0.23	0.19

5.5.2 Selecting dyes by curve fitting to minimise metamerism

The colorimetric tristimulus value matching method described in section 5.5.1 will deal with up to six dyes to minimise metamerism under two illuminants. An alternative approach using the full spectrophotometric match has been proposed by McGinnis [11].

In the spectrophotometric match the spectrophotometric curve of the target to be matched is reproduced by suitable combination of the colorants. Since colorimetric tristimulus matching also accounts for both observer and light source characteristics, it use is virtually mandatory when the colorants available for making the match are not similar to those in the target. Spectrophotometric matching is, however, a viable alternative when the same or quite similar colorants are used.

From a computational point of view the spectrophotometric approach is more direct and straightforward than colorimetric matching. In addition, the least-squares technique can be applied in spectrophotometric matching, so that both selection and combination computations are carried out simultaneously.

Consider the standard prediction equation extended to deal with n dyes (Eqn 5.40):

$$f(R_\lambda) = f(R_{s,\lambda}) + a_{\lambda,1}c_1 + a_{\lambda,2}c_2 + \ldots + a_{\lambda,n}c_n \quad (5.40)$$

To obtain a spectrophotometric match Eqn 5.40 must be satisfied at each wavelength – in practice, at the usual 16 wavelengths from 400 nm to 700 nm at intervals of 20 nm. There are therefore 16 equations of the form of Eqn 5.40 that must be satisfied.

McGinnis pointed out that in certain practical colour-matching situations only three colorants, or at most four or five, are required to obtain an acceptable match. In such cases it is impossible to satisfy identically all 16 equations. Even if we did use 16 dyes, there would still be some residual differences between the measured target reflectance and the predicted reflectance because of limitations in the theory, and also because of minor errors in compounding the match. What must be done is to satisfy the 16 equations as nearly as possible at all wavelengths and the least-squares technique is a method of doing this.

When applied to Eqn 5.40, the least-squares technique requires that the sums of the squares of the residuals for each equation be at a minimum (Eqn 5.41):

$$\min \sum \left(f(R_\lambda) - f(R_{s,\lambda}) - a_{\lambda,1}c_1 - a_{\lambda,2}c_2 - \ldots - a_{\lambda,n}c_n \right)^2 \quad (5.41)$$

McGinnis used standard matrix algebra techniques to enable calculation of the colorant concentrations c_i for a colour match based on least-squares fitting to the target reflectance curve.

Occasionally, some of the calculated concentrations have negative values. In these cases the colorants are removed from the calculations, which is in effect the same as limiting the minimum concentration of the colorant to zero.

Tables 5.4 and 5.5 show results obtained by McGinnis from least-squares prediction using a list of nine dyes to match the three target colours, which had been previously dyed with three of the dyes from the list. In successive cycles those dyes which gave negative concentrations were removed before the next cycle was started. In four cycles it can be seen that the program selected the correct original three dyes and gave recipes very close to those of the original targets.

Table 5.4 Least-squares prediction: percentage concentrations of direct dyes on cotton

Target recipes

Dye	T1	T2	T3
Y1	0.4	0.2	0.6
B1	0.4	0.2	0.2
R1	0.2	0.6	0.4

Predicted recipes

Cycle 1

Dye	T1	T2	T3
Y1	0.43	0.29	0.63
Y2	0.16	0.05	0.08
Y3	−0.12	−0.09	−0.11
B1	0.45	0.23	−0.03
B2	−0.06	−0.05	0.01
B3	0.03	0.03	0.02
R1	0.14	0.43	0.41
R2	0.17	0.26	0.21
R3	−0.07	−0.10	−0.10

Cycle 2

Dye	T1	T2	T3
Y1	0.44	0.21	0.63
Y2	−0.04	−0.03	−0.09
Y3	0	0	0
B1	0.60	0.40	0
B2	0	0	−0.05
B3	−0.12	−0.13	0.03
R1	0.21	0.56	0.49
R2	0.02	0.01	0.02
R3	0	0	0

Cycle 3

Dye	T1	T2	T3
Y1	0.38	0.17	0.50
Y2	0	0	0
Y3	0	0	0
B1	0.43	0.21	0
B2	0	0	0
B3	0	0	−0.01
R1	0.31	0.65	0.65
R2	−0.07	−0.08	−0.13
R3	0	0	0

Cycle 4

Dye	T1	T2	T3
Y1	0.40	0.18	0.53
Y2	0	0	0
Y3	0	0	0
B1	0.43	0.20	0.20
B2	0	0	0
B3	0	0	0
R1	0.22	0.56	0.49
R2	0	0	0
R3	0	0	0

5.5.3 Minimising metamerism colorimetrically

Sluban has investigated the strategy of programming recipe prediction to minimise directly the colour difference from target over three illuminants, rather than allowing the formulation program to iterate to zero colour difference in daylight and simply accepting the resulting colour differences in tungsten light and in fluorescent light [12]. The colour-difference minimisation technique can reduce the large variation in colour

Table 5.5 Least squares prediction: percentage concentration of wool dyes

Target recipe

Y1	0.10
R1	0.05
B1	0.02

Predicted recipe

Dye	Cycle 1	Cycle 2	Cycle 3	Cycle 4
Y1	0.120	0.102	0.105	0.104
Y2	−0.019	0	0	0
R1	0.048	0.045	0.049	0.048
R2	−0.001	0	0	0
B1	0.022	0.019	0.023	0.021
B2	−0.068	0	0	0
B3	0.001	−0.024	0	0
V1	−0.001	0	0	0
01	−0.001	0	0	0
G1	0.004	0.063	−0.001	0
N1	0.001	0.003	−0.001	0

differences sometimes obtained in the alternative illuminants but at the expense of accepting a small colour difference in daylight. The target for a colour match is shown in Eqn 5.42:

$$w_{D_{65}}^2 (\Delta E_{D_{65}})^2 + w_A^2 (\Delta E_A)^2 + w_{WWF}^2 (\Delta E_{WWF})^2 \to \min \quad (5.42)$$

where w = weighting applied to colour difference in daylight (D_{65}), tungsten light (A) and warm white fluorescent light (WWF) respectively. Typical examples are given in Table 5.6.

In certain cases spectrophotometric curve fitting gave larger colour differences in all three illuminants than the colorimetric fit to illuminant D_{65}. The colour-difference minimisation technique could be especially useful where an initial colour standard is being produced. In this case if a small change in shade in daylight could be accepted, a more colour-constant standard could be produced.

5.6 DEVIATIONS FROM LINEARITY IN RECIPE PREDICTION EQUATIONS

In the mathematics of recipe prediction in textiles, considerable problems are caused by deviations from linearity in the equations relating f(R) to dye concentration, particularly at higher dye concentrations. The deviations from linearity are due to:

Table 5.6 Colour-difference minimisation in three illuminants

	Predicted colour difference /CIELAB units		
Colour	D_{65}	A	WWF
Weighted spectrophotometric curve fitting			
Pink	0.35	0.52	0.20
Grey-violet	1.06	1.29	0.19
Green	0.62	1.38	1.09
Colorimetric fit to illuminant D_{65}			
Pink	0.0	0.51	0.38
Grey-violet	0.0	1.85	1.01
Green	0.0	1.26	1.33
Colour-difference minimisation			
Pink	0.24	0.33	0.17
Grey-violet	0.81	1.13	0.36
Green	0.75	0.76	0.78

(a) deficiencies in the theory relating concentration of dye in fibre with reflectance
(b) nonlinear uptake of dye by the fibre as the applied dye concentration is increased
(c) interference with dye uptake due to interaction or competition between dyes during the dyeing process.

Figure 5.10 shows a plot of $f(R) - f(R_s)$ against dye applied in the dyebath for a typical disperse dye on polyester yarn [13]. It is clear that Eqn 5.7 does not hold over a wide concentration range. This result is not unexpected since for all dyes exhaustion decreases with increasing amount of dye applied. In particular the linearity falls off at low reflectance values (high $f(R)$ values).

One explanation for the nonlinearity is that Eqn 5.7 does not take account of the light reflected from the surface of the fibre. This factor becomes more important at those wavelengths where the reflectance is very low (in the blue wavelengths for yellows and in the red wavelengths for blues, for instance; in certain cases the reflectance of a yellow can be lower than that of a black dyed on the same substrate). Amendments to the $f(R)$ function, such as those of Pineo, have been used in an attempt to provide more linear relationships (Eqn 5.43):

$$f(R) = \frac{(1-R)^2}{2(1-b)(R-b)} \quad (5.43)$$

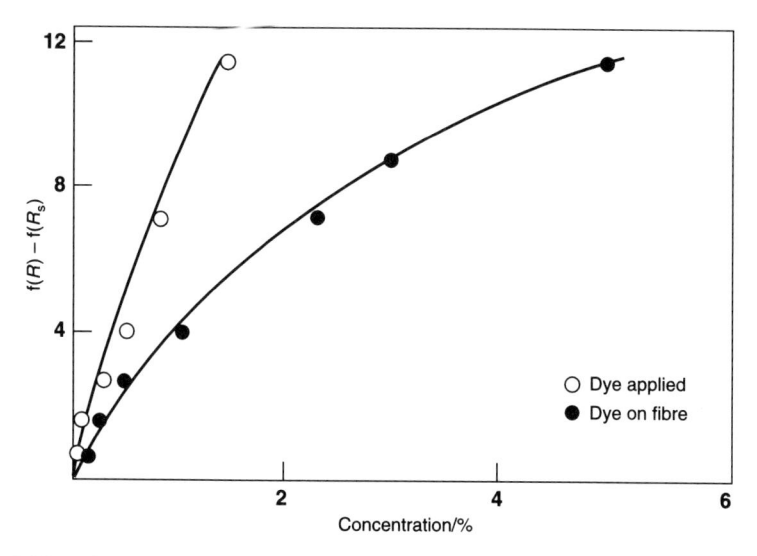

Figure 5.10 Reflectance function of CI Disperse Blue 128 on polyester at a wavelength of 540 nm

where b is the Pineo factor determined by experimentation. The introduction of a power term to the f(R) function has been tried (Eqn 5.44):

$$f(R_\lambda)^n = a_{\lambda,1}c_1 + a_{\lambda,2}c_2 + a_{\lambda,3}c_3 + f(R_{s,\lambda}) \tag{5.44}$$

where n is greater than unity. Unfortunately different dyes require different values of n, so the equation cannot be used in relation to mixtures.

Gailey et al. applied a power term to the dye-in-dyebath concentration term (Eqn 5.45):

$$f(R) = (a_{\lambda,1}c_1^{b_1})c_1 + (a_{\lambda,2}c_2^{b_2})c_2 + (a_{\lambda,3}c_3^{b_3})c_3 + f(R_{s,\lambda}) \tag{5.45}$$

where b is a negative constant with value between 0 and −0.5 [14]. Eqn 5.45, though successful for single dyes, was less satisfactory in mixtures, the residual errors being considered to be due to dye interaction and competition effects in the dyebath.

5.6.1 Interpolation solutions to nonlinearity

Two simple practical approaches to the problem of nonlinearity in Eqn 5.7 are widely used.

Use of overall best-fit absorption coefficients
If the nonlinearity is not too great, as is the case when high dye exhaustions can be achieved (with acid dyes on wool, for example), the simplest approach is to calculate

the average absorption coefficient over a range of calibration dye concentrations, typically six concentrations. In the early days of computer recipe prediction this was often done manually by converting the equation to its logarithmic form (Eqn 5.46):

$$\log[f(R_\lambda) - f(R_{s,\lambda})] = \log c + \log a_\lambda \qquad (5.46)$$

Plotting $f(R) - f(R_s)$ against c using log–log paper should give a plot of gradient 1, and an intercept $\log a$.

Nowadays a least-squares fitting technique is used. In least-squares fitting the sum of the squares of the differences between the measured values and the values predicted by the proposed equation is minimised. Thus if:

$$f(R) = f(R_{s,\lambda}) + a_\lambda c$$

then the sum of the squares is (Eqn 5.47):

$$\min \sum \left(f(R) - f(R_{s,\lambda}) - a_\lambda c\right)^2 \qquad (5.47)$$

Unfortunately this measure of fit tends to be biased towards higher dye concentrations which give larger $f(R)$ values, and so a weighted sum of squares is better (Eqn 5.48):

$$\sum \left(\frac{f(R_\lambda) - f(R_{s,\lambda}) - a_\lambda c}{f(R_\lambda)}\right)^2 \qquad (5.48)$$

This expresses the difference in percentage terms of the magnitude of the measured $f(R)$ values and so tends to give equal accuracy to all levels of dye concentration. If desired the weighting function can be introduced so that, for example, the median dye concentrations give more emphasis by using a manually selected weighting factor w, which is different for each dye concentration in the range defined by Eqn 5.49:

$$\min \sum \left(f(R_\lambda) - f(R_{s,\lambda}) - a_\lambda c\, w\right) \qquad (5.49)$$

Use of individual concentration-dependent coefficients
Instead of calculating the mean absorption coefficient for a single dye at a particular wavelength, which gives in some cases a very poor description of the variation of $f(R)$ with dye concentration, it is more usual to store on the computer disc the individual absorption coefficients calculated from each of, say, six dyeings in the calibration range using Eqn 5.50:

$$f(R_\lambda) = f(R_{s,\lambda}) + a_{\lambda,i} c_i$$
$$a_{\lambda,i} = \frac{f(R_\lambda) - f(R_{s,\lambda})}{c_i} \qquad (5.50)$$

DEVIATIONS FROM LINEARITY IN RECIPE PREDICTION EQUATIONS 237

The computer is first instructed to select the absorption coefficients from a medium-concentration dye curve for the initial attempt at prediction. At the end of the first iteration loop the computer assesses the initial recipe it has calculated and selects, for each dye in the recipe, the absorption coefficients whose dye concentration most nearly matches the initial recipe concentration; it then enters the correction loop of the process with these new coefficients. This process is repeated until a satisfactory computer match has been obtained.

A disadvantage of the above approach is shown in Figure 5.11. Here the absorption coefficients (gradients) specific to each dye concentration level are illustrated. Obviously, the gradients through each dye concentration point do not represent the rate of change of f(R) with dye concentration ($\partial [f(R)]/\partial c_i$) according to the curve drawn through the points. The correction matrix computed during the iteration process will therefore be inaccurate: the computer will overshoot or undershoot the target during convergence and the final recipe will be in error by some small amount.

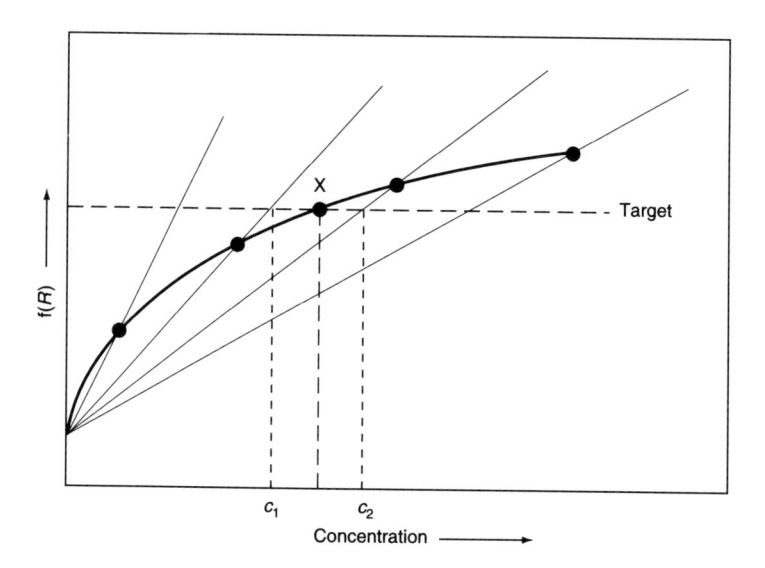

Figure 5.11 Plots of f(R) against dye concentration for several dyeings

For example, if the target f(R) is indicated by X in Figure 5.11 then the absorption coefficient from the dyeing with lower concentration than the target would yield a lower than true dye concentration c_1, whereas the absorption coefficient from the dyeing with higher concentration than the target would yield a higher than true dye concentration c_2. This problem has been partly overcome by using linear interpolation between the range of calibration dye concentrations [15]. A factor F is calculated from the formula (Eqn 5.51):

$$F = \frac{c_t - c_l}{c_h - c_l} \tag{5.51}$$

where c_t = target concentration for which the absorption coefficient is needed
c_l = concentration of the next lower level of calibration dyeings
c_h = concentration level of the next higher level of calibration dyeings.

The target absorption coefficient a is obtained from Eqn 5.52:

$$a = a_l + F(a_h - a_l) \tag{5.52}$$

where a_l = absorption coefficient from the next lower concentration dyeing
a_h = absorption coefficient from the next higher concentration dyeing.

Even if a more accurate single-constant absorption coefficient is obtained which could be drawn as a gradient through the point X in Figure 5.11, this gradient would still not represent the rate of change of f(R) with c at this point. In some commercial systems a more accurate estimate of $\partial[f(R)]/\partial c_i$, for use in calculating the correction matrix, is obtained by linear interpolation between the f(R) and c values of the calibration dyeings on either side of the target concentration.

The new absorption coefficient $a = [f(R) - f(R_s)]/c$ and $\partial[f(R)]/\partial c_i$ calculated from linear interpolation would still not exactly describe the variation of reflectance function with dye concentration about the desired target concentration if the relationship is not linear. The approximation will be much better, however, and the error between target and predicted dye concentrations will be considerably lessened; a more satisfactory predicted recipe will be produced.

Quadratic and cubic equations have also been used for interpolation in commercial systems, as giving a more accurate fit to the calibration data. Another refinement used in some commercial systems is to correct the predicted recipe dye concentration to compensate for the nonlinear relationship between $f(R) - f(R_s)$ and dye concentration (Eqn 5.53):

$$c_t = a_0 + a_1 c_p + a_2 c_p^2 \tag{5.53}$$

where c_t = true concentration of the dye to be applied to match the target
c_p = dye concentration computed from the absorption coefficients.

The polynomial coefficients are calculated from calibration dyeings at different concentrations. Saving in computer storage (if required) may also be achieved since it is now only necessary to hold a single set of absorption coefficients from one dye concentration for the dye (or an averaged set), together with the polynomial coefficients. This technique makes it easy to adjust the predictions on the basis of practical experience. It can also be used to adjust coefficients prepared on one substrate for use on another having slightly different dye absorption and reflectance characteristics. The polynomial coefficients can be recalculated as required using the concentrations originally calculated by the computer and the concentrations that were actually required in practice to produce the target shade.

5.6.2 Simple physical chemistry solutions to nonlinearity

Approaches such as those described in section 5.6.1 are widely used in commercial colorant formulation systems to improve the accuracy of first predictions. However, another more fundamental approach has been used by at least one textile company [16,17]. In this approach, developed in 1968, it is assumed that most of the nonlinearity between dye applied to the dyebath and the reflectance function is caused by problems of dye uptake. In single dyes this is the result of gradual fall-off in dye exhaustion as the concentration of dye applied from the dyebath is increased; a limit is reached beyond which no further increase in dye uptake can take place regardless of the amount of dye in the dyebath.

In dye mixtures the uptake of each dye can be affected by the presence of other dyes and by changes in the auxiliary chemicals added to the dyebath, such as salt or acid. It was postulated that there was a fairly good linear relationship between dye actually contained in the fibre and the reflectance function f(R), as is expected in prediction theory, but that all the dye added to the dyebath did not go on to the fibre. An equation relating dye-in-dyebath to dye-in-fibre was needed in the reflectance function relationship (Eqn 5.54):

$$f(R_\lambda) = f(R_{s,\lambda}) + a_\lambda c_f \qquad (5.54)$$

where c_f is dye concentration in the fibre.

There are two generally used equations to describe the relationship between dye in the dyebath and dye taken up by the fibre, namely the Freundlich and the Langmuir isotherms. Both of these describe the relationship between the amount of dye in the fibre $[D]_f$ and that in the dyebath $[D]_s$ at equilibrium, i.e. after dyeing has taken place for some time. In recipe prediction, however, we are dealing with the dye concentrations in the dyebath at the beginning of the dyeing cycle. Nevertheless the two equations can be used equally well with the initial concentrations of dye in the dyebath c, to give two empirical equations that describe the nonlinear relationship between dye in the dyebath and that taken up by the fibre, c_f.

The Freundlich isotherm is defined by Eqn 5.55:

$$[D]_f = k[D]_s^x \qquad (5.55)$$

where k is a constant and x is a fractional power, or its empirical equivalent (Eqn 5.56):

$$c_f = gc^b \tag{5.56}$$

in which g and b replace k and x respectively. Substituting in Eqn 5.54 we obtain Eqn 5.57:

$$f(R_\lambda) = f(R_{s,\lambda}) + a_\lambda \, gc^b \tag{5.57}$$

The Freundlich isotherm does not have any upper limit to dye absorption. The Langmuir isotherm, however, does imply an upper limit, having been developed initially for dyeing processes such as acid dyes on wool, where dyes are attached to specific sites and where chemical saturation occurs at high dye concentrations (Eqn 5.58):

$$[D]_f = \frac{k[S]_f[D]_s}{1+k[D]_s} \tag{5.58}$$

where $[S]_f$ = concentration of sites in the fibre, or its empirical equivalent (Eqn 5.59):

$$c_f = \frac{gc}{1+kc} \tag{5.59}$$

which substitutes into the reflectance function equation to give Eqn 5.60:

$$f(R_\lambda) = f(R_{s,\lambda}) + a_\lambda \left(\frac{gc}{1+kc} \right) \tag{5.60}$$

Eqn 5.59 can be extended to deal with, say, three dyes in a dye mixture in two ways (Eqns 5.61 and 5.62):

$$c_f = \frac{g_1 c_1}{1+k_1 c_1} + \frac{g_2 c_2}{1+k_2 c_2} + \frac{g_3 c_3}{1+k_3 c_3} \tag{5.61}$$

$$c_f = \frac{g_1 c_1 + g_2 c_2 + g_3 c_3}{1+k_1 c_1 + k_2 c_2 + k_3 c_3} \tag{5.62}$$

Eqn 5.61 implies that each dye, although it has an individual saturation limit, is taken up by the fibre independently, while Eqn 5.62 can be rewritten as Eqn 5.63:

$$c_f = \frac{g_1 c_1}{1+k_1 c_1 + k_2 c_2 + k_3 c_3} + \frac{g_2 c_2}{1+k_1 c_1 + k_2 c_2 + k_3 c_3} + \frac{g_3 c_3}{1+k_1 c_1 + k_2 c_2 + k_3 c_3} \tag{5.63}$$

and it can be seen that the uptake of each dye is affected by the concentration of each of the other dyes in the mixture, i.e. competition effects are taken into consideration. Mathematically this form is the preferable one, since the equation is additive, unlike Eqn 5.61. Suppose the equations represent one dye being added to the dyebath in three parts, then $g_1 = g_2 = g_3 = g$, and $k_1 = k_2 = k_3 = k$; therefore Eqn 5.62 reduces satisfactorily to that of a single dye added in three parts (Eqn 5.64):

$$c_f = \frac{g(c_1 + c_2 + c_3)}{1 + k(c_1 + c_2 + c_3)} \tag{5.64}$$

For this approach to prediction, the dye content in the fibre of the calibration ranges must first be estimated by standard analytical techniques. Details for most dye classes have been given by McDonald et al. [17]. These methods usually involve extraction of dye from the fibre and estimation by spectrophotometric techniques, or dissolving the dyed fibre in solvent, or estimation of dye uptake from exhaust dyebath liquors. Once the dye-in-fibre content of the calibration ranges are known, then absorption coefficients are calculated in the normal way, but relating the reflectance function to dye-in-fibre content.

A two-stage prediction procedure is then adopted. In the first stage the dye-in-fibre content required to match the target is predicted from the absorption coefficients. In the second stage the dye required to be added to the dyebath is calculated from the simplified dye uptake equations described above.

The non-additive form (Eqn 5.61) has been used successfully for match prediction in such systems as disperse dyes on polyester, where dyes tend to add on independently. The additive form (Eqn 5.62) has been used equally successfully for vat dyes, where some dye blocking is normally encountered especially at extremely high dye concentrations [18]. The results of the application of the above approach to prediction have been reported by McDonald et al. [17].

Where the extra effort of determining dye-in-fibre contents is considered too great, a simple modification of the equations can be used (Eqn 5.60). Combining constants gives Eqn 5.65:

$$f(R_\lambda) = f(R_{s,\lambda}) + \frac{g_\lambda c}{1 + kc} \tag{5.65}$$

In practice the k constants are also found to vary with wavelength (Eqn 5.66):

$$f(R_\lambda) = f(R_{s,\lambda}) + \frac{g_\lambda c}{1 + k_\lambda c} \tag{5.66}$$

and so we have two constants at each wavelength, giving a total of 32 constants to define the dye. This compares with 18 constants (16 reflectance constants and two dye uptake constants) required with the original forms of the equations. On the other hand there is no need for dye-in-fibre estimation, and the number of absorption coefficients is still less than the six sets of 16 coefficients held for the interpolation approach.

An equation identical to Eqn 5.66 can also be derived from Eqn 5.4 if both numerator and denominator are divided by S_s. This would imply that dyes may have scattering power different from that indicated by standard theory. Certainly when yellow dyes are applied to unbleached cotton the reflectance of the dyed material is often higher than that of the undyed substrate at the red–yellow end of the spectrum, even when the dyes are not fluorescent.

In terms of recipe prediction accuracy, recipes determined by dye-in-dyebath two-constant theory are more accurate on average than those determined by one-constant dye-in-dyebath theory using interpolation techniques to correct for nonlinearity. Table 5.6 shows the prediction errors obtained when recipes to match target shades were predicted using Eqn 5.66 extended as follows (Eqn 5.67):

$$f(R_\lambda) = \frac{g_{1,\lambda} c_1}{1 + k_{1,\lambda} c_1} + \frac{g_{2,\lambda} c_2}{1 + k_{2,\lambda} c_2} + \frac{g_{3,\lambda} c_3}{1 + k_{3,\lambda} c_3} + f(R_{s,\lambda}) \qquad (5.67)$$

and by three commercial prediction systems employing interpolation techniques. The same single-dye ranges were used for each approach. The average colour differences show the improvement obtained using Eqn 5.67. In effect, the use of two constants per wavelength is similar to the interpolation techniques described earlier, especially when the independent dye-uptake equation (Eqn 5.67) is used. Both approaches give good fits to single-dye data. The better results obtained using Eqn 5.67 may be due to a deficiency in the interpolation technique in calculating the absorption coefficients or $\partial[f(R)]/\partial c_i$ values for convergence in the prediction routines discussed in section 5.6.1 (Table 5.7).

Care is required in the calculation of the g and k constants in Eqns 5.59 and 5.66. For example, in Eqn 5.66 one method is to write the equation in the form of Eqn 5.68:

$$\frac{1}{f(R_\lambda) - f(R_{s,\lambda})} = \left(\frac{1}{g_\lambda}\right)\frac{1}{c} + \frac{k_\lambda}{g_\lambda} \qquad (5.68)$$

and carry out linear regression to determine the values of $1/g$ and k/g, and hence g and k. This method produces values of k and g that tend towards infinity, however [19–21]. Direct search techniques such as those described by Box *et al.* are used to determine g and k by minimising the sums of squares (Eqn 5.69 or 5.70) [22]:

$$\min \sum \left(\frac{R \text{ measured} - R \text{ predicted from equation}}{R \text{ measured}}\right)^2 \qquad (5.69)$$

Table 5.7 Comparative performance of recipe prediction systems: colour differences (CMC(3:1) units)

Colour	Disperse dyes on polyester			Reactive dyes on cotton			
	System 1	System 2	Eqn 5.67	System 1	System 2	System 3	Eqn 5.67
1	1.0	0.9	1.0	5.7	4.9	3.8	2.5
2	5.2	4.6	2.9	1.9	1.7	1.5	1.4
3	0.7	1.2	0.6	5.2	5.3	3.0	3.1
4	3.2	3.5	1.8	3.0	2.0	3.5	1.6
5	0.9	0.6	0.5	3.6	3.2	3.7	1.7
6	2.0	2.9	1.6	3.2	2.8	3.3	1.1
7	1.9	1.5	0.9	0.7	0.9	0.5	2.5
8	0.8	0.8	1.1	1.4	3.1	1.5	1.6
9	4.2	4.8	1.5	1.5	1.5	0.8	1.2
10	3.2	3.8	1.5	3.6	1.5	2.8	0.3
11	1.0	1.3	1.1	2.2	3.1	2.1	1.6
12	1.2	1.0	1.4	1.9	1.5	1.6	1.5
13	2.2	1.5	2.7	3.0	0.2	1.2	0.7
14	1.9	2.3	0.7	0.5	2.9	0.4	1.5
15	1.7	2.0	0.4	2.8	4.4	6.5	0.8
16	1.3	1.1	1.1	1.8	1.2	1.0	1.3
17	1.2	1.1	1.3	5.2	5.1	2.7	2.4
18	0.9	0.5	1.2	4.3	3.6	3.1	3.3
19	2.4	2.6	1.7	1.1	1.8	1.0	2.0
20	3.6	4.3	2.1	2.8	3.0	2.5	1.7
21	0.5	0.7	1.2				
22	2.3	3.2	1.6				
23	0.5	0.6	1.1				
24	1.0	0.8	1.4				
Mean	1.9	2.0	1.4	2.8	2.7	2.3	1.7
<2.0	62%	58%	88%	40%	40%	45%	65%
<1.5	50%	50%	63%	20%	15%	30%	35%

$$\min \sum \left(\frac{f(R) \text{ measured} - f(R) \text{ predicted from equation}}{f(R) \text{ measured}} \right)^2 \quad (5.70)$$

Minimising reflectance differences gives the better results.

One useful technique described by Box *et al.* is the Simplex method of Nelder and Meade. In the Simplex iteration technique the computer is fed with three sets of starting g and k values, and these are plotted to produce a triangle, shown as triangle A in Figure 5.12(a). Eqn 5.66 is evaluated for each dye concentration in a given dye range to produce a set of predicted f(R) values. The difference between the predicted values from Eqn 5.66 and the actual values determined from the measurements on the dyeings are used to obtain the sums of squares measure of goodness of fit indicated above.

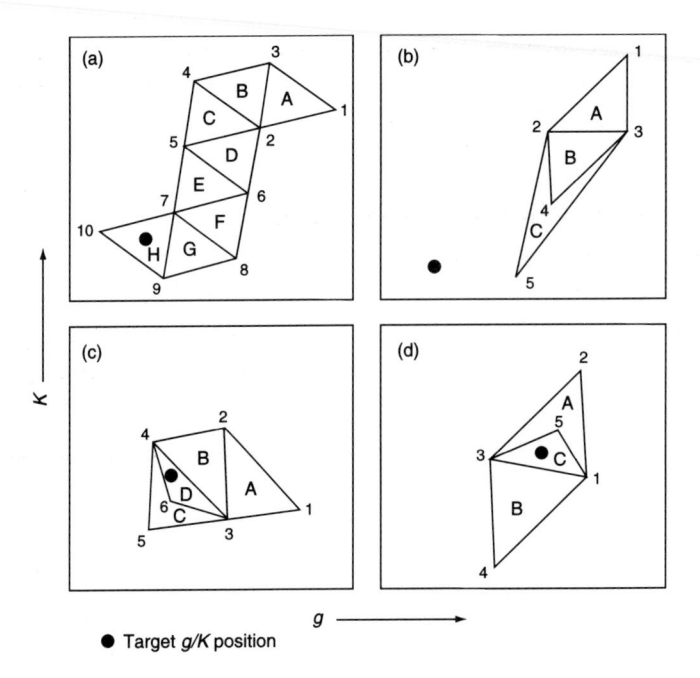

Figure 5.12 Simplex method of absorption coefficient calculation (for explanation see text)

By this method a sums of squares fit is calculated for each of the three sets of g and k values represented by the vertices of triangle A in Figure 5.12(a). The computer then reflects the triangle in a direction opposite to the vertex with the worst fit, producing a new triangle (B) and a new pair of g and k values. It is anticipated that one of the vertices in triangle B will provide a better fit to the measured data. If so, the procedure continues until the reflecting triangles eventually home towards the g and k values which give an optimum fit to the measurements of the dyeings. This is indicated by triangle H in Figure 5.12(a).

Procedures are included in the Simplex program to accelerate and improve the degree of convergence to the target values. For example, in Figure 5.12(b), if triangle B produces a vertex 4 with better fit than all three vertices of triangle A, then triangle B is extended to triangle C by doubling the vertex reflection to vertex 5. This accelerates convergence towards the target area.

Conversely, once the Simplex has reached the target area, provision must be made to reduce the size of the triangles so that convergence to the desired target position is possible. In Figure 5.12(c) reflection from triangle B to triangle C produces a fit for vertex 5, which is better than vertex 2 but worse than either 4 or 3. In this case the triangle reflection is contracted to vertex 6 to produce a smaller triangle D. In Figure 5.12(d) reflectance to triangle B produces vertex 4, which is a worse fit than vertices 1,

2 and 3; the Simplex program is therefore going in the wrong direction, and triangle C is contracted to give vertex 5, whose fit should lie between the worst vertex 1 and the better fits of vertices 2 and 3.

5.6.3 Piecewise linear solutions to nonlinearity

The nonlinear equations described in section 5.6.2 assume a regular decrease in dye uptake as the concentration of dye increases. Unfortunately, this is not the case with all dyes. Hoffenberg plotted the f(R) values of different dyes and pigments against concentration and showed that uptake can follow many different patterns (Figure 5.13) [23].

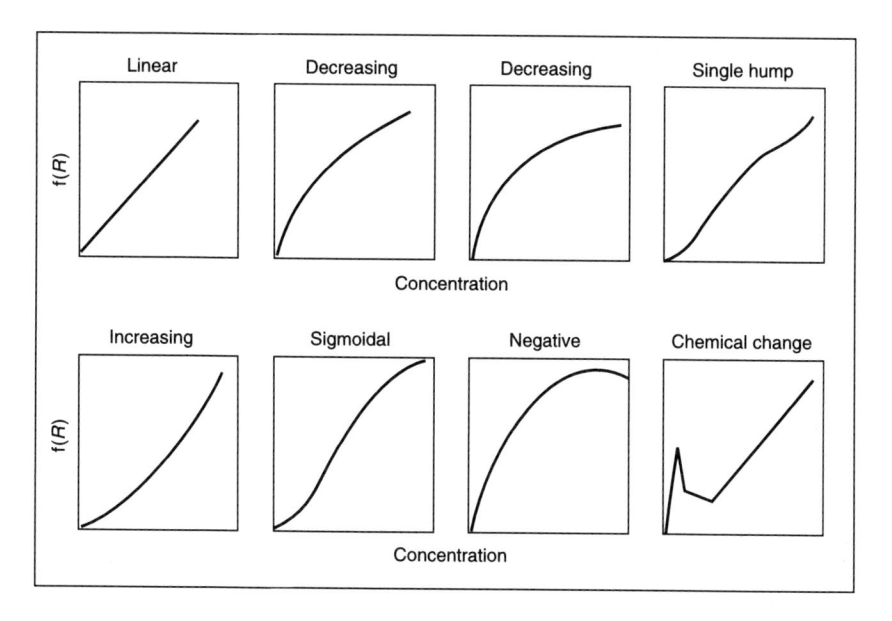

Figure 5.13 Relationships between f(R) and dye concentration

Some colorants show different uptake forms at different wavelengths. Hoffenberg's examples include a dye that shows linear uptake at the wavelength of maximum absorption and decreasing uptake at other less strongly absorbing wavelengths, and another that shows decreasing uptake at the wavelength of maximum absorption but continues to build linearly at other wavelengths. Yet another shows increasing uptake at all wavelengths. Some dyes show increasing uptake at the wavelength of maximum absorption and sigmoid uptake at other wavelengths, some show single-hump uptake, while others show negative uptake at high concentrations (usually associated with bronzing on the sample). The unexpected changes in uptake due to chemical effects are exemplified by a quinacridone violet in dope-dyeing when, at some concentrations, the pigment dissolves in the fibre polymer and changes chemical species. The sample therefore changes colour from violet to neutral grey at the lower concentrations.

Hoffenberg examined various nonlinear equations, including second- and third-order polynomial equations, exponential equations, and the Langmuir-type equation discussed in section 5.6.2. Even where these equations define dye uptake accurately, it is difficult to decide which equation-type is most applicable to a given colorant, or across wavelengths within a single colorant. Hoffenberg argued that the relationship between f(R) and dye concentration is too complex for predetermined nonlinear equations to be used for recipe formulation. Universal application of the approach using nonlinear equations would require highly complex software and, on the part of the user, substantial mathematical insight in order to choose the correct equations for every case.

Hoffenberg described a piecewise linear approximation to relate f(R) to concentration: the perceived smooth curve of whatever shape is approximated by a series of straight-line segments in which each segment is described by Eqn 5.71:

$$f(R) = m_j c + b_j \tag{5.71}$$

where m_j and b_j are the slopes and intercepts respectively of the f(R) curve for the jth line segment on which the concentration c falls. The colorant is treated as a series of 'mini-colorants', each of which builds linearly. When a substrate is coloured with three components using line segments j, k and l the total f(R) would be given by Eqn 5.72:

$$f(R)_{total} = m_{1,j} c_1 + m_{2,k} c_2 + m_{3,l} c_3 + b_{1,j} + b_{2,k} + b_{3,l} + f(R_s) \tag{5.72}$$

For convergence in recipe formulation the same approach as that described by Allen in section 5.4 is used. It is assumed that the difference in reflectance between the reflectance curve of the target and that of the substrate (or current mathematical match) can be approximated to by Eqn 5.28. The only difference between the piecewise linear approach and Allen's approach is the presence of the nonzero intercepts in Eqn 5.72. The final matrix for the first approximation set of concentrations is similar to Eqn 5.37 but includes the nonzero intercepts. To illustrate the differences from the Allen approach the matrix for the piecewise approach is shown in Eqn 5.73:

$$\begin{vmatrix} \Sigma(dE\,\bar{x}\,m_{1,j}) & \Sigma(dE\,\bar{x}\,m_{2,k}) & \Sigma(dE\,\bar{x}\,m_{3,l}) \\ \Sigma(dE\,\bar{y}\,m_{1,j}) & \Sigma(dE\,\bar{y}\,m_{2,k}) & \Sigma(dE\,\bar{y}\,m_{3,l}) \\ \Sigma(dE\,\bar{z}\,m_{1,j}) & \Sigma(dE\,\bar{z}\,m_{2,k}) & \Sigma(dE\,\bar{z}\,m_{3,l}) \end{vmatrix} \times \begin{vmatrix} c_1 \\ c_2 \\ c_3 \end{vmatrix}$$

$$= \begin{vmatrix} \Sigma(dE\,\bar{x}\,[f(R)_t - f(R)_p - b_{1,j} - b_{2,k} - b_{3,l}]) \\ \Sigma(dE\,\bar{y}\,[f(R)_t - f(R)_p - b_{1,j} - b_{2,k} - b_{3,l}]) \\ \Sigma(dE\,\bar{z}\,[f(R)_t - f(R)_p - b_{1,j} - b_{2,k} - b_{3,l}]) \end{vmatrix} \tag{5.73}$$

where $f(R)t$ is the target and $f(R)_p$ is the substrate of the current prediction in the recipe formulation iteration process. By analogy with Eqn 5.38, the concentrations are determined by inverting the coefficient matrix and multiplying it into the error vector.

In recipe formulation using the piecewise linear approach, an initial set of mini-colorants (generally the middle mini-colorant for each complete colorant) is chosen; if at any time a first approximation (or subsequent iteration) moves a concentration outside the domain of its mini-colorant, the algorithm starts all over again with a new set of mini-colorants. Thus the final match formulation contains only colorant concentrations predicted from the finally selected mini-colorants closest to the final recipe.

The piecewise linear approximation will fit exactly the dye ranges of the single dyes and has been used extensively in commercial systems for many years, but it does not compensate for the interaction between dyes in dye recipes. Hoffenberg reports a 'search and correct' technique to deal with this problem. A palette of known recipes is searched for one that is close in colour to the target, and this colour is corrected in the computer to provide an initial formulation.

In contrast to the piecewise linear equations described by Hoffenberg for describing the varying relationships between reflectance function and dye concentration, Jeler and Golob used one of five two-constant formulae for fitting the relationship between the reflectance function and dye concentration for each individual wavelength [24]:

(a) $f(R) = x/(ax + b)$
(b) $f(R) = ax \exp(xb)$
(c) $f(R) = ax + bx^2$
(d) $f(R) = a \ln(x + 1) + bx$
(e) $f(R) = ax + b$.

For each individual wavelength the equation from this list which best fits the relationship between function of reflectance and dye concentration is chosen for use in the recipe prediction program.

5.6.4 More complex physical chemistry solutions

Unfortunately, even if we can derive absolutely perfect absorption coefficients (whether by the use of two or more constants per wavelength or by interpolation) to describe the relationship between reflectance and the amount of dye added to the dyebath for single dyes, we still do not always get perfect predictions when these coefficients are applied to mixtures of dyes used in normal colorant formulations.

For this reason systems have been developed in which the concept of two-stage recipe prediction has been extended to include a full physical chemistry treatment of the relationship between dye-required-in-fibre and dye-required-in-dyebath. In this approach the amount of dye taken up by the fibre is deduced from equations incorporating a dye affinity term. Manual determination of the dye affinity is tedious and complicated, but the task is easy if properly designed computer programs are used.

In the case of ionic dyes Weedall *et al.* examined three alternative dye-uptake equations [13,18,25]. These were :
(a) Donnan diffuse adsorption model
(b) Langmuir-type adsorption model
(c) two-constant empirical model (section 5.6.2).

In particular Weedall investigated vat dyes, and found that the Donnan model gave the best fit to experimental data. In the case of single vat dyes the simple empirical two-constant dye uptake model developed for disperse dyes (Eqn 5.61) gave equally good predictions as the much more complicated Donnan model. However, when vat dyes were applied under dyeing conditions different from those used for the original calibration dyeings, the full Donnan model performed much more consistently. Table 5.8 shows the colour differences obtained between a set of ternary vat dye mixtures, applied first under standard conditions and then under completely different conditions of liquor ratio, salt and temperature. Set B predicted using the simple two-constant empirical model gave relatively large colour differences, whereas set A predicted using the Donnan adsorption model gave relatively small differences when the dyeing conditions were changed.

Table 5.8 Assessment of vat dyeings completed under different conditions (colour difference in ΔE (ANLAB(40) units)

Dyeing condition	Set A Donnan model	Set B Empirical model
1	2.77	7.06
2	2.22	4.69
3	1.87	2.76
4	0.90	2.88
5	0.56	3.98
6	2.39	3.22
7	4.30	6.46
8	3.09	5.86
9	7.11	11.65
10	3.69	4.89
11	1.98	3.23
12	1.69	4.67
Mean	2.71	5.11

The results of applying this physical chemistry approach to vat dye recipe prediction and the simpler two-constant empirical equations to disperse recipe prediction have been reported [17,25]. Prediction systems using the physical chemistry approach have also been developed for other dye classes.

5.7 RECIPE PREDICTION USING FLUORESCENT DYES

The measurement of fluorescent colours poses special problems, since light is absorbed by the fluorescent dye or pigment at one wavelength and then re-emitted in part as fluorescence at a longer wavelength.

There are two methods by which we can measure reflectance on a spectrophotometer. In the monochromatic illumination method, light from the source enters the monochromator and the monochromatic light emerging at a given wavelength irradiates the sample, is reflected by it, and is detected by the photodetector of the instrument. In the polychromatic illumination method, light from the source irradiates the sample and the reflected light then passes through the monochromator to be detected by the photodetector.

With nonfluorescent samples, the reflectance curves obtained by the two methods are identical. When a fluorescent sample is measured on a monochromatic illumination system, however, an incorrect reflectance curve is obtained that does not represent the true colour of the sample.

Figure 5.14 shows the reflectance curves obtained when a fluorescent red dye was measured under polychromatic and under monochromatic illumination conditions.

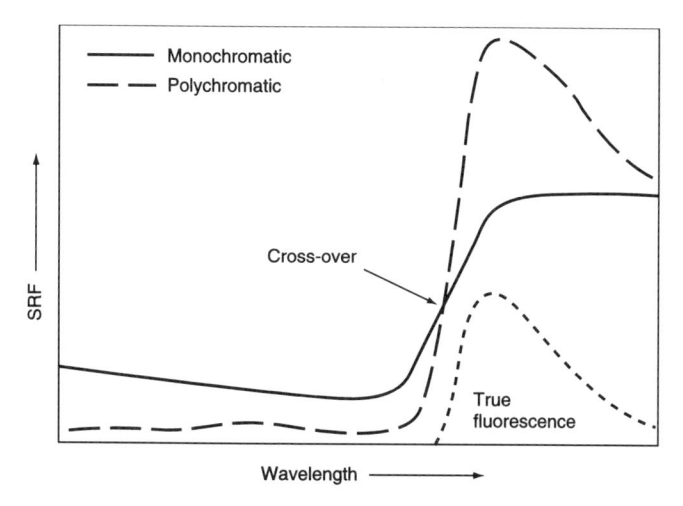

Figure 5.14 Reflectance curves of a fluorescent red dye under polychromatic and monochromatic illumination

Consider first the monochromatic illumination technique. If the sample is irradiated with monochromatic light (at 500 nm, say) which excites fluorescence, then some of this light will be reflected back to be detected by the photodetector, but some will be converted into fluorescent radiation and emitted at longer wavelengths. The instrument records both the true reflectance at 500 nm and the fluorescent emission at the longer wavelengths, and so gives an incorrectly high reading at 500 nm. On the other hand, at a longer wavelength (e.g. at 650 nm) where fluorescent emission occurs, if the sample is irradiated with monochromatic light of 650 nm (which does not excite fluorescence) then no fluorescence will be generated and the reflectance measured at 650 nm will be incorrectly too low.

Consider now the polychromatic illumination technique. Since the irradiating light contains all wavelengths, including those which excite fluorescence, the expected fluorescence will be excited. If such an instrument is set at 500 nm, then only light of that wavelength will be allowed to pass through the monochromator to be detected by the photodetector. Since no fluorescence is emitted at 500 nm, only the true reflectance of the material is detected. When the instrument is set at 650 nm, however, both reflected light and fluorescent emission are detected and the sum of these is measured.

In Figure 5.14 we can combine the polychromatic curve from the lower wavelengths of the spectrum (below the cross-over) with the monochromatic curve from the higher wavelengths (above the cross-over) to obtain a true reflectance curve over the whole spectrum. The latter is then subtracted from the polychromatic curve to obtain the true fluorescence emission curve [26]. This combination of polychromatic and monochromatic illumination is known as the two-mode method of spectrophotometry.

Since the curve obtained with polychromatic illumination is made up of a reflectance and a fluorescent emission component, it is inaccurate to call the curve of a fluorescent sample measured on a spectrophotometer a reflectance curve. Instead such a curve is called the *spectral radiance factor* (SRF) and is given the symbol β (Eqn 5.74):

$$\beta_t = \beta_r + \beta_f \qquad (5.74)$$

where β_r = reflectance and β_f = fluorescence [27].

The results from this technique are subject to slight errors at the cross-over point of the two curves, because here there can be both absorption of light and re-emission of fluorescence at the same wavelength. Interpolation has been used to correct the curves at this point [28].

Since fluorescent emission is dependent on the spectral energy distribution of the illuminant, a light source similar to daylight, such as a xenon lamp, should be used in the spectrophotometer. The coating on the integrating sphere affects the colour of the light and therefore the amount of fluorescence generated. Finally the sample itself alters the system by emitting a little energy at longer wavelengths back into the sphere which increases the energy in the reference beam of the instrument, and thus decreases slightly the SRF measured at longer wavelengths. Sample apertures on integrating spheres should therefore be as small as possible.

Donaldson introduced a two-monochromator method for investigating fluorescence emission [29]. In this technique a second monochromator fitted with a high-intensity light source is used to illuminate the integrating sphere of the main spectrophotometer. The sample can therefore be illuminated by monochromatic light of a given wavelength and the spectrum of the fluorescence generated can be detected directly on the main spectrophotometer. The shape of the fluorescence emission curve and the amount of fluorescent emission generated can be investigated in relation to the wavelength of excitation.

In one approach to colorant formulation the practical solution to the influence of the light source is to use a source in the spectrophotometer which has a spectral distribution similar to daylight, such as the xenon lamp, and to endeavour to produce colour matches in this illumination [21]. Effects of other light sources which may lead to mismatches when the same samples are viewed under them are ignored.

Figure 5.15 shows the SRF curves of a range of concentrations of the fluorescent disperse dye CI Disperse Yellow 82. At the blue end of the spectrum, where no fluorescence is emitted, the reflectance of each dyeing decreases as the dye concentration increases, in the expected fashion. At the yellow end of the spectrum, however, where

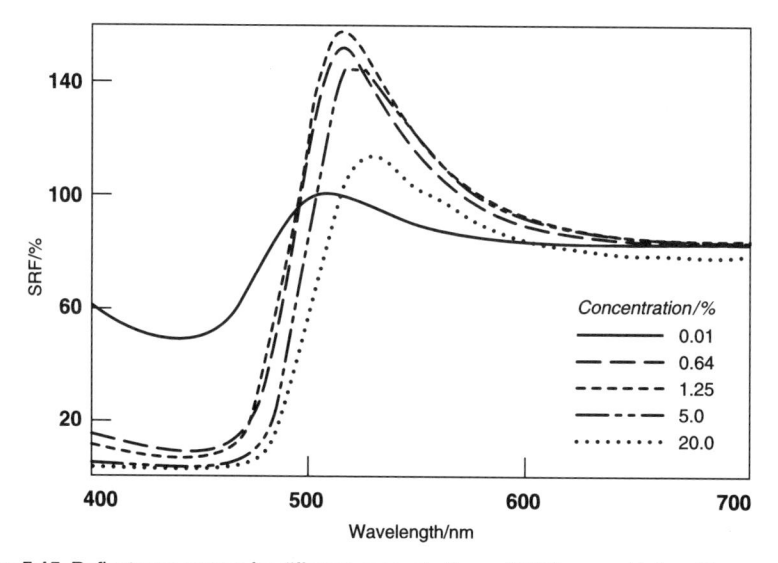

Figure 5.15 Reflectance curves for different concentrations of CI Disperse Yellow 82

fluorescent radiation is emitted, the SRF first rises to a maximum as the dye concentration is increased and then starts to fall again as the dye concentration is further increased.

At the blue end of the spectrum the dye behaves conventionally, so the SRF can be converted to f(R) values and conventional prediction theory will deal with the dye reflectance behaviour. At the fluorescing end of the spectrum, however, the behaviour of the dye is not at all conventional. There are two problems:
(a) the SRF increases and then decreases as the dye concentration is increased
(b) the SRF rises above 100%, so the Kubelka–Munk function cannot be used.

The SRF of a fluorescent dye can be treated as being made up of two components, a true-reflectance component β_r and a fluorescent component β_f (Eqn 5.74). The fluorescent component can be separated from the SRF by subtracting the true reflectance curve β_r as already explained, using both polychromatic and monochromatic illumination conditions if the spectrophotometer in use will permit. Alternatively the fluorescence-generating incident radiation can be removed to obtain β_r by placing cutoff filters either in front of the light source or in front of the samples at the measuring port of the spectrophotometer [21,30–32].

The simplest example illustrating the last-named techniques is with a fluorescent white. If we place a UV cutoff filter in front of the light source the blue-violet fluorescence of the fluorescent brightener is eliminated and we are left with the pale yellowish-white of the substrate.

In the case of fluorescent colours the problem is more complicated because the fluorescence is generated by exciting radiation of wavelengths which extend from the UV well across the lower half of the visible spectrum. In order to obtain the nonfluorescent true reflectance curve, therefore, we must use a range of cutoff filters to eliminate the fluorescence completely. Typical filters that have been used for this operation are given in Table 5.9. The transmission curves of these filters are shown in Figure 5.16.

In spectrophotometers permitting the insertion of the filter into the light path the process is as follows. The SRF of the fluorescent dye is first measured under direct xenon illumination using no filters. To remove the fluorescence-exciting wavelengths, the cutoff filters are then inserted in sequence into the light beam and the sample measured from 700 nm down to the wavelength at which the filter transmission has fallen to 50%, as indicated in Table 5.9. A series of curves is obtained (Figure 5.17). The difference between the curves for unfiltered and for filtered light gives a measure of fluorescence β_f at each wavelength. This technique has been used on the Beckman DK2A and Zeiss RFC3 spectrophotometers [21,30,31].

If the light source of the spectrophotometer in use is inside the integrating sphere, the cutoff filters cannot be positioned in front of the light source and so a different

Table 5.9 Wavelength range of various filters

Filter	Wavelength range of >50% transmittance/nm
Chance OY10	390–700
Schott FG10	420–700
Chance OY18	440–700
Chance OY8	470–700
Chance OY6	500–700
Schott GG495	510–700
Chance OY4	520–700
Chance OY3	530–700
Schott OG530	540–700
Schott OG550	560–700
Schott OY570	590–700
Chance OY1	600–700

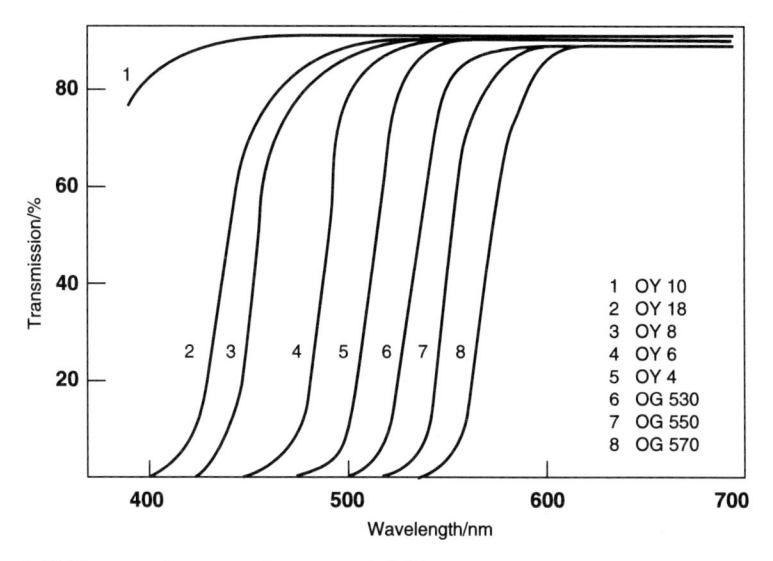

Figure 5.16 Transmission curves for some cutoff filters

technique must be used. In these cases the cutoff filters are placed directly over the sample to be measured. Light passes through the filter and the fluorescence-exciting radiation is removed. The light is then reflected by the sample and passes back through the cutoff filters a second time on its way into the integrating sphere to be measured. Thus both the reflecting and transmitting properties of the cutoff filters affect the reflectance recorded by the spectrophotometer.

The cutoff filters are calibrated by producing a five- to ten-step grey scale on the

254 RECIPE PREDICTION FOR TEXTILES

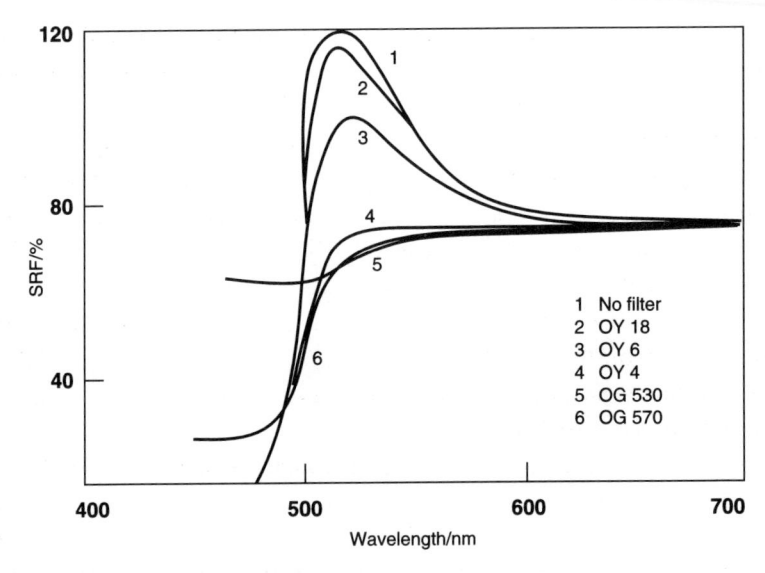

Figure 5.17 Reflectance curves for CI Disperse Yellow 82 (concentration 0.64%) measured in a spectrophotometer (Beckmann DK2A) with and without the presence of various filters

material involved using a nonfluorescent grey dye. Each of the samples is measured with and without cutoff filters. At each wavelength a linear relationship is found between with- and without-filter reflectance (R_f and R_w respectively) of the form shown in Eqn 5.75:

$$R_w = aR_f + K \qquad (5.75)$$

The value of the intercept K is probably attributable to surface reflection from the filter and compression of the textile samples behind each glass filter; both effects would add a constant reflectance to the without-filter reflectance values. The error introduced by the placing of the filters is less than 0.6% absolute reflectance.

Each with-filter measurement, within the working wavelength range of the filter, is converted to its without-filter measurement by using the factors obtained during the filter calibration. The lowest value at each wavelength is the nonfluorescent reflectance β_r. In practice a computer program is used to analyse the sets of reflectance curves to produce the two curves of interest, SRF and β_r. The computer also determines the difference between the curves to provide the measure of fluorescence β_f.

The model used for predicting the true reflectance β_r is either the two-stage approach, using Eqns 5.59–5.63 (in which the reflectance function is related linearly to dye-in-fibre concentration, followed by use of a two-constant equation to convert dye-in-fibre into dye-in-dyebath concentrations), or the simpler two-constant Eqn 5.66.

These cope with the nonlinearity found when the reflectance function f(β_r) is plotted against dye-in-dyebath concentration. In this respect the true reflectance of the dye behaves as it would with any normal nonfluorescent dye.

When the β_f values of a fluorescent component at a given wavelength are plotted against the dye-in-fibre concentration, curves similar to that shown in Figure 5.18 for CI Disperse Yellow 82 are obtained. Clearly, the fluorescent component is extremely large at certain wavelengths (up to 80% in a SRF of 150% at a wavelength of 520 nm); it first increases as the concentration increases and then decreases as the concentration is further increased. Therefore it is the fluorescence that accounts for the peculiar changes in SRF of the fluorescent dye with concentration as shown in Figure 5.15, and not the true reflectance, which behaves as it would with any normal dye.

The problem is how to define the changes in β_f with increasing dye concentration. When a dye molecule capable of producing fluorescence is subjected to exciting radiation of a frequency corresponding to a permitted energy change within the molecule, the molecule absorbs energy and is raised to a higher energy level. The molecule in its excited state is unstable and rapidly tends to revert back to a lower energy level. In doing so the absorbed energy is dissipated either in nonradiative processes or by fluorescence, according to Stokes' law. If, when in the excited state, the dye molecule collides with another molecule, then the energy is dissipated in nonradiative processes, and fluorescence in the visible region is suppressed. This is known as *quenching* of fluorescence and can take place when other dyes are present in the mixture or at high concentrations of the fluorescent dye alone, when it is known as *self-quenching* of fluorescence.

It has been found first by empirical methods [30], and later by study of the mechanism of generation of fluorescence [21,31], that the fluorescent component β_f can be defined for a single dye by Eqn 5.76:

$$\beta_f = \frac{a_{1,\lambda} c_1}{(1 + b_{1,\lambda} c_1)(1 + b_{11,\lambda} c_1)} \quad (5.76)$$

where $a_{1,\lambda}$ and $b_{1,\lambda}$ are constants defining the generation of fluorescence in dye 1 and $b_{11,\lambda}$ is the self-quenching coefficient of dye 1 at wavelength λ. All these constants can be derived by the method of least squares from the β_f values obtained from a range of dyeings using the fluorescent dye at various concentrations. Usually twelve concentrations are required for this operation to ensure coverage of the increase and decrease in fluorescence. Solving Eqn 5.76 gives two constants, one of which is always much larger than the other: the larger value is allocated to $b_{1,\lambda}$ and the smaller to $b_{11,\lambda}$.

A demonstration of the accuracy of these equations for a single dye is shown in Figure 5.18 for the fluorescent dye CI Disperse Yellow 82. It can be seen that the equation defines the increase and decrease of fluorescence with increase in dye concentration very satisfactorily. In Figure 5.19 the actual and predicted SRF values are given, where the true reflectance has been fitted by the two-constant equation developed earlier for nonfluorescent dyes (Eqns 5.59–5.61).

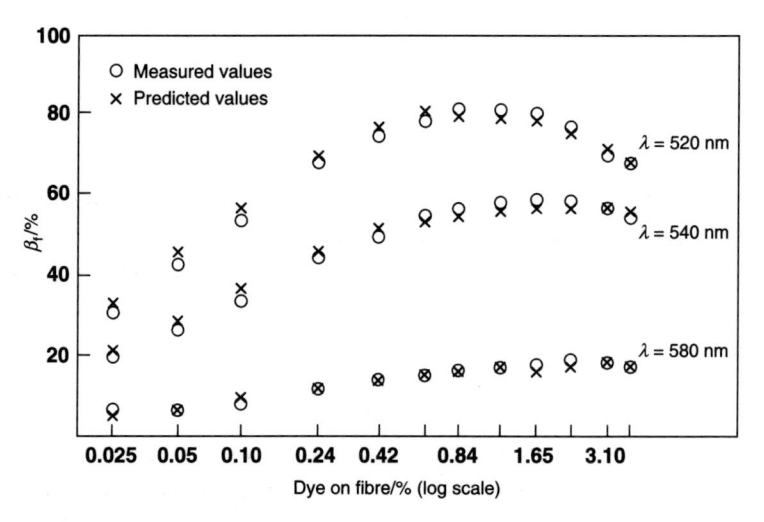

Figure 5.18 Measured and predicted β_f values plotted against concentration of CI Disperse Yellow 82 at different wavelengths

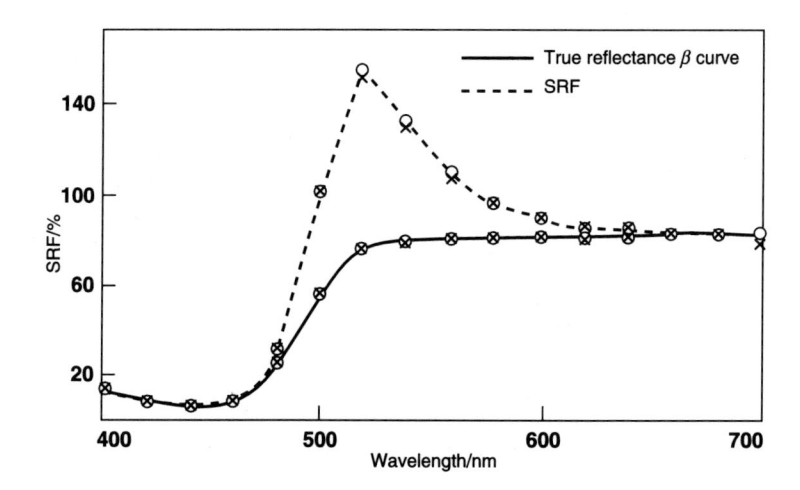

Figure 5.19 Measured and predicted SRF values plotted against wavelength for CI Disperse Yellow 82 (0.5% dye on fibre) (for key see Figure 5.18)

Eqn 5.76 has been extended to deal with up to three fluorescent dyes in admixture [21,31]. In the case of mixtures of fluorescent and nonfluorescent dyes quenching of the fluorescence of dye 1 by the other dyes present will occur as illustrated in Eqn 5.77:

$$\beta_f = \frac{a_{1,\lambda} c_1}{(1 + b_{1,\lambda} c_1)(1 + b_{11,\lambda} c_1 + b_{21,\lambda} c_2 + b_{31,\lambda} c_3)} \qquad (5.77)$$

where $b_{21,\lambda}$ = quenching effect of nonfluorescent dye 2 on dye 1
$b_{31,\lambda}$ = quenching effect of nonfluorescent dye 3 on dye 1.

The fluorescence quenching effects of other dyes are determined by dyeing binary mixtures of the fluorescent dye with each of the other dyes individually in turn. To determine the effect of nonfluorescent dye 2 on fluorescent dye 1 at wavelength λ we write the two-dye mixture equation (Eqn 5.78):

$$\beta_f = \frac{a_{1,\lambda} c_1}{(1 + b_{1,\lambda} c_1)(1 + b_{11,\lambda} c_1 + b_{21,\lambda} c_2)} \qquad (5.78)$$

Eqn 5.78 is solved for $b_{21,\lambda}$ by the least-squares method, and iterative procedures can be used when the equation is extended for mixtures of two fluorescent dyes [21,31]. The accuracy of the fit of these equations in describing the fluorescent behaviour of binary mixtures is shown in Figure 5.20 and extended for two fluorescent dyes in Figure 5.21.

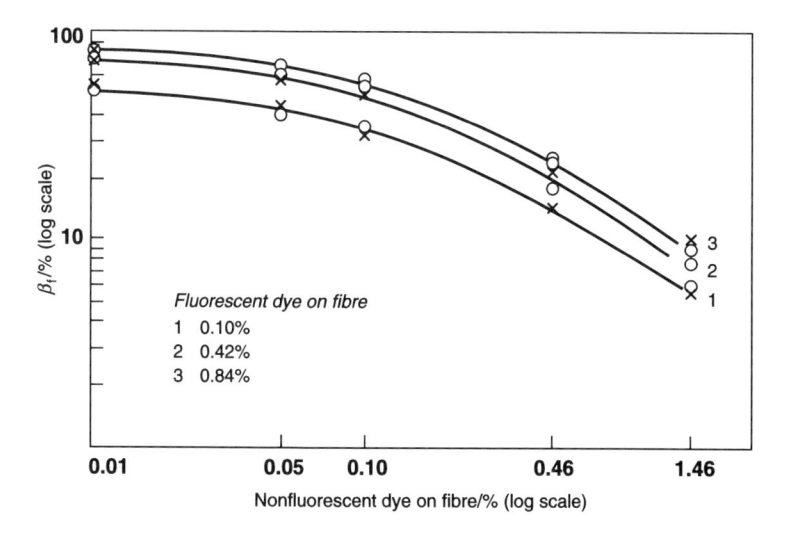

Figure 5.20 Quenching effect of a nonfluorescent dye (CI Disperse Blue 176) on a fluorescent dye (CI Disperse Yellow 82) at a wavelength of 520 nm (for key see Figure 5.18)

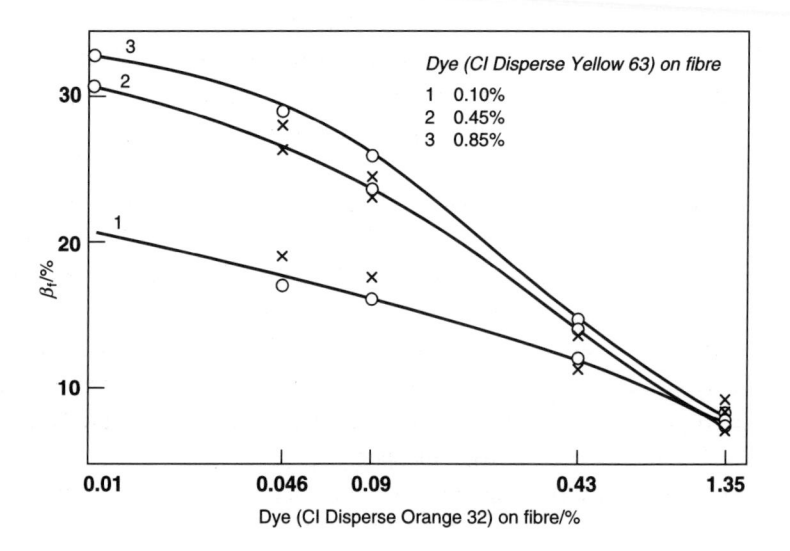

Figure 5.21 Quenching effect of a fluorescent dye on another fluorescent dye (CI Disperse Orange 32 and Yellow 63) at a wavelength of 560 nm (for key see Figure 5.18)

In order to carry out recipe prediction with fluorescent dyes we must derive a correction matrix that will enable the computer to minimise δX, δY and δZ from the target. From Eqn 5.74 we know that $\beta_t = \beta_r + \beta_f$. Therefore for dye i we can write (Eqn 5.79):

$$\begin{aligned}\frac{\partial X}{\partial c_i} &= \sum \left(\frac{\partial \beta_t}{\partial c_i}\right) E_\lambda \bar{x}_\lambda \\ &= \sum \left(\frac{\partial \beta_r}{\partial c_i} + \frac{\partial \beta_f}{\partial c_i}\right) E_\lambda \bar{x}_\lambda \\ &= \sum \left(\frac{\partial [f(\beta_r)]}{\partial c_i} \times \frac{\partial \beta_r}{\partial [f(\beta_r)]} + \frac{\partial \beta_f}{\partial c_i}\right) E_\lambda \bar{x}_\lambda\end{aligned} \quad (5.79)$$

and similarly for Y and Z, and for the other dyes in the mixture. Eqn 5.66 defines $f(R_\lambda)$ for the nonfluorescent component. Therefore Eqn 5.80 can be written:

$$\frac{\partial [f(\beta_r)]}{\partial c_i} = \frac{g_{i,\lambda}}{(1+k_{i,\lambda} c_i)^2} \quad (5.80)$$

and $\partial \beta_r / \partial [f(\beta_r)]$ is calculated from the Kubelka–Munk function of the predicted true reflectance.

We still need to obtain $\partial \beta_f / \partial c_i$. Because of the complexity of differentiating the fluorescence equations, this is done numerically by calculating β_f from the recipe concentrations and then making a small addition (0.0001%) to the recipe concentration of each dye in turn, calculating the effect on the change in β_f, and then $\partial\beta_f/\partial c_i$ for each dye directly. McKay, using disperse dyes on spun polyester yarn, demonstrated that this approach to recipe prediction is equivalent in accuracy to prediction with nonfluorescent dyes [21].

Another approach to the problem of predicting fluorescent dye mixtures has been developed by Funk *et al.* [27,33], who used the two-mode method of spectrophotometry to separate out β_r and β_f. Funk additionally investigated the mechanism of fluorescent emission using a Diano Matchscan spectrophotometer which normally used polychromatic illumination in the integrating sphere. This instrument was modified by the addition of a second monochromator so that the polychromatic light source could be replaced by a second grating monochromator and a xenon lamp, to produce monochromatic illumination in the integrating sphere.

To investigate the relationship between excitation energy and fluorescent emission, the second monochromator is set up at a given excitation wavelength and a scan is made of the sample by the main spectrophotometer. Measurements are made in single-beam mode, and the fluorescent curve measured must therefore be factorised at each emission wavelength by the response of the instrument (Eqn 5.81):

$$F_\lambda = \frac{F'_\lambda}{r_\lambda} \qquad (5.81)$$

where F_λ = corrected emission
F'_λ = uncorrected emission
r_λ = response of the instrument.

Measurements made by this technique are shown in Figure 5.22. The response is obtained by measuring a white calibration standard at fixed slit width in single-beam mode. At each wavelength the value obtained is the product of:
(a) the energy of the light source
(b) the light-transmitting power of the monochromator at the particular slit width
(c) the electrical response of the photodetector system to the amount of light detected
(d) the reflectance of the white standard.

If we divide this value by the known reflectance of the white standard and by the relative energy of the light source, we obtain the response of the instrument to light of a given wavelength. In double-beam mode the effect of the instrument response is cancelled because all measurements are made relative to an almost simultaneous measurement of the white standard at each wavelength.

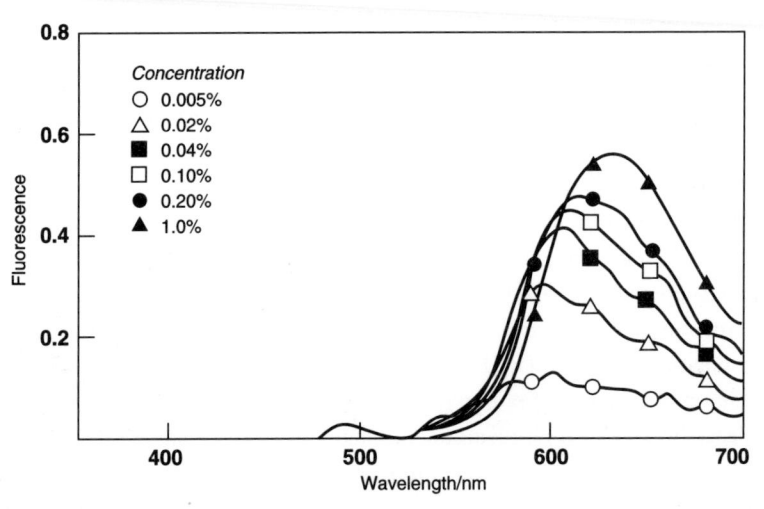

Figure 5.22 Fluorescence emission of a red dye uncorrected for absorption by the substrate and the dye

The assumption is made that, at the molecular level, the distribution of fluorescent emission has a constant shape. That is, that if a molecule of a fluorescent dye is irradiated by light of varying wavelength (varying energy) or intensity (varying energy), the amount of fluorescence produced will vary but the relative shape of the emission curve will remain constant. It is also assumed that the intensity of fluorescence is directly proportional to the quantity of excitation energy absorbed by the fluorescing molecule. These assumptions have been confirmed experimentally. In practice the absorption of the substrate and dyes present affects the amount of energy available for the excitation of fluorescence and the amount of emission that escapes from the sample.

The curves shown in Figure 5.22 are not constant in shape as predicted by the theory. This is because both the substrate and the fluorescent dye itself are absorbing some of the emitted fluorescence. The measured values are therefore divided by the true reflectance of the dye (which includes the substrate absorbance) at each wavelength, to get the true fluorescence as generated at the molecular level, $\beta_{f,mol}$ (Eqn 5.82):

$$\beta_{f,mol} = \frac{\beta_{f,\lambda}}{\beta_{r,\lambda}} \tag{5.82}$$

This gives emission curves of constant shape, as predicted by the theory (Figure 5.23).

To define the shape of emission curves in a manner independent of dye concentration, fluorescent factors (FFAC) were determined by summing the fluorescent emission of each concentration over the entire emission band, and dividing the fluorescence at each wavelength by this total (Eqn 5.83):

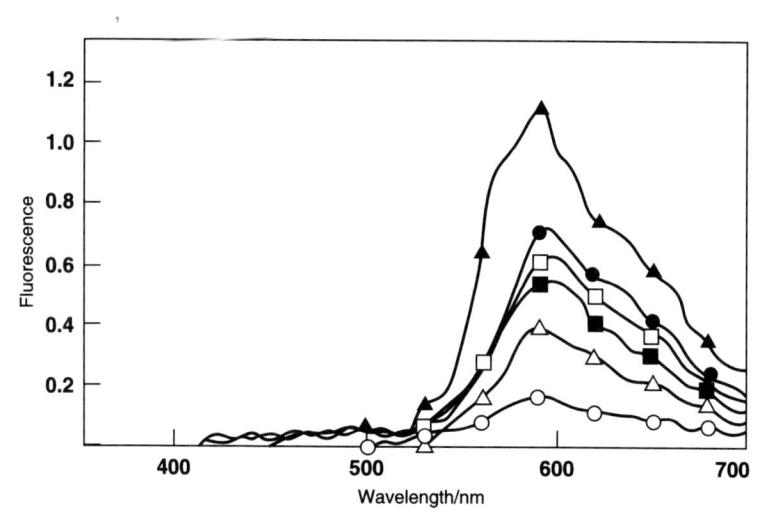

Figure 5.23 Fluorescence emission of a red dye corrected for absorption by the substrate and the dye (for key see Figure 5.22)

$$(\text{FFAC}) = \frac{\beta_{f,\text{mol}}}{F_{\text{total}}} \quad (5.83)$$

where $F_{\text{total}} = \Sigma \beta_{f,\text{mol},\lambda}$.

The FFAC values should be constant for each wavelength of emission regardless of concentration. In practice this holds reasonably well and the value at each wavelength, averaged over the concentration range, is used. Thus if the total fluorescence emission F_{total} of a given dye can be determined, the FFAC values enable calculation of the distribution of the emission curve (Eqn 5.84):

$$\beta_{f,\text{mol},\lambda} = (\text{FFAC}) \times F_{\text{total}} \quad (5.84)$$

The total fluorescent emission F_{total}, generated by light of a given excitation wavelength, can be related to the total amount of energy absorbed by the fluorescent dye. This depends on the spectral energy distribution of the irradiating light source and the absorption of the substrate and dyes in the mixture, including the fluorescent dye itself. Using the two-monochromator set-up, the energy passing through the equipment in single-beam mode is corrected for the response of the system according to Eqn 5.85:

$$E_\lambda = \frac{E'_\lambda}{r_\lambda} \quad (5.85)$$

where E_λ = corrected value of excitation
E'_λ = uncorrected value of excitation.

Correction for absorption by the substrate and dyes must also be made. For example, if the undyed substrate has a true reflectance β_{rs} of 80% at a given wavelength and the dyed substrate has a true reflectance β_r of 60%, then 20% of the available irradiating energy at that wavelength has been absorbed by the dye and substrate and is not available to excite fluorescence. Therefore at a given wavelength, for the given concentration of dye, the energy available to excite fluorescence becomes $E_\lambda(\beta_{rs,\lambda} - \beta_{r,\lambda})$.

We can establish a relationship between the excitation energy available at a given wavelength and the total fluorescent emission F_{total} produced by excitation by that wavelength (Eqn 5.86):

$$h_\lambda = \frac{F_{total}}{E_\lambda(\beta_{rs,\lambda} - \beta_{r,\lambda})} \qquad (5.86)$$

The term h_λ is the relative amount of fluorescence produced over the entire emission band when a unit amount of energy is absorbed at a given wavelength; it is unique for each dye and is independent of dye concentration. Graphs of relative fluorescent efficiency against irradiating wavelength can be prepared for each fluorescent dye. Figure 5.24 shows a typical graph, with the values normalised so that the highest value = 1.0. Hence Eqn 5.86 enables us to define the total fluorescent emission generated by the excitation energy absorbed at a given wavelength. The fluorescence generated by the incident wavelength is related to the dye concentration, since the total excitation absorbed depends on the true reflectance of the dye at the excitation wavelength, which is in turn dependent on the concentration.

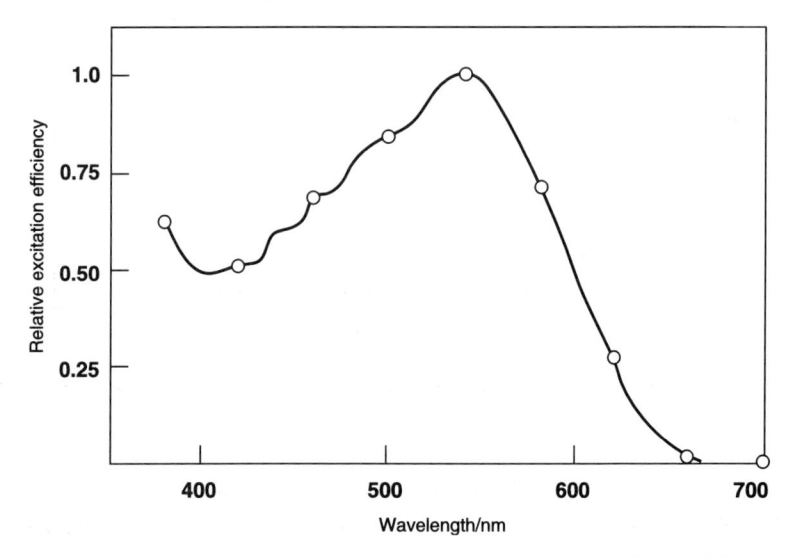

Figure 5.24 Relative excitation efficiency curve for the red dye shown in Figures 5.22 and 5.23 (concentration 0.2% dye on fibre)

The total excitation for a given concentration of dye generated by a given irradiating wavelength is $h_\lambda E_\lambda(\beta_{rs,\lambda} - \beta_{r,\lambda})$, which yields an equivalent F_{total} emission. Energy is absorbed over a range of wavelengths, however, each making a different contribution to the total fluorescence emitted. Summing Eqn 5.86 over all wavelengths of excitation gives the total excitation energy, weighted by the relative efficiency at generating fluorescence of each excitation wavelength (Eqn 5.87):

$$E_{total} = \sum h_\lambda E_\lambda(\beta_{rs,\lambda} - \beta_{r,\lambda}) \tag{5.87}$$

Since the h_λ curve does not change with concentration, plots of E_{total} against F_{total} should be linear: this has been confirmed experimentally (Figure 5.25). This type of relationship also holds for different light sources. If we derive the equation for the straight line fitted through the points we have a relationship between total excitation (i.e. dye concentration) and total fluorescence (Eqn 5.88):

$$F_{total} = m\, E_{total} \tag{5.88}$$

where m is the gradient of the fitted line.

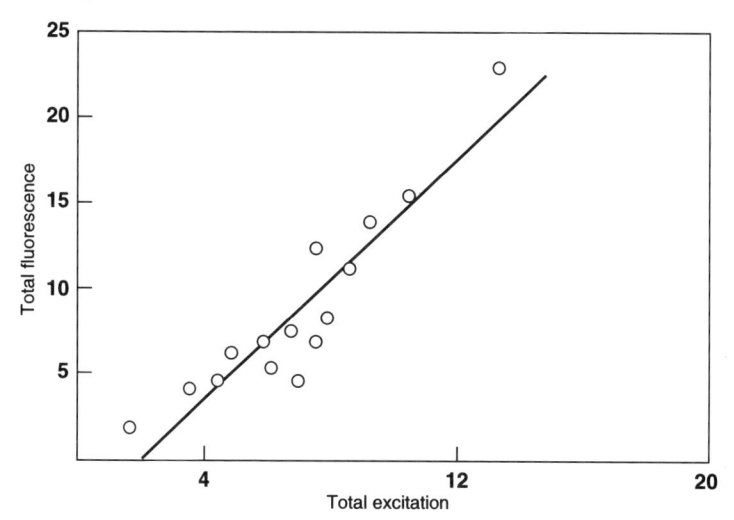

Figure 5.25 Total excitation versus total fluorescence for the red dye shown in Figures 5.22 and 5.23

The steps in recipe prediction using the above theory are then as follows:
1. Predict the true reflectance curve β_r of the mixture of dyes using standard nonfluorescent prediction theory.
2. Predict the total energy available for excitation, E_{total}, using Eqn 5.87.

3. Use Eqn 5.88 to calculate the total energy emitted, F_{total}.
4. Use the FFAC values in Eqn 5.84 to calculate the fluorescent emission distribution $\beta_{f,mol}$.
5. Correct the emission curve $\beta_{f,mol}$ at each wavelength for absorption by other dyes in the mixture by multiplying determined emission by the true reflectance of the mixture of dyes at these wavelengths to produce β_f (Eqn 5.82).
6. Add this resultant emission to the true reflectance curve to obtain the spectral radiance factor for the mixture (Eqn 5.74).

For higher accuracy in prediction, Eqn 5.87 is be modified by the addition of a factor f_λ to give Eqn 5.89:

$$E_{total} = \sum h_\lambda E_\lambda (\beta_{rs,\lambda} - \beta_{r,\lambda}) f_\lambda \qquad (5.89)$$

where f is that fraction of the total light absorbed by the mixture which is represented by the fluorescent dye (Eqn 5.90):

$$f_\lambda = \frac{a_{1,\lambda} c_1}{a_{1,\lambda} c_1 + a_{2,\lambda} c_2 + a_{3,\lambda} c_3} \qquad (5.90)$$

a_1, a_2, and a_3 being the nonfluorescent absorption coefficients used in single-constant prediction theory in Eqn 5.6.

The two methods of dealing with the prediction of recipes containing fluorescent dyes described above require separation of the fluorescence and the true reflectance from the SRF. This is a difficult and tedious procedure, requiring special techniques and equipment. A third method has been developed by Man in an attempt to overcome these problems [34]. It defines fluorescence β_f in a single dye (Eqn 5.91):

$$\beta_f = Mc \; \exp(Lc^k) \qquad (5.91)$$

where M, L and k are constants for a particular dye at a particular wavelength, determined by computer iterative techniques.

Man divided the SRF curve into two parts. The first part included those wavelengths where the SRFs of the dyeings are always greater than the reflectance of the substrate. (It is assumed that the dye does not absorb any light at these wavelengths and that its true reflectance equals that of the substrate. Therefore the difference in reflectance between the SRF of the dyed material and the reflectance of the substrate is fluorescence.) The second part consisted of those wavelengths where the SRF of the dyed material first of all increases with increasing dye concentration and then decreases to a value which may be well below that of the substrate. At these wavelengths the dye is absorbing light as well as emitting fluorescence. In order to obtain the true reflectance curve β_r for these wavelengths, it is assumed that at the highest dye concentration no fluorescence occurs because of self-quenching. Therefore absorption coefficients are calculated from the reflectance curve of the heaviest dye concentration and these are used to predict the true reflectance curve for the dyeing, and by subtracting β_r from SRF the fluorescence β_f can be calculated. Thus β_f and β_r do not actually have to be measured.

For mixtures of a fluorescent dye with a nonfluorescent dye, the quenching effect is defined by Eqn 5.92:

$$\beta_f = p \times \frac{\beta'_f}{1+qc_n} \times c_f^{-0.2} \qquad (5.92)$$

where β'_f = fluorescence emitted by the fluorescent dye alone
p, q = constants
c_f = concentration of the fluorescent dye
c_n = concentration of the nonfluorescent quenching dye.

Shah et al. have also proposed that, at longer wavelengths where fluorescent emission is strong, the fluorescent emission can be defined as the difference between the total SRF and the substrate (Eqn 5.93) [53]:

$$\beta_f = \beta_t - \beta_s \qquad (5.93)$$

Shah et al. suggested that fluorescent dye can be characterised from polychromatic illumination spectrophotometric measurements only – that is, without the necessity for separate measurement of the fluorescent and nonfluorescent components of SRF. They assume that fluorescent emission begins at the wavelength where the SRF curve of the fluorescent dye first crosses above that of the substrate (Figure 5.15). They observed that, with several fluorescent dyes at low wavelengths, the SRF values of the dye range are all below the substrate and suggest that at these wavelengths the variation in SRF with concentration is due entirely to light absorption, some of which will excite fluorescence (the excitation domain). At longer wavelengths the SRF starts to rise above the substrate. There is a gradual increase in the wavelength at which cross-over takes place as the dye concentration increases (Figure 5.15). At some concentration level corresponding to high fluorescence this wavelength shift ceases and it is suggested that this is the wavelength above which there is no longer any light absorption but only emission of fluorescence (the emission domain). The wavelength range between the point where the dye curve first crosses the substrate curve and that where wavelength shift ceases is known as the overlap domain.

Shah et al. found that β_f can be defined by Eqn 5.94:

$$\beta_f = \frac{mc_f}{k} \times \exp\left(\frac{-wc_f(\ln \lambda - \ln \lambda_{max})^2}{2k^2}\right) \quad (5.94)$$

where c_f = concentration of fluorescent dye
   m, w, k = constants
   λ_{max} = wavelength of maximum fluorescence
   λ = wavelength for calculating β_f.

Eqn 5.94 is claimed to give good prediction of fluorescent dye SRF in the fluorescence emission domain and in the overlap domain for different dye concentrations.

Doring has attempted to bypass most of the difficulties associated with the measurement and prediction of fluorescent colour recipes by preparing mixtures of a single fluorescent pigment with up to three nonfluorescent pigments in an enamel paint system [35]. The database of single fluorescent pigment and the mixtures are prepared, and the CIE xyY values for each sample are obtained. The difference Δx, Δy and ΔY between the pure fluorescent dye and each mixture is assumed to vary linearly with change in pigment concentration over a restricted area of colour space and can be expressed in a matrix as Eqn 5.95:

$$\Delta x = \left(\frac{\partial x}{\partial c_1}\right)\Delta c_1 + \left(\frac{\partial x}{\partial c_2}\right)\Delta c_2 + \left(\frac{\partial x}{\partial c_3}\right)\Delta c_3$$

$$\Delta y = \left(\frac{\partial y}{\partial c_1}\right)\Delta c_1 + \left(\frac{\partial y}{\partial c_2}\right)\Delta c_2 + \left(\frac{\partial y}{\partial c_3}\right)\Delta c_3 \quad (5.95)$$

$$\Delta Y = \left(\frac{\partial Y}{\partial c_1}\right)\Delta c_1 + \left(\frac{\partial Y}{\partial c_2}\right)\Delta c_2 + \left(\frac{\partial Y}{\partial c_3}\right)\Delta c_3$$

Using the pure fluorescent pigment and a nearby colour-mixture sample, Eqn 5.95 can be solved to obtain the coefficients in brackets. Knowing the coefficients, Eqn 5.93 can then be used to calculate the concentrations of the three nonfluorescent dyes to match the target colour. It is claimed that only two or three corrections are required to produce a match within 0.8 CIELAB units of standard. It is suggested that the technique should be equally suitable for dye systems.

5.8 RECIPE CORRECTION

Recipes predicted by computer formulation may fail in actual dyeing to give a satisfactory match to the target for a variety of reasons, including:

(a) failure of the prediction theory to deal adequately with the nonlinear relationship between dye applied from the dyebath and reflectance
(b) interaction between dyes, leading to dye-uptake behaviour which is different from the uptake of the dyes when applied individually
(c) scaling-up problems, when recipes based on laboratory-scale dyeing ranges are applied in bulk machinery, and deliberate or accidental variations in the dyeing process compared to the calibration dyeings
(d) variations in substrate physical structure, compared with the calibration ranges
(e) mistakes in processing, such as incorrect dye weighing or wrong temperature.

The failure to match the target may be detected either in a check dyeing carried out in the laboratory or in the dye lot processed under bulk conditions. In either case it is necessary to have some method of correcting the defective recipe to bring the colour closer to target.

If the colour difference between dyeing and target is fairly small (up to about 3 CIELAB units), the correction matrix derived during the recipe-correction loop in the formulation program can be used very effectively. The difference between ΔX, ΔY and ΔZ values of the target and dyeing are obtained from colour measurement and entered into the correction matrix (Eqn 5.23) to obtain the corrections to the original recipe to improve the match, for example:

Actual dyeing: X = 30.50   Y = 27.60   Z = 16.50
Target: X = 30.70   Y = 27.55   Z = 16.90
Difference: ΔX = 0.20 ΔY = –0.05 ΔZ = 0.40

Hence we have (Eqn 5.96):

$$\Delta c_1 = 0.20 \left(\frac{\partial c_1}{\partial X}\right) - 0.05 \left(\frac{\partial c_1}{\partial Y}\right) + 0.40 \left(\frac{\partial c_1}{\delta Z}\right)$$

$$\Delta c_2 = 0.20 \left(\frac{\partial c_2}{\partial X}\right) - 0.05 \left(\frac{\partial c_2}{\partial Y}\right) + 0.40 \left(\frac{\partial c_2}{\partial Z}\right) \quad (5.96)$$

$$\Delta c_3 = 0.20 \left(\frac{\partial c_3}{\partial X}\right) - 0.05 \left(\frac{\partial c_3}{\partial Y}\right) + 0.40 \left(\frac{\partial c_3}{\partial Z}\right)$$

The corrected recipe is therefore:
new c_1 = original c_1 + Δc_1
new c_2 = original c_2 + Δc_2
new c_3 = original c_3 + Δc_3.

This technique has been used for correcting laboratory dyeings and bulk dye lots from the earliest days of computer colorant formulation. It was formerly the practice to output the correction matrix at the time of computer prediction, but it is now more usual to recalculate the matrix automatically from the original recipe at the time of correction.

If the colour difference between dyeing and target is large, the effectiveness of the correction matrix deteriorates and the correction may overshoot or undershoot the target (although still reducing the colour difference) and not produce an exact match. The reasons are the same as that applicable to the computer correction loop used during the original recipe calculation process: that $dR/d[f(R)]$ and $d[f(R)]/dc_i$ in the correction matrix are linear approximations to a nonlinear phenomenon and only apply reliably over fairly small changes in dye concentration or in reflectance values. For example, if the target concentration was, say, 2% and the test-dyeing dye concentration was only 1.5%, then the $d[f(R)]/dc_i$ value would be calculated at a concentration of 1.5%; if, as is usually the case, the relationship between $f(R)$ and c is nonlinear, this would not be the same as that applicable for the target concentration.

It is therefore now more common to measure the test dyeing and use the computer to calculate a recipe to match it. The concentrations computed for the test dyeings are then subtracted from the original concentrations computed for the target to give the dye corrections to be applied to the test-dyeing recipe. For small colour differences this is almost identical to applying the correction matrix as originally used for correcting off-shade dye lots. This method allows the computer to incorporate corrections for nonlinear dye-uptake effects and colour-space effects by using either interpolation, or multiconstant empirical or physicochemical equations, in the prediction program as already described. It has been reported by McDonald et al. as being particularly effective in calculating redyes (in fresh dyebaths) of off-shade dyeings, when the correction is based on dye-in-fibre concentrations in conjunction with physical chemistry equations [17]. In this case, the technique gives the amount of additional dye in the fibre needed to match the target dye-in-fibre. The physicochemical equations are then used to calculate the amount of each dye required in the dyebath to give the necessary level of dye on the fibre for the target and the sample. The difference between the two is the amount of dye needed for a fresh dyebath to allow for bleed-off of dye already on the substrate.

In a different approach to recipe correction, it is assumed that the dyeing is correct and the inaccuracy in prediction is due to errors in the absorption coefficients, for example because the substrate being dyed is not exactly the same as that on which the calibration dyeings were prepared [15]. If the computed concentrations for the original target recipe are t_1, t_2 and t_3 and those calculated for the test dyeing are d_1, d_2 and d_3, then correction factors can be calculated from the two sets of concentrations which represent the apparent inaccuracies in the calibration data of the original dye ranges.

For example, suppose that the target concentration for dye 1 is $t_1 = 2.5\%$ and the dye concentration actually taken up by the fibre is found to be equivalent to $d_1 = 2.0\%$. This implies that in the particular dyeing process or substrate concerned the extent of dye uptake is only 2/2.5 (= 80%) of that found in the original calibration ranges. To compensate for this difference in uptake, we should divide the original target concentration by 0.8 to give a corrected target concentration, which in practice would give a dye uptake equal to the desired 2.5%. Correction factors are therefore determined for each dye as follows (Eqn 5.97):

$$f_1 = \frac{t_1}{d_1} \qquad f_2 = \frac{t_2}{d_2} \qquad f_3 = \frac{t_3}{d_3} \tag{5.97}$$

By applying the correction factors to both target and dyeing concentrations t_i and d_i, we obtain new recipes with dye concentrations corrected for the observed variations in uptake, as follows (Eqn 5.98):

$$\begin{aligned} e_1 &= f_1 t_1 & e_2 &= f_2 t_2 & e_3 &= f_3 t_3 \\ g_1 &= f_1 d_1 & g_2 &= f_2 d_2 & g_3 &= f_3 d_3 \end{aligned} \tag{5.98}$$

From these new recipes the concentration adjustments, corrected for dye-uptake effects specific to the process, can be calculated as follows (Eqn 5.99):

$$\begin{aligned} \Delta c_1 &= e_1 - g_1 & \Delta c_2 &= e_2 - g_2 & \Delta c_3 &= e_3 - g_3 \\ &= f_1(t_1 - d_1) & &= f_2(t_2 - d_2) & &= f_3(t_3 - d_3) \end{aligned} \tag{5.99}$$

If the results of several dyeings of the same recipe are available, then the correction factors can be averaged and used to adjust the absorption coefficients for future predictions under these particular dyeing conditions or substrate type. Experience will show whether the correction factors obtained are reasonable. Serious errors in dye application would be shown by unreasonably high or low correction factor values.

Corrections computed by the above methods may be used to amend a dye recipe for use in future dye lots, or to correct the off-shade dye lot currently being processed. Negative corrections imply that the dye lot already contains too much of the dye and will have to be stripped or otherwise reduced in depth of shade before correction procedures can be carried out (although it is sometimes possible to ignore very small negative corrections).

An alternative approach reported by Kuehni is to change the standard to a slightly deeper shade and attempt to match to this. Standard and dyeing reflectance data are converted into the additive reflectance function values $f(R_t)$ target and $f(R_d)$ dyeing. A strength factor P is calculated at the wavelength where $f(R_d) - f(R_t)$ is at a maximum (Eqn 5.100):

$$P = \frac{f(R_d)}{f(R_t)} \tag{5.100}$$

All 16 $f(R_t)$ values are multiplied by this same factor and then converted back to reflectance values. This curve represents a new target which is stronger than the original, but attainable by correcting the current dyeing. If the colour difference between this amended curve and the original target curve is acceptable, then a prediction to match this curve can be made.

If the substrate currently being processed is known to have different absorptivity from that used for the calibration dyeings, then it is frequently useful to determine an average correction factor by dyeing a set of ternary mixture shades on the current substrate. Usually a dyeing of medium concentration with roughly equal amounts of each dye will be sufficient. The reflectance of this check dyeing is measured and a recipe is predicted to match this curve using the original calibration data absorption coefficients. The predicted concentrations c_{d1}, c_{d2} and c_{d3}, are used to calculate correction factors by comparing them with the concentrations used for the original target shade c_{t1}, c_{t2} and c_{t3} (Eqn 5.101):

$$f_1 = \frac{c_{t1}}{c_{d1}} \qquad f_2 = \frac{c_{t2}}{c_{d2}} \qquad f_3 = \frac{c_{t3}}{c_{d3}} \tag{5.101}$$

If the correction factors from the check mixtures do not differ substantially from each other then the variation in uptake is uniform for all dyes and the average of the correction factors can be assumed to apply approximately to all other dyes in the dye class. This method is useful for improving the accuracy of first predictions in cases where substrate variability is a factor in prediction errors, and where it is impracticable to prepare fresh calibration ranges for each dye. More complex nonlinear equations (e.g. quadratic) are also used in some commercial systems to correct for varying dye uptakes by different substrates.

5.9 USING HISTORICAL DATA TO IMPROVE ACCURACY OF PREDICTION

The accuracy of prediction in recipe formulation depends not only on the underlying mathematical theory but also on the appropriateness of the database from which the coefficients for prediction are derived. One of the problems with prediction systems is that the database of single-dye ranges can become out of date due to changes in the dye application technology, such as additions of dyebath chemicals or changes of substrate. Even when the mathematics can exactly describe the relationship between dye concentration and reflectance for a single dye, it may not work completely satisfactorily with dye mixtures due to extraneous factors such as dye interaction. There are often significant differences between the results from a laboratory dyeing and the same recipe dyed in bulk machinery. For these reasons attempts have been made to incorporate these changes in the recipe prediction process.

The most used technique is to search the file of established recipes for the required substrate to find one or more recipes for similar shades to the target colour, which have been dyed with the same dye combination. The nearest shade or group of shades is then used to establish the concentration correction factors, either of the type described in section 5.8 or more sophisticated equations which can be used to adjust recipes for this area of colour space. The degree of sophistication of the interpolation between surrounding recipes depends on the system designer. The recipe for the target colour is predicted and adjusted with these factors or equations to produce a more accurate recipe.

If enough previously dyed recipe mixtures with the same dye combinations are available then computer direct-search programs can be used to recalculate, from the mixture dyeings, the absorption coefficients for each of the dyes in prediction equations such as Eqn 5.67. This approach has been used since 1980 in the Coats Viyella colour physics system, which was marketed externally for a few years between 1989 and 1991. The system is claimed to produce more accurate recipes than can be obtained from single-dye absorption coefficients. In continuous use, any changes that affect dye uptake will be present in the most recently dyed mixtures and are therefore incorporated into the revised prediction coefficients.

5.10 SELECTION OF OPTIMUM DYE COMBINATIONS FOR RECIPE PREPARATION

We have already considered how a typical computer formulation program can be designed to predict all possible recipes from a given dye selection to match a given colour target. The resultant recipes can be sorted in order of cost or metameric index. Cost and metamerism are not of course the only factors to be considered in choosing the optimum dye combination. Before a final recipe is selected from the computer shortlist the colourist must take into consideration other factors such as compatibility of dyes, fastness characteristics, level-dyeing properties and stability of recipe. Factors such as these have been described in detail by Mackin [36].

Having satisfied the initial requirements of fastness and cost effectiveness, each dye which is in the dye selection must be tested for:
(a) colour yield and build-up performance

(b) compatibility, i.e. rate of dyeing, build-up, blocking effects on other dyes, response to different conditions of pH, temperature, electrolyte concentration, all of which must be assessed by reference to manufacturers' literature or by practical dyeing tests
(c) level-dyeing behaviour assessed by strike–migration tests
(d) colour constancy in different illuminants, the object being to select where possible only those dyes which show minimal shade alteration in such circumstances
(e) selection of homogeneous dyes; where dye makers do not supply information on this aspect, the dye samples are subjected to chromatographic analysis and blow testing.

These tests enable unsatisfactory dyes to be deleted from a prospective selection. A typical selection table for vat dyes is given in Figure 5.26 [36].

Once satisfactory dyes have been selected, their absorption coefficients can be used to compute the position in CIELAB colour space of each dye at a series of different lightness levels. The problem of dye combination is then reduced to finding a suitable enclosing triangle of three dyes on an appropriate lightness plane. The criteria of suitability must be practical in nature and take into account the compatibility of the three dyes with reference to levelness, dyeing method and so on. The actual positions of the dyes and binary lines are calculated by computer, while prospective three-dye selections are made by the colourist. A typical dye map illustrating positions of various dyes and predetermined ternary dye combinations is shown in Figure 5.27.

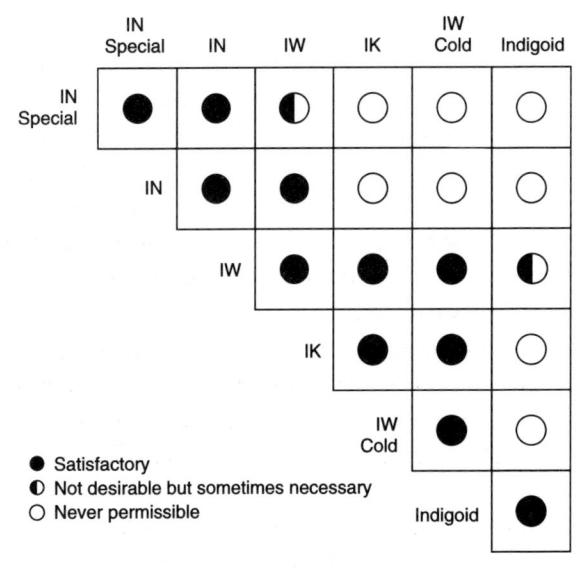

Figure 5.26 Permissible combinations of dyeing methods with different groups of vat dyes

SELECTION OF OPTIMUM DYE COMBINATIONS FOR RECIPE PREPARATION   273

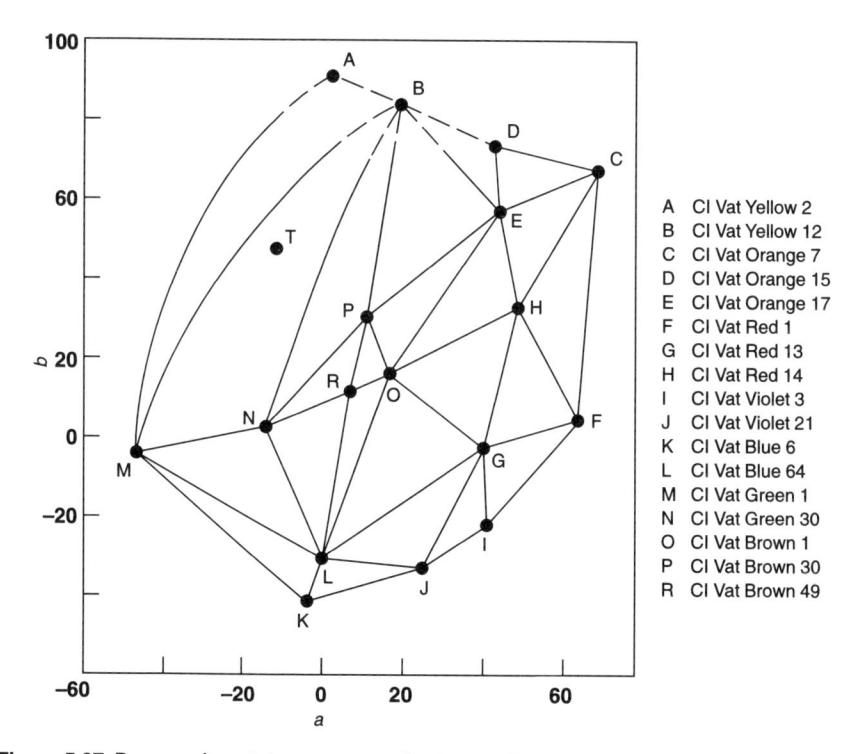

Figure 5.27 Dye map for vat dyes on mercerised cotton (*L* planes 48–74)

A	CI Vat Yellow 2
B	CI Vat Yellow 12
C	CI Vat Orange 7
D	CI Vat Orange 15
E	CI Vat Orange 17
F	CI Vat Red 1
G	CI Vat Red 13
H	CI Vat Red 14
I	CI Vat Violet 3
J	CI Vat Violet 21
K	CI Vat Blue 6
L	CI Vat Blue 64
M	CI Vat Green 1
N	CI Vat Green 30
O	CI Vat Brown 1
P	CI Vat Brown 30
R	CI Vat Brown 49

No triangulation of a particular lightness plane is unique, and a particular point may be enclosed within an inner and several outer triangles, all of which will give the required colour. This raises the question of whether recipes using primary colours are more satisfactory than recipes using internal primary colours. A recipe of primary colours is one that employs three dissimilar saturated dyes, whereas a recipe of internal primary colours is one employing three dyes of the same general hue: for example, three reds to produce a red target shade, or two browns and a yellow to produce a brown shade. Mackin has shown that the most stable recipes, i.e. those which are least affected by variations in application method or drugstore weighing variation, are those produced by internal primaries [36]. Mackin's conclusions were therefore as follows:
(a) Dye combination for a point in colour space should be by means of the nearest enclosing internal primary dyes, provided a suitable dye triangulation has been made.
(b) This method does not break down even when the point in colour space is near a boundary (i.e. a binary dye combination), showing that numerical equality of concentrations in a recipe does not significantly aid its stability.

(c) A method of dyeing based on primary dyes will tend to magnify any errors in strength, weighing and preparation, and the use of internal primaries is to be preferred, whether or not their concentrations are numerically equal.

In general, no more than 25 dyes are required to give good coverage of the colour gamut and sound recipe construction for any dye–fibre system, and in many dye classes a selection of about 15 dyes is adequate. The brightest technically acceptable dyes must always be included or covered in any dye selection. Selected dyes should be as evenly spaced as possible throughout the chromaticity gamut; for example, the inclusion of a homogeneous green dye is always preferable to a binary combination of a blue and yellow dye giving approximately the same colour (Figure 5.27).

In practice the dye triangulation may change at different lightness levels. For pale colours, factors of prime importance are good level-dyeing properties, good light fastness and good compatibility with other dyes, while cost and wet fastness are of relatively little significance (wet fastness is usually satisfactory at these depths). Moreover, at these depths the hues of dyes of roughly similar colour (browns and oranges, for instance) come much closer together, and so it is often practicable to restrict the number of dyes used. For medium colours, cost and wet fastness are the most important characteristics for dye selection. In heavy depths, all these factors are significant but cost considerations are of paramount importance; dyes must also possess outstanding build-up properties. As with pale depths, it is also possible to restrict the number of dyes required. Frequently the properties of dye mixtures are quite significantly different from what might be indicated by the properties of single dyes (catalytic fading is an example), and such considerations can lead to further modification of the dye selection.

Once a dye selection has been made, the dye maps similar to those in Figure 5.27 can be made for the different lightness levels. When the final dye selections for each lightness level have been made, a 'triangular' selection table is made up so that the computer can select the correct dye combinations for any area of colour space entirely automatically. The computer logic can be set out in the form of a table for the dyes in Figure 5.27 (Table 5.10).

The computer is programmed to go to the alternative triangle indicated in Table 5.10 if any dye is computed to have a negative concentration. For example, in triangle BMN if dye M is negative the computer switches to alternative triangle BNP; if dye B is negative the computer switches to triangle LMN; and if dye N is negative triangle ABM is tried. Midpoint CIE a^*b^* values are also computed so that the computer can select the nearest midpoint of the target coordinates as a starting triangle indicator. It can happen that this is not the correct dye selection to give the target colour; for example, to match the target T in Figure 5.27, the nearest midpoint is in triangle ABM, whereas the correct triangle for predicting the colour is triangle BMN. In this case a

Table 5.10 'Triangular' dye selection (Figure 5.27)

Dye triangle	Alternative triangles	CIELAB coordinates of triangle midpoint	
		a*	b*
ABM	None	−18	65
BMN	BNP, LMN, ABM	−20	25

start is made in triangle ABM: dye A is negative, so the computer moves to triangle BMN to complete the prediction procedure.

The triangulation approach has been criticised for not always giving the cheapest recipe. Mackin and Purves have tested its performance on 235 polyester recipes [37]. They found that a properly prepared triangulation map will produce recipes which are on average only 3.3% dearer than the cheapest recipes. On the other hand the most expensive recipes produced by the computer by conventional permutation methods were 900% dearer than those from the triangulation approach. Figure 5.28 shows that relatively few recipes were significantly cheaper than those from a properly prepared triangulation system, whereas Figure 5.29 shows that many of the alternative recipes produced by permutation methods were excessively expensive.

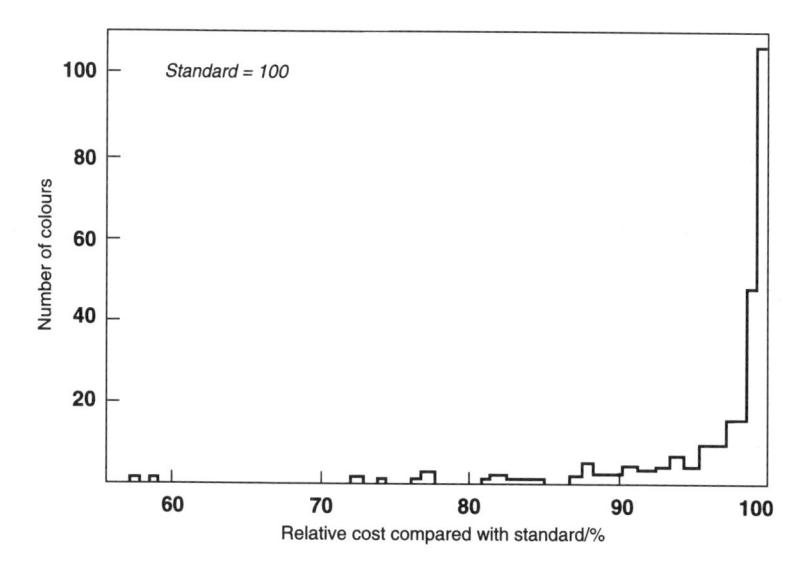

Figure 5.28 Cost distribution of cheaper recipes compared with standard recipe for disperse dyes on spun polyester fibre (235 colours from Coats Viyella card)

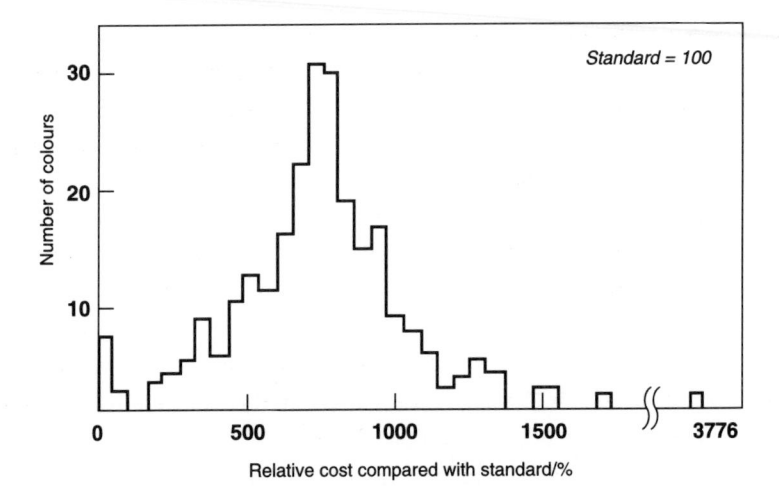

Figure 5.29 Cost distribution of dearer recipes compared with standard recipe for disperse dyes on spun polyester fibre (235 colours from Coats Viyella card)

Mackin and Purves also describe the use of prediction methods for comparing the cost-effectiveness of new dyes. The first method is for comparing dyes having the same *Colour Index* generic name but supplied by different manufacturers. The new dye is first applied in different amounts to give a calibration range. Using the existing dye selection, predictions are made to match each of the new dye reflectance curves. The cost of the resulting ternary mixture recipes are then compared with the cost of the single-dye recipes at each concentration in the range to demonstrate the cost-effectiveness of the new dye.

Mackin and Purves also describe the prediction of recipes to match a large number of CIELAB colour coordinates evenly spaced across colour space in the form of a cubic lattice or grid. Alternatively, all the current standard colours in the colour range of a company can be matched. This method is particularly suitable for assessing whether a new product is worthy of incorporation into the current dye selection, or whether it can be used to substitute a currently used dye. Single or simple dye comparisons do not necessarily provide an accurate basis for overall dye recipe cost-saving calculations; moreover, popular and seasonal colours must be costed with a production weighting before any accurate assessment of dye cost saving can be established. Dyehouse production disruption due to dye additions, redyes, administration and stock control may not be recouped from the dye savings if the shade mix does not include large amounts of heavy- and medium-depth dyeings.

Another advantage of the triangulation prediction method is that recipes can be formulated by computer by personnel unskilled in dyeing technology. The permutation method always requires a technologist to make the final recipe selection from the short-list produced by the computer, sometimes under the stress of production requirements, and to assess the dyeing characteristics of all dyes in use before making a decision. The triangulation method allows the colourist to determine optimum dye combinations on cost and technological grounds once and for all, after which the computer can be allowed to proceed automatically with reliable recipe formulation.

Some caution should be exercised over the interpretation of recipe costs from alternative predicted formulations. Predicted recipes are only as reliable as the accuracy of prediction will permit. It is an enlightening experience to dye up a selection of alternative predicted recipes made to match a given target. Some combinations will be found to be good matches and others mediocre. The subsequent adjustments to bring the latter recipes on target may affect their position in the cost table. Where small cost differences are important, this factor should be borne in mind.

5.11 RECIPE FORMULATION ON FIBRE BLENDS

5.11.1 Recipe formulation for fibre mixtures

The simplest technique for recipe formulation for mixtures of fibres, where each component fibre requires a different class of dyes, consists of predicting a recipe for each fibre independently, based on calibration ranges for pure fibre already held in the computer. Adjustment is made for the proportions of different fibres in the mixture, and the dyeing proceeds in the normal way. In practice, however, this simple approach does not always work well because cross-staining of the fibres can take place [38].

It is often not possible to obtain samples of the pure component fibres with which to prepare calibration ranges. Chong developed a technique for polyester/cotton in which the blend is dyed separately with dyes from each relevant class: for example, one sample may be dyed with disperse dyes for the polyester and another sample with reactive dyes for the cotton [39]. The respective undyed fibres are then dissolved out chemically from the two samples using potassium hydroxide in ethanol for the polyester and sulphuric acid for the cotton, and calibration absorption coefficients are calculated for the disperse dye/polyester and for the reactive dye/cotton components.

5.11.2 Recipe formulation for loose-stock blends

Computer formulation has also been used in the area of blending stock-dyed fibres together to produce a given colour in the blended fibre. The first prediction technique used a version of the standard single-constant recipe prediction theory and was described by Guthrie *et al.* [40,41]. They used a formula developed by Friele [42] in which the total light absorbance of the blended fibres is the sum of the component light absorbance values (Eqn 5.102 and 5.103):

$$f(R) = af(R_1) + bf(R_2) + cf(R_3) \qquad (5.102)$$

$$f(R) = \exp\left(\frac{-S(1-R)^2}{2R}\right) \qquad (5.103)$$

where a, b, c = proportions of fibres present ($a + b + c = 1$)
R_1, R_2, R_3 = reflectance values of the three individual component fibres
S = scattering factor optimised experimentally (for viscose fibre S = 0.28).

Only a restricted range of colours can be blended if solid shade effects are to be obtained. Acceptable ranges of colours were determined by Guthrie et al. and about 20 basic colours of dyed stock were found necessary to produce a good gamut of blended fibre colours.

Burlone found that the conventional single-constant Kubelka–Munk theory normally used for single fibres gave inaccurate results when applied to fibre blends [43]. Significantly more accurate results were obtained when the two-constant theory normally applied to paints or plastics is employed (Eqn 5.4). In this approach each coloured fibre is assigned two constants, a pseudo-absorption constant K and a pseudo-scattering constant S. In the case of fibre blends these constants are essentially dependent on the physical form and composition of the sample rather than on the presence of any true Kubelka–Munk absorbing and scattering particles.

Burlone measured coloured fibres in the form of carded puffs and used nine stock-dyed samples of nylon fibre as 'primary' colours for blend formulation. By analogy with the paint formulation techniques described in Chapter 6, a sample of 100% coloured fibre (in paint formulation called the 'mass-tone') and a blended sample consisting of known percentages of coloured fibre and white fibre were used to solve the simultaneous equations to obtain the K and S values for each of the coloured fibres. Mixtures of coloured fibre with 25%, 50% and 75% white fibre were used. The values of K and S obtained were independent of the composition of the mixtures and the average K and S obtained from the range of samples were adopted for formulation. Again, by analogy with paint formulation, it was found that for yellow and orange colours it was better to use a mixture of 90% coloured fibre and 10% black fibre to obtain satisfactory K and S values for these colours. The scattering coefficients varied greatly from dye to dye, which helps to explain why single-constant theory, which assumes constant S, cannot describe the colour of a fibre blend.

Eqn 5.4 was used by Burlone to predict matches to 44 targets in the form of blends of four 'primary'-coloured fibres. In these cases the most satisfactory predicted recipes contained the same four 'primary' colours as the predicted recipe. Use of alternative coloured fibre mixtures was not so useful for fibre blends because, even if a colorimetric match was obtained, it did not necessarily result in a satisfactory visual match. For example, the visual appearance of a grey sample consisting of a mixture of black and white fibres is not the same as that of a self-dyed grey sample, although on measurement they may have the same XYZ values. Those alternative colour mixtures that contained 'primary' colours similar in colour to those in the original target blend were more likely to produce a recipe that would be considered to be an acceptable match in colour and appearance to the target sample. Sample preparation and measurement errors were found to be the most significant contributors to colour-matching inaccuracy in formulation of fibre blends. Spectrophotometric readings are sensitive to the pressure applied to the fibre puffs during sample measurement, and sample appearance and measurement results can be altered by variations in the blending operation.

Burlone also predicted the same target shades using Friele's reflectance function (Eqn 5.103) and an earlier reflectance function by Stearns and Noechel (Eqn 5.104) [44]:

$$f(R) = \frac{1-R}{M(R-0.01)+0.01} \qquad (5.104)$$

where M is a factor that is optimised experimentally ($M = 0.15$ for wool blends).

He found that the two-constant recipe prediction equations gave significantly more accurate results than single-constant theory using Eqns 5.103 and 5.104. In addition, the two-constant approach does not require the preparation of experimental mixtures to optimise the parameters S and M for the given substrate in the reflectance functions shown in Eqn 5.102 and Eqn 5.103.

Walowit et al. proposed a method for improving the accuracy of the two-constant theory for predicting four-component fibre blends [45]. This involves the use of a least squares technique to solve directly for the K and S values of the constituent coloured fibres from existing blends of four coloured fibres. By analogy with Eqn 5.4, for a four-component blend there will be four K values and four S values, i.e. eight constants to be determined. Theoretically this would require eight mixtures to create an exactly determined system of eight linearly independent equations in the eight unknowns. In practice, however, more than eight mixtures are employed to produce an overdetermined system of equations, thereby increasing the reliability of the calculated values of K and S.

Walowit et al. have also adapted this least squares technique to produce a spectrophotometric curve recipe prediction technique for fibre blends using Eqn 5.4, which for a four-dye blend is written as Eqn 5.105 [46]:

$$(K/S)_{\text{std},\lambda} = \frac{c_1 K_{1,\lambda} + c_2 K_{2,\lambda} + c_3 K_{3,\lambda} + c_4 K_{4,\lambda}}{c_1 S_{1,\lambda} + c_2 S_{2,\lambda} + c_3 S_{3,\lambda} + c_4 S_{4,\lambda}} \quad (5.105)$$

Extracting the concentrations and rearranging gives Eqn 5.106:

$$c_1 \left[K_{1,\lambda} - S_{1,\lambda} (K/S)_{\text{std},\lambda} \right] + c_2 \left[K_{2,\lambda} - S_{2,\lambda} (K/S)_{\text{std},\lambda} \right]$$
$$+ c_3 \left[K_{3,\lambda} - S_{3,\lambda} (K/S)_{\text{std},\lambda} \right] + c_4 \left[K_{4,\lambda} - S_{4,\lambda} (K/S)_{\text{std},\lambda} \right] = 0 \quad (5.106)$$

Since the K_λ and S_λ values of the n colorants and $(K/S)_{\text{std},\lambda}$ are all constants, they can be written as a coefficient matrix, making the following substitutions (Eqn 5.107):

$$\begin{aligned} X_{1,\lambda} &= K_{1,\lambda} - S_{1,\lambda} (K/S)_{\text{std},\lambda} & X_{3,\lambda} &= K_{3,\lambda} - S_{3,\lambda} (K/S)_{\text{std},\lambda} \\ X_{2,\lambda} &= K_{2,\lambda} - S_{2,\lambda} (K/S)_{\text{std},\lambda} & X_{4,\lambda} &= K_{4,\lambda} - S_{4,\lambda} (K/S)_{\text{std},\lambda} \end{aligned} \quad (5.107)$$

The coefficient matrix at 10 nm intervals from 400 to 700 nm is then written as Eqn 5.108:

$$[\text{COEFFS}] = \begin{vmatrix} X_{1,\lambda_1} & X_{2,\lambda_1} & X_{3,\lambda_1} & X_{4,\lambda_1} \\ \cdots & \cdots & \cdots & \cdots \\ X_{1,\lambda_{31}} & X_{2,\lambda_{31}} & X_{3,\lambda_{31}} & X_{4,\lambda_{31}} \\ 1 & 1 & 1 & 1 \end{vmatrix} \quad (5.108)$$

The matrix also includes a set of coefficients = 1 as a constraint.

The observation vector is formed from the right side of the 32 equations, namely zeros and a value = 1 as a constraint in Eqn 5.109:

$$[\text{OBS}] = \begin{vmatrix} 0_1 \\ \cdots \\ \cdots \\ 0_{31} \\ 1 \end{vmatrix} \quad (5.109)$$

Without the constraints, the problem would be indeterminate since the observation matrix would contain all zeros for which there is an infinite number of solutions.

The solution vector has the following form (Eqn 5.110):

$$[C] = \begin{vmatrix} C_1 \\ C_2 \\ C_3 \\ C_4 \end{vmatrix} \qquad (5.110)$$

Using matrix linear regression the colorant concentrations can now be computed using Eqn 5.111:

$$[C] = \left\{ [COEFFS]' [COEFFS] \right\}^{-1} [COEFFS]' [OBS] \qquad (5.111)$$

where [COEFFS]' is the transpose of [COEFFS].

Since the entire calculation is performed with only one set of matrix multiplication and a single matrix inversion, this algorithm should run faster than tristimulus matching which generally involves multiple iterations. Because up to 31 wavelengths are used in the calculation to determine if there is a match, rather than the three differences in XYZ or *Lab* for tristimulus matching, higher numerical precision should result. This should make the technique particularly useful if matching is done with colorants that are very similar to each other. In fibre blends Walowit *et al.* found that the spectrophotometric matches were closer to standard than with tristimulus matching. They suggest that the accuracy of technique for matching with similar colours should be particularly useful in the production of an on-shade product by mixing previously produced unsatisfactory off-shade batches.

One other advantage of this technique is that it can also be used for recipe prediction in the UV or IR regions (for camouflage, for example) where tristimulus matching is not possible.

5.12 USE OF NEURAL NETWORKS IN RECIPE FORMULATION

The conventional approaches to recipe formulation depend on equations resulting from the analytical treatment of the relationship between reflected light and colorant concentration. In general these equations are only approximations to the physical processes actually taking place and this leads to inaccuracies in recipe formulation. The relationship between dyes, dyeing procedures, substrates, physical structure and reflectance is extremely complex and it is unlikely that a completely accurate analytical solution can ever be obtained. Similar problems arise in many sciences, and the need to tackle them has led to the application of the branch of artificial intelligence known as *artificial neural network* (ANN) theory.

In a conventional computer program, the computer carries out a set of specific instructions to complete a given task. ANN programs represent a different approach to problem-solving which has strong parallels to the way the human brain is thought to operate. The brain is believed to learn to react to given problems (gain experience) by modifying the information signals received in the synapses (junctions) between neurons, in time altering the structure of brain to enable it to carry out certain tasks. Unlike a conventional computer program a neural network is designed to adapt and acquire knowledge over time in order to complete a certain task.

There are many types of ANN. One of the simplest and most successful which has been described used colour problem-solving [47,48]; this is the multi-layer perceptron (MLP, Figure 5.30). The MLP consists of simple processing units arranged in layers. Each unit receives input, modifies this in a simple way, and produces an output. In an ANN each unit receives an input from all the units in the previous layer, modifies these, and produces an output which is then passed on to each unit in the subsequent layer. The units in the first layer receive their input from the outside world at the start of the system and the units in the final layer send their outputs to the outside world at the other end of the system.

Figure 5.30 shows a network for the conversion of reflectance values to cyan, magenta, yellow and black (CMYK) values for use in a paper printer. The connections

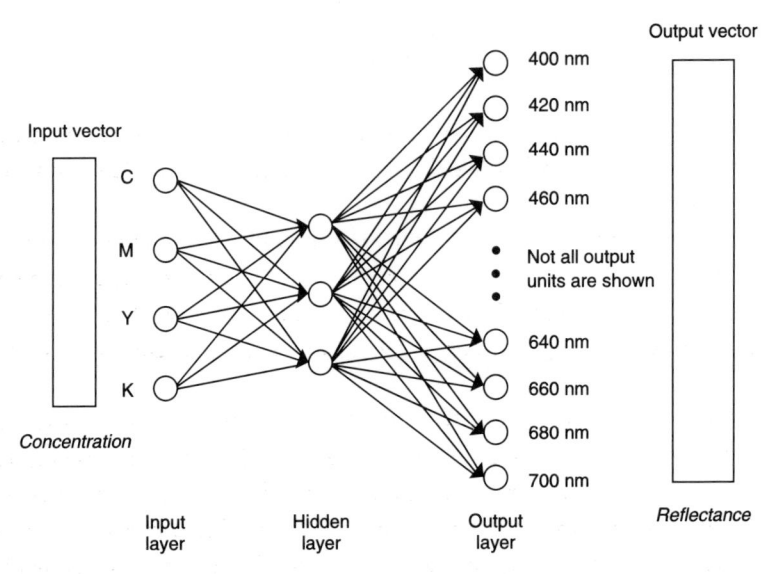

Figure 5.30 CMYK neural architecture (bias units not shown)

between the units are represented by straight lines. The input for each unit is computed from the sum of the outputs of the previous units in the system, after each output has been modified by a weighting factor specific to the connection between the two units. In addition (not shown in Figure 5.30) each unit receives an input from a bias unit, whose output is always unity. Long-term knowledge is stored in the network in the form of the interconnection weights linking these units.

Depending on the problem to be solved, networks of varying complexity can be selected. The ANN will contain one or more hidden layers of units between the input and output layers, the number of units in each layer also depending on the complexity of the problem. If the output of the jth unit in the previous layer is represented by O_j, and the weight between the jth unit in the previous layer and the ith unit in the current layer is represented by $W_{j,i}$, then the input I_i to the ith unit in the current layer is given by Eqn 5.112:

$$I_i = \sum (W_{j,i} O_j + W_i) \qquad (5.112)$$

where W_i is the bias weight.

Unit i receives an input I_i which is the sum of the weighted inputs and modifies this input, usually by computing a simple nonlinear function to produce an output. For example, if unit i computes the nonlinear sigmoid logistic activation function, the output O_i of unit i is given by Eqn 5.113:

$$O_i = \frac{1}{1 + \exp(-I_i)} \qquad (5.113)$$

Linear output functions are not used and would not lead to satisfactory results in these networks because the functional composition of several linear functions is itself a linear function.

Before the network can be used to solve a given task it must first be trained using known pairs of input and output vectors. For example, to train the system to convert a given reflectance curve to CMYK values, the output vectors would be a selection of CMYK values and the input vectors the measured reflectance curves of the samples printed with these values.

The learning process in an ANN consists of adjusting the weights W in the network so that the units in the output layer produce the desired output when a certain input vector is presented at the input layer. Pairs of input and output vectors are presented to the input and output layers of the network respectively. The input vector is used to generate an output for each unit in the network, layer by layer, until an output is produced at the final output layer. The weights are then modified so as to reduce the error between the calculated output of the last output layer and the desired output. This is repeated for a series of training pairs until a set of weights is found that can accurately predict the correct output for all the training input vectors. The process of presenting all the training pairs to the system and assessing the goodness of prediction is known as an *epoch*. To train the system to satisfactory accuracy may take several thousand epochs. Mathematical techniques such as 'back propagation of the generalised delta rule' are used for systematic optimisation of the weights to minimise error [48]. The trained network can then be used to compute output for inputs that have not been presented to it before. Kang and Anderson have applied neural networks to colour scanner and printer calibrations [49].

Westland *et al.* have applied neural networks to recipe prediction with some degree of success [47,48]. The network consisted of three input units (CIE $L^*a^*b^*$ values), 24 hidden units arranged in two layers of 8 and 16 units, and three output units (three dye concentrations). The training input consisted of a series of two-dye recipe samples and single-dye recipe samples and the output vector was the measured CIE $L^*a^*b^*$ values of the samples. The training consisted of 55 000 epochs using 30 dyed samples, consisting of 12 single-dye samples and 18 binary mixtures. Predictions were then made to match these 30 samples plus another 21 samples not included in the training set. In the 54 predictions, 33 of the targets were binary mixtures and for these 78.8% gave errors less than 0.8 CMC(2:1) unit. Predictions of the remaining 21 targets dyed with single dyes were not so accurate, leading to an overall accuracy for the complete data set of 60% with errors less than one CMC(2:1) unit.

Westland claims that the use of neural networks offers several potential advantages over the conventional recipe prediction approach using absorption coefficients:

(a) It is not necessary to prepare a special database of single dyes in order to use the neural network method. The network can be trained on actual production samples.
(b) The network can continue to learn after the initial training period, since future production samples can be presented to the system and this knowledge incorporated into the network weights. This gives the network the potential to adapt to changes in factors such as water supply, change of substrate, change of dye strength and so forth.
(c) The network may be able to learn the behaviour of colorants for which the mathematical descriptions are complex. For example, fluorescent dyes and metallic paint systems are currently difficult to treat using standard Kubelka–Munk theory.

The ANN technique has also been applied to other types of colour problem, including the characterisation of colour printers, pass/fail colour assessment, colour notation transformations, and reproduction of colour generated on computer screens by colour printers. These have been reviewed by Westland [50]. The development of neural networks may have a significant effect on the approach to solving colour problems in the future.

5.13 EFFECT OF EXTRANEOUS FACTORS ON RECIPE FORMULATION ACCURACY

Mathematical formulation of recipes may not be an exact technique, but there are many extraneous factors which produce errors greatly in excess of those introduced by the defects of the colour physics equations. In particular, calibration and stability of the spectrophotometer, accuracy in dyeing of calibration ranges, and reproducibility in sample preparation and presentation are extremely important. Lack of attention to these details will significantly reduce the accuracy of the formulation system.

The stability of the spectrophotometer is critical. If the instrument drifts then absorption coefficients and other constants calculated from initial measurements will become worthless in time. In most colour control systems the computer monitors the spectrophotometer and compensates for any drift that may occur. The reference values of calibration tiles are stored in the computer and the instrument is recalibrated with these tiles every few hours, correction factors being calculated from these measurements and used to adjust the instrument back to its initial state. These techniques have almost eliminated the once-familiar drift-related deterioration in prediction performance.

The correction factors stored in the computer should deal with changes in the photometric scale and zero offset. In some systems the black tile is replaced by a 'black box' which is lined with either black velvet or a suitable arrangement of black glass that almost totally absorbs incident light.

Reproducible sample preparation and presentation to the instrument is important. In general, care must be taken that the sample is thick enough to be completely opaque to light, otherwise the background on which it is mounted will influence the colour. A fabric sample should be folded several times to ensure opacity. Yarn should be wound to a sufficient thickness on a suitable mounting, using a parallel lay to ensure a flat surface. Cardboard is often used for the mount, but a better technique is to use a metal or plastic card with a hole in the centre so that light penetrating the layers of yarn on one side of the card is given the chance to be reflected by the layers on the other side, before reaching the background mount. The hole must therefore be larger than the aperture of the spectrophotometer. Such a card saves yarn and produces a flatter surface since fewer layers need to be wound to attain opacity. Suitable card winders for yarn are made by Ayrton, Manchester, and Zweigle, Switzerland.

With all textile surfaces, texture plays a significant part in the reflectance results obtained. Samples exhibiting textural effects are best measured several times, usually at each of four orientations, and the average value input to the computer. Yarns show large orientation effects, and it is best to standardise on one orientation for best reproducibility.

Experience has shown that with textiles optimum reproducibility is obtained by measurement with the specular component included, with the possible exception of very heavily textured samples or carpets, where exclusion of the specular component gives better results.

5.14 HARDWARE FOR COMPUTER COLORANT FORMULATION

Equipment for computer colorant formulation is marketed by several specialist companies. It usually consists of:
(a) a recording spectrophotometer operating over the visible spectrum, normally in the range 400–700 nm
(b) a computer for carrying out the necessary calculations
(c) a disc data-storage system for holding the prediction software and the basic data.

The spectrophotometer is generally interfaced to the computer so that the reflectance measurements can be transferred directly to the computer for processing; some instruments, however, contain a dedicated microprocessor to process the reflectance data and convert it into colour coordinates and to maintain calibration of the instrument. In some spectrophotometers the spectrum is produced by prism or grating methods, and in other models by interference filter wedges or wheels, coupled with either traditional photomultipliers or arrays of solid-state photodetectors. By these means reflectance scans can be completed in from 2 to 9 s. All these systems give reliable and reproducible colour measurements.

The computer system will contain a visual display unit (VDU) to provide instructions and information to the operator, and a printer for output of results. The modern personal computer is powerful enough to carry out all necessary calculations required for a colour physics control system. In some situations a computer network may be required to allow multi-user access to computer files for purposes such as recipe information, stock control, scaling-up recipes and control of dye- and chemical-dispensing systems. To ensure data integrity in the event of a computer failure or other accident a suitable secure backup storage system is an essential part of the computer hardware.

The size of the storage disc depends on the requirements of the system and software and the amount of information (for example, on colour standards and calibration dyeings) which is to be retained. It can easily be determined from knowledge of these factors coupled with knowledge of the file structure used to hold the basic information.

There are now well-established requirements for colour control systems which all specialist system manufacturers will endeavour to satisfy. The variation between packages reflects each manufacturer's attempts to produce a reliable, effective and user-friendly system. Regardless of the facilities offered by a commercial package it is likely that each customer will require some facilities tailored to meet specific requirements. These requirements can usually be met either by minor system alterations by the supplier or by the addition of special software written to suit the user's particular needs.

Commercial systems usually have provision for most of the following facilities:
(a) measurement of reflectance
(b) automatic calibration and compensation for drift of the spectrophotometer, either by spectrophotometer design or by computer compensating software
(c) pass/fail quality control software, including calculation of colour differences between samples in several different colour-difference formulae
(d) calculation of absorption coefficients, including techniques to compensate for a nonlinear relationship between reflectance and dye concentration
(e) recipe-prediction programs with preselected combinations of dyes or with permutation of dyes to give minimum cost or minimum metamerism
(f) display on the monitor screen of a simulation of the approximate colour of a given recipe in different illuminants, as an aid to optimum recipe selection where metamerism is a problem
(g) ability to use previously dyed samples to improve the accuracy of recipe prediction
(h) recipe-correction programs for off-shade dye lots, and programs to allow conversion to a different colour for serious dyehouse errors
(i) bulking-up programs to convert predicted recipes into actual dye weights for dyeing a particular weight of material
(j) Storage of bulk proven recipes with file search and recall of recipes; general file interrogation programs
(k) downloading of recipes to computer-controlled dye- and chemical-dispensing equipment
(l) accumulation of management statistics to monitor dyehouse performance.

The advantage of commercially designed colour-control systems is that the cost of preparing the necessary programs and assembling the complete package are included in the purchase price. Specialist companies can spread the cost of preparing the package over many installations and can maintain specialist programming staff and colour experts to keep abreast of new developments. The commercially designed package allows a dyehouse to introduce computer colour control quickly and without the necessity of lengthy training of staff in colour physics. Generally, customers do not require a complete understanding of the operating software of the system and require only sufficient knowledge to enable its effective operation.

However well a general-purpose colour physics system is designed, it will usually be found to be deficient in some aspect when used in any particular dyehouse situation. All users want programs tailored to their specific requirements. In this respect the commercial package must be capable of adaptation to meet the requirements of the user.

5.15 COST-EFFECTIVENESS OF COMPUTER PREDICTION SYSTEMS

Park and Thompson have summarised the influence on costs of the installation of a colour physics system into a dyehouse engaged in dyeing acrylic and polyester yarn [51]. One of the advantages of computer match prediction is that it provides a means of computing recipes quickly, including the minimum-cost dye combination from the stock-list of dyes available. Dye savings of 10–20% are often claimed by producers of match-prediction systems. Park and Thompson confirmed these figures when they rematched existing shade ranges of 50 shades on acrylic and 100 on polyester using the cheapest predicted dye combinations, providing that the dye combination was technically satisfactory. The average dye cost was reduced by 10% for the acrylic dyeings and by 30% for the polyester dyeings.

While such an exercise could be carried out without a match-prediction system, it would be much more difficult and time-consuming. The exact cost savings of course depend on the cost-effectiveness of the recipes in use before the introduction of computer match prediction.

Other cost-saving techniques are also possible. For example, Park and Thompson also state that the convenience of computer recipe file storage enables the dyehouse to purchase dyes on a spot-lot/best-price basis, provided that batches of dyes are sufficiently large (say 500 kg). The colour physics system is then used to determine the strength of the dye and update every recipe containing that dye in the recipe storage file.

Apart from the possible reduction of dye costs in preparation of new recipes, computer-based colour control systems can provide other potential cost-saving facilities, depending on the requirements of the individual dyehouse. These are discussed below.

5.15.1 Recipe storage and retrieval

This assists in reducing redyes because it is possible to reproduce recipes accurately and to take account of changes in dye strength. File sorting and interrogation enable rapid assessment to be made of the possible effects on the recipe file of factors such as dye substitution and changes in dye strength.

5.15.2 Colour correction
Where the shade difference is considerable, the correction program enables the dyeing to be matched more accurately than is possible by visual means, enabling rapid correction of redyes, or the conversion of old dyed stock into saleable material.

5.15.3 Instrumental colour-difference measurement
Coupled with a reliable pass/fail formula, this provides more consistent quantitative assessment of colour difference than is possible by visual assessment, usually leading to less redyeing. Records can be kept of the colour of each dye lot, providing a history of the colour for recall before recipe adjustment is made to standard file recipes. This reduces unnecessary and erratic adjustment of file recipes, but facilitates correct adjustment when appropriate. To assist in pass/fail decision-making, most systems now have a graphical colour space display of the distribution around the colour standard of the current dye lot and of previous dye lots.

5.15.4 Inventory control of dyes and drugs
In many commercial systems this function is coupled with automatic weighing and/or dispensing equipment. On some systems the bulk recipe weights can be passed by the computer to the dye-dispensing equipment, to ensure that the required dye weight is correctly dispensed. Automatic deduction of dye and chemical weights from stock can be carried out, with printing of re-order instructions when the stock is reduced to a predetermined level.

5.15.5 Statistics
Production of summary statistics of recipe performance can be specified by the dyehouse management.

In 1983 Gailey summarised the interacting facilities which can be provided by a well-designed colour control system (Figure 5.31) [52]. Supplemented by automatic dye/chemical dispensing systems and dyeing machine control systems, the instrumental approach to colour control provides a most important contribution to efficient production in the dyehouse.

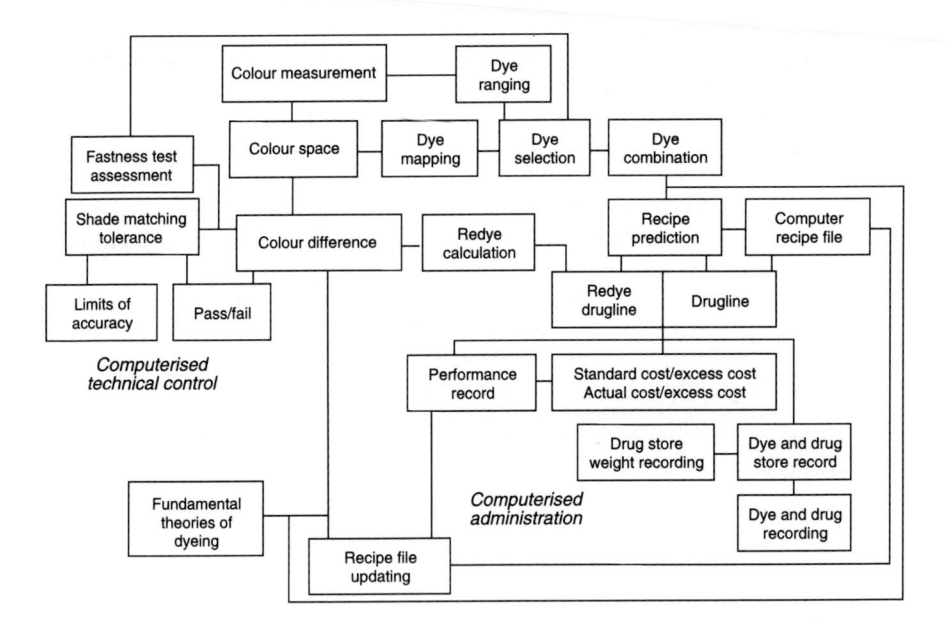

Figure 5.31 Modern dyeing system using a reflectance spectrophotometer and a computer with disc storage

REFERENCES

1. J V Alderson, E Atherton and A N Derbyshire, *J.S.D.C.*, **77** (1961) 657.
2. R H Park and E I Stearns, *J. Opt. Soc. Amer.*, **34** (1944) 112.
3. *J.S.D.C.*, **79** (1963) 573.
4. H R Davidson, H Hemmendinger and J L R Landry, *J.S.D.C.*, **79** (1963) 577.
5. A E Cutler, *J.S.D.C.*, **81** (1965) 601.
6. L Gall, *Colour 73* (Bristol: Adam Hilger, 1973) 153.
7. E Allen, *J. Opt. Soc. Amer.*, **56** (1966) 1256.
8. E Allen, *Color 77* (Bristol: Adam Hilger, 1978) 153.
9. H W Holdaway, *Col. Res. Appl.*, **5** (2) (1980) 93.
10. H W Holdaway, *Aust. Text.*, **1** (6) (1968) 28.
11. P W McGinnis, *Color Eng.*, **15** (6) (1967) 22.
12. B Sluban, *Col. Res. Appl.*, **18** (2) (1993) PAGE?.
13. A H Liddell, D McKay and P J Weedall, *J.S.D.C.*, **92** (1976) 53.
14. R B Love, S Oglesby and I Gailey, *J.S.D.C.*, **81** (1965) 609.
15. R G Kuehni, *Computer colorant formulation* (Lexington, Massachusetts: D C Heath, 1975)
16. R McDonald and D Shaw, J & P Coats internal research report (5 Sept 1968).
17. R McDonald, D McKay and P J Weedall, *J.S.D.C.*, **92** (1976) 39.
18. P J Weedall, PhD thesis, University of Bradford (1981).
19. D Colquhoun, *Appl. Statistics*, **18** (2) (1969) 130.
20. J E Brown and D S Riggs, *J. Biol. Chem.*, **240** (1965) 863.

21. D McKay, PhD thesis, University of Bradford (1976).
22. M J Box, D Davies and W H Swann, *Non-linear optimisation techniques* (Edinburgh: Oliver and Boyd, 1969).
23. Paul Hoffenberg Proc., AATCC Ann. Conf., Philadelphia (1989).
24. S Jeler and V Golob, *Color 89*, Proc. AIC Conf., Buenos Aires (1989).
25. A H Liddell, D McKay and P J Weedall, *J.S.D.C.*, **90** (1974) 164.
26. F T Simon, *J. Color Appear.*, **1** (4) (1972) 5.
27. R A Funk, PhD thesis, Clemson University (1980)
28. G E Beebe, MSc thesis, Clemson University (1974).
29. R Donaldson, *J. Appl. Phys.*, **5** (1954) 210.
30. R McDonald and D McKay, J & P Coats internal research report (5 Nov 1970).
31. D McKay, Preprints SDC symposium, Nottingham (1981).
32. J E Tyler and F P Callahan, *J. Opt. Soc. Amer.*, **41** (1951) 997.
33. F T Simon, R A Funk and A C Laidlaw, *Col. Res. Appl.*, **19** (6) (1994) 461.
34. Tak Min Man, PhD thesis, University of Bradford (1985).
35. G Doring, *Col. Res. Appl.*, **11** (4) (1986) 287.
36. J F Mackin, *J.S.D.C.*, **91** (1975) 75.
37. J F Mackin and J A Purves, *J.S.D.C.*, **96** (1980) 177.
38. R St John, *Colour 73* (Bristol: Adam Hilger, 1973) 431.
39. T F Chong, Proc. Textile Institute Congress, Hong Kong (1985).
40 J C Guthrie, J Moir and P H Oliver, *J.S.D.C.*, **78** (1962) 27.
41. A Miller et al., *J.S.D.C.*, **79** (1963) 604.
42. L F C Friele, *J. Text. Inst.*, **43** (1952) P604.
43. D A Burlone, *Col. Res. Appl.*, **8** (2) (1983) 114.
44. D A Burlone, *Col. Res. Appl.*, **9** (4) (1984) 213.
45. E Walowit, C J McCarthy and R S Berns, *Col. Res. Appl.*, **12** (6) (1987) 340.
46. E Walowit, C J McCarthy and R S Berns, *Col. Res. Appl.*, **13** (6) (1988) 358.
47. S Westland, J M Bishop, M J Bushnell and A L Ushert, *J.S.D.C.*, **107** (1991) 235
48. J M Bishop, M J Bushnell and S Westland, *Col. Res. Appl.*, **16** (1) (1991) 3.
49. H R Kang and P G Anderson, *J. Electron. Imaging*, **12** (Apr 1992) 125.
50. S Westland, *J.S.D.C.*, **110** (1994) 370.
51. J Park and T M Thompson, *J.S.D.C.*, **98** (1982) 74.
52. I Gailey, *J.S.D.C.*, **99** (1983) 253.
53. H S Shah, *Color 89*, Proc. AIC Conf., Buenos Aires (1989).

FURTHER READING

J H Nobbs, *Rev. Prog. Col.*, **15** (1985) 66.

J C Guthrie and J Moir, *Rev. Prog. Col.*, **9** (1978) 1.

J Park, *Instrumental colour formulation* (Bradford: SDC, 1994).

CHAPTER 6

Colour-match prediction for pigmented materials

James H Nobbs

6.1 INTRODUCTION

6.1.1 Historical development

The aim of recipe match prediction is to produce a formulation that when applied to a substrate by a standard method results in a colour identical to that of the standard under specified conditions of illumination and observation. Two models are needed to enable such calculations to be carried out:

(a) an optical model that relates the reflectance spectrum of the surface to the concentration of the components in the formulation, the thickness of coating layer and the reflectance spectrum of the substrate, and
(b) a model of colour vision that relates the reflectance spectrum of a surface to colour appearance.

Although the Schuster–Kubelka–Munk theory and the CIE XYZ method of colour specification fulfilled these requirements and were available by the mid-1930s, the technique of match prediction took many years to develop. Park and Stearns in 1944 were the first to describe equations that calculate an approximate match and an iterative technique for improving upon the match [1]. They used a mechanical calculator and each recipe took several hours to determine. During the next 20 years further progress was hindered by the lack of suitable methods of performing the large number of calculations required for each prediction.

In 1955 an interesting demonstration of the technique was made by Atherton, who made use of ingenious analogue computing methods based on a device constructed from bent wires, cardboard and adhesive tape [2]. It was not until the late 1950s that the advantages of using a programmable digital computer were exploited, when Atherton and Cowgill showed how recipes could be determined in less than a minute [3]. Commercial application of the technique now became a realistic proposition. The availability of mainframe programmable digital computers stimulated progress during the early 1960s, particularly by Alderson, Atherton and Derbyshire at ICI Dyestuffs Division [4]. The ICI group soon developed the technique to the stage where it could be applied to a wide range of textile dyeing and printing systems [5]. Not long afterwards ICI introduced instrumental match prediction as a technical service to customers, the calculations being carried out on a Ferranti Pegasus computer.

The high cost of mainframe computers at this time led to the development of analogue computer methods of solving the equations. Two analogue systems are of note: the COMIC system developed in the USA by Davidson and Hemmendinger [6] and the Redifon [7] system developed in the UK. The COMIC system stored the optical properties of each dye as settings on 16 potentiometers mounted in a small plug-in module. The reflectance spectrum of the standard was registered by setting 16 potentiometers on the main console. The operator selected the modules that represented the dyes in the proposed recipe and plugged them into the computer. A cathode-ray tube displayed two traces, one corresponding to the spectrum of the standard and the other a calculated spectrum of the recipe. The operator adjusted potentiometers representing the concentration of each dye in the mixture until the traces coincided, a process that normally took about 30 minutes.

Modern personal computers are several times more powerful than the machines used by the ICI group and many more facilities are available in modern colour management software than just recipe match prediction. The match-prediction calculation often forms only a small part of colour management software, most of the programming being concerned with the additional facilities and the user interface.

The nature of computing has changed in other ways. The powerful mathematical tools available in spreadsheet programs can allow the algorithms of a new match-prediction program to be initially tested in spreadsheet format. Writing the final program in a high-level computer language is only undertaken when the quality of the matching method has been established via the spreadsheet. All the examples given in this chapter have been obtained with the aid of the Microsoft Excel 5.0 spreadsheet.

In this chapter the basic theory of the calculations is considered in sections 6.1 to 6.3, the way in which the database of optical constants is obtained is described in sections 6.4 to 6.6, and finally match prediction is dealt with in sections 6.7 and 6.8. Both practical and theoretical aspects are considered, following the advice of Patterson given over 95 years ago: 'Theoretical knowledge alone cannot make a successful colour mixer, but it certainly proves of great value in explaining the causes of failure and directing the conditions that lead to success' [8].

6.1.2 Schuster–Kubelka–Munk theory of diffuse reflectance

An important part of colour physics is the prediction of the amount of light that a surface will reflect or transmit at each wavelength from knowledge of the concentration of each colorant in the layer. A simple theory was presented by Kubelka and Munk in 1931 [9], based on considerations developed in 1905 by Arthur Schuster [10], an astrophysicist who studied the passage of light through scattering and absorbing layers. The application of the theory to prediction of reflectance has been reviewed by Nobbs [11]. The treatment developed here refers to coating materials that contain conventional pigments and excludes pigments with highly direction-dependent optical properties, such as metal flake pigments and the mica-based pigments that create colour by exploiting interference effects.

Two fluxes

Schuster simplified the problem by considering the interaction of light with the layer in terms of two fluxes of energy (Figures 6.1 and 6.2). The *i* energy flux includes all light directions that have a component travelling towards the illuminated surface and the *j* energy flux includes all light directions that have a component travelling away from the illuminated surface. For most viewing conditions objects are illuminated from above and as a result *i* and *j* are often termed the down flux and up flux respectively.

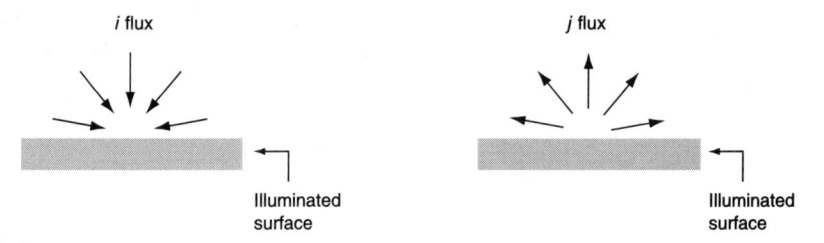

Figure 6.1 Directions of light included in the *i* energy flux

Figure 6.2 Directions of light included in the *j* energy flux

Absorption and scattering coefficients

The interaction of the light within the layer is described by an absorption coefficient K and a scattering coefficient S defined as follows. We consider a thin section of thickness dx at a distance x from the illuminated surface of the layer. The intensity of the two fluxes at this plane are i_x and j_x (Figure 6.3). Then $K\,dx$ is the fractional amount of light lost from a flux by absorption as it passes through the thickness dx, and $S\,dx$ is the fractional amount lost from a flux by scattering as it passes through the thickness dx.

The coefficients defined in this way are not the absorption and scattering extinction coefficients of conventional spectroscopy. In particular the term 'scattering' is used to imply a change in direction of a photon or light beam in a flux so that it is no longer

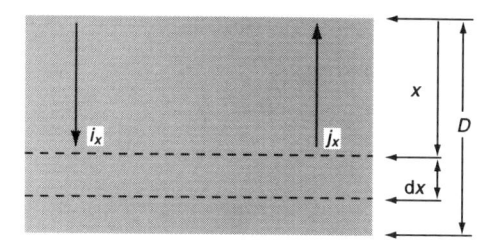

Figure 6.3 Fluxes at a thin section inside a layer

counted in that flux. It follows that energy lost by scattering from one flux appears as energy gained by the other flux.

The net change in each flux is obtained by adding the energy lost and gained as the flux passes through the thin section dx (Eqns 6.1 and 6.2):

$$di_x = -Ki_x\,dx - Si_x\,dx + Sj_x\,dx \tag{6.1}$$

$$-dj_x = -Kj_x\,dx - Sj_x\,dx + Si_x\,dx \tag{6.2}$$

$$\uparrow \qquad \uparrow \qquad \uparrow \qquad \uparrow$$
net change · loss by absorption · loss by scattering · gain by scattering

The negative sign for the change in the j_x flux arises because the direction of j_x is opposite to that of x. Collecting terms together and dividing through by dx leads to two coupled differential equations (Eqns 6.3 and 6.4):

$$\frac{di_x}{dx} = -(K+S)i_x + Sj_x \tag{6.3}$$

$$\frac{dj_x}{dx} = -Si_x + (K+S)j_x \tag{6.4}$$

The solution to Eqns 6.3 and 6.4 relate measurable terms such as reflectance and transmittance of the layer to the coefficients K, S and the thickness of the layer D.

Reflectance ratio of an opaque layer
There are several ways of solving Eqns 6.3 and 6.4 to obtain R_∞, the reflectance ratio at the top surface of an opaque layer. One of the simplest depends on consideration of the reflectance ratio r_x at a point x from the illuminated surface (Eqn 6.5):

$$r_x = \frac{j_x}{i_x} \tag{6.5}$$

The value of R_∞ is just the value of r_x when $x = 0$. The way in which r_x changes with position in the layer is found by differentiating Eqn 6.5 with respect to x (Eqn 6.6):

$$\frac{dr_x}{dx} = \frac{i_x \frac{dj_x}{dx} - j_x \frac{di_x}{dx}}{i_x^2} \tag{6.6}$$

Substitution of Eqns 6.3 and 6.4 for di_x/dx and dj_x/dx respectively and r_x for j_x/i_x gives Eqn 6.7:

$$\frac{dr_x}{dx} = -Sr_x^2 + 2(K+S)r_x - S \tag{6.7}$$

At and near to the top surface of an opaque layer the value r_x does not change with x (Eqn 6.8):

$$\frac{dr_0}{dx} = -Sr_0^2 + 2(K+S)r_0 - S = 0 \tag{6.8}$$

This is a quadratic equation which can be solved to give r_0, which is identical to R_∞ (Eqn 6.9):

$$R_\infty = 1 + K/S - [(1 + K/S)^2 - 1]^{1/2} \tag{6.9}$$

The inverse equation is Eqn 6.10:

$$K/S = \frac{(1 - R_\infty)^2}{2R_\infty} \tag{6.10}$$

which is known as the Kubelka–Munk function and which shows that the reflectance of an opaque layer depends only on the ratio of K to S and not on their absolute values. Figure 6.4 illustrates the dependence of R_∞ on K/S, and clearly indicates the difficulty of achieving a pure, clean white: the slightest amount of absorption reduces the reflectance dramatically.

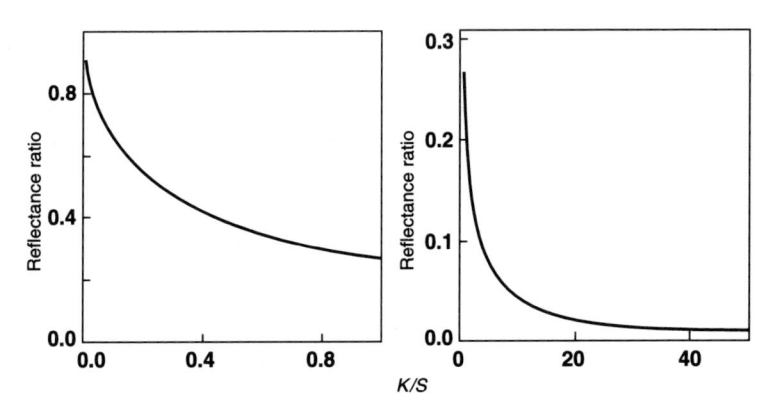

Figure 6.4 Plots of reflectance against K/S for an opaque layer

Units of the coefficients

The K and S coefficients refer to the absorption and scattering of a coating per unit thickness of the layer. The probability that a photon of light will be absorbed or scattered by a particular particle depends on the effective cross-sectional area it presents to the light beam (Figure 6.5). The effective cross-sectional area is the product of the physical cross-sectional area and a dimensionless efficiency factor. The efficiency factors for the absorption and the scattering processes are different, and depend on several factors that include the optical properties of the particle material, the size and shape of the particle, and the properties of the medium in which the particle is immersed.

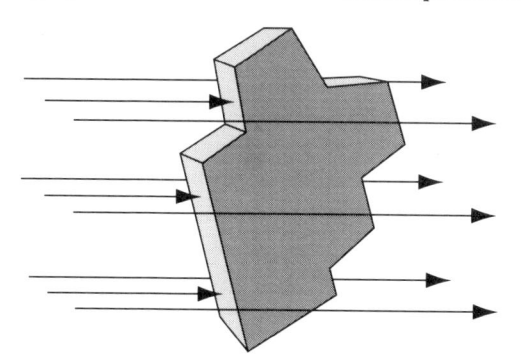

Figure 6.5 Interaction of light with a particle surface

K is the overall probability of absorption as the flux travels through unit thickness of a layer: it is the product of the particle 'absorption' cross-sectional area and the number of particles per unit volume. It follows that the units of K are reciprocal length (m^{-1}, for example). Analogously, the scattering coefficient S has similar units.

The K and S coefficients can be related to the fundamental optical properties of the layer ε and σ. To a first approximation it can be shown that (Eqn 6.11):

$$K = 2\varepsilon \qquad S = \sigma \qquad (6.11)$$

where ε is the linear absorption extinction coefficient and σ is the linear scattering extinction coefficient.

Additivity of the coefficients

Under ideal conditions the absorption coefficient of a mixture is the sum of the contributions from each component present. In a similar way the scattering coefficient is the sum of the contributions from each component present. For example, the K_M and S_M values for a mixture of components A and B, are given by Eqn 6.12:

$$K_M = K_A + K_B \qquad S_M = S_A + S_B \qquad (6.12)$$

The number of particles per unit volume depends on the volume fraction of pigment in the coating. Thus the contributions to K_M and S_M by a component A in a mixture should be a linear function of the volume concentration of A, C_A (Eqn 6.13):

$$K_A = C_A \hat{K}_A \qquad S_A = C_A \hat{S}_A \qquad (6.13)$$

where $\hat{K}$ and $\hat{S}$ are the absorption and scattering coefficients per unit thickness for unit volume of component A.

6.1.3 Calculated and measured values of reflectance

The reflectance ratio calculated by the theory is defined by Eqn 6.14:

$$R_\infty = \frac{\text{energy flux reflected from top surface}}{\text{energy flux incident on top surface}} \qquad (6.14)$$

In general this is not the parameter measured by a reflectance spectrophotometer. These instruments normally report the ratio of intensity detected when the sample is placed against the measurement port to that detected when a diffusely reflecting ideal white material is placed against the same measurement port. The precise nature of the measurement and how it is related to the reflectance ratio depends on the optical geometry of the instrument and reflection characteristics of the sample.

To a good approximation an instrument containing an integrating sphere making measurements on nonglossy surfaces, such as textile materials and matt-finish paints, measures the reflectance ratio of the samples.

6.1.4 Limitations of the Schuster–Kubelka–Munk theory

Certain conditions need to be satisfied for the theory to be a reasonable approximation to the measured values.

Diffuse illumination
The sample should be illuminated with diffuse light, that is, the light incident on the surface should be of the same intensity from all directions. The closest we may come to experiencing diffuse illumination is on a foggy day or when looking out of the window of an aircraft as it flies through thick cloud. Diffuse illumination conditions can be created in a measuring instrument by an integrating sphere.

Diffuse reflection
The sample should be a diffuse reflector. A diffusely reflecting layer reflects an equal intensity of light along each direction, whether the incident light is diffuse or not.

Such a surface is known as a *Lambertian reflector* after the physicist Charles Lambert, who first described its properties. Matt white surfaces such as white paper and paint are good approximations to diffuse reflectors. Very good diffuse reflectors can be made from lightly compressed finely powdered materials such as titanium dioxide or magnesium carbonate. A layer of fresh powdery snow is an excellent Lambertian reflector.

Layer composition
Certain conditions apply to the layer itself. The surface should be smooth, have a large illuminated area compared to its thickness and be homogeneous in composition. The large area : thickness ratio required ensures that only a negligible fraction of the light is lost from the edges of the layer.

Not taken into account
The theory does not take into account the polarisation of the light, interference effects and the partial reflection and refraction of the light at any air-to-layer interface that may be present.

6.2 SEMITRANSPARENT LAYERS

One of the most common uses of pigments is in generating the colour of printing inks. Printed layers are usually semitransparent and the light reflected from the printed surface often includes light directly scattered from the pigments in the ink layer as well as light that has been transmitted through the layer, reflected by the substrate and transmitted back out of the system (Figure 6.6).

This is a more complex situation than the opaque layer described in section 6.1. The Kubelka–Munk theory does provide a solution and it can be developed by first establishing the optical properties of a layer in isolation (Figure 6.7) and then determining the effect when a substrate is placed beneath the layer.

The optical properties of the an isolated layer can be described by the reflectance R_o, transmittance T and absorbance A of energy as defined in Eqns 6.15–6.17:

$$R_o = \frac{\text{energy flux reflected by layer}}{\text{energy flux incident on layer}} \quad (6.15)$$

$$T = \frac{\text{energy flux transmitted through layer}}{\text{energy incident on layer}} \quad (6.16)$$

$$A = 1 - R_o - T \quad (6.17)$$

Provided the assumptions of the Kubelka–Munk theory are satisfied, then the way in which a light energy flux interacts with a plane parallel layer is described by Eqns 6.3

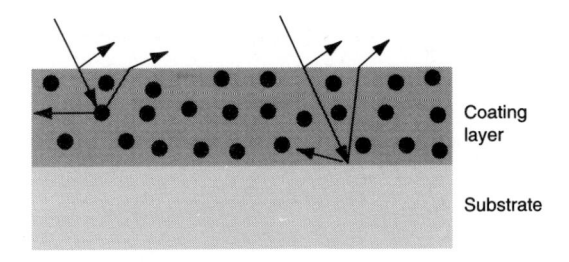

300 COLOUR-MATCH PREDICTION FOR PIGMENTED MATERIALS

Figure 6.6 Light reflection at a semitransparent layer

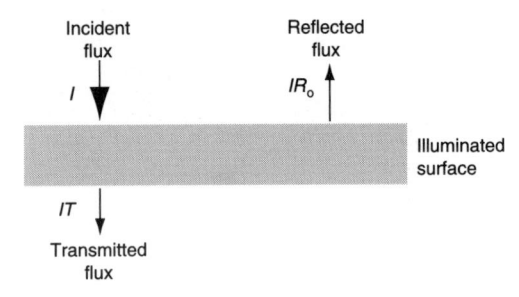

Figure 6.7 Reflectance and transmission of an isolated layer

and 6.4. The general solution provides expressions for the two fluxes i_x and j_x at the position a distance x from the illuminated surface of the layer. Of more practical interest is the reflectance R_o and transmittance T in terms of the Kubelka–Munk absorption and scattering coefficients K and S and the reflectance of an opaque layer R_∞ (Eqn 6.18 and 6.19):

$$R_o = R_\infty \left[\frac{1 - \exp(-2Z)}{1 - R_\infty^2 \exp(-2Z)} \right] \tag{6.18}$$

$$T = (1 - R_\infty^2) \left[\frac{\exp(-Z)}{1 - R_\infty^2 \exp(-2z)} \right] \tag{6.19}$$

where Z (which can be thought of as a measure of the optical thickness of the layer) is defined by Eqn 6.20 in which D is the physical thickness of the layer:

$$Z = D[K(K + 2S)]^{1/2} \tag{6.20}$$

In order to determine the properties of the coating on a substrate, two more reflectance expressions need to be defined (Eqns 6.21 and 6.22):

$$R_g = \frac{\text{energy flux reflected by substrate}}{\text{energy flux incident on substrate}} \tag{6.21}$$

SEMITRANSPARENT LAYERS 301

$$R = \frac{\text{energy flux reflected by system}}{\text{energy flux incident on system}} \quad (6.22)$$

The reflectance of the composite is obtained by imagining the coating layer split from the substrate (Figure 6.8). A flux I is assumed to be incident on the top surface of the layer, the amount IR_o is reflected by the layer and IT is transmitted. A second flux J is assumed reflected back from the substrate and is incident on the bottom surface of the layer. The amount JR_o is reflected and JT transmitted though the layer.

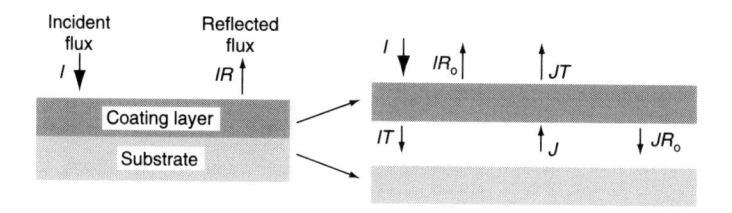

Figure 6.8 Imaginary splitting of the coating from a substrate to form an isolated layer

From the definition of R, Eqn 6.23 follows:

$$R = \frac{IR_o + JT}{I} \quad (6.23)$$

The flux J can be related to I via R_g (Eqn 6.24):

$$J = R_g(IT + JR_o) \quad (6.24)$$

Rearrangement gives Eqn 6.25:

$$J = \frac{IT R_g}{1 - R_g R_o} \quad (6.25)$$

and substitution for J in Eqn 6.23 gives Eqn 6.26:

$$R = R_o + \frac{T^2 R_g}{1 - R_g R_o} \quad (6.26)$$

In principle the problem is now solved. The relationship between K, S and the reflectance R_o and transmittance T of an isolated layer is expressed by Eqns 6.18–6.20. Eqn 6.26 relates these values to the reflectance R of the layer coated on to a substrate of reflectance R_g. This relationship, defined by a lengthy single equation, can be expressed more simply by defining two new terms α and β (Eqn 6.27):

$$R = \frac{\alpha R_g + \beta R_\infty}{\alpha + \beta} \qquad (6.27)$$

where α is a function of the opaque reflectance R_∞ and the reflectance of the substrate R_g (Eqn 6.28):

$$\alpha = \frac{1 - R_\infty^2}{1 - R_g R_\infty} \qquad (6.28)$$

and β is a function of K, S (and D) via the variable Z defined in Eqn 6.20 (Eqn 6.29):

$$\beta = \exp(2Z) - 1 \qquad (6.29)$$

The changes in R_o, T and R as the thickness of the coating increases are shown in Figures 6.9, 6.10 and 6.11. The diagrams correspond to weakly ($K = 0.2$), moderately ($K = 1.0$) and strongly ($K = 5.0$) absorbing coating layers respectively with the scattering coefficient the same for all three examples, $S = 2.0$. The behaviour of the three layers is similar, with the transmittance starting at 100% and rapidly decreasing with increasing thickness. Conversely the reflectance of the isolated layer starts at zero and increases with increasing thickness to approach the opaque layer value. The net reflectance starts at the value for the substrate (R_g) and as the layer thickness increases tends towards the opaque layer reflectance R_∞.

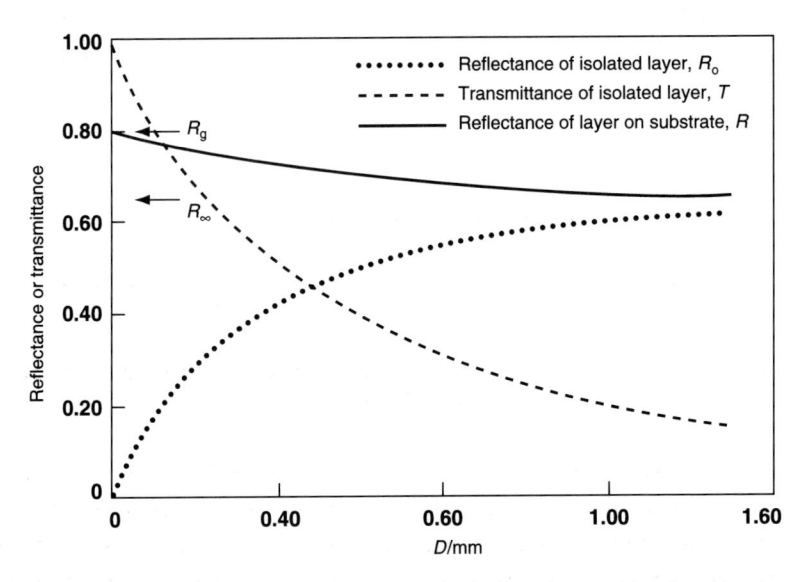

Figure 6.9 Dependence of reflectance and transmission of a weakly absorbing layer on thickness; $K = 0.2$, $S = 2.0$, $R_\infty = 0.642$, $R_g = 0.800$

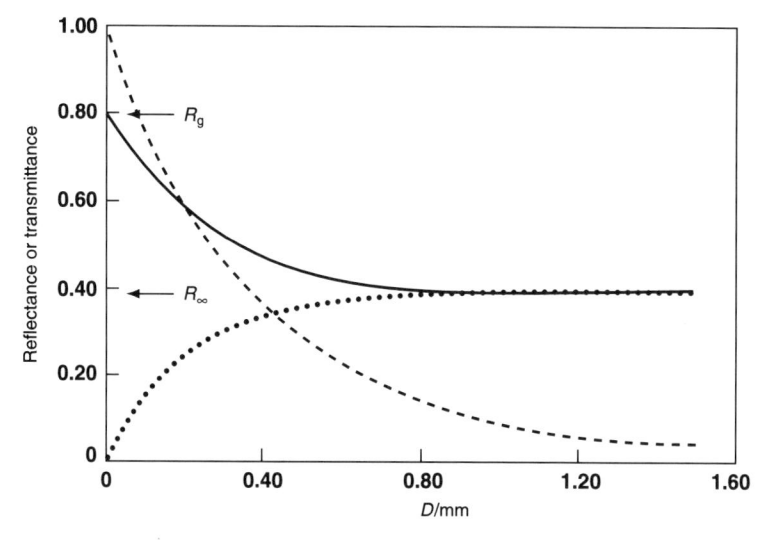

Figure 6.10 Dependence of reflectance and transmission of a moderately absorbing layer on thickness; $K = 1.0$, $S = 2.0$, $R_\infty = 0.382$, $R_g = 0.800$; for key see Figure 6.9

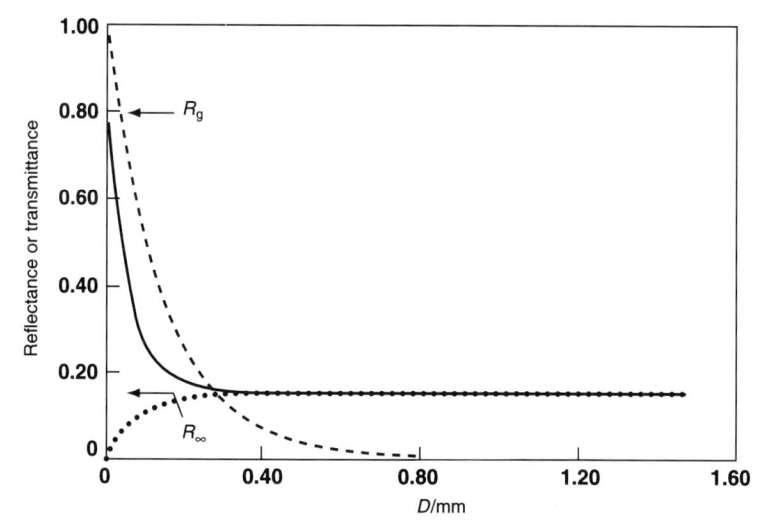

Figure 6.11 Dependence of reflectance and transmission of a strongly absorbing layer on thickness; $K = 5.0$, $S = 2.0$, $R_\infty = 0.146$, $R_g = 0.800$; for key see Figure 6.9

Although the general trend shown in the three figures is similar, the relative layer thickness at which the coating becomes opaque ($R \approx R_\infty$) is strongly influenced by the value of the absorption coefficient (Table 6.1). The strongly absorbing layer is approximately opaque when D is 0.45 units, whereas the weakly absorbing layer needs to be over 4 times as thick ($D = 1.90$ units) before opacity is achieved.

304 COLOUR-MATCH PREDICTION FOR PIGMENTED MATERIALS

Table 6.1 Relative layer thickness for opacity $S = 2.00$

K	D at $R \approx R_\infty$
0.2	1.90
1.0	1.10
5.0	0.45

Table 6.2 Sample calculation values for a layer with $K = 1.00$, $S = 2.00$, $R_g = 0.800$ and $R_\infty = 0.3820$

D	Z (Eqn 6.20)	R_o (Eqn 6.18)	T (Eqn 6.19)	R (Eqn 6.26)
0.0000	0.0000	0.0000	1.0000	0.8000
0.2000	0.4472	0.2401	0.5808	0.5741
0.4000	0.8944	0.3261	0.3579	0.4647
0.6000	1.3416	0.3594	0.2255	0.4166
0.8000	1.7889	0.3728	0.1433	0.3962
1.0000	2.2361	0.3782	0.0914	0.3878

Table 6.3 Sample calculation values for a layer with $K = 1.00$, $S = 2.00$, $R_g = 0.800$, $R_\infty = 0.3820$ and $\alpha = 0.8130$

Z (Eqn 6.20)	β (Eqn 6.29)	R (Eqn 6.27)
0.0000	0.000	0.8000
0.4472	1.446	0.5741
0.8944	4.983	0.4647
1.3416	13.633	0.4166
1.7889	34.791	0.3962
2.2361	86.544	0.3878

Tables 6.2 and 6.3 show the values obtained from a typical calculation and can be used to check the accuracy of computer programs or spreadsheets.

6.3 PARTIAL REFLECTION AT AIR/COATING INTERFACE

As mentioned in section 6.1, the Schuster–Kubelka–Munk theory does not take into account the partial reflection of light at the air-to-coating interface as the light enters and leaves the layer. The reflection arises from the difference in the refractive index of air to that of the coating material. The process is illustrated in Figure 6.12(a), where a parallel light beam of intensity I is shown incident on an ideally flat boundary between air of refractive index n_1 and a surface coating material of refractive index n_2. A fraction t of the incident beam is transmitted through the boundary in a direction closer to the normal to the boundary surface and a fraction r is reflected away from the surface at the same angle relative to the surface normal as the incident beam. The partially reflected light has had no opportunity to interact with the colorants present in the coating layer and will have roughly the same spectral characteristics as the incident light.

A similar interaction for a light beam approaching the interface from within the coating layer is illustrated in Figure 6.12(b). Part of the beam will be reflected back into the coating material and the remainder will be transmitted. In this case the transmitted beam direction will be refracted to a greater angle relative to the surface normal than that of the incident beam. When the angle of incidence is greater than a critical angle then total internal reflection takes place and no light is transmitted through the boundary.

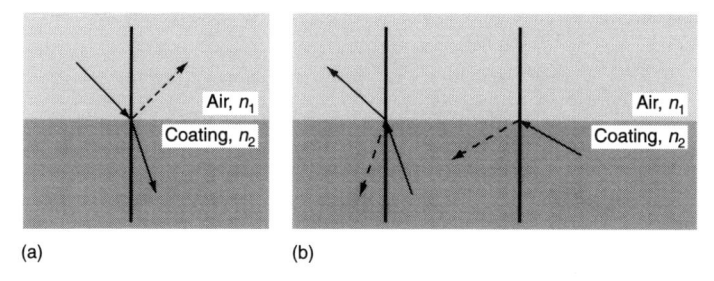

Figure 6.12 Partial reflection and transmission; (a) at air/coating interface, (b) for both a low and a high angle of incidence when total internal reflection occurs

The amounts of light reflected and transmitted can be calculated from Fresnel's equation and depends on the angle of incidence and the refractive indices of the two layers. At normal incidence the equation is very simple and the fraction reflected is given by Eqn 6.30:

$$r = \frac{(n_2 - n_1)^2}{(n_2 + n_1)^2} \tag{6.30}$$

The effect of the boundary has to be included in the calculation of the measured reflectance of a material irrespective of whether the coating layer is opaque or semitransparent. The extent of the effect will depend on the refractive index of the coating, the angular distribution of the incident light and the roughness of the surface.

A correction equation can be derived by considering the interaction of two energy fluxes with the boundary. An energy flux I is imagined travelling towards the boundary

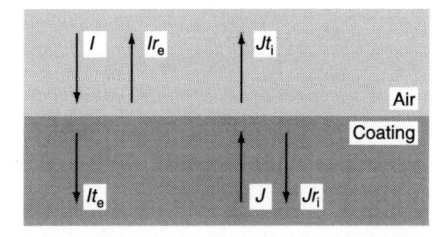

Figure 6.13 Partial reflection of the fluxes *I* and *J* at air/coating interface

from the air and flux *J* travelling towards the boundary from within the layer (Figure 6.13). The fractions of the incident flux *I* that are reflected and transmitted by the interface are r_e and t_e respectively. The fractions of the incident flux *J* that are reflected and transmitted by the boundary are r_i and t_i respectively.

The measured reflectance ρ of the layer is the ratio of the total reflected energy flux to the total incident flux (Eqn 6.31):

$$\rho = \frac{r_e I + t_i J}{I} \tag{6.31}$$

The flux *J* is linked to flux *I* via the true or body reflectance *R* of the layer, which is the reflectance modelled by the Schuster–Kubelka–Munk theory, by Eqn 6.32:

$$J = R(t_e I + r_i J) \tag{6.32}$$

Rearrangement gives Eqn 6.33:

$$J = \frac{t_e I R}{1 - r_i R} \tag{6.33}$$

Substitution of Eqn 6.33 for *J* in Eqn 6.31 provides the relation between ρ and *R* (Eqn 6.34):

$$\rho = r_e + \frac{t_e t_i R}{1 - r_i R} \tag{6.34}$$

This equation is often known as the Saunderson equation [12]. Given that glass and many of the polymeric materials used in paints and printing inks have refractive indices of about 1.50, the correction coefficients when flux *I* is a collimated beam and flux *J* is a diffuse beam are $r_e = 0.040$ and $r_i = 0.600$. Then, since $t_e = 1 - r_e$ and $t_i = 1 - r_i$, it follows that $t_e = 0.960$ and $t_i = 0.400$.

The large value of r_i compared with r_e arises because a large fraction of the diffuse flux *J* will approach the air-to-coating boundary at an angle of incidence greater than the critical angle of about 40°. This light will be totally reflected back into the coating layer. No such effect occurs for the light in flux *I*.

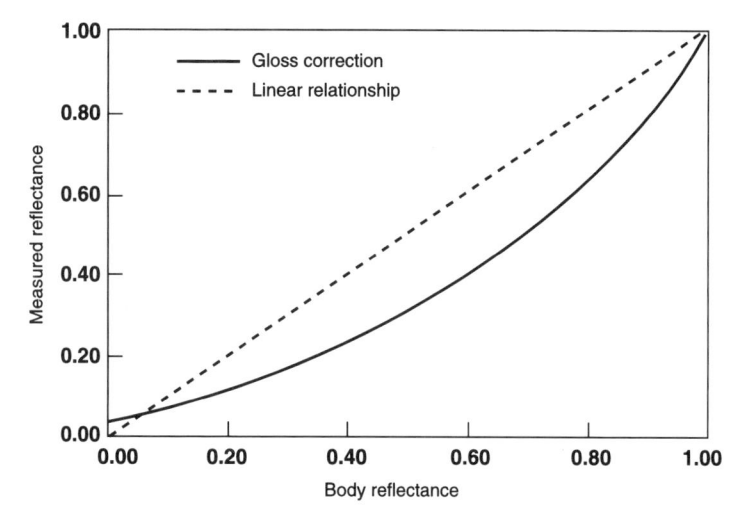

Figure 6.14 Calculated relationship between measured and body (or true) reflectance for a coating of refractive index 1.5

The relationship between the true and the measured reflectance for these values of the coefficients is plotted in Figure 6.14. This shows that the partial reflection at the surface can have a considerable effect on the measured reflectance, the effect being greatest at a true reflectance of about 60% which would be reduced by the interface to a measured value of 40%.

Eqn 6.34 can be rearranged to allow the calculation of the true or body reflectance from the measured value (Eqn 6.35):

$$R = \frac{\rho - r_e}{t_e t_i + r_i(\rho - r_e)} \tag{6.35}$$

It is the body reflectance R that is used in all calculations involving the Schuster–Kubelka–Munk theory.

6.4 DATABASE CALIBRATION OF OPAQUE LAYERS

Practical application of the Schuster–Kubelka–Munk theory depends on the ability to determine K and S values. This can be done by the preparation of calibration panels and their measurement on an appropriate reflectance spectrophotometer.

The assumptions of the Schuster–Kubelka–Munk theory are most closely met by instruments that use an integrating sphere measurement head. These devices provide either diffuse illumination or diffuse observation, depending on the details of the optical arrangement. The two-flux theory assumes both a diffuse illumination and a diffuse

6.4.1 Determination of K and S values

It was shown in section 6.1 that the reflectance of an opaque layer depends on the ratio of K to S and not on the absolute values of the coefficients. Thus the calibration process can be simplified by determining values relative to those of some standard material. It is normal practice to take the scattering power of a standard white formulation S_W as the standard property. It is reasonably practical to make a white paint with consistent properties because of the excellent properties of modern titanium dioxide pigments.

First consider a sample panel of a paint containing a volume concentrations C_A of pigment A and C_W of the standard white. The reflectance spectrum of the panel is measured and a value of ρ obtained at a particular wavelength. The measured reflectance is converted to the body or true reflectance using Eqn 6.35 above, where typical values of the coefficients are $r_e = 0.040$, $r_i = 0.600$, $t_e = 0.960$, $t_e = 0.400$.

The ratio of K to S is then calculated from Eqn 6.36:

$$K/S = \frac{(1-R)^2}{2R} \tag{6.36}$$

This can be expressed in terms of contributions from the individual components by Eqn 6.37:

$$\frac{K_M}{S_M} = \frac{K_A + K_W}{S_A + S_W} \tag{6.37}$$

The optical coefficients are made relative to the scattering power of white by dividing the numerator and denominator terms by S_W (Eqn 6.38):

$$\frac{K_M}{S_M} = \frac{\dfrac{K_A}{S_W} + \dfrac{K_W}{S_W}}{\dfrac{S_A}{S_W} + 1} \tag{6.38}$$

The equation is now becoming cumbersome and further development is simplified by defining a quantity ω to represent the ratio of K to S (Eqn 6.39):

$$\omega_M = \frac{K_M}{S_M} \qquad \omega_A = \frac{K_A}{S_A} \qquad \omega_W = \frac{K_W}{S_W} \tag{6.39}$$

Noting that:
$$K_A = S_A \omega_A$$

Eqn 6.39 can be re-expressed as Eqn 6.40:

$$\omega_M = \frac{\omega_A \left(\dfrac{S_A}{S_W}\right) + \omega_W}{\dfrac{S_A}{S_W} + 1} \tag{6.40}$$

Eqn 6.40 can be rearranged to extract the term of interest S_A/S_W, the scattering coefficient of the coloured pigment relative to the scattering power of the white (Eqn 6.41):

$$\frac{S_A}{S_W} = \frac{\omega_M - \omega_W}{\omega_A - \omega_M} \tag{6.41}$$

The relative absorption coefficient is obtained via Eqn 6.42:

$$\frac{K_A}{S_W} = \omega_A \left(\frac{\omega_M - \omega_W}{\omega_A - \omega_M}\right) \tag{6.42}$$

Of the three terms on the right-hand side of Eqns 6.41 and 6.42, only ω_M is known from measurement of the mixture panel. The unknown terms ω_A and ω_W represent the ratio K/S for component A and the white standard respectively. If it is assumed that the value of K/S of a colorant is constant, the ratio has the same value irrespective of whether the colorant is a component in a mixture or is present in a formulation on its own. The value of ω_A can be determined from the reflectance of an opaque panel with only component A in the formulation and that of ω_W from the reflectance of an opaque panel with only the white standard in the formulation.

6.4.2 Sample calculation for Pigment Red (oxide red)

The following example refers to paint panels of formulations containing mixtures of predispersed colorants (Sandosperse) in a heat-set resin (Synthese). In each case the total volume of pigment in the dry layer is no more than 25%. The test colorant is Sandosperse S-GAJ which contains red iron oxide and the reference white is Sandosperse S-CBJ which contains titanium dioxide. Predispersions of this type are manufactured to controlled colour and strength and contain additives that promote the dispersion of the colorant into the formulation.

Figure 6.15 displays the true reflectance spectra of panels containing white only (R_W), red only (R_A) and a mixture of red and white (R_M). Figure 6.16 displays the values of K_A/S_W and S_A/S_W determined from these spectra, showing absorption in the

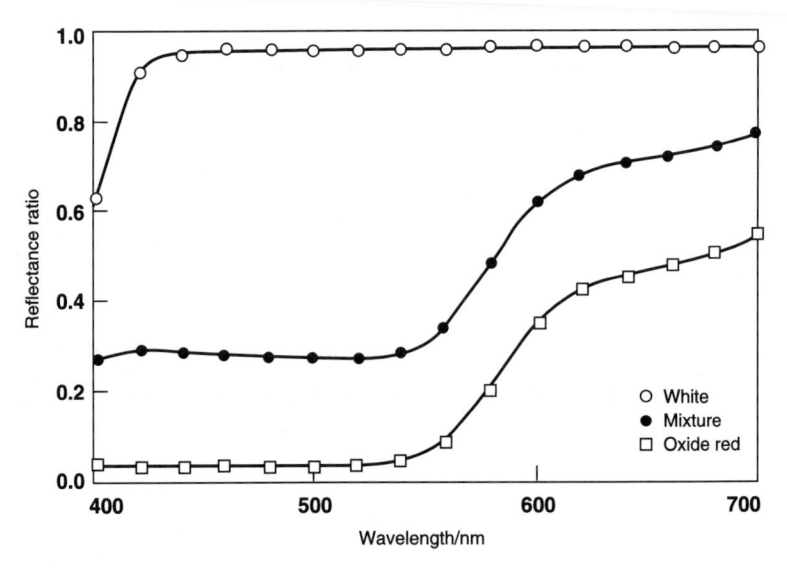

Figure 6.15 True reflectance of panels containing a reference white, a pigment (oxide red) and a test mixture

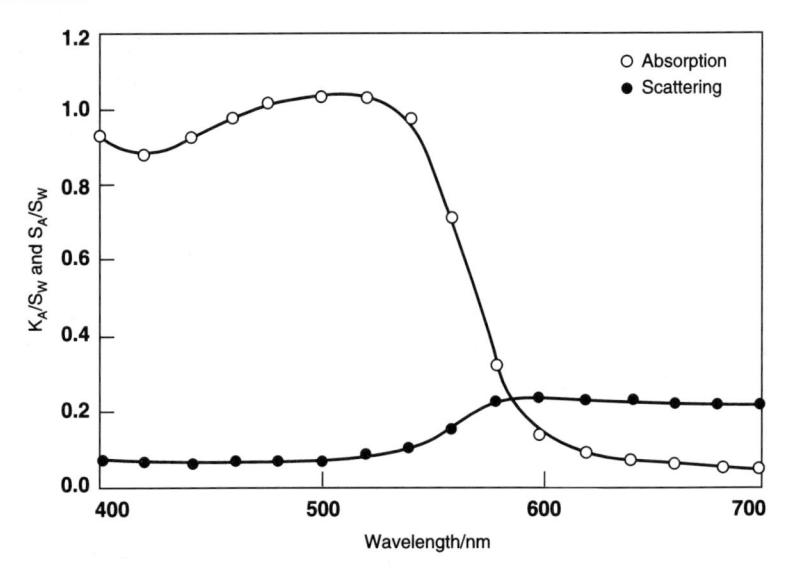

Figure 6.16 Relative absorption and scattering values for the oxide red component in a test mixture (see Figure 6.15)

blue and green regions of the spectrum and scattering in the red region; this pattern is typical for a red inorganic pigment.

Table 6.4 summarises results calculated at five wavelengths that cover a range of absorption and scattering values. The table can be used to check the correct operation

DATABASE CALIBRATION OF OPAQUE LAYERS 311

Table 6.4 Results from a calculation of relative absorption and scattering values from the true reflectance of panels containing white only, oxide red only and a mixture of both.

Wave-length /nm	Reflectance ratio			K/S values (ω) (Eqn 6.36)			Eqn 6.42 K_A/S_W	Eqn 6.41 S_A/S_W
	White	Oxide red	Mixture	White	Oxide red	Mixture		
400	0.6240	0.0372	0.2721	0.11330	12.4693	0.9733	0.9328	0.0748
500	0.9611	0.0329	0.2732	0.00079	14.2002	0.9665	1.0362	0.0730
600	0.9679	0.3558	0.6247	0.00053	0.5832	0.1127	0.1391	0.2385
700	0.9671	0.5527	0.7763	0.00056	0.1811	0.0322	0.0385	0.2128

of programs and spreadsheets. The value of S_W depends on the amount of white present in the mixture. Further assumptions have to be made to determine the absorption and scattering coefficients relative to a fixed reference value. If the properties of the reference white material are satisfactory then S_W will be a linear function of the volume concentration of white in the test mixture. By defining the fixed reference parameter as $\hat{S}_W$, the scattering coefficient of unit concentration of white per unit layer thickness, Eqn 6.42 becomes Eqn 6.43:

$$\frac{K_A}{C_W \hat{S}_W} = \omega_A \left(\frac{\omega_M - \omega_W}{\omega_A - \omega_M} \right) \tag{6.43}$$

where C_W is the concentration of the reference white in the mixture. The left-hand side of Eqn 6.43 includes the ratio of an optical coefficient to $\hat{S}_W$; this type of ratio is often needed and the notation in Eqn 6.44 is used:

$$K_{AW} = \frac{K_A}{\hat{S}_W} \qquad \hat{K}_{AW} = \frac{\hat{K}_A}{\hat{S}_W} \tag{6.44}$$

where $\hat{K}_A$ is the absorption coefficient of unit concentration of colorant A per unit layer thickness.

Rearrangement of Eqn 6.43 gives the absorption of the oxide red relative to the reference scattering value (Eqns 6.45 and 6.46):

$$K_{AW} = C_W \omega_A \left(\frac{\omega_M - \omega_W}{\omega_A - \omega_M} \right) \tag{6.45}$$

$$S_{AW} = \frac{K_{AW}}{\omega_A} \qquad (6.46)$$

The colorant concentrations in these equations can be expressed as fractions of the volume of the colorants present, $(C_A + C_W)$, rather than as a fraction of the total volume. This follows from the dependence of the reflectance of an opaque layer on the ratio of K to S and not on the absolute values.

A match-prediction database has to characterise the absorption and scattering of each component at any concentration. The concentration dependence is determined by preparing mixtures of the test colorant with the reference white at a range of relative concentrations. If the colorant is well behaved then the values of K_{AW} obtained from the mixtures will be a linear function of concentration C_A (Eqn 6.47):

$$K_{AW} = C_A \hat{K}_{AW} \qquad (6.47)$$

A plot of K_{AW} against C_W should be a straight line with slope $\hat{K}_{AW}$, the linear calibration constant of absorption for colorant A. It represents the absorption of unit amount of colorant A relative to the scattering of unit amount of the standard white. The scattering calibration constant can be obtained from Eqn 6.48:

$$\hat{S}_{AW} = \frac{\hat{K}_{AW}}{\omega_A} \qquad (6.48)$$

A separate plot is needed at each wavelength of the spectrum. Figure 6.17 shows the results obtained for the oxide red mixtures described earlier. Plots at five wavelengths

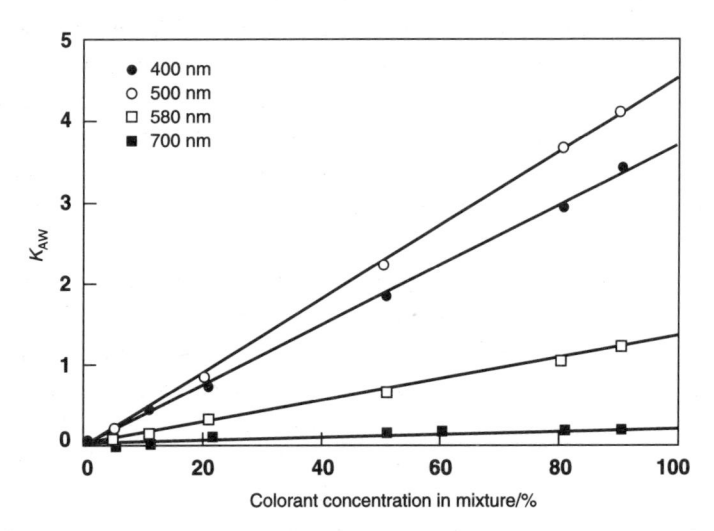

Figure 6.17 Concentration dependence of absorption of the oxide red component in mixtures with a reference white

are shown; the lines are linear regression fits to the data points. A linear relationship between absorption and volume concentration is seen throughout the spectrum. Oxide red pigment is easy to disperse, especially in the predispersed form used to produce these panels; other pigments will not necessarily show a linear relationship at each wavelength in the spectrum. The small deviations of the data points from the line indicate that errors arising from reflectance measurement and sample preparation are small.

6.4.3 Estimates of error of the optical constants

The precision of the measured values of K_{AW} and S_{AW} will depend on the precision of measurement of the reflectance values. This in turn is determined by the method of measurement and the precision of the instrument itself. To a first approximation it can be assumed that the error in the reflectance measurement ($\Delta\rho$) for a given type of panel is constant and is usually better than ± 1%. Because the values of K_{AW} and S_{AW} are determined from the reflectance of three different panels, however, the associated error in K_{AW} depends on the reflectance values of those panels. It can be expected that the error in K_{AW} will be large when the mixture reflectance ρ_M is close to the value of either the colorant-only panel ρ_A or the reference white ρ_W. This is confirmed by the curves in Figure 6.18, which plots the percentage relative error of K_{AW} as a function of the difference in reflectance ($\rho_M - \rho_A$). The values have been calculated assuming a reflectance error of ±0.1%, keeping ρ_W and ρ_A constant and varying the reflectance of the mixture ρ_M. The plot shows a rapid increase in error as ρ_M approaches to within 2.5% of ρ_A or ρ_W. The plot shows that ρ_M must be at least 2.5% different from both ρ_W and ρ_A if absorption values accurate to within ± 10% are to be obtained.

The relationship between $\Delta\rho$ and $\Delta K_{AW}/K_{AW}$ arises from the form of Eqn 6.45 in that when either ($\omega_M - \omega_W$) or ($\omega_A - \omega_M$) is small, small changes in ω_M, ω_W or ω_A will strongly affect the value of K_{AW}. One way this can occur is when the measured reflectance of the mixture panel is similar to that of the reference white or of the colorant-only panel.

Figure 6.18 suggests that the most precise values of K_{AW} are obtained when ρ_M is the arithmetic mean of ρ_W and ρ_A; in Figure 6.19 the relative error in K_{AW} is plotted against ($\rho_W - \rho_A$) when ρ_M fulfilled this criterion. The figure shows a rapid increase in error as ρ_A approaches ρ_W and that a difference of at least 5% is needed to provide absorption values accurate to better than ±10%. The lower limit of ($\rho_M - \rho_A$) or ($\rho_W - \rho_M$) needed to achieve a relative error in K_{AW} of less than ±10% is roughly in proportion to the value of $\Delta\rho$ as shown in Table 6.5.

The magnitude of the errors shown in Figures 6.18 and 6.19 have been estimated using differential coefficients to determine how a small change in measured reflectance

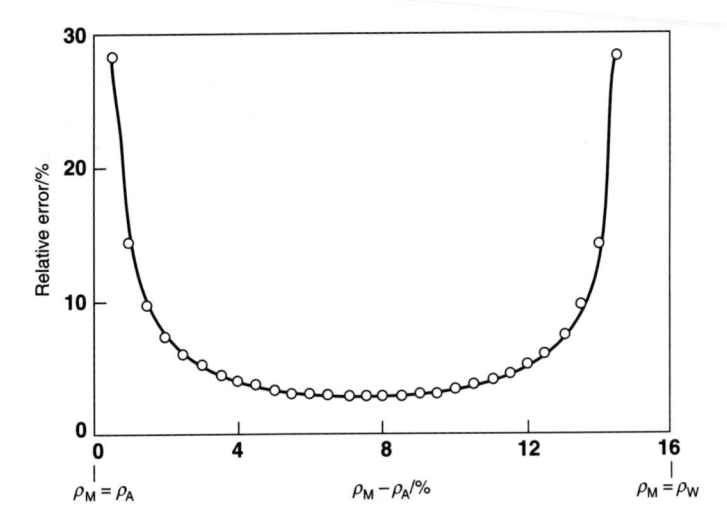

314 COLOUR-MATCH PREDICTION FOR PIGMENTED MATERIALS

Figure 6.18 Relative error in K_{AW} for various values of $\rho_M - \rho_A$; $\rho_A = 75\%$, $\rho_W = 81\%$, $\Delta\rho = 0.1\%$

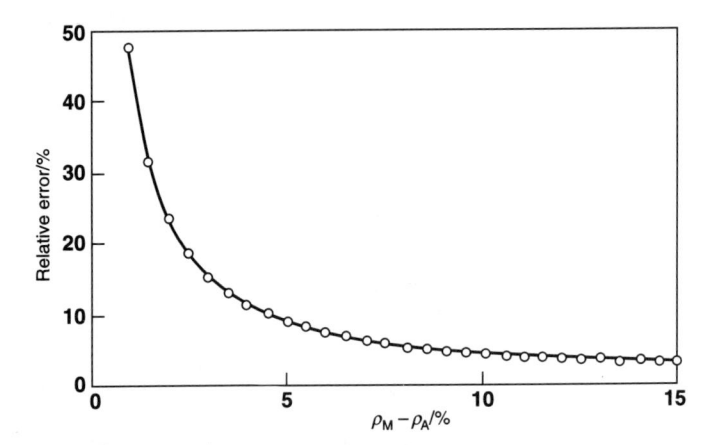

Figure 6.19 Relative error in K_{AW} for various values of $\rho_W - \rho_A$; $\rho_M = (\rho_W + \rho_A)/2$, $\Delta\rho = 0.1\%$

Table 6.5 Lower limit of reflectance values to achieve a relative error in K_{AW} of less than 10%

Measurement error $\Delta\rho$ /%	Limit of $(\rho_M - \rho_A)$ and $(\rho_W - \rho_M)$ /%
0.20	5.00
0.10	2.50
0.05	1.25

$\Delta\rho$ produces a corresponding change in K_{AW}. Separate relationships are needed to determine how the error in each reflectance value affects the value of K_{AW}. The expressions are simplified by using the expression ΔQ to denote the change in K_{AW}.

The error arising from a small change in the reflectance of the white only panel is ΔQ_W given by Eqn 6.49:

$$\Delta Q_W = \frac{\partial Q}{\partial \omega_W} \Delta \omega_W$$
$$= -\frac{C_W \omega_A}{\omega_A - \omega_M} \Delta \omega_W \quad (6.49)$$

where $\Delta\omega$ is given by Eqn 6.50:

$$\Delta \omega_W = \frac{1}{2}\left(1 - \frac{1}{R_W^2}\right)\left[\frac{(1 - r_i R_W)^2}{t_e t_i}\right] \Delta \rho_W \quad (6.50)$$

and r_i, t_e, and t_i are the gloss correction coefficients. Eqns 6.51 and 6.52 give ΔQ_A and ΔQ_M:

$$\Delta Q_A = \frac{\partial Q}{\partial \omega_A} \Delta \omega_A$$
$$= -\frac{C_W \omega_M (\omega_M - \omega_W)}{(\omega_A - \omega_M)^2} \Delta \omega_A \quad (6.51)$$

$$\Delta Q_M = \frac{\partial Q}{\partial \omega_M} \Delta \omega_M$$
$$= \frac{C_W \omega_A (\omega_A - \omega_W)}{(\omega_A - \omega_M)^2} \Delta \omega_M \quad (6.52)$$

The subscripts W, A and M denote values that refer to the white-only, colorant-only and mixture panels respectively. The average error in K_{AW} for independent random errors in each of the reflectance measurements is found by summing the squares of all the ΔQ terms and then taking the square root (Eqn 6.53):

$$\Delta Q = (\Delta Q_W^2 + \Delta Q_A^2 + \Delta Q_M^2)^{1/2} \quad (6.53)$$

6.4.4 Analysis of very bright colorants

The long-wavelength reflectance of single-colorant panels of very bright yellow, orange and red colorants is so close to that of the reference white that the optical properties at these wavelengths cannot be accurately determined from mixtures with white.

316 COLOUR-MATCH PREDICTION FOR PIGMENTED MATERIALS

A useful method of obtaining additional calibration data in these cases involves preparing a panel of a mixture of the test colorant with a small amount of a standard black colorant. A panel containing a mixture of white with the standard black colorant is also required. As in the previous method of analysis, single-colorant panels of the white, black and test colorant are needed.

The reflectance of the black-only, white-only and mixture of black with white are converted to K/S to give ω_B, ω_W and $\omega_{M,BW}$ respectively. The absorption coefficient of the black relative to the scattering of white is determined from Eqns 6.45 and 6.47, as shown in Eqn 6.54:

$$\frac{C_{BW}\hat{K}_B}{\hat{S}_W} = (1 - C_{BW})\,\omega_B \left(\frac{\omega_{M,BW} - \omega_W}{\omega_B - \omega_{M,BW}} \right) \qquad (6.54)$$

where C_{BW} is the volume fraction of black relative to the total amount of colorant in the mixture.

The reflectance of the colorant-only and the mixture of black with colorant are converted to ω_A and $\omega_{M,BA}$ respectively and used to determine the absorption of the black relative to the scattering of the colorant (Eqn 6.55):

$$\frac{C_{BA}\hat{K}_B}{\hat{S}_A} = (1 - C_{BA})\,\omega_B \left(\frac{\omega_{M,BA} - \omega_A}{\omega_B - \omega_{M,BA}} \right) \qquad (6.55)$$

Dividing Eqn 6.54 by Eqn 6.55 eliminates K_B, and rearrangement gives Eqns 6.56 and 6.57:

$$S_{AW} = \frac{\hat{S}_A}{\hat{S}_W} = \frac{C_{BA}(1 - C_{BW})}{C_{BW}(1 - C_{BA})} \left(\frac{\omega_{M,BW} - \omega_W}{\omega_B - \omega_{M,BW}} \right) \left(\frac{\omega_B - \omega_{M,BA}}{\omega_{M,BA} - \omega_A} \right) \qquad (6.56)$$

$$\hat{K}_{AW} = \omega_A \hat{S}_{AW} \qquad (6.57)$$

Reflectance values for a set of panels for calibration of Sandosperse Yellow EHJ are shown in Figure 6.20; the black concentration in the mixtures is 1% by mass. The results of the analysis are shown in Figure 6.21, which also shows error bars assuming an error of ±0.1% in reflectance measurement. Figure 6.21 shows small errors in the calibration values at wavelengths where the reflectance of the Yellow EHJ panel is very different from that of the standard black panel (540–700 nm) and very large errors when the reflectance of the two panels are similar (400–520 nm).

The error bars show that mixtures with white are needed to calibrate the optical properties of the yellow colorant at the short wavelengths and mixtures with black to

DATABASE CALIBRATION OF OPAQUE LAYERS 317

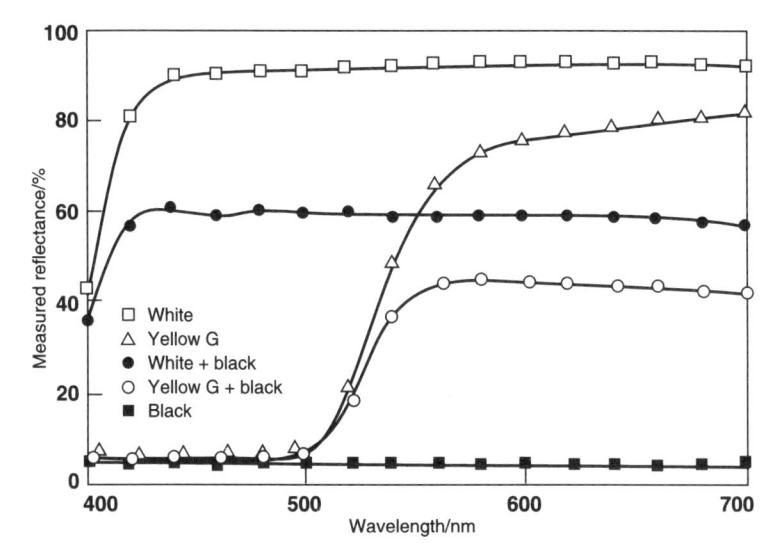

Figure 6.20 Reflectance of panels for calibrating the scattering of Sandosperse Yellow EHJ; mixtures containing 1% black

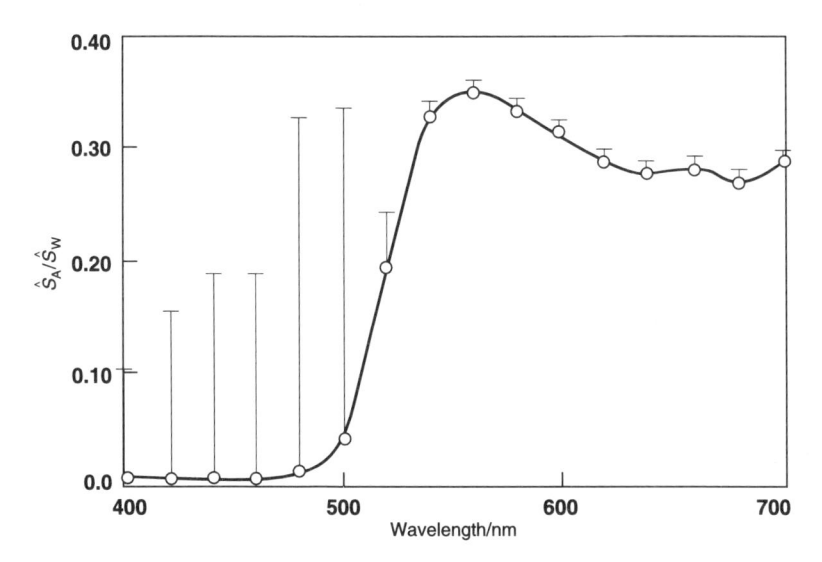

Figure 6.21 Relative scattering coefficient of Sandosperse Yellow EHJ obtained from mixtures with a standard black

calibrate at the long wavelengths. Table 6.6 contains the numerical values for four wavelengths and can be used to check the operation of programs and spreadsheets designed to perform the calculations.

318 COLOUR-MATCH PREDICTION FOR PIGMENTED MATERIALS

Table 6.6 Results from a calculation of the relative scattering values of Sandosperse Yellow EHJ from panels containing white only, black only, yellow only and mixtures of white with black and yellow with black[a]

Wavelength /nm	White ω_W	Black ω_B	Yellow ω_A	Black + white $\omega_{M,BW}$	Black + yellow $\omega_{M,BA}$	Eqn 6.56 $\hat{S}_{AW}$
400	0.1133	40.1571	8.6917	0.1664	15.4497	0.0040
500	0.0008	47.3052	5.1242	0.0306	5.6330	0.0426
600	0.0005	55.7750	0.0081	0.0330	0.0931	0.3142
700	0.0006	76.1032	0.0040	0.0383	0.1122	0.2874

[a] $C_{BW} = 1.68\%$; $C_{BA} = 1.39\%$

Table 6.7 Summary of opaque panels and their function in the analysis of the optical properties of a colorant relative to the scattering power of a standard white

Panel	Function	Equation
White only	Determination of $\omega_W = K_W/S_W$	6.36
Colorant only	Determination of $\omega_A = K_A/S_A$	6.36
Colorant and white mixture series	Determination of $\omega_M = K_M/S_M$ and the concentration dependence of K_{AW}	6.36, 6.45
Black only	Determination of $\omega_B = K_B/S_B$	6.36
White and black	Determination of $\omega_{M,BW} = K_M/S_M$	6.36
Colorant and black	Determination of $\omega_{M,BA} = K_M/S_M$ and $\hat{S}_{AW}$ for bright yellow and red colorants	6.36, 6.56

6.4.5 Summary of calibration panels

The functions of the various calibration panels are summarised in Table 6.7.

6.4.6 K and S analysis: a least-squares method

The method of analysis given in section 6.4.4 determines a value of the relative absorption coefficient K_{AW} from calibration panels by means of Eqn 6.45 or 6.56 and 6.57. This approach makes use of the reflectance of a colorant-only panel to determine the value of ω_A (Eqn 6.39). Producing a panel that is opaque at each wavelength in the spectrum can be difficult when the coating formulation contains only a bright pigment with low scattering power. Some types of bright yellow, red and orange organic pigments are almost non-absorbing at long wavelengths, and if the scattering power is also low then even relatively thick layers may be semitransparent at these wavelengths.

This produces a systematic error in the ratio K_A/S_A obtained from reflectance values and corresponding errors in the values of K_{AW} and S_{AW}.

An alternative method based on the principle of least-squares fitting has been described by Walowit et al. [13]. The method still requires that all the calibration panels are opaque, but it does not demand the production of a colorant-only panel (although this can be included if available). The method treats K_A and S_A as independent quantities and the values are adjusted automatically to produce an optimised (least-squares) fit to the information obtained from the set of calibration panels. A typical calibration panel set (Table 6.9, page 325) will include mixtures of the colorant with the standard white and with the standard black. The least-squares method automatically deals with both types of panel.

At any particular wavelength, the ratio of absorption to scattering of any one of the calibration mixtures in Table 6.9 is given by Eqn 6.58:

$$\frac{K_M}{S_M} = \frac{K_A + K_B + K_W}{S_A + S_B + S_W} = \frac{K_{AW} + K_{BW} + K_{WW}}{S_{AW} + S_{BW} + S_{WW}} \tag{6.58}$$

where the subscripts M, A, B and W denote values that refer to the mixture, the colorant A, the standard black and the standard white respectively. Eqn 6.58 is not in the linear form required by the fitting process and a conversion needs to be made. The following equations use the notation $\hat{K}_{JW}$ to represent the calibration constant $\hat{K}_J/\hat{S}_W$ of colorant J and w_J to represent K_J/S_J. In this notation the individual terms in Eqn 6.58 are as follows (Eqn 6.59):

$$\begin{aligned} K_{AW} &= C_A \hat{K}_{AW} & S_{AW} &= C_A \hat{S}_{AW} \\ K_{BW} &= C_B \hat{K}_{BW} & S_{BW} &= C_B \hat{S}_{BW} \\ K_{WW} &= C_W \hat{K}_{WW} = C_W \omega_W & S_{WW} &= C_W \hat{S}_{WW} = C_W \end{aligned} \tag{6.59}$$

Eqn 6.58 becomes Eqn 6.60:

$$\omega_M = \frac{C_A \hat{K}_{AW} + C_B \hat{K}_{BW} + C_W \omega_W}{C_A \hat{S}_{AW} + C_B \hat{S}_{BW} + C_W} \tag{6.60}$$

This equation is converted to a linear form (Eqn 6.61):

$$C_W(\omega_M - \omega_W) = C_A \hat{K}_{AW} - C_A \hat{S}_{AW}\omega_M + C_B \hat{K}_{BW} - C_B \hat{S}_{BW}\omega_M \tag{6.61}$$

The object of the fitting process is to obtain values for the optical coefficients $\hat{K}_{AW}$, $\hat{S}_{AW}$, $\hat{K}_{BW}$ and $\hat{S}_{BW}$ such that when the concentrations in a calibration mixture are used in Eqn 6.60 the value of ω_M matches the measured value of ω for that mixture. The ratio ω is obtained from the gloss-corrected reflectance R of the calibration panel using

Eqn 6.36. The fitting equation (Eqn 6.62) is obtained by substituting ω for ω_M in Eqn 6.61:

$$C_W(\omega - \omega_W) = C_A \hat{K}_{AW} - C_A \hat{S}_{AW}\omega + C_B \hat{K}_{BW} - C_B \hat{S}_{BW}\omega \quad (6.62)$$

In this treatment the value of ω_W is obtained from measurement of the white-only panel and is not adjusted during the fitting process. The known quantities in Eqn 6.62 are ω, ω_W, C_A, C_B and C_W, while the optical coefficients $\hat{K}_{AW}$, $\hat{S}_{AW}$, $\hat{K}_{BW}$, and $\hat{S}_{BW}$ are unknown. At every wavelength there is a set of values of ω, C_A, C_B and C_W for each of the calibration panels. In the example given in Table 6.9, eight mixtures are included in the analysis. The least-squares process calculates the values of the four optical coefficients that gives the best match over the eight sets of data.

The fitting process

The fitting process can be explained by considering a more general problem. Consider a group of n sets of data where the ith set consists of values of a dependent variable V_i and of the independent variables f_j. V is linked to f_j via a function F (Eqn 6.63):

$$F = a_1 f_1 + a_2 f_2 + a_3 f_3 + a_4 f_4 \quad (6.63)$$

where the a_j are constant coefficients. The value of the expression for the ith set of data is given by Eqn 6.64:

$$F_i = a_1 f_{1,i} + a_2 f_{2,i} + a_3 f_{3,i} + a_4 f_{4,i} \quad (6.64)$$

The deviation of the ith value of F from the value of V is ε_i (Eqn 6.65):

$$\varepsilon_i = V_i - F_i \quad (6.65)$$

The least-squares method calculates the constants a_j that produce a minimum in E, the sum of the squares of the deviations over the n panels in the calibration set (Eqn 6.66):

$$E = \sum_{i=1}^{n} \varepsilon_i^2 \quad (6.66)$$

Normally each deviation is equally weighted in the fitting process. The equation of interest has however been changed to obtain a form suitable for the fitting process, and the least square of the deviations ε_i does not necessarily represent the best fit to the initial equation. In such a case the deviations can be weighted so that the minimum in the sum of the squares of the weighted deviations coincides with the best fit to the initial equation (Eqns 6.67 and 6.68):

$$W_i \varepsilon_i = W_i(V_i - F_i) \quad (6.67)$$

$$E = \sum_{i=1}^{n} W_i^2 \varepsilon_i^2 \tag{6.68}$$

For example, in this analysis the overall aim is to obtain the optical coefficients that would best fit the reflectance values of the calibration panels, and the appropriate weighted deviation would be equivalent to $(R - R_M)$.

The least-squares method leads to a set of four simultaneous equations (Eqns 6.69–6.72). In these equations the symbol Σ is used to represent a summation from $i = 1$ to $i = n$. The value of each of the summations is calculated from the known values of f, W and V. The coefficients a_1 to a_4 may be found by solving the four simultaneous equations:

$$\sum W_i^2 f_{1,i} V_i =$$
$$a_1 \sum W_i^2 f_{1,i} f_{1,i} + a_2 \sum W_i^2 f_{1,i} f_{2,i} + a_3 \sum W_i^2 f_{1,i} f_{3,i} + a_4 \sum W_i^2 f_{1,i} f_{4,i} \tag{6.69}$$

$$\sum W_i^2 f_{2,i} V_i =$$
$$a_1 \sum W_i^2 f_{2,i} f_{1,i} + a_2 \sum W_i^2 f_{2,i} f_{2,i} + a_3 \sum W_i^2 f_{2,i} f_{3,i} + a_4 \sum W_i^2 f_{2,i} f_{4,i} \tag{6.70}$$

$$\sum W_i^2 f_{3,i} V_i =$$
$$a_1 \sum W_i^2 f_{3,i} f_{1,i} + a_2 \sum W_i^2 f_{3,i} f_{2,i} + a_3 \sum W_i^2 f_{3,i} f_{3,i} + a_4 \sum W_i^2 f_{3,i} f_{4,i} \tag{6.71}$$

$$\sum W_i^2 f_{4,i} V_i =$$
$$a_1 \sum W_i^2 f_{4,i} f_{1,i} + a_2 \sum W_i^2 f_{4,i} f_{2,i} + a_3 \sum W_i^2 f_{4,i} f_{3,i} + a_4 \sum W_i^2 f_{4,i} f_{4,i} \tag{6.72}$$

Comparing Eqn 6.62 with Eqn 6.64 shows that the V functions may be defined by Eqn 6.73:

$$V_i = C_W(\omega - \omega_W) \tag{6.73}$$

and the f functions by Eqn 6.74:

$$f_{1,i} = C_A \qquad f_{2,i} = -C_A \omega \qquad f_{3,i} = C_B \qquad f_{4,i} = -C_B \omega \tag{6.74}$$

Using an appropriate weighting function, the solution provides a least-square fit, where the optical coefficients are given by Eqn 6.75:

$$\hat{K}_{AW} = a_1 \qquad \hat{S}_{AW} = a_2 \qquad \hat{K}_{BW} = a_3 \qquad \hat{S}_{BW} = a_4 \tag{6.75}$$

The white standard

In this treatment the value of ω_W is treated as a fixed term at each wavelength. As a result each colorant can be analysed independently of the others in the set, but they all refer to the same standard white. New colorants may be added to the database without having to re-analyse the complete set of data.

The weighting function

When the functions defined in Eqns 6.73–6.75 are substituted in Eqns 6.64 and 6.65, then the deviation is found to be given by Eqn 6.76:

$$\varepsilon = \omega S_M - K_M \tag{6.76}$$

This can be rewritten in terms of ΔK and ΔS, the deviations of K_M and S_M from the actual values for the panel K and S (Eqns 6.77 and 6.78):

$$K_M = K + \Delta K \qquad S_M = S + \Delta S \tag{6.77}$$

$$\varepsilon = \omega \Delta S - \Delta K \tag{6.78}$$

Partial differentials can be used to express small deviations in K and S in terms of corresponding small deviations in the calculated reflectance values from the measured value ($\Delta R = R_M - R$). The right-hand side of Eqn 6.78 can be re-expressed in terms of ΔR (Eqn 6.79):

$$\Delta K = \frac{\partial K}{\partial R} \Delta R_K \qquad \Delta S = \frac{\partial S}{\partial R} \Delta R_S \tag{6.79}$$

where the differentials are given by Eqn 6.80:

$$\frac{\partial K}{\partial R} = -\frac{K(1+R)}{R(1-R)} \qquad \frac{\partial S}{\partial R} = \frac{S(1+R)}{R(1-R)} \tag{6.80}$$

Substituting these equations for ΔK and ΔS in Eqn 6.78 gives Eqn 6.81:

$$\varepsilon = \frac{K(1+R)}{R(1-R)} (\Delta R_K + \Delta R_S) \tag{6.81}$$

The deviation term is a complicated function of K and R. The unweighted fit produces optical coefficients that minimises the sum of the squares of these deviations. This fit does not necessarily coincide with a minimum in the sum of the squares of the deviation of the calculated reflectance from the actual panel values. The purpose of the weighting term is to make the weighted deviation a more appropriate function of ΔR.

The term K in Eqn 6.81 refers to the absorption coefficient of the calibration panel mixture. At this stage of the analysis the value of K is unknown, and the only information available is the ratio $\omega = K/S$ (defined by Eqn 6.36). It is necessary to make an assumption about the nature of the calibration panel before an appropriate weighting can be determined. The first of three possible assumptions is that K is constant and only S varies across the spectrum. The appropriate weight function is given by Eqn 6.82:

$$W = \frac{R(1-R)}{(1+R)} \quad \text{which gives} \quad W\varepsilon = K(\Delta R_K + \Delta R_S) \quad (6.82)$$

The second possible assumption is that S is constant. Re-expressing Eqn 6.81 in terms of S gives Eqn 6.83:

$$\varepsilon = S\left(\frac{1-R^2}{2R^2}\right)(\Delta R_K + \Delta R_S) \quad (6.83)$$

The appropriate weight function is now given by Eqn 6.84:

$$W = \frac{2R^2}{1-R^2} \quad \text{which gives} \quad W\varepsilon = S(\Delta R_K + \Delta R_S) \quad (6.84)$$

A third possible assumption is that the optical thickness of the layer Z (defined by Eqn 6.20) is constant. Re-expressing Eqn 6.81 in terms of Z gives Eqn 6.85:

$$\varepsilon = \frac{Z}{R}(\Delta R_K + \Delta R_S) \quad (6.85)$$

The 'constant Z' assumption provides the simplest weight function (Eqn 6.86):

$$W = R \quad \text{which gives} \quad W\varepsilon = Z(\Delta R_K + \Delta R_S) \quad (6.86)$$

A further modification can be made to take account of the action of the gloss correction equation. The weight function shown in Eqn 6.87 acts to minimise the sum of the squares of the deviation of the equivalent measured reflectances $(\rho_M - \rho)$:

$$W = R\left[\frac{t_e t_i}{(1-r_i R)^2}\right] \quad \text{which gives} \quad W\varepsilon = Z(\Delta \rho_K + \Delta \rho_S) \quad (6.87)$$

where r_i, t_i and t_e are the gloss correction coefficients defined in section 6.3.

The performances of the three weighting functions are compared in Table 6.8, where the standard deviations of the calculated reflectances from the measured values over the calibration set are given for four different colorants. It appears that the 'constant Z' method (Eqn 6.87) gives the lowest standard deviation overall.

Table 6.8 Standard deviation of the calculated reflectance of the mixture ρ_M from the measured reflectance ratio ρ

Pigment	Z constant	K constant	S constant
Yellow 4G	0.0034	0.0043	0.0066
Oxide Red GAJ	0.0018	0.0025	0.0016
Blue 2GLS	0.0024	0.0031	0.0036
Green 3GLS	0.0027	0.0038	0.0041

Matrix methods of least-squares fitting

The various summations and the solution of the simultaneous equations involved in a least-squares fit can be made by matrix methods. Many spreadsheet programs include matrix operations as standard functions allowing the optimised fit method of analysis to be carried out.

At the chosen wavelength, each calibration panel has an associated weighted version of Eqn 6.64 (Eqn 6.88):

$$WV = Wa_1 f_1 + Wa_2 f_2 + Wa_3 f_3 + Wa_4 f_4 \tag{6.88}$$

The set of n weighted equations can be represented in matrix notation by Eqn 6.89:

$$[WV] = [Wf][a] \tag{6.89}$$

where $[a]$ and $[WV]$ are (4×1) and $(n \times 1)$ matrices, and $[Wf]$ is an $(n \times 4)$ matrix (Eqn 6.90):

$$[a] = \begin{pmatrix} a_1 \\ a_2 \\ a_3 \\ a_4 \end{pmatrix} \quad [WV] = \begin{pmatrix} W_1 V_1 \\ W_2 V_2 \\ \ldots \\ W_n V_n \end{pmatrix}$$

$$[Wf] = \begin{pmatrix} W_1 f_{1,1} & W_1 f_{2,1} & W_1 f_{3,1} & W_1 f_{4,1} \\ W_2 f_{1,2} & W_2 f_{2,2} & W_2 f_{3,2} & W_2 f_{4,2} \\ \ldots & \ldots & \ldots & \ldots \\ W_n f_{1,n} & W_n f_{2,n} & W_n f_{3,n} & W_n f_{4,n} \end{pmatrix} \tag{6.90}$$

The matrix form of least-squares solution is given by Eqn 6.91:

$$[a] = \left\{ [Wf]^T [Wf] \right\}^{-1} [Wf]^T [WV] \tag{6.91}$$

where the superscripts T and –1 denote taking the transpose and the inverse of a matrix respectively.

Sample calculation

The set of opaque panels described in Table 6.9 were prepared using Sandosperse predispersed colorants in Synthese, a high-gloss heat-setting resin. Each panel was measured in the specular-included mode on a reflectance spectrophotometer with an integrating sphere. The predispersions used were White CBJ, Black BNLS and Yellow 4G. The calculated values of ω, ω_W, C_A, C_B and C_W at 500 nm are shown in Table 6.10, and the functions V and f_1 to f_4 in Table 6.11. The fitted optical coefficients are reported in Table 6.12.

Table 6.9 Calibration panel set used in the least-squares method of analysis; predispersions used were White CBJ, Black BNLS and Yellow 4G

Panel		Mass fraction/%		
No.	Description	Colorant	Black	White
0	White only	0.0	0.0	100.0
1	Black only	0.0	100.0	0.0
2	White and black	0.0	5.0	95.0
3	Colorant and white	5.0	0.0	95.0
4	Colorant and white	10.0	0.0	90.0
5	Colorant and white	20.0	0.0	80.0
6	Colorant and white	50.0	0.0	50.0
7	Colorant and white	80.0	0.0	20.0
8	Colorant and black	99.0	1.0	0.0

Table 6.10 Formulation in volume fraction units of the calibration panels; gloss-corrected reflectance R and ω (K/S ratio) at 500 nm also shown

	Panel numbers								
Function	0	1	2	3	4	5	6	7	8
C_W	1.0000	0.0000	0.9182	0.9189	0.8430	0.7047	0.3736	0.1298	0.0000
C_B	0.0000	1.0000	0.0818	0.0000	0.0000	0.0000	0.0000	0.0000	0.0101
C_A	0.0000	0.0000	0.0000	0.0811	0.1570	0.2953	0.6264	0.8702	0.9899
R	0.9651	0.0225	0.5577	0.8210	0.7620	0.6722	0.4940	0.3343	0.1935
ω	0.0006	21.2335	0.1754	0.0195	0.0372	0.0799	0.2591	0.6628	1.6807

Table 6.11 Least-squares fit function of Eqn 6.89

	Least-squares fit for panel numbers							
Function	1	2	3	4	5	6	7	8
V	0.0000	0.1605	0.0174	0.0309	0.0559	0.0966	0.0859	0.0000
f_1	0.0000	0.0000	0.0811	0.1570	0.2953	0.6264	0.8702	0.9899
f_2	0.0000	0.0000	−0.0016	−0.0058	−0.0236	−0.1623	−0.5768	−1.6637
f_3	1.0000	0.0818	0.0000	0.0000	0.0000	0.0000	0.0000	0.0101
f_4	−21.2335	−0.0143	0.0000	0.0000	0.0000	0.0000	0.0000	−0.0170
W	0.0089	0.4862	1.2309	0.9983	0.7288	0.3852	0.2019	0.0956

Table 6.12 Least-squares optical coefficients for Yellow 4G at 500 nm

Function	Coefficient
$\hat{K}_{AW}$	0.1968
$\hat{S}_{AW}$	0.1372
$\hat{K}_{BW}$	1.9792
$\hat{S}_{BW}$	0.0932
$\hat{K}_{WW}$	0.0006
$\hat{S}_{WW}$	1.0000

The results of the weighted fit are shown in the reflectance spectra plotted in Figure 6.22: reasonable agreement has been obtained between the calculated and measured reflectance values of the calibration panels. The plots include the measured reflectance spectrum of a colorant-only panel together with lines calculated from the calibration data. The figure shows good agreement between the measured and predicted spectra of the colorant-only panel, even though no such panel was included in the analysis.

The least-squares fit method has the disadvantage that the assumption of linear dependence of K_{AW} on concentration is implicit in the analysis method. This assumption is valid for most systems but in a significant number a nonlinear dependence is shown.

6.4.7 Nonlinear concentration dependence

Nonlinear dependence of K_{AW} on concentration often appears as a saturation effect, a plot of K_{AW} against concentration forming a convex curve rather than a straight line.

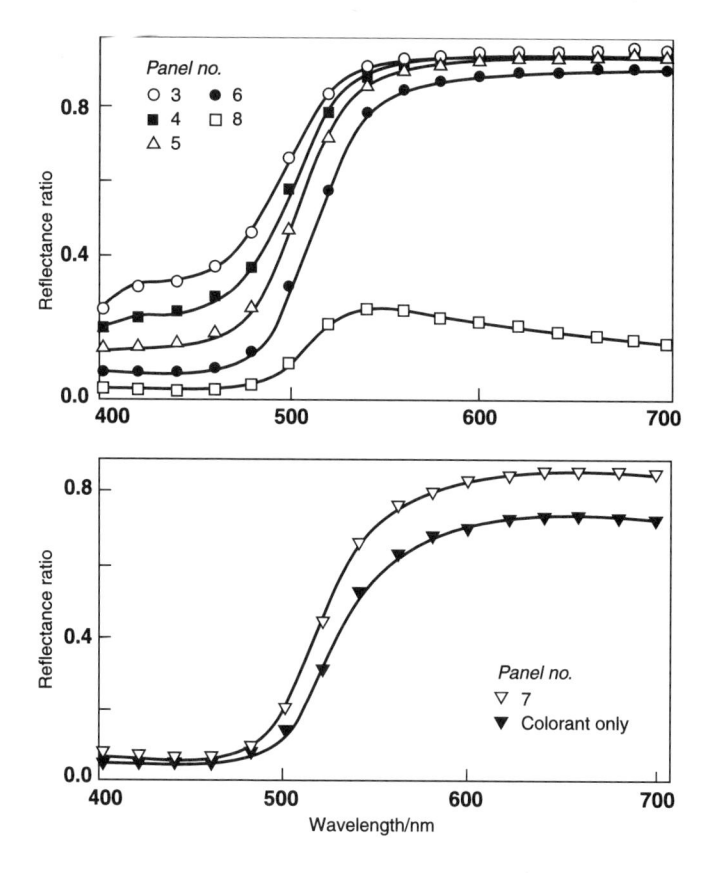

Figure 6.22 Reflectance of mixtures with Yellow 4G; measured values are shown as points, and the lines are calculated from the least-squares-fitted optical coefficients

The nonlinearity may be characterised by fitting the data to a power series, such as Eqn 6.92:

$$K_{AW} = aC_A + bC_A^2 + cC_A^3 \qquad S_{AW} = \frac{K_{AW}}{\omega_A} \qquad (6.92)$$

It is assumed that ω_A is constant and that the nonlinearity is a property of the test colorant and not of the reference white. The database entry for this colorant would consist of the values of ω_A, a, b and c for each wavelength to be included in the match-prediction calculation.

The power series is only one form by which the nonlinearity may be characterised; it is used here and in the match prediction section simply as an illustration. It is important to the match-prediction algorithms that the chosen function is differentiable with respect to concentration.

6.4.8 Generalised method of analysis

A method that has general application combines the advantage of not requiring an opaque single-colorant panel with the ability to determine a value for K_{AW} at each calibration panel concentration. This is achieved by using the least-squares method to obtain $\hat{K}_{AW}$ and $\hat{S}_{AW}$, and hence w_A. The value of w_A is used in the direct method (Eqn 6.45) to determine K_{AW} at each concentration. The results are then fitted to Eqn 6.92 to obtain the database constants.

6.5 PREPARATION OF CALIBRATION PANELS (PAINT)

Careful preparation of the calibration panels is well rewarded by the accuracy of the predictions achieved. This stage often determines the performance of the match-prediction system, irrespective of the complexity of the optical model and the power of the computer. The link of the system to the real world is via the database panels. Time is well spent in considering the most reproducible method for carrying out the following operations:
(a) weighing
(b) mixing
(c) application
(d) drying
(e) reflectance measurement.

6.5.1 Concentration series

There are three basic methods of formulation for colour matching:
(a) Blending finished paints: in this case each component in the formulation is a finished paint in its own right.
(b) Tinting base paints: limited amounts of predispersions of colorants, also known as tinters or stainers, are mixed into a base paint. The base paint may be white or coloured.
(c) Full formulation: predispersions or dry colorants are dispersed into a transparent, colourless diluent. Care has to be taken to not to exceed the maximum pigment volume concentration allowed in the dry coating layer.

Blending finished paints
Colour matching by mixing together finished paints is one of the easiest methods of colour formulation. The pigment volume concentration has already been established in the formulation of the finished paints and only simple mixing is required to blend the components together. Database preparation is straightforward and the sequence of mixtures

Table 6.13 Recommended concentration sequence of calibration panels for a finished system

Test material (coloured paint)/%	Reference material (standard paint)/%
2.5	97.5 white
5.0	95.0 white
10.0	90.0 white
20.0	80.0 white
40.0	60.0 white
70.0	30.0 white
100.0	0.0 white
99.0	1.0 black

shown in Table 6.13 is often used. Notice how the concentration of the colorant paint roughly doubles between mixtures following the principle of having more calibration panels at low concentrations than at high. In most database analysis programs the mixtures can be based on either mass or volume measurement at the convenience of the user.

Tinting of a base paint
Colour matching by mixing colorant predispersions into a base paint is a common method of producing small volumes of custom-matched paints in do-it-yourself stores, particularly in the USA where several thousand systems of this type are installed. The base paint is a finished paint that is coloured by a limited addition of concentrated tinters or stainers, usually no more than 10% of the total volume. The paint scheme may contain several base paints, each designed to cover a range of strengths of shade or a particular hue range at a reasonable covering power. Most schemes contain three white bases (for pastel, medium and deep shades respectively) and a clear base. The pastel base contains the greatest amount of white pigment and the deep base the least. Some schemes include coloured base paints which may be tinted to a small range of highly chromatic colours that cannot be obtained by tinting the white bases. The coloured bases are normally red, yellow and blue and occasionally brown. The clear base is only used when a match cannot be obtained in any of the other bases.

In order for the system to give consistent results from stored or predicted formulations, the properties of the base paints and the tinters need to be controlled to very high standards of reproducibility. In principle this should allow the entire scheme to be calibrated from a single set of panels using the pastel base as the reference white. The clear base is used for the colorant-only and the colorant-plus-black panels. Whether this is possible depends on the nature of the match-prediction program; some systems require calibration panels for the tinters in each base paint in the scheme.

The concentration range of the calibration mixtures reflects the limited level of maximum colorant addition. For example, if the maximum addition is 10% then the panel sequence might be as shown in Table 6.14.

Table 6.14 Recommended concentration sequence of calibration panels for a base scheme system

Test material (tinter)/%	Reference material (standard base)/%
0.25	99.75 white
0.50	99.50 white
1.00	99.00 white
2.00	98.00 white
5.00	95.00 white
10.00	90.00 white
30.00	70.00 clear
29.50 + 0.50 black	70.00 clear

Full formulation
The calibration of systems based on the addition of colorants to a clear base requires more consideration than the two previous schemes. The numbers and types of panels needed are similar to those of the other schemes, but the limits on the pigment volume concentration in the dry coating must be carefully taken into account when preparing the mixtures. A time-efficient method of base production is first to prepare a finished paint for each of the colorants by mixture with the clear base. In order to obtain opaque layers at a reasonable layer thickness, it is tempting to use as much colorant as possible in the clear base consistent with the critical pigment volume concentration of the dry layer. On this principle, each finished paint would contain a different pigment concentration. This has the danger of possible calculation errors when determining the amounts of each paint needed in the calibration mixture. A more reasonable approach is to set a choice of two levels of colorant in the clear base: a fixed high level for colorants that are easy to disperse, such as predispersions and inorganic pigments, and a lower value for difficult-to-disperse colorants such as organic pigments. This would result in two possible dilution series.

Consider the following example: a white finished paint has 1 g white pigment per

2.5 g finished paint, while the test paint has 1 g colorant per 5 g finished paint. Calculation of the dilution sequence is illustrated in Table 6.15.

Table 6.15 Calculation of finished paint formulations needed to produce set colorant-to-white mixture concentrations (see text)

Colorant amount/g		Finished paint amount/g		Finished paint (100 g)/%	
Colorant	White	Colour	White	Colour	White
2.5	97.5	12.5	243.8	4.88	95.12
5.0	95.0	25.0	237.5	9.52	90.48
10.0	90.0	50.0	225.0	18.18	81.82
20.0	80.0	100.0	200.0	33.33	66.67
40.0	60.0	200.0	150.0	57.14	42.86
70.0	30.0	350.0	75.0	82.35	17.65
100.0	0.0	500.0	0.0	100.00	0.00

6.5.2 Panel production

Reference materials
The optical properties of the colorants are expressed relative to the scattering power of a reference white material. The reference white must be of consistent strength throughout the entire database panel set; it should have the highest possible reflectance values, and any contamination by poor working practices or yellowing with age should be avoided. The entry of extra colorants at a later stage is simple if they are prepared using the same reference white as the original base. Similar requirements of consistent properties also apply to the reference black.

Substrates
Specially prepared card is available, which is printed with a contrasting black and white pattern that indicates clearly if the coating layer is not opaque. Metal panels are also available for use with air-drying and high-temperature heat-set formulations.

Measuring out components
Great care is needed in measuring components for the calibration mixtures. Whether the method is gravimetric or volumetric, the resolution of the measurement must be better than 0.01% of the batch size: for example, the components for a 100 g batch must be weighed accurately to within 10 mg, preferably with a precision digital top-pan

balance standing in a draught-free position on a firm bench. Where many sets of calibration panels are to be produced for the same set of colorant predispersions (in a base scheme calibration, for example) an automatic dispensing system may be a more efficient method of mixture production.

It is good practice to measure the largest component into the mixing vessel first and then add the remaining components in descending order of amount. If a small component is placed first into a container a significant amount may be trapped in a corner or remain stuck to the bottom surface and not fully mix into the formulation. Strict cleanliness is required and a separate set of tools should be kept for handling the reference white material. The effects of accidental cross-contamination can be reduced by first preparing the calibration mixtures of the brighter colorants (yellow, orange and red) and finishing with the strongest, darkest colorants (green, blue, violet and black).

Mixing, dispersion, application and drying
The optical properties of a pigment dispersion depend on the distribution of sizes of the pigment particles. The tinting strength of a poor pigment dispersion is significantly less than that of a well-dispersed system. The smaller the average particle size of the dispersion, the higher is the colour strength that is exhibited. It follows that the method of mixing and dispersion of colorants into the formulation must be the same for each calibration mixture. For similar reasons the method of application of the coating can affect the strength developed by the colorants: application by a spray gun, for example, performs considerably more work on the coating formulation than spreading by a wire-wound bar or a slot coater. As a result different methods can produce slightly different colours from the same coating formulation.

The conditions under which the calibration panels dry also influence the colour developed. If the panels are to be dried by an accelerated process then it is necessary to check that there is no measurable colour change compared with panels dried by the normal method. Moreover, the colour of a panel can change by a small but significant amount during the first few hours following drying. It is necessary to check the time stability and if the colour change is appreciable, a time delay between drying and colour measurement may be appropriate. The delay should reflect the way in which the match-predicted formulations will be judged: for example, if a predicted recipe is mixed, applied, dried and measured in the shortest possible time then the database calibration panels should be prepared and measured in a similar way.

6.5.3 Additional data

A practical surface coating formulation has several specifications apart from colour and hiding power. The dry, solid coating is required to have good adhesion to the substrate and low permeability to moisture, and to be durable. Such properties are strongly influenced by the volume ratio of the particulate matter (the discontinuous phase) to the polymeric binder (the continuous phase) in the dry coating. This is often expressed in terms of the pigment volume concentration, the volume of pigment relative to the total volume of the dry coating. The application properties change with this parameter and rapidly worsen when it exceeds a certain value known as the *critical pigment volume concentration* (CPVC). The optical properties also change in this region, with a loss in scattering power and reduction in tinctorial strength so that the absorption and scattering values become nonlinear functions of concentration of colorant. The value of the CPVC usually lies between 25 and 50% and depends on the nature of both the continuous and discontinuous phases. The formulation must thus satisfy the combined requirements of matching the colour of the target panel, providing a given level of hiding and having a pigment volume concentration less than the critical value.

The most useful way of expressing the formulation of a coating material is in terms of the amount of each component in the form appropriate at the dispensing stage (as a liquid or a powder, for example). A simple format for keeping track of pigment volume concentration is to characterise each of the dispensed components by the solids content, the ratio of the volume of the dry component to the volume at the dispensing stage. In addition the dry solids part of the component is split into the fraction that contributes to the continuous or 'binder' phase and the remainder that contributes to the discontinuous or 'pigment' phase. This data, together with the density of the component, allows a dispense formulation to be given in mass or volume units and allows the calculation of pigment volume concentration and layer thickness of the dry coating. In addition a conversion between the wet layer thickness and dry layer thickness can be calculated. Example values for a clear diluent, a white pigment powder and a white predispersion are given in Table 6.16.

Table 6.16 Sample data that characterises a formulation component in its dispensed form

Component	'Wet' density /g cm^{-3}	Solids[a]/%	Binder[b]/%
Synthese	1.02	60	100
White Pigment	2.57	100	0
White S-CBJ	1.71	60	20

a Amount = (dry volume/wet volume) × 100
b Amount = (binder volume/dry volume) × 100

6.6 DATABASE CALIBRATION FOR SEMITRANSPARENT LAYERS

In most printing processes a semitransparent layer is coated on to a substrate. The colour is determined by the optical properties of the inks, the thickness of the layer and the reflectance of the substrate. The optical properties of the inks may be characterised from a series of prints at a fixed layer thickness on one or more substrates. The substrates used for the calibration prints play a similar role to the white paint in the calibration of an opaque paint database. In section 6.2 a description was given of how the Schuster–Kubelka–Munk theory may be used to relate the reflectance of a semitransparent layer to K and S (Eqn 6.27, where α is defined in Eqn 6.28).

Equations that allow the calculation of K and S from reflectance values are obtained by substituting for α in Eqn 6.27 and rearranging, making β the subject. The values of Z, K and S are then determined from Eqn 6.93:

$$Z = \frac{1}{2} \ln(\beta + 1) \qquad K = \frac{Z}{D}\left(\frac{1-R_\infty}{1+R_\infty}\right) \qquad S = \frac{Z}{D}\left(\frac{2R_\infty}{1-R_\infty^2}\right) \qquad (6.93)$$

It is normal practice to define unit layer thickness as the thickness of the calibration prints. The value of D is then 1 for the calibration database.

6.6.1 Determination of R_∞ for a semitransparent layer

The equations above include R_∞, the reflectance of an opaque layer of the ink. For many ink systems, however, it is difficult to obtain a satisfactory opaque layer. Thick layers of ink may take a long time to dry and often crack. In addition, because of the limitations of the Kubelka–Munk theory, the value of R_∞ measured from a real opaque layer may not be appropriate for predicting the reflectance of the semitransparent system.

A value of R_∞ may be determined from the reflectance of prints of the same thickness over a white and over a black substrate. Let R_w and $R_{g,w}$ represent the reflectance of the print over white and of the white substrate respectively and R_b, $R_{g,b}$ the corresponding values for the print on black and the black substrate respectively.

The value of β depends only on the values of K and S for the layer and is identical for both prints. The value of β from the print over the black and white substrates is given by Eqn 6.94:

$$\begin{aligned}\beta &= \left(\frac{R_{g,b} - R_b}{R_b - R_\infty}\right)\left(\frac{1-R_\infty^2}{1-R_{g,b}R_\infty}\right) \\ &= \left(\frac{R_{g,w} - R_w}{R_w - R_\infty}\right)\left(\frac{1-R_\infty^2}{1-R_{g,w}R_\infty}\right)\end{aligned} \qquad (6.94)$$

Rearranging produces a quadratic equation for R_∞ (Eqn 6.95):

$$R_\infty = B - (B^2 - 1)^{1/2} \tag{6.95}$$

where B is given by Eqn 6.96:

$$B = \frac{(1+R_b R_w)(R_{g,w} - R_{g,b}) - (1+R_{g,b}R_{g,w})(R_w - R_b)}{2(R_b R_{g,w} - R_{g,b} R_w)} \tag{6.96}$$

The ratio $\omega = K/S$ can be obtained directly from B ($\omega = B - 1$).

6.6.2 Sample calculation

The reflectance values of calibration panels for a yellow lithographic ink are shown in Figure 6.23, together with the calculated opaque layer reflectance. The prints were made at a layer thickness of 0.4 cm^3 m^{-2} on to coated card suitable for packaging. Figure 6.24 shows the calculated values of K and S plotted against wavelength: a strong absorption band appears in the region 400–480 nm and weak scattering is present throughout the spectrum. The reflectance values and the results of the calculation at five wavelengths across the spectrum are given in Table 6.17, which can be used to verify the operation of a spreadsheet or computer program.

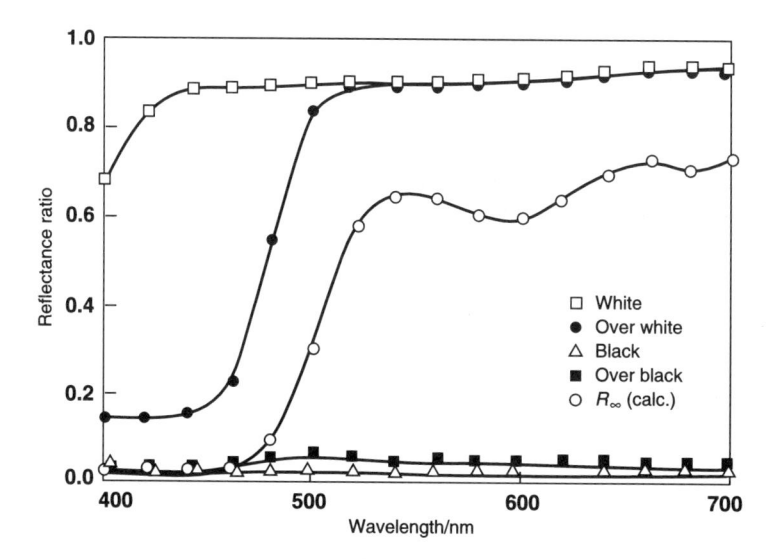

Figure 6.23 Corrected reflectance of yellow lithographic ink calibration prints (0.4 cm^3 m^{-2}) and calculated opaque reflectance R_∞.

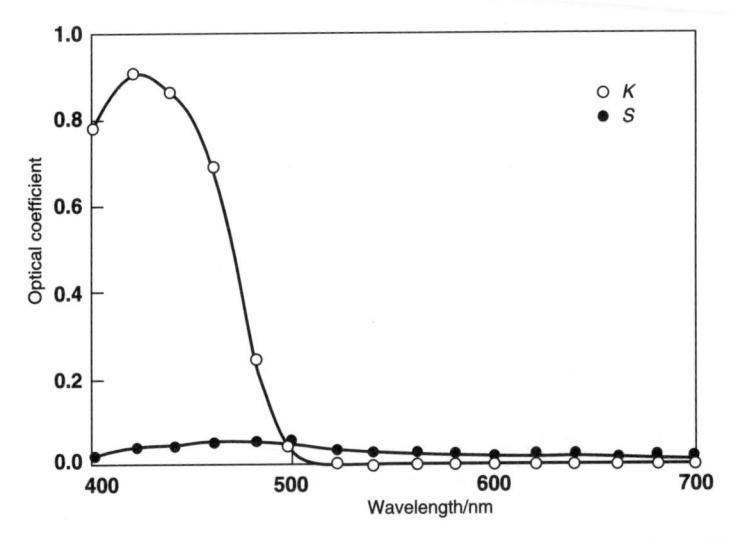

336 COLOUR-MATCH PREDICTION FOR PIGMENTED MATERIALS

Figure 6.24 Optical coefficients K and S for a print of a yellow lithographic ink (0.4 cm^3 m^{-2})

Table 6.17 Corrected reflectance ratios and results of calculations for calibration prints of a yellow lithographic ink on coated card at 0.4 cm^3 m^{-2}

Wave-length /nm	Corrected reflectance ratio					Calculated value			
	Substrate $R_{g,w}$	Substrate $R_{g,b}$	On white R_w	On black R_b	B Eqn 6.96	R_∞ Eqn 6.95	K Eqn 6.93	S Eqn 6.93
420	0.8404	0.0239	0.1465	0.0221	22.962	0.0218	0.9078	0.0413
460	0.8930	0.0246	0.2305	0.0347	13.348	0.0375	0.7026	0.0569
480	0.8973	0.0251	0.5546	0.0581	5.143	0.0982	0.2448	0.0591
500	0.9035	0.0249	0.8445	0.0619	1.780	0.3074	0.0342	0.0438
600	0.9158	0.0261	0.9114	0.0439	1.130	0.6041	0.0025	0.0192

6.6.3 Concentration dependence of K and S

The value of K depends on the volume concentration of the colorant in the layer, in a manner that is found by producing prints using a series of dilutions of the test ink with a clear diluent. The concentration series is similar to that for opaque layers, more mixtures being prepared at low colorant concentrations than at high. A typical set is shown in Table 6.18. Prints of the same layer thickness are produced and the reflectance spectrum measured. After application of the Saunderson correction, the true reflectance is used in Eqn 6.27 to calculate β, and then Eqn 6.93 is used to obtain K and S for that concentration.

Table 6.18 Typical dilution series for calibration of a printing ink

Coloured ink	Diluent ink	Substrate
2	98	White
5	95	White
10	90	White
20	80	White
30	70	White
50	50	White
70	30	White
100	0	White
100	0	Black

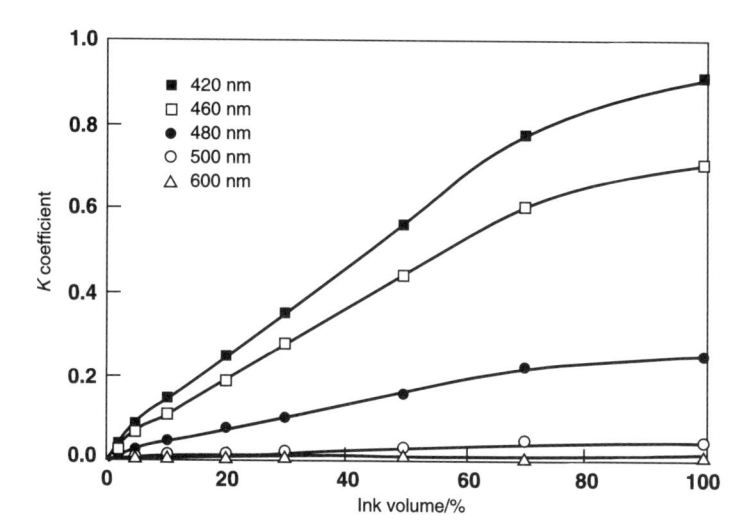

Figure 6.25 Concentration dependence of K for a yellow lithographic ink printed on coated card at 0.4 cm³ m⁻²

Figure 6.25 shows a plot of K against the volume concentration of a yellow lithographic ink in a mixture with clear diluent. Plots at five wavelengths are shown, illustrating a range of absorption strengths. The degree of linearity of the plots is typical for lithographic databases, with the onset of saturation indicated by a decrease in slope at high concentrations. The nonlinear dependence of K on C can be characterised by an equation of the general form of Eqn 6.97:

$$K = aC + bC^2 + cC^3 \tag{6.97}$$

The database will consist of values of the coefficients a, b, c and ω at each of the measured wavelengths in the spectrum. The coefficients a, b and c at each wavelength can be obtained by a regression fit of the K and C values for the set of database panels to Eqn 6.97. It is beneficial to use a weighted fit (section 6.4.6) to minimise the sum of the square deviation of the reflectance values rather than absorption values. The weight for the jth data pair in the fit is given by Eqn 6.98:

$$W_j = \frac{(1 - R_\infty^2)(R_\infty - R_g)}{1 - R_g R_\infty} \frac{2(1 + \beta_j)}{\alpha + \beta_j} \frac{t_e t_i}{(1 - r_i R_j)^2} \quad (6.98)$$

6.6.4 Production of calibration prints

The same detailed consideration of methods of measuring out, mixing, application and drying is needed for printing ink calibration as for opaque paints. The method of print production has to be carefully chosen to mimic the results obtained on production machinery. For example, consider the process of proofing lithographic paste inks. The ink is transferred to the substrate from a rubber-surfaced roller. The factors that influence the amount of ink transferred are the speed of printing, the pressure on the roller and the softness of the roller. The first two are normally controlled by adjustment of the proofing machine. The softness of the print roller depends on the cleaning process, which solvent is used and how thoroughly the solvent is removed after cleaning.

Ideally each calibration print should have the same layer thickness. This demands very precise measurement of the amount of ink on to the proofing apparatus and an accurate, reproducible proofing operation. It is also necessary for all the inks in the mixture to have approximately the same rheological properties, since when their properties differ significantly the amount of ink transferred to the substrate will vary systematically with the composition of the mixture. If the print thickness is known then a simple ratio technique can be used to adjust the absorption values from K' for a print of thickness D' to K corresponding to a standard thickness D (Eqn 6.99):

$$K = \frac{K'D}{D'} \quad (6.99)$$

Alternatively the K could be left at the analysed value and the volume concentration adjusted from C' to C to compensate for the change in thickness (Eqn 6.100):

$$C = \frac{C'D}{D'} \quad (6.100)$$

Cleaning the proofing apparatus can take a significant time and it is difficult to prevent cross-contamination of materials. In this context it is helpful to keep separate sets of proofing apparatus for yellows, reds, greens and blue/blacks.

Bronzing

Bronzing refers to the metal-like lustre that is exhibited by some prints at high concentrations of a coloured ink. The effect is not taken into account by match-prediction algorithms. Bronzing is due to an unusually high reflectance of the print at a narrow band of wavelengths at the edge of a strong absorption band. The additional reflectance has an angular dependence similar to surface gloss and has the complementary hue to the colour of the ink itself, giving a bronze-like appearance to the print at certain angles of view. It is most apparent in very strong pigments such as phthalocyanine blue, which exhibits a reddish-violet tint on bronzing. Bronzing is present in inks of other colours such as yellow, but is less obvious because the high lightness and chroma of, for example, a full-strength yellow print masks the visual impression of bronzing.

When bronzing occurs in a database dilution series the reflectance at the bronzing wavelengths initially behaves normally, decreasing with increasing colorant concentration, but at higher concentrations the reflectance starts to rise. It is not possible to accommodate bronzing into the calibration procedure but the effect can be dramatically reduced by overvarnishing the prints.

The optical behaviour of a bronzing print is consistent with an increase in the refractive index of the layer at the affected wavelengths. The refractive index of a material has a weak dependence on the absorption extinction coefficient. Normally the increase in absorption with colorant concentration has little effect on the index, but at high concentrations of strong colorants the increase in index at certain wavelengths is sufficient to cause a noticeable increase in surface gloss. Overvarnishing the print replaces the single large step in refractive index at the air/ink boundary with a two smaller steps, an air/varnish boundary and a varnish/ink boundary. The two intermediate steps in refractive index produce a greatly reduced 'bronze' gloss compared to the unvarnished surface.

Fluorescent whitened paper and board

Many white papers and carton-boards include fluorescent compounds to increase their impression of whiteness. The fluorescent emission can pose a problem to the calibration of a substrate for colour-prediction calculations. The compounds may be stimulated by the UV component in the light incident on the sample in the reflectance spectrophotometer. They would then re-emit most of the absorbed UV energy as light in the blue regions of the spectrum, creating artificially high reflectance values for the substrate at these wavelengths. Most printing ink materials, including the clear diluent, absorb UV light, reducing both the stimulation of the fluorescent compounds in the substrate and the measured reflectance at the fluorescence emission wavelengths. This apparent drop in reflectance will be falsely interpreted by the equations as an increase in absorption coefficient. The effect can be reduced by measuring the reflectance with the UV light filtered from the light incident on the sample and by calibration of the substrate by measurement of a sample overprinted with the clear diluent.

Changing substrates

A database prepared by prints on one substrate can be used to predict the colour of a mixture applied to a different substrate, provided the surface properties of the substrates are similar. It is often necessary to adjust the apparent strength of the inks to obtain the optimum match-prediction performance. Databases cannot be exchanged when the degree of penetration of the ink into the substrates is very different. The degree of penetration will influence the optical path of the light in the material. For example, a database prepared from prints on coated carton-board (low penetration) will usually produce poor predictions of formulations for uncoated board (high penetration), even with adjustments for strength.

Additional information

Some types of ink are made by mixing components such as a pigment concentrate, a binder to form the polymeric continuous phase and a solvent to adjust the viscosity of the ink to a standard value. The calibration process was described in terms of mixing together finished inks and normally the predicted formulation would also be in these terms. With additional information such as the volume fraction of each component in the finished ink and the respective densities, it is possible to re-express the formulation in terms of the component amounts of concentrate, binder and solvent.

6.7 MATCH PREDICTION OF AN OPAQUE LAYER

A match-prediction calculation can be broken down into seven steps:
1. Specify the colour of the standard.
2. Select the colorants to be used in the formulation from the database list. Usually between two and five colorants are chosen.
3. Calculate the amount of each colorant in the formulation.
4. At each wavelength in the spectrum, sum the contribution from each colorant in the formulation to determine the total absorption (K_M) and total scattering (S_M) coefficients.
5. Calculate the reflectance spectrum of the mixture from the inverse Kubelka–Munk equation.
6. Convert the reflectance R_M to the reflectance ρ_M that would be measured by the spectrophotometer by means of the Saunderson equation.
7. Calculate the colour coordinates of the recipe spectrum and compare with those of the standard.

The following text will concentrate on step 3. Unfortunately there are no explicit equations that link colour coordinates to the concentration of colorants in a mixture. This is illustrated in Figure 6.26, which shows a section of the surface in CIELAB colour space on which mixtures of a white, a yellow and a blue colorant lie. Even for this simple combination the surface is strongly curved, eventually folding backwards on itself for very strong dark blue and green shades.

A common method of determining a matching recipe is carried out in colour-difference space. Colour-difference space is illustrated in Figure 6.27, which shows the ΔE^* values between recipes containing a mixture of white, yellow and blue and a standard

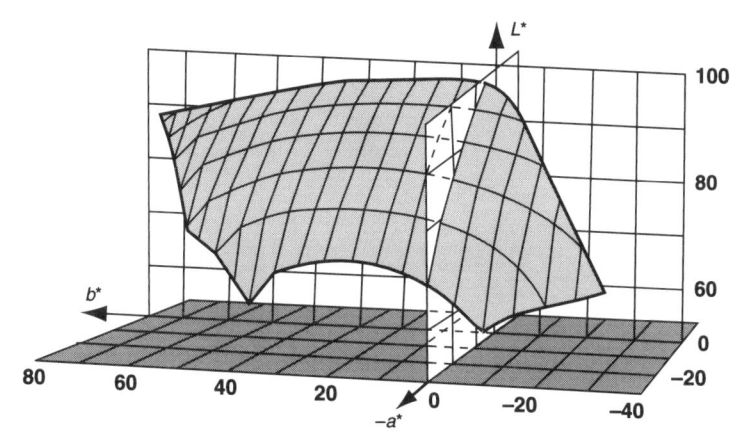

Figure 6.26 Part of the colour surface produced by mixtures of white, yellow and blue colorants in an opaque paint layer

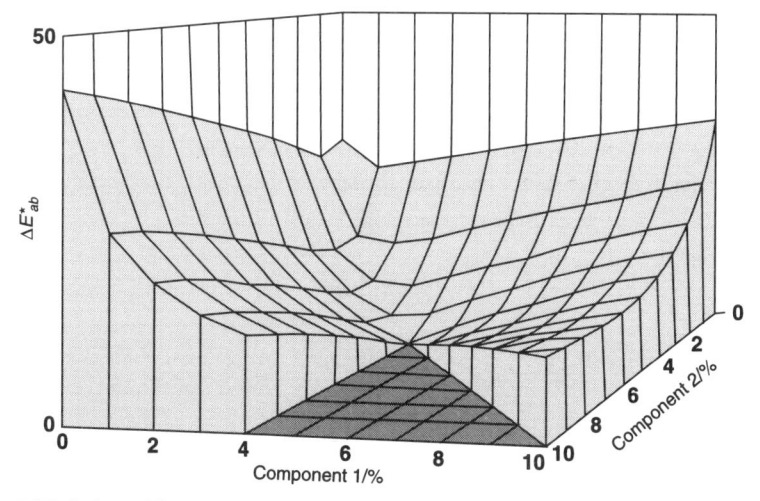

Figure 6.27 Colour-difference surface between recipes containing white, yellow (1) and blue (2) colorants and a standard sample; one section is cut away to show the $\Delta E^* = 0$ point

colour. The points lie on a surface which has a minimum height at the formulation that matches the standard. The colorimetric method of match prediction provides a way of systematically adjusting a recipe towards the minimum point in colour-difference space. The same principle applies when there are four, five or even six colorants in the recipe, even though it is not possible to imagine the shape of four-, five- or six-dimensional colour-difference space.

Two methods of match prediction are in common use: spectrophotometric curve matching and colorimetric matching. These have recently been compared by Sluban [14].

6.7.1 Calculation of reflectance spectrum of an opaque layer

The calculation of the reflectance spectrum for a known recipe is straightforward. The K/S value of a recipe that contains n colorants is obtained by summing the contributions to K and S from each component (Eqn 6.101):

$$\frac{K_M}{S_M} = \frac{K_1 + K_2 + \ldots + K_n}{S_1 + S_2 + \ldots + S_n} \tag{6.101}$$

where the subscripts M, 1, 2 and n denote values that refer to the mixture and colorants 1, 2 and n respectively. The information in the calibration database allows calculation of the absorption and scattering of each colorant relative to the scattering of the calibration white. Re-expressing Eqn 6.101 in terms of these values gives Eqn 6.102:

$$\omega_M = \frac{K_{1W} + K_{2W} + \ldots + K_{nW}}{S_{1W} + S_{2W} + \ldots + S_{nW}} \tag{6.102}$$

The equations are developed for the general case of a nonlinear relationship between concentration (C_j) and the optical properties (K_{jW}, S_{jW}). For example, if a power series has been used to relate the optical constant K_{jW} to C_j then Eqn 6.103 can be written:

$$K_{jW} = a_j C_j + b_j C_j^2 + c_j C_j^3 \qquad S_{jW} = \frac{K_{jW}}{\omega_j} \tag{6.103}$$

Once the K_{jW} and S_{jW} have been determined, the value of ω_M at each wavelength is obtained from Eqn 6.102 and the reflectance calculated from the inverse Kubelka–Munk equation (Eqn 6.104):

$$R_M = (1 + \omega_M) - [(1 + \omega_M)^2 - 1]^{1/2} \tag{6.104}$$

Finally the effect of gloss (section 6.3) is added to obtain the reflectance value of the mixture panel, ρ_M (Eqn 6.105):

$$\rho_M = r_e + \frac{t_e t_i R_M}{1 - r_i R_M} \quad (6.105)$$

The colour coordinates XYZ and $L^*a^*b^*$ are calculated in the usual way.

6.7.2 Spectrophotometric curve matching

This method considers data from all the wavelengths simultaneously. At each wavelength in the spectrum the K/S value of the standard is calculated from the gloss-corrected reflectance values (Eqn 6.36). The object of the fitting process is to obtain values of K_{jW} and S_{jW} in Eqn 6.102 such that ω_M matches the value of ω obtained from the standard panel at each wavelength. A least-squares method is used to calculate the concentrations of colorants that minimise the sum over all wavelengths of the square of the deviation of the K/S value of the recipe from that of the standard. The method has been described by McGinnis for dye coloration (single constant) [15], and further developed by Walowit for pigments (two constant) [16].

The first step towards a fitting equation is to rearrange Eqn 6.102 and group together the terms for each colorant (Eqn 6.106):

$$0 = (K_{1W} - \omega_M S_{1W}) + (K_{2W} - \omega_M S_{2W}) + \ldots + (K_{nW} - \omega_M S_{nW}) \quad (6.106)$$

The fitting equation is obtained by substituting ω for ω_M (Eqn 6.107):

$$F = (K_{1W} - \omega S_{1W}) + (K_{2W} - \omega S_{2W}) + \ldots + (K_{nW} - \omega S_{nW}) \quad (6.107)$$

Assume that an initial recipe exists with colorant concentrations C_j, and let F be the value calculated from Eqn 6.107. The deviation ε of F from a target value V is given by Eqn 6.108:

$$\varepsilon = V - F \quad (6.108)$$

When the value of ε is zero at each wavelength the recipe is a spectral match to the standard. In most cases this will not be true and the concentration of each component j will need to be adjusted by an amount ΔC_j to correct for the deviation according to Eqn 6.109:

$$\varepsilon = \Delta C_1 \frac{\partial F}{\partial C_1} + \Delta C_2 \frac{\partial F}{\partial C_2} + \ldots + \Delta C_n \frac{\partial F}{\partial C_n} \quad (6.109)$$

The function $\partial F/\partial C_j$ is the partial derivative of F with respect to C_j and describes how the value of F will change for a small change in C_j, with the concentrations of the other components held constant. It is given by Eqn 6.110:

$$\frac{\partial F}{\partial C_j} = \frac{dK_{jw}}{dC_j} - \omega \frac{dS_{jw}}{dC_j} \quad \text{since} \quad \frac{dS_{jw}}{dC_j} = \frac{1}{\omega_j} \frac{dK_{jw}}{dC_j} \quad (6.110)$$

$$\frac{\partial F}{\partial C_j} = \frac{dK_{jw}}{dC_j}\left(1 - \frac{\omega}{\omega_j}\right)$$

For example, when there is a linear relationship between concentration and K_{jW} or S_{jW} the derivative is given by Eqn 6.111:

$$\frac{\partial F}{\partial C_j} = \hat{K}_{jW}\left(1 - \frac{\omega}{\omega_j}\right) \quad (6.111)$$

If the relationship is a power series then the derivatives are given by Eqn 6.112:

$$\frac{\partial F}{\partial C_j} = (a_j + 2b_j C_j + 3c_j C_j^2)\left(1 - \frac{\omega}{\omega_j}\right) \quad (6.112)$$

It follows that the only unknown terms in Eqn 6.109 are the concentration change values. There are typically 16 equations, one for each of the wavelengths used to characterise the samples.

It is convenient at this point to add a 17th equation to impose a constraint on the total concentration of the colorants. Using the nomenclature already defined, the concentration equation, target value and deviation are F_C, V_C and ε_C respectively (Eqns 6.113–6.115):

$$F_C = C_1 + C_2 + \ldots + C_n \quad (6.113)$$

$$\varepsilon_C = V_C - F_C \quad (6.114)$$

$$\varepsilon_C = \Delta C_1 + \Delta C_2 + \ldots + \Delta C_n \quad (6.115)$$

Eqns 6.109 and 6.115 have the same general form (Eqn 6.116):

$$\varepsilon = \Delta C_1 f_1 + \Delta C_2 f_2 + \ldots + \Delta C_n f_n \quad (6.116)$$

The least-squares method can be used to solve the 17 simultaneous equations for the values of ΔC_j that produce a minimum in S, the sum of the squares of the deviations over n equations (Eqn 6.117):

$$S = \sum_{i=1}^{n} \varepsilon_i^2 \quad (6.117)$$

Weighted deviations

As in section 6.4.6, the unweighted deviation defined in Eqn 6.109 can be expressed in terms of ΔK and ΔS, the deviations of K_M and S_M from the values needed to match the standard (Eqn 6.118):

$$\varepsilon = \Delta K - \omega \Delta S \qquad (6.118)$$

where $\Delta K = K - K_M$ and $\Delta S = S - S_M$

A weighted form of the deviation can be used and the weights chosen so that the minimum in S coincides with the best fit of the reflectance of the recipe panel to the reflectance of the standard panels (Eqn 6.67 and Eqn 6.119):

$$S = \sum_{i=1}^{n} W_i^2 \varepsilon_i^2 \qquad (6.119)$$

The form of the appropriate weighting function for this case has been discussed in section 6.4.6; if the 'constant Z' assumption is made then the weighting is the same as Eqn 6.87. In order to ensure that the volume condition of Eqn 6.115 is satisfied it is necessary to weight the terms in this equation at least a hundred times greater than any weight given by Eqn 6.87 ($W_C = 100$).

Method of least-squares fitting

Both the methods described in section 6.4.6 can be used to solve the least-squares fitting equations for concentrations. The advantages of matrices in providing a short-hand method of representing groups of equations in colorimetry has long been recognised [17], and is adopted here.

The set of 16 wavelength equations plus the concentration equation can be represented in matrix notation by Eqn 6.120:

$$[W\varepsilon] = [Wf][\Delta C] \qquad (6.120)$$

where $[\Delta C]$ and $[W\varepsilon]$ are the $(n \times 1)$ and (17×1) matrices given in Eqn 6.121:

$$[\Delta C] = \begin{pmatrix} \Delta C_1 \\ \Delta C_2 \\ \ldots \\ \Delta C_n \end{pmatrix} \qquad [W\varepsilon] = \begin{pmatrix} W_1 \varepsilon_1 \\ W_2 \varepsilon_2 \\ \ldots \\ W_{16} \varepsilon_{16} \\ W_C \varepsilon_C \end{pmatrix} \qquad (6.121)$$

and $[Wf]$ is a $(17 \times n)$ matrix (Eqn 6.122):

$$[Wf] = \begin{pmatrix} W_1 f_{1,1} & W_1 f_{2,1} & \cdots & W_1 f_{n,1} \\ W_2 f_{1,2} & W_2 f_{2,2} & \cdots & W_2 f_{n,2} \\ \cdots & \cdots & \cdots & \cdots \\ W_{16} f_{1,16} & W_{16} f_{2,16} & \cdots & W_{16} f_{n,16} \\ W_C & W_C & \cdots & W_C \end{pmatrix} \quad (6.122)$$

Each column of [Wf] refers to one colorant. The first 16 rows refer to the 16 wavelength equations (Eqn 6.109) so that $f_{2,1}$ is the value of $\partial F/\partial C_2$ at wavelength 1. The 17th row refers to the concentration (Eqn 6.115). The least-squares solution is given by Eqn 6.123:

$$[\Delta C] = \left\{ [Wf]^T [Wf] \right\}^{-1} [Wf]^T [W\varepsilon] \quad (6.123)$$

where the superscripts T and −1 again denote taking the transpose and the inverse of a matrix respectively. The new recipe is obtained by adding the changes to the initial concentration values (Eqn 6.124):

$$C_{j,1} = C_j + \Delta C_j \quad \text{for } j = 1 \text{ to } n \quad (6.124)$$

The first recipe

Rather surprisingly, a recipe can be determined from this set of equations by setting the concentration of each colorant in the initial recipe to zero (Eqn 6.125):

$$C_j = 0 \quad \text{for } j = 1 \text{ to } n \quad (6.125)$$

The values of F_i and F_C are all zero in this case. The target values are given by Eqn 6.126:

$$V_j = 0 \quad \text{for } i = 1 \text{ to } 16 \quad \text{and } V_C = 1 \quad (6.126)$$

The values of ε are determined and the equations are solved as described. The first formulation is given by Eqn 6.127:

$$C_{j,1} = \Delta C_j \quad (6.127)$$

If the optical constants are a linear function of concentration then no further improvement will be obtained from a correction step.

The correction step

For systems with nonlinear concentration relationships the match may be improved by applying the equations again. C_j is set to the values of the current recipe and the target values are unchanged (Eqn 6.128):

$$C_j = C_{j,1} \quad \text{for } j = 1 \text{ to } n \quad (6.128)$$

The values of F_i and F_C are calculated and the deviations ε determined. The equations are solved as described and the next formulation is given by Eqn 6.129:

$$C_{j,2} = C_{j,1} + \Delta C_j \tag{6.129}$$

The correction step is applied repeatedly until the value of S, the sum of the squares of the deviations, either falls below a preset limit value or does not change significantly as a result of the iteration.

Sample calculation

The properties of spectrophotometric curve matching can be illustrated by considering an example. In general, the method produces an excellent formulation when it is possible to match the spectrum of the standard using the selected colorants, but problems arise when an exact spectral match cannot be made using the chosen colorant combination. Figure 6.28 shows the reflectance spectrum of a green colour standard based on the Natural Colour System (NCS) 2030 B90G, together with the results of two spectrophotometric curve matches using four Sandosperse predispersions. The dashed curve was obtained by an unweighted fit ($W_i = 1$) and the solid curve by a fit using weighting equation 6.87. Neither method produced a perfect match (in fact a perfect match is not possible using these pigments). The better fit by the formulation calculated using weighted equations is clearly apparent. This is reflected in turn in a lower $L^*a^*b^*$ colour difference, as shown in Table 6.19. On average the weighted method produces colour differences between a quarter and a half of those obtained by the unweighted method.

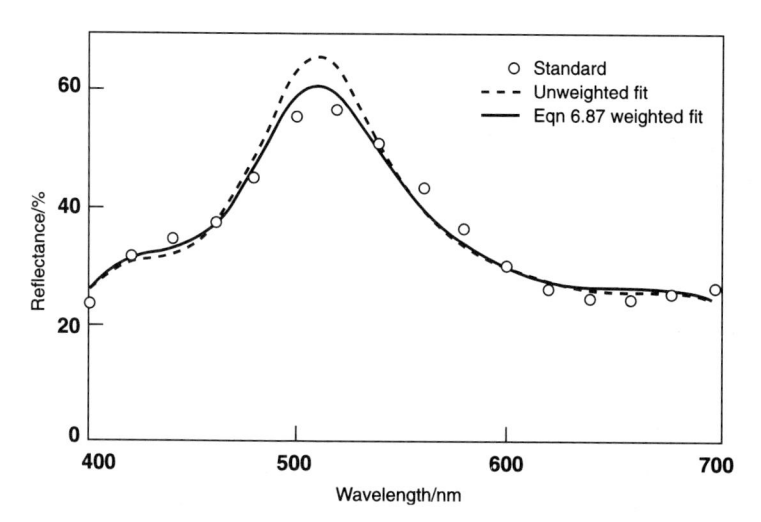

Figure 6.28 Spectrophotometric matches to a green standard using four colorants (Sandosperse predispersions White CBJ, Black BLNS, Yellow 4G and Blue 2GLS)

Table 6.19 CIELAB colour difference for different types of a four-colorant colour match to NCS 2030 B90G

Illuminant[a]	Spectrophotometric match		Colorimetric match under illuminant D_{65}
	Unweighted fit	Weighted fit	
D_{65}	5.77	1.54	0.04
A	2.72	0.79	1.64
TL84	2.31	0.93	1.86

a 10° observer

The spectrophotometric matching method can be applied to two, three, four, five or more colorants. When a good spectral match is not possible, however, there is often a significant colour difference between the predicted recipe and the standard. In many cases of this type, a more acceptable recipe would have a good match under the primary illuminant and a poorer match under other illuminants. The colorimetric method offers this type of solution, as shown in the third column of Table 6.19.

6.7.3 Colorimetric (XYZ) match prediction

Colorimetric matching was one of the first colour-matching methods to be established. The method was put on a formal basis for the two-constant Kubelka–Munk theory by Allen [18,19]. The method starts from an initial or first-guess recipe and determines the changes in concentration ΔC_j that will make the colour coordinates of the recipe more closely match those of the standard.

Colorimetric matching with four colorants
Consider four colorants in an initial recipe mixture with the respective concentrations C_1, C_2, C_3 and C_4. In colorimetric matching the parameters of interest are the tristimulus values X_M, Y_M and Z_M, which are related to concentration through the database entries as described in section 6.7.1. The method can be developed in a similar way to the spectrophotometric equations in section 6.7.2, but instead of 16 wavelength equations there are three tristimulus equations (Eqn 6.130):

$$\begin{aligned} F_1 &= X_M = f_X(C_1,C_2,C_3,C_4) \\ F_2 &= Y_M = f_Y(C_1,C_2,C_3,C_4) \\ F_3 &= Z_M = f_Z(C_1,C_2,C_3,C_4) \end{aligned} \quad (6.130)$$

MATCH PREDICTION OF AN OPAQUE LAYER 349

The total colorant volume constraint is again introduced (Eqn 6.131):

$$F_C = C_1 + C_2 + C_3 + C_4 \tag{6.131}$$

If the tristimulus values of the standard are set as the target values for Eqn 6.130, then the deviations are given by Eqn 6.132:

$$\varepsilon_1 = X - X_M \quad \varepsilon_2 = Y - Y_M \quad \varepsilon_3 = Z - Z_M \quad \varepsilon_C = V_C - F_C \tag{6.132}$$

The colorimetric method determines the changes in concentration ΔC_1, ΔC_2, ΔC_3 and ΔC_4 that are needed to reduce the deviations to zero.

The method makes use of partial derivative terms such as $\partial X_M / \partial C_1$ to describe the change in X_M, the tristimulus value of the mixture, for a small change in the concentration of colorant 1. The changes in X_M are assumed to be additive, so that the total effect of the changes in concentration is the sum of the individual contributions. The total change should act to correct the deviation (Eqn 6.133):

$$\begin{aligned}
\varepsilon_1 &= \Delta C_1 \frac{\partial X_M}{\partial C_1} + \Delta C_2 \frac{\partial X_M}{\partial C_2} + \Delta C_3 \frac{\partial X_M}{\partial C_3} + \Delta C_4 \frac{\partial X_M}{\partial C_4} \\
\varepsilon_2 &= \Delta C_1 \frac{\partial Y_M}{\partial C_1} + \Delta C_2 \frac{\partial Y_M}{\partial C_2} + \Delta C_3 \frac{\partial Y_M}{\partial C_3} + \Delta C_4 \frac{\partial Y_M}{\partial C_4} \\
\varepsilon_3 &= \Delta C_1 \frac{\partial Z_M}{\partial C_1} + \Delta C_2 \frac{\partial Z_M}{\partial C_2} + \Delta C_3 \frac{\partial Z_M}{\partial C_3} + \Delta C_4 \frac{\partial Z_M}{\partial C_4}
\end{aligned} \tag{6.133}$$

The constraint of a constant total concentration gives rise to a fourth equation (Eqn 6.134):

$$\varepsilon_C = \Delta C_1 + \Delta C_2 + \Delta C_3 + \Delta C_4 \tag{6.134}$$

We now describe how the partial derivatives in Eqn 6.133 may be calculated from the database constants so that the only unknown terms in the four simultaneous equations are the changes in concentration. The equations may then be solved to obtain these values.

Expressing the four equations in matrix form gives Eqn 6.135:

$$[\varepsilon] = [IT][\Delta C] \tag{6.135}$$

where $[\varepsilon]$ and $[\Delta C]$ are (4×1) matrices (Eqn 6.136):

$$[\varepsilon] = \begin{pmatrix} \varepsilon_1 \\ \varepsilon_2 \\ \varepsilon_3 \\ \varepsilon_4 \end{pmatrix} \quad [\Delta C] = \begin{pmatrix} \Delta C_1 \\ \Delta C_2 \\ \Delta C_3 \\ \Delta C_4 \end{pmatrix} \tag{6.136}$$

and [IT] is the (4 × 4) matrix often termed an influence matrix (Eqn 6.137):

$$[IT] = \begin{pmatrix} \frac{\partial X_M}{\partial C_1} & \frac{\partial X_M}{\partial C_2} & \frac{\partial X_M}{\partial C_3} & \frac{\partial X_M}{\partial C_4} \\ \frac{\partial Y_M}{\partial C_1} & \frac{\partial Y_M}{\partial C_2} & \frac{\partial Y_M}{\partial C_3} & \frac{\partial Y_M}{\partial C_4} \\ \frac{\partial Z_M}{\partial C_1} & \frac{\partial Z_M}{\partial C_2} & \frac{\partial Z_M}{\partial C_3} & \frac{\partial Z_M}{\partial C_4} \\ 1 & 1 & 1 & 1 \end{pmatrix} \quad (6.137)$$

The solution is straightforward; since [IT] is a square matrix it can be inverted to give the solution (Eqn 6.138):

$$[\Delta C] = [IT]^{-1} [\varepsilon] \quad (6.138)$$

The adjusted formulation is given by Eqn 6.139:

$$C_{j,1} = C_j + \Delta C_j \quad \text{for } j = 1 \text{ to } 4 \quad (6.139)$$

Determination of the partial derivatives
The partial derivative terms describe how the tristimulus values of the mixture will change for a small change in the amount of one of the components. These values may be determined using either a numerical method or an analytical method. The numerical method involves calculating the colour of a mixture with one component changed by a small amount ΔC_j. It follows that Eqn 6.140 can be written:

$$\frac{\partial X_M}{\partial C_j} \approx \frac{\Delta X_M}{\Delta C_j} \quad \frac{\partial Y_M}{\partial C_j} \approx \frac{\Delta Y_M}{\Delta C_j} \quad \frac{\partial Z_M}{\partial C_j} \approx \frac{\Delta Z_M}{\Delta C_j} \quad (6.140)$$

Analytical equations may be developed in a step-by-step fashion (Eqn 6.141):

$$\frac{\partial X_M}{\partial C_j} \approx \sum E \bar{x} \frac{\partial \rho_M}{\partial C_j} \quad (6.141)$$

where E is the spectral power of the illuminant and $\bar{x}$ is the observer weighting term. The summation is over the measured wavelengths in the spectrum. Similar equations are used to determine the other derivatives, replacing $\bar{x}$ with $\bar{y}$ or $\bar{z}$ as appropriate.

The derivative of reflectance can be expressed as the product of three terms (Eqn 6.142):

$$\frac{\partial \rho_M}{\partial C_j} = \frac{d\rho_M}{dR_M} \frac{dR_M}{d\omega_M} \frac{\partial \omega_M}{\partial C_j} \quad (6.142)$$

The first two derivatives are obtained directly from Eqns 6.104 and 6.105 (Eqns 6.143 and 6.144):

$$\frac{d\rho_M}{dR_M} = \frac{t_e t_i}{(1 - r_i R_M)^2} \tag{6.143}$$

$$\frac{dR_M}{d\omega_M} = -\frac{2R_M^2}{1 - R_M^2} \tag{6.144}$$

The third term describes the way in which ω_M changes for a small change in the concentration of component j, and is obtained from Eqn 6.102 (Eqn 6.145):

$$\frac{\partial \omega_M}{\partial C_j} = \omega_M \left(\frac{1}{K_M} \frac{\partial K_M}{\partial C_j} - \frac{1}{S_M} \frac{\partial S_M}{\partial C_j} \right) \tag{6.145}$$

Taking into account that if C_j is increased, then the amounts of the other components are decreased to maintain the constant-volume constraint, the derivatives are given by Eqn 6.146:

$$\frac{\partial K_M}{\partial C_j} = \frac{dK_{jW}}{dC_j} - K_M \qquad \frac{\partial S_M}{\partial C_j} = \frac{dS_{jW}}{dC_j} - S_M \tag{6.146}$$

For example, where there is a linear relationship between K_{jW}, S_{jW} and concentration then we may write Eqn 6.147:

$$\frac{\partial K_M}{\partial C_j} = \omega_M \left(\frac{\hat{K}_{jW}}{K_M} - \frac{\hat{S}_{jW}}{S_M} \right) \tag{6.147}$$

so that Eqn 6.148 can be written:

$$\frac{\partial \rho_M}{\partial C_j} = -\frac{R_M(1 - R_M)}{1 + R_M} \frac{t_e t_i}{(1 - r_i R_M)^2} \left(\frac{\hat{K}_{jW}}{K_M} - \frac{\hat{S}_{jW}}{S_M} \right) \tag{6.148}$$

Application of colorimetric matching

The method relies on the assumption of additivity expressed by Eqn 6.133, which is only valid for small changes in concentration and small deviations. It is often applied to correct a recipe produced by another method, as shown for the weighted spectrophotometric method in Table 6.19. When the colour of the initial recipe is moderately different from that of the standard then a single application of the colorimetric match calculation may not produce the optimum recipe. The adjusted recipe is then used as the initial recipe and a repeat calculation made. This cycle is continued (iterated) until either a match is achieved or no further reduction in colour difference is obtained.

It is possible to use the colorimetric method even with an arbitrary starting formulation, such as equal amounts of each component. In this case the colour of the starting recipe may be very different from that of the standard and the additivity condition is not satisfied for such large deviations. Direct application of the colorimetric calculation may produce concentrations that are negative or greater than 1.0. A solution may be obtained by reducing the colour deviation in a series of steps using an iterative technique, the maximum colour deviation allowed in each step being set so that the additivity assumption is still valid. The values of ε_1, ε_2 and ε_3 are calculated as before, except that if the magnitude of any one is greater than a preset limiting value then only a fraction of the deviations is used in the calculation. The fraction is set so that the magnitude of the largest deviation is equal to the preset limit. Finally, if the calculated concentration of any component at any stage is less than zero then it is set to zero and the other concentrations are adjusted in proportion to maintain a total colorant volume of unity.

This process is illustrated in Table 6.20, where the same standard (NCS 2030 B90G) and colorants are used as in the earlier example (Table 6.19). The initial recipe has equal amounts of each colorant and the maximum colour deviation is set as 10 units. The value of ΔE^* is reduced from 41.56 units to 0.05 units after four iterations, which is sufficiently fast to allow the colorimetric method to be routinely used without an initial recipe calculation. The method is then a true colorimetric match in the sense that only the values of X, Y and Z of the standard are required by the equations and not the full reflectance spectrum.

Table 6.20 Colorimetric match prediction from an arbitrary initial recipe; limited values of ε_1, ε_2 and ε_3 of recipe 1 are used to determine recipe 2, and so on

Parameter	Recipe 1	Recipe 2	Recipe 3	Recipe 4	Recipe 5
White	0.2500	0.6198	0.8419	0.8476	0.8483
Black	0.2500	0.0000	0.0115	0.0021	0.0037
Blue	0.2500	0.2140	0.0864	0.0872	0.0860
Yellow	0.2500	0.1662	0.0602	0.0631	0.0621
ε_1	25.95	13.12	1.65	−0.17	0.00
ε_2	34.87	13.98	3.20	−0.58	−0.02
ε_3	31.49	15.92	1.53	−0.16	0.00
Limited ε_1	7.44	8.24	1.65	−0.17	0.00
Limited ε_2	10.00	8.78	3.20	−0.58	−0.02
Limited ε_3	9.03	10.00	1.53	−0.16	0.00
ΔE^*_{ab}	41.56	14.00	4.64	1.24	0.05

Colorimetric matching with two or three colorants
The colorimetric match-prediction method can be adapted for use with formulations that contain two or three colorants. In the three-colorant case Eqns 6.133 and 6.134 apply, except that the right-hand side now contains only three concentration terms. The corresponding influence matrix [IT] now has four rows and three columns; the direct inversion method of solution (Eqn 6.138) cannot be applied as [IT] is not a square matrix.

It is possible to use a least-squares method of solution that calculates the changes in concentration that minimise the sum of the squares of the deviation values (Eqn 6.149):

$$S = \varepsilon_1^2 + \varepsilon_2^2 + \varepsilon_3^2 + \varepsilon_C^2 \tag{6.149}$$

As with the spectrophotometric match, the equations can be weighted to establish the relative importance of the deviation arising from that equation. In particular Eqn 6.134 should be weighted at least 100 times more heavily than Eqn 6.133 in order to maintain the constant-volume condition. The matrix form of the equations becomes Eqn 6.150:

$$[W\varepsilon] = [WI][\Delta C] \tag{6.150}$$

where $[W\varepsilon]$ is a (4×1) and $[\Delta C]$ is a (3×1) matrix as follows (Eqn 6.151):

$$[W\varepsilon] = \begin{pmatrix} W\varepsilon_1 \\ W\varepsilon_2 \\ W\varepsilon_3 \\ W\varepsilon_4 \end{pmatrix} \qquad [\Delta C] = \begin{pmatrix} \Delta C_1 \\ \Delta C_2 \\ \Delta C_3 \end{pmatrix} \tag{6.151}$$

[WIT] is the (4×3) influence matrix (Eqn 6.152):

$$[WIT] = \begin{pmatrix} W\dfrac{\partial X_M}{\partial C_1} & W\dfrac{\partial X_M}{\partial C_2} & W\dfrac{\partial X_M}{\partial C_3} & W\dfrac{\partial X_M}{\partial C_4} \\ W\dfrac{\partial Y_M}{\partial C_1} & W\dfrac{\partial Y_M}{\partial C_2} & W\dfrac{\partial Y_M}{\partial C_3} & W\dfrac{\partial Y_M}{\partial C_4} \\ W\dfrac{\partial Z_M}{\partial C_1} & W\dfrac{\partial Z_M}{\partial C_2} & W\dfrac{\partial Z_M}{\partial C_3} & W\dfrac{\partial Z_M}{\partial C_4} \\ W_C & W_C & W_C & W_C \end{pmatrix} \tag{6.152}$$

The least-squares solution is given by Eqn 6.153:

$$[\Delta C] = \left\{[WIT]^T [WIT]\right\}^{-1} [WIT]^T [W\varepsilon] \tag{6.153}$$

As before, the superscripts T and −1 denote taking the transpose and the inverse of a matrix respectively; typically $W = 1$ and $W_C = 100$.

Colorimetric matching with five colorants

In order for the method to produce a unique formulation for recipes containing five colorants an additional equation needs to added to the four in Eqns 6.133 and 6.134. This may be either a colour equation (the most common choice) or an additional constraint on the concentration. Although only one extra equation is needed, there is an advantage in adding three equations and describing the colour under a second illuminant. The method is then similar to the equilibrate match system described by Sluban [14]. The spectral power distribution of the second illuminant should be very different from that of the first; for example, illuminants D_{65} and A would form an appropriate pair whereas illuminants D_{65} and C would not. The weighted form of the equations as expressed in Eqn 6.150 is used. If desired the weights can be set to emphasise the importance of the colour deviation under the first illuminant relative to that under the second. As before, Eqn 6.153 is used to obtain the solution; however, the matrices need to be redefined to take into account the extra illuminant. The method can be used for any number of colorants between two and five, and hence the definitions will be made in a general form.

In order to save space the matrix definitions have been written in the transpose form, where the rows are exchanged for columns. The matrix [$W\varepsilon$] is the (7 × 1) deviation matrix shown in transpose form in Eqn 6.154. The additional subscripts a and b are used to denote X, Y and Z and other values referring to the first and second illuminants respectively:

$$[W\varepsilon] = [\; W_a\varepsilon_{1,a} \;\; W_a\varepsilon_{2,a} \;\; W_a\varepsilon_{3,a} \;\; W_b\varepsilon_{1,b} \;\; W_b\varepsilon_{2,b} \;\; W_b\varepsilon_{3,b} \;\; W_C\varepsilon_C \;]^T \quad (6.154)$$

[$W\Delta C$] is the (n × 1) concentration change matrix, where n is the number of colorants in the formulation. It is shown in transpose form in Eqn 6.155:

$$[\Delta C] = [\Delta C_1 \;\; \Delta C_1 \; ... \; \Delta C_n]^T \quad (6.155)$$

[WIT] is the (7 × n) influence matrix shown in transpose form in Eqn 6.156. The matrix method of obtaining a least-squares solution is given by Eqn 6.153. The equations are general, since they can be applied to two, three, four or five colorants in the formulation.

In the generalised form shown here the equations look daunting, but many aids are available to help with the task of organising the information and calculating the results. The solution is obtained by common mathematical operations which are available in standard spreadsheet programs. Since the solution is a least-squares fit in the

MATCH PREDICTION OF AN OPAQUE LAYER 355

$$[WIT] = \begin{pmatrix} W_a \dfrac{\partial X_{M,a}}{\partial C_1} & W_a \dfrac{\partial Y_{M,a}}{\partial C_1} & W_a \dfrac{\partial Z_{M,a}}{\partial C_1} & W_b \dfrac{\partial X_{M,b}}{\partial C_1} & W_b \dfrac{\partial Y_{M,b}}{\partial C_1} & W_b \dfrac{\partial Z_{M,b}}{\partial C_1} & W_C \\ W_a \dfrac{\partial X_{M,a}}{\partial C_2} & W_a \dfrac{\partial Y_{M,a}}{\partial C_2} & W_a \dfrac{\partial Z_{M,a}}{\partial C_2} & W_b \dfrac{\partial X_{M,b}}{\partial C_2} & W_b \dfrac{\partial Y_{M,b}}{\partial C_2} & W_b \dfrac{\partial Z_{M,b}}{\partial C_2} & W_C \\ \ldots & \ldots & \ldots & \ldots & \ldots & \ldots & \ldots \\ W_a \dfrac{\partial X_{M,a}}{\partial C_n} & W_a \dfrac{\partial Y_{M,a}}{\partial C_n} & W_a \dfrac{\partial Z_{M,a}}{\partial C_n} & W_b \dfrac{\partial X_{M,b}}{\partial C_n} & W_b \dfrac{\partial Y_{M,b}}{\partial C_n} & W_b \dfrac{\partial Z_{M,b}}{\partial C_n} & W_C \end{pmatrix}^T$$

(6.156)

same sense as fitting a regression line through a set of data points on a chart, direct matrix methods need not necessarily be used. For example, in the Microsoft Excel spreadsheet the seven elements of $[W\varepsilon]$ are placed in a column and treated as the dependent variable (the y values). The elements of $[WIT]$ are placed in n columns each of seven elements and treated as the independent variables (the x values), then application of the LINEST function returns the values of ΔC_j as the fitted coefficients.

Figure 6.29 shows the reflectance spectrum of the green colour standard based on NCS 2030 B90G. The solid curve shows the results of a reflectance-weighted spectrophotometric curve match followed by a colorimetric correction using five Sandosperse

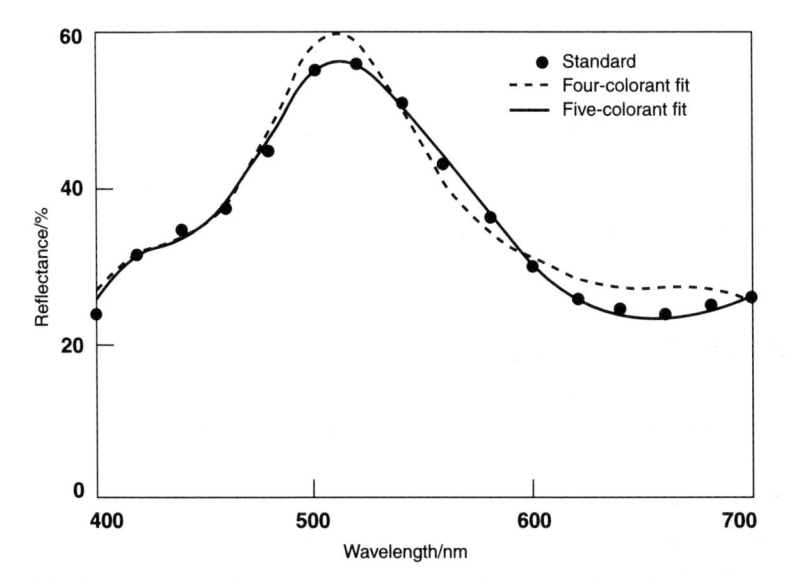

Figure 6.29 Colorimetric matches to a green standard using four and five colorants

predispersions, White CBJ, Black BLNS, Yellow 4G, Blue 2GLS and Green 3GLS. The dashed curve was obtained using four colorants: white, black, yellow and blue. In both calculations the weights were W_a = 1.0, W_b = 0.1 and W_C = 100. The formulations and the $L^*a^*b^*$ colour differences are shown in Table 6.21.

Table 6.21 Comparison between four-colorant (tinter) and five-colorant matches to a green standard

	Volume fraction			ΔE^*_{ab}	
Colorant	Four tinters	Five tinters	Illuminant	Four tinters	Five tinters
White CBJ	0.8485	0.8670	D_{65}	0.02	0.01
Black BLNS	0.0037	0.0103	A	1.62	0.15
Blue 2GLS	0.0858	0.0333	TL84	1.84	0.12
Yellow 4G	0.0620	0.0439			
Green 3GLS	0.0000	0.0455			

Both formulations provide a good match under the primary illuminant, but the colour difference under the secondary illuminant is reduced from 1.62 to 0.15 when the number of colorants is increased from four to five. The improved quality of the five-colorant fit is also clear from Figure 6.29.

6.7.4 Colorimetric ($L^*a^*b^*$) match prediction

The colorimetric (XYZ) method adjusts the recipe to minimise the sum of the squares of the weighted errors in the tristimulus values. With relatively little extra effort it is possible to generate equations that minimise the CIELAB colour difference ΔE^*. The

$$[WIE] = \begin{pmatrix} W_a \dfrac{\partial L^*_{M,a}}{\partial C_1} & W_a \dfrac{\partial a^*_{M,a}}{\partial C_1} & W_a \dfrac{\partial b^*_{M,a}}{\partial C_1} & W_b \dfrac{\partial L^*_{M,b}}{\partial C_1} & W_b \dfrac{\partial a^*_{M,b}}{\partial C_1} & W_b \dfrac{\partial b^*_{M,b}}{\partial C_1} & W_C \\ W_a \dfrac{\partial L^*_{M,a}}{\partial C_2} & W_a \dfrac{\partial a^*_{M,a}}{\partial C_2} & W_a \dfrac{\partial b^*_{M,a}}{\partial C_2} & W_b \dfrac{\partial L^*_{M,b}}{\partial C_2} & W_b \dfrac{\partial a^*_{M,b}}{\partial C_2} & W_b \dfrac{\partial b^*_{M,b}}{\partial C_2} & W_C \\ \ldots & \ldots & \ldots & \ldots & \ldots & \ldots & \ldots \\ W_a \dfrac{\partial L^*_{M,a}}{\partial C_n} & W_a \dfrac{\partial a^*_{M,a}}{\partial C_n} & W_a \dfrac{\partial b^*_{M,a}}{\partial C_n} & W_b \dfrac{\partial L^*_{M,b}}{\partial C_n} & W_b \dfrac{\partial a^*_{M,b}}{\partial C_n} & W_b \dfrac{\partial b^*_{M,b}}{\partial C_n} & W_C \end{pmatrix}$$

(6.158)

deviation terms ε and the influence matrix [WIT] are replaced by the equivalent $L^*a^*b^*$ versions (Eqn 6.157) (in transpose form the matrix [WIE] is given by Eqn 6.158 on page 356):

$$\varepsilon_1 = L^* - L^*_M \quad \varepsilon_2 = a^* - a^*_M \quad \varepsilon_3 = b^* - b^*_M \quad \varepsilon_C = V_C - F_C \quad (6.157)$$

The individual partial derivatives can be obtained from the corresponding tristimulus terms (Eqns 6.159–6.161):

$$\frac{\partial L^*}{\partial C} = \frac{116}{3Y}\left(\frac{Y}{Y_0}\right)^{1/3}\frac{\partial Y}{\partial C} \quad (6.159)$$

$$\frac{\partial a^*}{\partial C} = \frac{500}{3X}\left(\frac{X}{X_0}\right)^{1/3}\frac{\partial X}{\partial C} - \frac{500}{3Y}\left(\frac{Y}{Y_0}\right)^{1/3}\frac{\partial Y}{\partial C} \quad (6.160)$$

$$\frac{\partial b^*}{\partial C} = \frac{200}{3Y}\left(\frac{Y}{Y_0}\right)^{1/3}\frac{\partial Y}{\partial C} - \frac{200}{3Z}\left(\frac{Z}{Z_0}\right)^{1/3}\frac{\partial Z}{\partial C} \quad (6.161)$$

The least-squares solution follows from Eqn 6.153 (Eqn 6.162):

$$[\Delta C] = \left\{[WIE]^T [WIE]\right\}^{-1} [WIE]^T [W\varepsilon] \quad (6.162)$$

The benefits of using the extra calculations involved in the $L^*a^*b^*$ method arise when it is not possible to produce a good match with the chosen set of colorants. The method can provide a recipe that is visually closer to the standard than that obtained from the tristimulus method.

Sensitivity and robustness of a recipe
The influence matrix [WIE] can be used to provide additional useful information describing the properties of the recipe, in particular the sensitivity and the related parameter of robustness [20]. The *sensitivity* s_j of component j in a recipe describes the change in colour (ΔE^*_{ab}) that would arise from a dispensing error of ΔC_j in the volume concentration of component j (Eqn 6.163):

$$\Delta E^*_{ab} = s_j |\Delta C_j| \quad (6.163)$$

s_j is obtained from the partial derivatives used in the weighted influence matrix [WIE] (Eqn 6.164):

$$s_j = \left[\left(\frac{\partial L^*_M}{\partial C_j}\right)^2 + \left(\frac{\partial a^*_M}{\partial C_j}\right)^2 + \left(\frac{\partial b^*_M}{\partial C_j}\right)^2\right]^{1/2} \quad (6.164)$$

The values of s_j for the example formulation are shown in Table 6.22; this indicates that the colour of the recipe is around 50 times more sensitive to an error in dispensing the black component than to an error in dispensing the white.

Table 6.22 Sensitivity of the colour of recipe to individual colorants in a match to NCS 2030 B90G[a]

Colorant	Volume	Sensitivity
White CBJ	0.8670	11.97
Black BLNS	0.0103	606.80
Blue 2GLS	0.0333	155.19
Yellow 4G	0.0439	211.18
Green 3GLS	0.0455	151.59

a Sensitivity and robustness of the recipe are 678.24 and 0.0015 respectively

The average sensitivity of the entire recipe can be developed by assuming that the errors are truly random in nature and the dispensing precision is the same for each component in the formulation. Under these conditions the laws of combination of errors can be used to obtain the recipe sensitivity $\langle s \rangle$ (Eqn 6.165):

$$\langle s \rangle = \left(\sum_{j=1}^{n} s_j^2 \right)^{1/2} \qquad (6.165)$$

The term $\langle s \rangle |\Delta C|$ can be interpreted as the root mean square value of ΔE^*_{ab} that would be expected from repeated preparations of the same formulation, with a dispensing error of $\pm \Delta C$. The value may be used to select between recipes that are otherwise similar in quality of colour matching.

The *robustness* of a recipe is defined as the dispensing error that would produce a colour difference of one unit between the correct recipe and the incorrect recipe. Robustness is the reciprocal of recipe sensitivity, and it follows that highly sensitive recipes are not very robust. For practical purposes the precision of dispensing should be at least twice as good as the robustness. Under these conditions the colour of 95% of repeat preparations of a recipe will be within $\Delta E^*_{ab} = 1$ of the target recipe. The robustness of the recipe shown in Table 6.22 is 0.0015, and the average dispensing precision needs to be better than one part in 1000 to satisfy the above criteria. The sensitivity/robustness can also be expressed in terms of other colour-difference equations such as the CMC(l:c) equation [21].

6.7.5 Hiding power of a coating formulation

A paint is intended to hide the colour of the underlying substrate completely. Several national standards describe methods of specifying the hiding of a paint layer by reflectometry. Typically the coating is applied over standard white and standard black substrates. The colour of each sample is measured and the hiding expressed as the *contrast ratio* C_R (ASTM D2805-88) or the colour difference ΔE^*_{ab} (DIN 55987) at the layer thickness (Eqn 6.166):

$$C_R = \frac{Y_B}{Y_W}$$

$$\Delta E^*_{ab} = \left[(L^*_W - L^*_B)^2 + (a^*_W - a^*_B)^2 + (b^*_W - b^*_B)^2 \right]^{1/2}$$

(6.166)

where the subscripts W and B denote the colour of the coating over the white and the black substrate respectively.

Both standards define a specific value to be regarded as just-hiding: $C_R = 0.98$ and $\Delta E^*_{ab} = 1.00$ respectively. An opaque layer would have $C_R = 1$ and $\Delta E^*_{ab} = 0$.

The ability of a particular paint to hide or mask the colour of the substrate depends on the absorption and scattering coefficients per unit layer thickness. The database for opaque layers provides information from which K_{jW} and S_{jW}, the relative absorption and scattering coefficients of each component, can be determined. These values are relative to the scattering of the calibration white. In order to obtain the absolute values of K and S for the mixture, the sum of the contributions from the colorants must be multiplied by further factors. At each wavelength the coefficients are given by Eqn 6.167:

$$K = V\hat{S}_W \sum_{j=1}^{n} K_{jW} \qquad S = V\hat{S}_W \sum_{j=1}^{n} S_{jW}$$

(6.167)

where V is the total volume fraction of colorants in the coating formulation, and $\hat{S}_W$ is the scattering coefficient of unit concentration of the calibration white colorant at unit layer thickness.

The reflectance spectrum of a formulation coated on to a substrate of reflectance R_g can be calculated from Eqns 6.27–6.29, developed in section 6.2 for a semitransparent layer. The reflectance spectrum is corrected for surface effects using Eqn 6.34. The reflectance spectrum and the colour are calculated for the layer over the white ($R_g = 0.80$) and black ($R_g = 0.01$) substrates, and finally the hiding parameter C_R or ΔE^*_{ab} is determined.

Evaluation of $\hat{S}_W$

The scattering power of the calibration white may be determined from two calibration panels made with a paint containing only the standard white colorant and a clear diluent. An opaque layer is prepared (R_∞) and a semitransparent layer (R) applied over a black substrate (R_g). The volume fraction V of calibration white in the paint and the thickness D of the layer of the semitransparent panel are set to obtain a reflectance that is approximately halfway between R_g and R_∞. The method of calculation has been explained in section 6.6 (Eqn 6.168):

$$\hat{S}_W = \frac{Z}{VD} \frac{2R_\infty}{1-R_\infty^2} \qquad (6.168)$$

where Z is defined by Eqn 6.169:

$$\beta = \frac{R_g - R}{R - R_\infty} \frac{1-R_\infty^2}{1-R_g R_\infty} \qquad Z = \frac{1}{2}\ln(\beta + 1) \qquad (6.169)$$

Care must be taken to ensure that the units of V and D are consistent between the calibration and the hiding parameter calculation: for example, they should both refer to the wet layer or both to the dry layer. Clearly the values can be converted from wet to dry or vice versa when the volume content of solids in each component in the formulation is known.

Spreading rate of a coating formulation

In practice, the ability of a paint to hide the colour of the underlying substrate can be expressed in several ways. One of the most useful is the spreading rate H, the area (in m^2) that one litre of the test paint will cover at a layer thickness that gives the limiting value of C_R or ΔE^*_{ab}. The thickness of the coating (D in μm) that satisfies the just-hiding limit is linked to the spreading rate via Eqn 6.170:

$$H = \frac{1000}{D} \qquad (6.170)$$

For many formulations the spreading rate is almost linearly related to ΔE^*_{ab} (Figure 6.30).

Gall [22] and Cairns [23] described computer-based methods of predicting the hiding power from the K and S coefficients of the colorants in the formulation. The principle has been reviewed and added to by Volz [24]. In the method described below an iterative technique is used to determine the value of the spreading coefficient $\hat{H}$ for a hypothetical coating with $V = 1$ that satisfies the just-hiding criteria.

Let the just-hiding value of ΔE^*_{ab} be P. The value of H is set to a start value of H_n and the layer thickness determined from Eqn 6.171:

$$D_n = \frac{1000}{H_n} \tag{6.171}$$

The values of K and S corresponding to $V = 1$ are determined from Eqn 6.167 and the reflectance spectra and P_n (ΔE^*_{ab}) are calculated. A slightly different value, H_{n+1}, is assigned to H and the process repeated to obtain P_{n+1}. If the value of P_{n+1} is not close enough to the target value P then a corrected value H_{n+2} is estimated from Eqn 6.172 and P_{n+2} calculated:

$$H_{n+2} = H_{n+1} + (P - P_{n+1})\left(\frac{H_{n+1} - H_n}{P_{n+1} - P_n}\right) \tag{6.172}$$

The iteration is repeated by assigning the values of H_{n+1} and P_{n+1} to H_n and P_n, H_{n+2} and P_{n+2} to H_{n+1} and P_{n+1}. This is continued until the value of P_{n+2} is within preset limits of the target value when $\hat{H} = H_{n+2}$.

The results of the calculation for the five-colorant formulation described in Table 6.21 are shown in Figure 6.30. The plot shows the nearly linear relationship between H and ΔE^*_{ab} which leads to a rapid convergence of the iteration calculation. In this case the calculated ΔE^*_{ab} value is within the range limits of 1.00 ± 0.01 after three iterations. The iteration also converges rapidly when contrast ratio is used instead of ΔE^*_{ab}.

The resulting value of $\hat{H}$ can be interpreted in several ways. For example, if there is a lower limit of coating thickness D_{min}, then the thickness is set to this value and colorant loading in the paint formulation is given by Eqn 6.173:

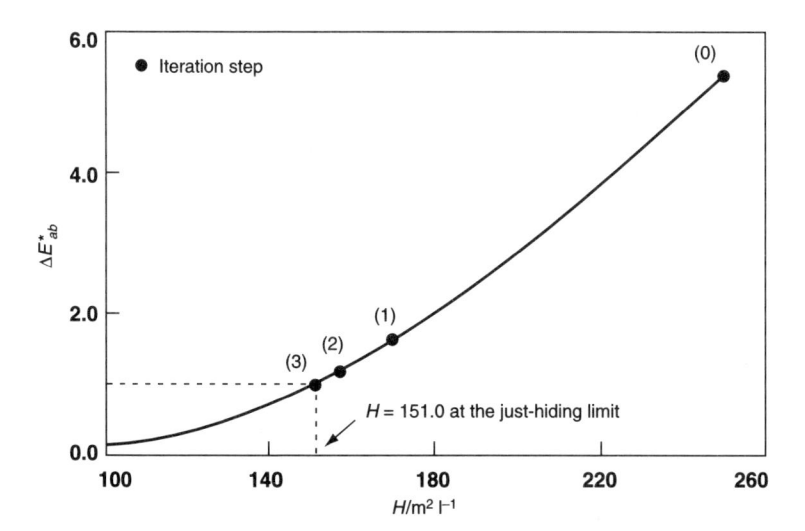

Figure 6.30 Iterative calculation of the hiding power of a five-colorant formulation for the green standard NCS 2030 B90G

$$V = \frac{1000}{\hat{H}D_{min}} \tag{6.173}$$

Suppose, for example, the minimum thickness may be taken as 25 μm. Then the colorant volume fraction of the green paint that just satisfies the hiding limit is:

$$V = \frac{1000}{151.0 \times 25} = 0.265$$

There may be an upper limit to the volume fraction of colorant that can be incorporated into the paint, V_{max}. If V exceeds this limit, then the volume fraction is set at the limit and the layer thickness increased to satisfy the hiding criteria (Eqn 6.174):

$$H = \hat{H}V_{max} \qquad D = \frac{1000}{H} \tag{6.174}$$

Taking as an example a maximum colorant loading of 10.0%, then the spreading rate and thickness of the green paint to achieve hiding are respectively:

$$H = 151.0 \times 0.10 = 15.1 \text{ m}^2 \text{ l}^{-1} \qquad D = \frac{1000}{15.1} = 66.3 \text{ μm}$$

6.7.6 Including a fixed component into the recipe

The match-prediction methods described so far allow each colorant in the formulation to take any fraction of the total colorant volume between 0 and 1. On occasions, a particular colorant is required to make up a specific amount of the total. The most common situations are as follows:

(a) Match-predicting the amounts of tinting colorants to add to a can already partly filled with a base paint. For example, a one-litre can is prefilled with 900 ml of a white base paint, leaving 100 ml for the addition of a mixture of tinters to produce the target shade. The base paint is a fixed component in the recipe and the tinter volume is restricted to the remaining 10% of the total volume.
(b) The colour correction of an off-shade batch of paint where the correction method calculates the mixture of colorants to add to an existing batch to correct the colour.
(c) The forced inclusion into the match-prediction recipe of a limited amount of a waste paint of similar shade to the target colour.

The equations described in section 6.7.1 are rewritten to include the terms for the fixed component in the recipe; these are denoted by the subscript P and the variable components by subscripts 1 to n (Eqn 6.175):

$$\omega_M = \frac{K_M}{S_M} = \frac{K_{PW} + K_{1W} + K_{2W} + \ldots + K_{nW}}{S_{PW} + S_{1W} + S_{2W} + \ldots + S_{nW}} \quad (6.175)$$

The total colorant volume is given by Eqn 6.176:

$$1 = C_P + C_1 + C_2 + \ldots + C_{nW} \quad (6.176)$$

The spectrophotometric and colorimetric prediction equations make use of partial differential equations to calculate changes in concentration to correct the deviations ε of the recipe values of the parameters F from the target values V. The method of calculating the changes in concentration ΔC_j is identical to that described earlier; since C_P is fixed, however, there is no partial derivative term and no ΔC_P term.

Spectrophotometric matching with a fixed component
The terms of Eqn 6.175 are separated as in Eqn 6.106 and the fitting equation (Eqn 6.177) obtained by substituting ω (the K/S value of the standard) for ω_M:

$$0 = (K_{PW} - \omega_M S_{PW}) + (K_{1W} - \omega_M S_{1W}) \\ + (K_{2W} - \omega_M S_{2W}) + \ldots + (K_{nW} - \omega_M S_{nW}) \quad (6.177)$$

The fitting functions F and F_C are defined in the same way as in section 6.7.2 (Eqn 6.178):

$$F = (K_{1W} - \omega S_{1W}) + (K_{2W} - \omega S_{2W}) + \ldots + (K_{nW} - \omega S_{nW}) \\ F_C = C_1 + C_2 + \ldots + C_n \quad (6.178)$$

and the target values V and V_C of F and F_C respectively are changed to take into account the presence of the fixed component in the formulation (Eqn 6.179):

$$V = -(K_{PW} - \omega S_{PW}) \qquad V_C = 1 - C_P \quad (6.179)$$

The initial recipe is made by setting the fixed component concentration to the desired value and the concentrations of each of the variable components to zero (Eqn 6.180):

$$C_P = \text{fixed value} \qquad C_j = 0 \quad \text{for } j = 1 \text{ to } n \quad (6.180)$$

The values of K_{PW} and S_{PW} are determined from the database entries and the target values V and V_C are calculated. The values of F and F_C are calculated by Eqn 6.182 and the deviations ε calculated. Finally the values of the partial derivatives are determined (Eqn 6.110). The weighted equations are solved by the methods described earlier to give the corrected recipe (Eqn 6.181):

$$C_{P1} = C_P \qquad C_{j1} = C_j + \Delta C_j \quad (6.181)$$

Colorimetric matching with a fixed component

Fewer changes are needed to include a fixed component into the colorimetric matching method (section 6.7.3) than with the spectrophotometric method. The tristimulus (or $L^*a^*b^*$) values of an initial formulation that includes the fixed component are determined (Eqn 6.182):

$$F_1 = X_M = f_X(C_P, C_1, C_2, \ldots C_4)$$
$$F_2 = Y_M = f_Y(C_P, C_1, C_2, \ldots C_4) \quad (6.182)$$
$$F_3 = Z_M = f_Z(C_P, C_1, C_2, \ldots C_4)$$

The total colorant volume constraint is given by Eqn 6.178. The tristimulus (or $L^*a^*b^*$) values of the standard are the target values for Eqn 6.182, and once again $V_C = 1 - C_P$ is the target value of the volume constraint.

The deviations ε of the colour parameters of the initial recipe from the target values are calculated, and the values of the partial derivatives are determined using modified forms of Eqn 6.146 that take into account the presence of the fixed component (Eqn 6.183):

$$\frac{\partial K_M}{\partial C_j} = \frac{dK_{jW}}{dC_j} - \frac{K_M - K_{PW}}{1 - C_P} \qquad \frac{\partial S_M}{\partial C_j} = \frac{dS_{jW}}{dC_j} - \frac{S_M - S_{PW}}{1 - C_P} \quad (6.183)$$

The weighted equations are solved by the methods described earlier to give the corrected recipe from Eqn 6.181.

Matching with a minimum amount of a component

It is possible to select the same colorant both for the fixed component and as one of the variable components. In this case there will be a minimum concentration of this colorant in the formulation and the match-prediction calculation may include more if the colour match requires it.

6.7.7 Correction of an off-shade formulation

There are several reasons why the colour of a test panel made from a match-predicted paint might not match the target colour sufficiently well to be accepted as a match, such as the following:

(a) a change in the tinctorial strength of one or more of the colorants; a similar effect can arise from a change in the apparent tinctorial strength caused by an alteration in the efficiency of mixing the colorant into the paint, or a change in the method of paint application, or a systematic error in dispensing the colorant

(b) a database prepared from an inaccurate set of calibration panels

(c) an optical model that does not provide an accurate description of the properties of the system (for example, the test panel or the database panels are not opaque)
(d) a nonsystematic error in the dispensing of one or more of the colorants in the formulation
(e) contamination of the paint.

There are two types of correction process: recipe correction and batch correction.

Recipe correction can be thought of as a laboratory-based procedure for testing a predicted recipe by producing a panel from a small sample batch of paint. The colour of the trial panel is measured, and if it is off-shade a corrected recipe is calculated. The corrected formulation is redispensed as a new sample of paint.

Batch correction is the process of adjusting an off-shade batch of paint by adding small amounts of colorants to that batch. The cause of the mismatch will influence the degree of success of the correction procedure. For example, recipe correction is only effective if the error is systematic, i.e. type (a) to (c) above; random errors of type (d) or (e) will not be corrected. A batch correction procedure should be effective on all five types of error.

Both correction methods are based on making use of the reflectance spectrum of the trial paint panel to calculate an adjustment to the recipe of the trial paint.

Recipe correction

The correction method found by McDonald et al. to be effective in the correction of off-shade textile materials also provides a simple system for the correction of a trial paint that is reasonably close in colour to the standard ($\Delta E^* \leq 3$) [25]. A match-prediction recipe is calculated for the colour of the trial panel using the same colorants as in the paint. This provides the predicted colorant fraction P_j of each component in the paint recipe. These values are subtracted from the colorant fractions actually used (T_j) to produce a difference ΔC_j (Eqn 6.184):

$$\Delta C_j = T_j - P_j \quad \text{for } j = 1 \text{ to } n \qquad (6.184)$$

The concentration difference between the two recipes is associated with the difference in colour between the trial panel and the standard. It follows that the concentration differences can be used to adjust the recipe of the trial to produce a corrected colorant recipe (Eqn 6.185):

$$C_j = T_j + \Delta C_j \quad \text{for } j = 1 \text{ to } n \qquad (6.185)$$

If a colorant loading V was determined for the first predicted recipe (section 6.7.5) then this value can still be used for the corrected recipe.

One correction method assumes that the inaccuracy in the prediction arises from errors in the optical coefficients K_{jW} and S_{jW} determined from the database [26]. A correction factor t_j is obtained from the values of P_j and T_j (Eqn 6.186):

$$t_j = \frac{P_j}{T_j} \quad \text{for } j = 1 \text{ to } n \qquad (6.186)$$

The absorption and scattering terms are determined from the database taking this factor into account, for example Eqn 6.187:

$$K_{jW} = t_j C_j \hat{K}_{jW}$$
$$\text{or} \quad K_{jW} = t_j \left(a_j C_j + b_j C_j^2 + c_j C_j^3 \right) \qquad (6.187)$$
$$\text{and} \quad S_j = \frac{K_{jW}}{\omega_j}$$

The recipe for the standard colour is repredicted using the corrected optical coefficients. The correction factors are valid only for the trial under consideration. The constant total colorant volume constraint in the prediction calculations will produce a non-unity correction factor for at least two colorants in the recipe, even though only one of them may be the cause of the inaccurate prediction.

Batch correction
Calculation of a batch correction is more complex than that of a recipe correction. A method that has been found effective by the author involves creating a temporary database entry to represent the optical coefficients of the trial batch of paint.

A match prediction is made to the colour of the trial panel using the same colorants as in the formulation. The recipe is used to provide values of K_M and S_M, the absorption and scattering of the predicted recipe of the trial batch of paint (Eqn 6.188):

$$K_M = \sum_{j=1}^{n} K_{jW} \qquad S_M = \sum_{j=1}^{n} S_{jW} \qquad (6.188)$$

In the case of a contaminated batch the predicted trial recipe may not be a good match to the colour of the trial panel. In order to force the optical constants of the trial paint to match the colour of the trial panel perfectly, the values are adjusted to obtain $\hat{K}_{TW}$ and $\hat{S}_{TW}$, the linear calibration constants of the trial paint (Eqn 6.189):

$$\hat{K}_{TW} = \frac{V\omega_T(K_M + S_M)}{1 + \omega_T} \qquad \hat{S}_{TW} = \frac{\hat{K}_{TW}}{\omega_T} \qquad (6.189)$$

where $\omega_T = \frac{(1 - R_T)^2}{2R_T}$

and R_T is the gloss-corrected reflectance of the trial panel and V the total volume fraction of colorants in the trial paint formulation. Since it is not possible to determine whether the error is in the absorption or the scattering term, a compromise is made by assuming that the value of $(K_M + S_M)$ remains constant; this tends to assign most of the error to the smaller of the two values.

The recipe for the colour of the standard shade is repredicted, forcing a minimum (usually 80%) of the trial paint into the correction formulation (section 6.7.6). The colorants included in the prediction calculation are not restricted to those used in the trial; any suitable colorant in the database can be used to correct the trial paint batch.

6.8 MATCH PREDICTION OF A SEMITRANSPARENT LAYER

Most nontextile printing processes can be described as the coating of a transparent or semitransparent layer of coloured material on to the surface of a substrate. The reflectance spectrum of the coated surface is determined by the optical properties of the coating material, the thickness of the layer and the reflectance of the substrate. The general principles of the colour match-prediction methods developed for an opaque layer can be applied to semitransparent layers, though the equations linking reflectance to concentration are more complex.

The K_M and S_M values of the layer are obtained by adding the contributions from each component in the formulation, taking into account the relative thickness D of the layer, to that used for the calibration prints (Eqn 6.101). The R_M reflectance of the layer depends on K_M, S_M and the reflectance of the substrate R_g via the equations derived in section 6.2 (Eqn 6.190):

$$R_M = \frac{\alpha R_g + \beta R_\infty}{\alpha + \beta} \qquad (6.190)$$

where α (Eqn 6.28) is a function of both the opaque reflectance R_∞ and R_g, and R_∞ is given by Eqn 6.191:

$$R_\infty = 1 + \frac{K_M}{S_M} - \left[\left(1 + \frac{K_M}{S_M}\right)^2 - 1\right]^{1/2} \qquad (6.191)$$

and β is a function of the absorption (K_M) and scattering (S_M) values of the layer and the thickness D of the layer relative to that used for the calibration prints (Eqn 6.20 and 6.29). The prediction of a reflectance spectrum is completed by taking into account the effect of the coating-to-air interface (section 6.3) (Eqn 6.192):

$$\rho_M = r_e + \frac{t_e t_i R_M}{1 - r_i R_M} \qquad (6.192)$$

6.8.1 Spectrophotometric curve matching

The spectrophotometric matching method determines the concentration of colorants in a formulation that will have, at each wavelength in the spectrum, the same value of a fitting parameter as the standard panel. In the case of an opaque layer the fitting parameter is related to the ratio K/S, and the target values to be matched by the formulation are directly determined from the reflectance spectrum of the standard (section 6.7.2).

For a semitransparent system the reflectance of the coated substrate depends on both K and S and not just on their ratio (K/S). Unfortunately target values for either K or S cannot be determined from the reflectance spectrum of the standard alone; the reflectance of the substrate on which the coating will be made (R_g) and the opaque layer reflectance of the formulation (R_∞) are also required. While R_g is known, R_∞ is not. It follows that spectrophotometric matching equations equivalent to those of the opaque layer cannot be developed.

6.8.2 Colorimetric match prediction

The colorimetric method of match prediction (section 6.7.3) is not specific to opaque layers or even to materials coloured with pigments. All that is required to apply the technique to a particular coloration system is a method of determining the influence matrix, that is, the matrix that describes how small changes in a colorant recipe will influence the colour coordinates of the test print (XYZ or $L*a*b*$). Once this is established the match-prediction calculation is identical to that already described for opaque layers, including the use of more than four or less than three components in the formulation, incorporation of a fixed amount of a component and the calculation of the sensitivity and robustness of the recipe.

Compared to the opaque layer prediction, the main difference lies in the direct inclusion of a colourless, transparent diluent component into the formulation. In an opaque layer the amount of diluent is calculated after the colorant recipe has been determined and used to adjust the hiding power (spreading rate) of the formulation. In a semitransparent system the amount of a clear diluent will directly influence the colour of the coated surface by determining the 'strength' of the ink. The diluent is included in the recipe calculation in an identical manner to a white or coloured component, even though the K and S coefficients for the diluent may be zero.

Determination of the partial derivatives

Analytical equations for the partial derivatives can be developed in a step-by-step fashion in a similar way to the opaque layer case. For example, the way in which the X value of the recipe print will change for a small change in the concentration of component j is given by Eqn 6.141. The summation is over the measured wavelengths in the spectrum. The derivative of measured reflectance is determined from the true reflectance in the usual way (Eqn 6.193):

$$\frac{\partial \rho}{\partial C_j} = \frac{d\rho}{dR}\frac{dR}{dC_j} \quad \text{where} \quad \frac{d\rho}{dR} = \frac{t_e t_i}{(1-r_i R)^2} \qquad (6.193)$$

A change in the amount of component j in the recipe will simultaneously alter the values of R_∞ (Eqn 6.191) and Z (Eqn 6.194):

$$Z = D[K_M(K_M + 2S_M)]^{1/2} \qquad (6.194)$$

The effect of both changes must be taken into account in determining the change in R_M. Fortunately it is possible to treat these as two separate, additive terms. The first term calculates the change in R_M assuming Z is kept constant, for example by a compensating change in the thickness D. The second term calculates the change in R_M assuming that only Z changes and R_∞ is kept constant (Eqns 6.195–6.197):

$$\frac{\partial R_M}{\partial C_j} = \left(\frac{\partial R_M}{\partial C_j}\right)_Z + \left(\frac{\partial R_M}{\partial C_j}\right)_{R_\infty} \qquad (6.195)$$

$$\left(\frac{\partial R_M}{\partial C_j}\right)_Z =$$
$$-\beta\left[\frac{1}{\alpha+\beta} + \frac{(R_g - R_\infty)(\alpha R_g - 2 R_\infty)}{(\alpha+\beta)^2(1 - R_g R_\infty)}\right] \frac{R_\infty(1-R_\infty)}{1+R_\infty}\left(\frac{1}{K_M}\frac{\partial K_M}{\partial C_j} - \frac{1}{S_M}\frac{\partial S_M}{\partial C_j}\right)$$
$$(6.196)$$

$$\left(\frac{\partial R_M}{\partial C_j}\right)_{R_\infty} = 2(1+\beta)\left[\frac{\alpha(R_\infty - R_g)}{(\alpha+\beta)^2}\right]\frac{1+R_\infty}{1-R_\infty}\frac{\partial K_M}{\partial C_j} \qquad (6.197)$$

The derivatives $\partial K_M/\partial C_j$ and $\partial S_M/\partial C_j$ are obtained from the database information in the manner given in section 6.7.3.

Once the derivatives are known, then Eqns 6.141 and 6.193 are used to determine the elements in the influence matrix. The equations are then solved by the methods described in section 6.7.3 and the change in concentrations added to the initial amounts to produce the corrected recipe.

An arbitrary starting formulation (such as equal amounts of each component) is chosen and the stepwise method of solution described in section 6.7.3 is used. The maximum colour deviation allowed at each step in the calculation is set so that the additive assumption is still valid.

Sample calculation

The results of stepwise colorimetric match prediction to a brown standard are shown in Table 6.23. The prediction is for lithographic inks printed on to a white-coated cartonboard. The initial recipe contains equal amounts of each of the five components and has a colour difference of over 23 units from the standard. After four iterations the colour difference has been reduced to zero, indicating the efficiency of the matching algorithms.

Table 6.23 Five-component colorimetric (*XYZ*) match prediction of a brown standard[a]

Parameter	Recipe 1	Recipe 2	Recipe 3	Recipe 4	Recipe 5
Clear	0.2000	0.3481	0.4473	0.4271	0.4267
Green	0.2000	0.0992	0.1040	0.1056	0.1053
Red	0.2000	0.1616	0.1395	0.1448	0.1449
Yellow	0.2000	0.3141	0.2711	0.2814	0.2818
Black	0.2000	0.0770	0.0380	0.0411	0.0413
ΔE^*_{ab}[b]	23.35	5.43	0.94	0.05	0.00
ΔE^*_{ab}[c]	25.19	5.36	0.88	0.09	0.10

a Starting from an arbitrary initial recipe using weightings $W_a = 1.0$, $W_b = 0.1$, $W_c = 100$
b Illuminant D_{65}, 10° observer
c Illuminant A, 10° observer

Figure 6.31 shows the reflectance spectrum of the standard, the predicted reflectance spectra of recipes 1 and 5, and the spectrum of a four-component recipe (no green) matched under D_{65} only.

6.8.3 Opaque white and transparent white

In the example calculation the clear diluent was used to adjust the strength of the ink. It is interesting that a white, pigmented diluent such as a white base ink could also be used for this purpose. The names 'transparent white' and 'opaque white' have come

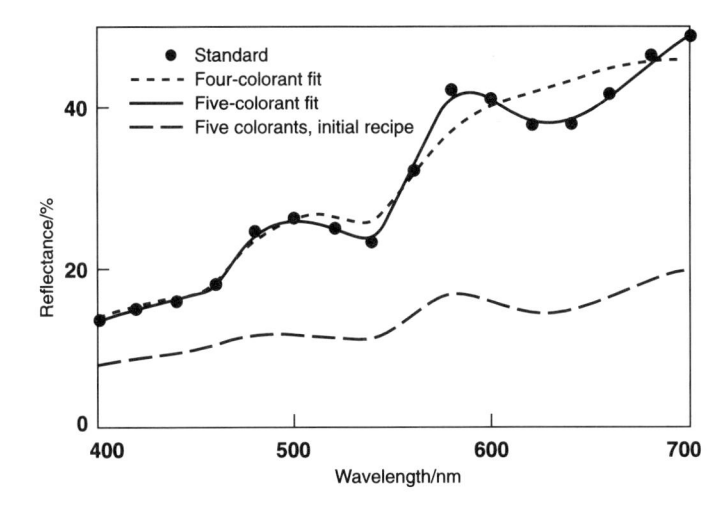

Figure 6.31 Colorimetric matches to a brown standard using four and five colorants, including a clear diluent; formulations shown in Table 6.23

Table 6.24 Colour difference between two five-component matches to a brown standard (10° observer)

Formulation 1		Formulation 2		Illuminant	ΔE^*_{ab}
Name	Volume	Name	Volume	D_{65}	0.00
				A	0.05
				TL84	0.12
White	0.2736	Clear	0.4267		
Green	0.1131	Green	0.1053		
Red	0.1677	Red	0.1449		
Yellow	0.4029	Yellow	0.2818		
Black	0.0427	Black	0.0413		

into use for a clear diluent and a white base ink respectively. A formulation containing a white base ink has greater hiding power than one with a clear diluent. This can have the advantage of the printed layer partially masking the colour of a substrate, reducing the visibility of any small colour variations.

The match prediction of recipes printed on a white substrate where both white and clear are included in a formulation can pose a problem. Table 6.24, for example, shows two formulations which both match the brown standard, as shown by the near-zero colour difference between prints made from the two formulations. Any blend of the two formulations would also match the standard. It follows that the colorimetric match-prediction method cannot determine a unique recipe under these conditions. If such a formulation is required then one of these components must be entered as a fixed amount.

6.8.4 Correction of an off-shade formulation

The recipe correction methods described in section 6.7.6 can be applied to ink formulations, but the method of batch correction must be changed to take into account the semitransparent nature of the trial print layer. A match prediction is made to the colour of the trial print and the values of K_M, S_M and R_∞ are determined for the predicted recipe of the trial print. The optical constants are adjusted to values that give a perfect match to the reflectance spectrum of the trial print by assuming that R_∞ remains constant. R_∞, R_g and R, obtained from the measured reflectance of the print, are used in Eqn 6.93 to obtain K_T and S_T, the adjusted optical constants of the batch of ink. The colour of the standard is then repredicted forcing a minimum amount of the trial batch in the formulation (section 6.7.7).

REFERENCES

1. R H Park and E I Stearns, *J. Opt. Soc. Amer.*, **34** (1944) 112.
2. E Atherton, *J.S.D.C.*, **71** (1955) 389.
3. E Cowgill, ICI Dyestuffs Division Technical Note (Sept 1963).
4. J V Alderson, E Atherton, and A N Derbyshire, *J.S.D.C.*, **77** (1961) 657.
5. J V Alderson, E Atherton, C Preston and D Tough, *J.S.D.C.*, **79** (1963) 723.
6. H R Davidson, H Hemmendinger and J L R Landry, *J.S.D.C.*, **79** (1963) 577.
7. A E Cutler, *J.S.D.C.*, **81** (1965) 601.
8. D Patterson, *The science of colour mixing* (London: Scott Greenwood, 1900).
9. P Kubelka and F Munk, *Z. Tech. Phys.*, **12** (1931) 593.
10. A Schuster, *Astrophys. J.*, **21** (1905) 1.
11. J H Nobbs, *Rev. Prog. Col.*, **15** (1985) 66.
12. J L Saunderson, *J. Opt. Soc. Amer.*, **32** (1942) 727.
13. E Walowit, C J McCarthy and R S Berns, *Col. Res. Appl.*, **12** (1987) 340.
14. B Sluban, *Col. Res. Appl.*, **18** (2) (1993) 74.
15. P W McGinnis, *Col. Eng.*, **15** (6) (1967) 22.
16. E Walowit, C J McCarthy and R S Berns, *Col. Res. Appl.*, **13** (6) (1988) 358.
17. E Allen, *Col. Eng.*, **4** (4) (1966) 24.
18. E Allen, *J. Opt. Soc. Amer.*, **64** (7) (1974) 991.
19. E Allen in *Optical radiation measurements*, Vol. 2, Ed. F Grum and C J Bartleson (New York: Academic Press, 1980).
20. B Sluban, *Die Farbe*, **39** (1993) 247.
21. B Sluban and J H Nobbs, *Col. Res. Appl.*, (1995) in press.
22. L Gall, *Farbe Lacke*, **71** (1966) 955, 1058.
23. E L Cairns, *J. Paint Technol.*, **44** (1972) 76.
24. H G Volz, *Prog. Org. Coatings*, **15** (1987) 99.
25. R McDonald, D McKay and P J Weedall, *J.S.D.C.*, **92** (1976) 39.
26. R G Kuehni, *Computer colorant formulation* (Lexington, MA: D C Heath, 1975).

CHAPTER 7

Colour in visual displays

Lindsay W MacDonald

7.1 INTRODUCTION

This chapter deals with displays from the point of view of the colorimetrist, considering only those aspects of the technology that relate to how the devices produce colour and how it can be controlled and measured. Of the vast array of display types that have been developed for a wide range of differing applications, we concentrate here on only two: the cathode-ray tube (CRT) and the backlit liquid-crystal display (LCD). These predominate in television and computer-based workstations and are the types of display that the reader is most likely to meet in everyday working practice. The principles of colorimetry set out in this chapter are, however, applicable to emissive displays of all types.

7.2 CRT DISPLAYS

7.2.1 Cathode-ray tube construction

The cathode-ray tube (CRT) is a specialised and highly developed form of the thermionic valve. It is based on a principle invented by Karl Ferdinand Braun in 1897, but owes its refinement mainly to the development of television. The experiments of John Logie Baird (1925), carried out when he was working on mechanical systems for the sensing and reproduction of images, stimulated others to devise electronic solutions.

The Kinoscope display tube was invented by Vladimir Zworykin of Westinghouse in 1929, and over the next ten years various designs competed against one another, trading size versus brightness versus replacement cost. The first public service television transmissions in Europe were made by the BBC in 1936, with receivers incorporating either 7–8.5-inch direct-view CRTs or 3–4-inch projection tubes. By the 1950s direct-view CRTs had gained the dominant market position and the attractiveness of television as a medium generated the substantial R&D funding needed to fuel competitive development, resulting in bigger and more rectangular screens, all-glass envelopes and larger deflection angles to shorten the tubes and hence reduce overall cabinet size.

From that point until the recent development of flat-panel displays, the CRT has been the ubiquitous display device. It is found in television sets, computer terminals, oscilloscopes, industrial control consoles, radar and many other types of information display. Strengthened versions are used in marine, avionic, automotive and military instrument panels and even in stereoscopic head-mounted displays, with very demanding specifications for luminance range, compactness and reliability. Large-screen models with high resolution and negligible distortion serve for the photo-realistic display of colour images in computer-based workstations for graphic arts applications, such as desktop publishing, creative design, multimedia databases and animation. Without the CRT none of these applications would have enjoyed its present success, because the CRT has provided such a commercially viable and visually effective window into the world of electronic information processing.

The main features of a monochrome CRT of the type used in television are shown in Figure 7.1. A cathode is coated with a suitable electron-emitting material so that when its temperature is raised by a heater a cloud of electrons is emitted. Because the glass envelope is evacuated, the electrons can be freely accelerated by applying a high positive voltage (typically 16 to 25 kV) to an anode attached to a phosphor-coated screen. The electrons are focused into a narrow beam by the geometry of the first anode plates, and the strength of the electron beam is controlled by the signal voltages applied to the grid electrodes. The position at which the beam strikes the screen is

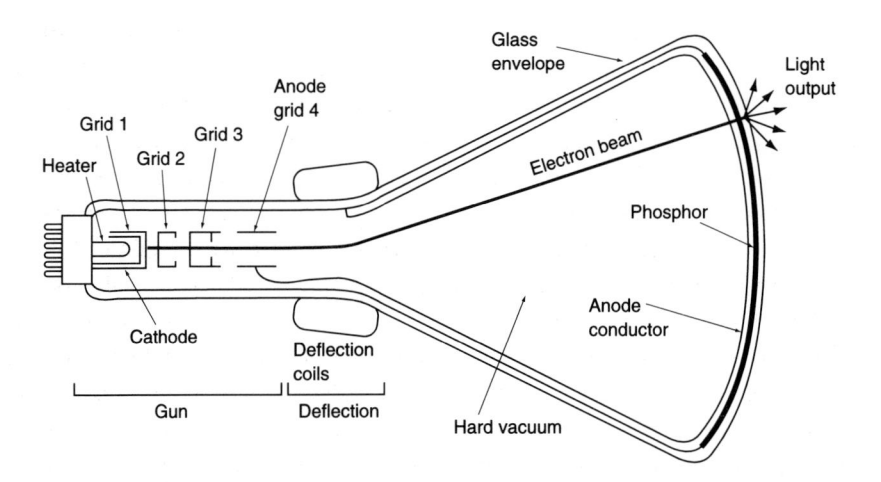

Figure 7.1 Construction of a monochrome cathode-ray tube, showing its resemblance to the thermionic valve

controlled by the magnetic field generated by the deflection coils wound in a yoke around the neck of the tube. The electron-generating apparatus in the narrow part of the CRT is known collectively as the *electron gun*. When the electrons strike the phosphor-coated screen their kinetic energy, proportional to the accelerating voltage, is partially converted into heat and partially transferred to the phosphor, resulting in the stimulation of light output.

7.2.2 Electron-beam characteristics

Great refinement is needed in the beam formation and scanning mechanisms of a CRT. The inside surfaces of the conical part of the glass envelope, for example, are coated with a conductive layer connected to (and therefore held at the same voltage as) the anode. Thus the electron beam moves through a space that is virtually free from any electrostatic fields. The inner surface of the phosphor is also coated by a very thin conductive layer of aluminium which is connected to the anode. This not only assists in eliminating electrostatic fields but also prevents secondary emission of electrons from the phosphor [1].

Focusing of the electron beam is dynamic, whereby the focal length is adjusted according to the coordinate position on the screen. This is necessary because the faceplate of a modern CRT screen is nearly flat, so that not all points are equidistant from the centre of deflection. The cross-sectional density of the beam is Gaussian, and the intensity of the light spot generated on the phosphor has a constant distribution, often referred to as the *pixel point spread function* (Figure 7.2). The spot thus has no distinct edge and its size is conventionally specified as the diameter at which the luminance is 50% of the peak value. In a typical office visual display unit the spot size is about 0.3 mm, but in a high-resolution monochrome CRT it can be as little as 0.15 mm. The fineness of the beam is bounded by both electrostatic repulsion among the electrons, which limits achievable electron density, and the inherent aberrations and nonlinearities of the electron optical lenses and deflection systems. Design of a CRT is a fine compromise among conflicting requirements for high luminance, speed, small uniform spot size, low geometric distortion and low power consumption [2].

The phosphor is deposited as layers of powdered material on the inside of the faceplate of the tube. Most phosphors for monochrome CRTs are based on zinc sulphide and they can be obtained with a wide variety of characteristics, principal among which are spectral power distribution, luminous efficiency, decay time and usable life. The luminous efficiencies are generally in the range 10 to 50 lumens per watt (lm W^{-1}), with the higher values being for emissions in the green wavelengths where the eye has greatest sensitivity (section 8.7.3). Phosphor decay time, known as *persistence*, can range from less than a microsecond to several seconds. The life of a phosphor

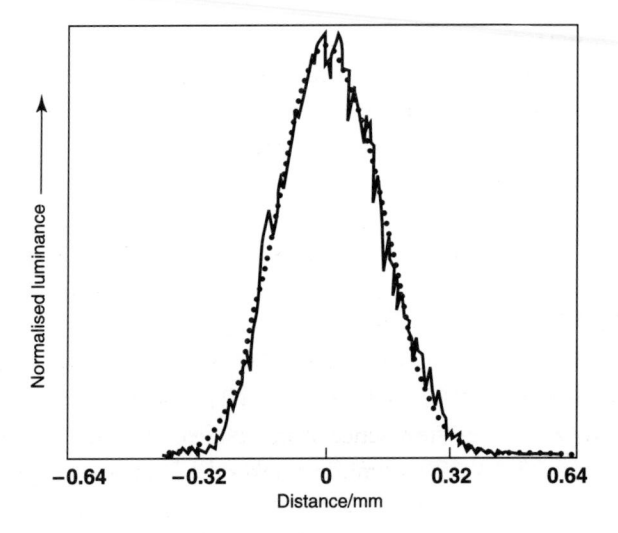

Figure 7.2 Horizontal profile of a monochrome pixel (the dotted line represents the best-fitting Gaussian curve (from ref. 3, reproduced by permission of Macmillan Press)

depends on the total energy imparted to it by the electron beam, but by avoiding excessive beam current lifetimes of several thousand hours can easily be achieved.

The luminance Y from the phosphor can be expressed in terms of the electron beam current and anode voltage by Eqn 7.1:

$$Y = \frac{k_b I_b V_b^{n}}{A} \qquad (7.1)$$

where k_b = phosphor luminous efficiency, in lm W^{-1}
I_b = beam current
V_b = accelerating voltage
n = constant (value approximately 2)
A = area of phosphor excited by electron beam.

The electron beam current is related to the signal voltage applied to the control grid by a complex equation that is dependent on the relative geometry of the various electrodes and other factors. In practice, the transfer function that relates the output luminance Y to the input signal voltage takes the form of Eqn 7.2:

$$Y = Y_0 + (Y_{max} - Y_0)\left(\frac{\alpha V_i}{V_{max}} + \beta\right)^{\gamma} \qquad (7.2)$$

where Y_0 = residual luminance when $V_i = 0$
Y_{max} = maximum luminance when $V_i = V_{max}$
V_i = input signal voltage
V_{max} = maximum input signal voltage
α, β, γ = constants.

The constants α and β represent the gain and offset applied to the signal voltage by the display's contrast and brightness controls respectively. The exponent γ is known as the *gamma* of the display and is typically in the range 2 to 3. It must be measured individually for each display unit as it depends critically on the internal structure of the CRT and can vary from one tube to another, even within batches of the same model from the same manufacturer. In the absence of any residual light (i.e. $Y_0 = 0$), the transfer function when plotted on logarithmic axes takes the form of a straight line of slope γ (Figure 7.3). By convention the display's white point ($V_i = V_{max}$ and $Y = Y_{max}$) is chosen as the origin in the top right-hand corner.

7.2.3 Shadow-mask colour CRTs

In order to display colour on a CRT three independent electron guns are required, together with three phosphor types to produce the primary coloured lights of red, green and blue (RGB). The problem of ensuring that each electron beam excites only its corresponding phosphor is solved through the introduction into the CRT of a shadow

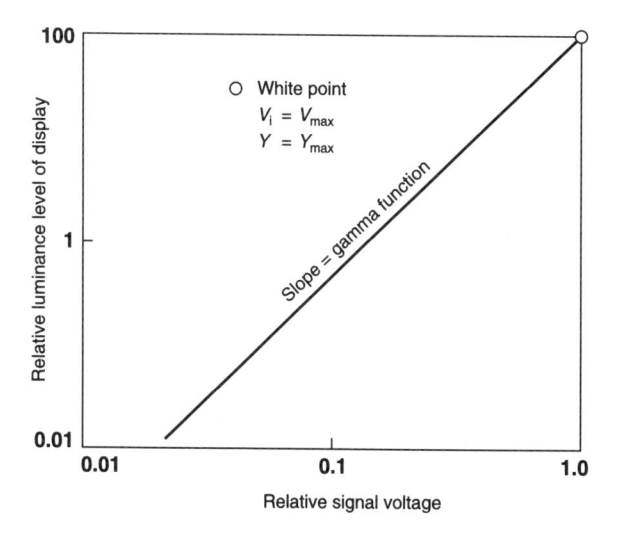

Figure 7.3 Typical CRT transfer function is linear when plotted on logarithmic axes

378 COLOUR IN VISUAL DISPLAYS

mask, a metal plate behind the screen pierced with holes or slots through which the focused electron beams pass. The shadow mask is positioned close behind the faceplate, and there is a very wide angle of deflection of the electron beam, as much as 110° between its extremities.

Two types of shadow mask and phosphor geometry are commonly employed: the delta and the linear types. In the delta type, originally developed by RCA, the circular holes in the shadow mask are uniformly spaced, and the phosphor dots are deposited in a hexagonal array on the screen. Each hole serves a group of three adjacent RGB phosphor dots and the angle of incidence of each electron beam as it passes through the hole determines the location (and hence the colour) of the phosphor dot it excites. Two different groupings of the phosphor excitation pattern are possible, depending on the relative positions of the electron guns (Figure 7.4(a) and (b)). When these are arranged in an equilateral triangle or delta formation (Figure 7.4(a)) then an inverted triangle of phosphor dots is excited. When they are arranged in a straight line (Figure 7.4(b)) then the phosphor dots excited also lie in a straight line.

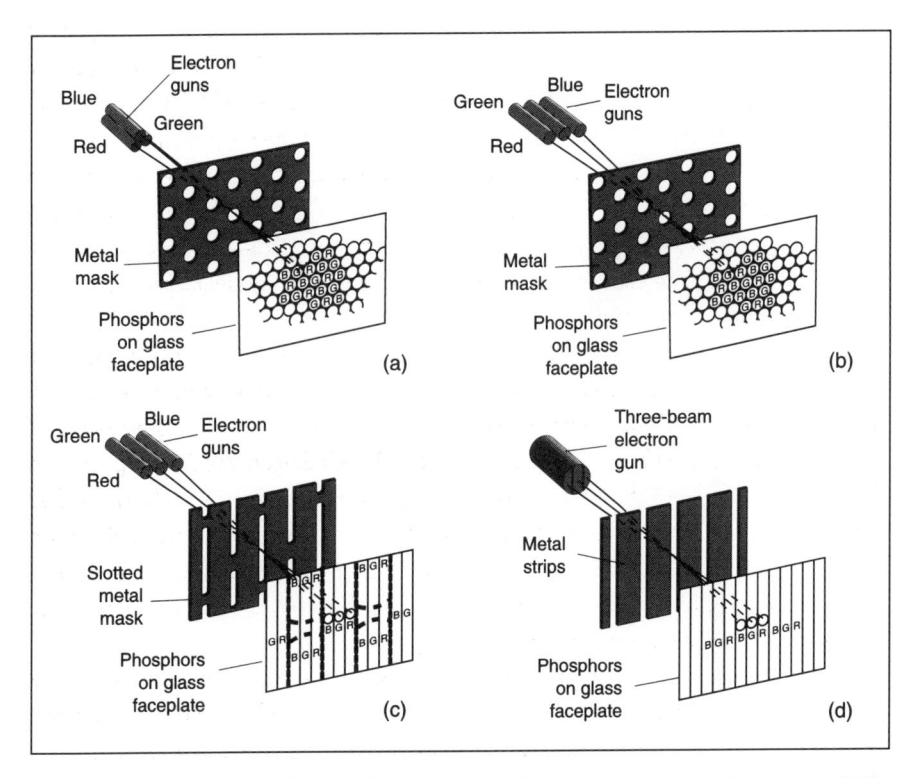

Figure 7.4 The four principal types of electron gun and shadow mask geometry in colour CRTs (from ref. 4, reproduced by permission of Academic Press)

The linear type has vertical slots in the shadow mask with the phosphors deposited in vertical stripes on the screen. The guns are arranged in line so that the phosphors are excited as three points in a straight line. The slotted shadow mask (Figure 7.4(c)) has greater mechanical strength, but is unable to excite all phosphor dots uniformly. The metal-strip construction (Figure 7.4(d)), developed by Sony for its popular Trinitron range, allows the vertical apertures to extend the full height of the screen, with two thin horizontal wires to hold the strips in place, thus improving the electron transparency by about 30% over the delta-type shadow mask. Because the mask must be held under tension in the vertical direction, both mask and screen have a cylindrical rather than a spherical surface, that is, they are curved only in the horizontal plane.

Each of the basic types of construction has its advantages. Delta gun tubes have better resolution because the three colour phosphor dots cluster closer together, but they are particularly difficult to maintain in alignment and have been largely superseded. In-line guns are easier to arrange for convergence over the full screen surface, though they cause a fall-off in image sharpness at the edges of the display due to the aspherical profile of the electron beam. CRTs with the delta type of shadow mask cannot produce high luminance levels because the shadow mask intercepts about 80% of the electron beam current. Increasing the current results in heating of the shadow mask, causing it to expand outwards (known as *doming*), thereby shifting the landing positions of the electron beams and degrading the purity of the colour. Tubes with vertical phosphor stripes can achieve higher luminance levels and do not suffer from the inherent moiré patterning of the delta formation phosphor dots. Sony has continued over the past 25 years to make improvements to every aspect of the Trinitron design. A guard grille at the phosphor surface aligned with the aperture grille, for example, now virtually eliminates colour misregistration [2].

The pitch of the phosphor dots or stripes varies from about 0.6 mm for domestic television screens down to about 0.2 mm in the highest-resolution graphics displays. A standard graphics CRT typically has a pitch of about 0.3 mm. For an understanding of how this relates to pixel size, consider a CRT of 20-inch diagonal in 5 : 4 aspect ratio displaying an image of 1280 by 1024 pixels. The image size is approximately 375 by 300 mm and each pixel is therefore 0.29 mm square, well matched to the basic phosphor dot resolution. Displays for Apple Macintosh computers (irrespective of screen size) always provide 72 pixels per inch, each pixel being equivalent to one printer's point, which translates to a pixel size of 0.353 mm.

The resolution of the displayed image is often dependent more on the spot size and video bandwidth of the electronics than on the phosphor pitch. Because the electron beam has a Gaussian distribution (Figure 7.2) its diameter is necessarily larger than that of a single shadow mask hole, so that in addition to the primary target phosphor dot it also excites neighbouring phosphor dots, albeit at a lower intensity. Figure 7.5

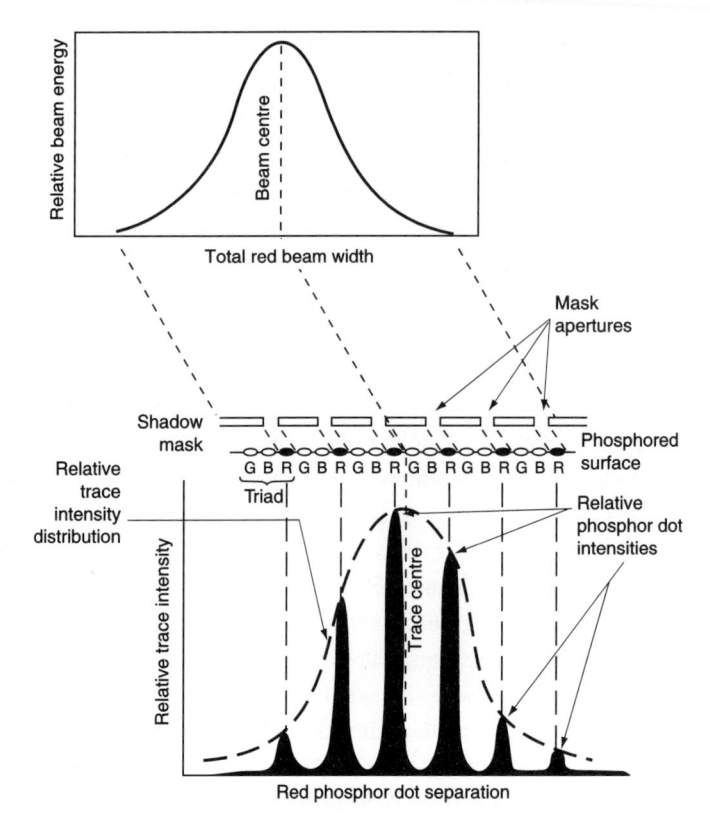

Figure 7.5 Intensity distribution caused by the electron beam passing through a metal-strip shadow mask (from ref. 4, reproduced by permission of Academic Press)

shows the luminance distribution for the red electron gun projected through the shadow mask. This effect is disguised because as the electron beam sweeps across the screen it excites each pixel location in turn so that the total light output is integrated over all the elements, and also because the eye cannot resolve the individual phosphor dots at typical viewing distances. Making the spot size larger increases luminance at the expense of spatial resolution (image sharpness), and is one of the necessary trade-offs in CRT design. The size of the blue spot is less critical for sharpness than red and green, because the eye has a lower spatial sensitivity to blue [5]. In general the luminance profile of two horizontally adjacent pixels cannot be predicted as the sum of the luminance profiles measured when the pixels are illuminated separately, because an isolated pixel never achieves its full output intensity [3].

In conventional delta mask displays careful design minimises the spill-over of the electron beam on to the neighbouring phosphor dots, which are of different colours and whose light output would dilute that of the target phosphor and hence reduce the colour purity. One common method is to separate the phosphor dots by a guard band which provides a tolerance for the beam mis-landings that can occur through production variations or thermal distortion. If the guard band is made from a light-absorbing material such as carbon black, it also reduces internal reflections from the phosphors and the reflection of ambient light. Such black matrix screens have slightly lower luminance but greatly enhanced contrast and are therefore preferred for text and image displays in office environments.

7.2.4 Phosphor properties

Phosphors are inorganic crystalline solids that emit light through a process called *cathodoluminescence*. When an electron beam strikes the phosphor coating of the screen kinetic energy is transferred from the electrons in the beam to the electrons in the phosphor atoms, causing them to jump to higher quantum energy levels. When subsequently they return to their ground state these excited electrons give up their extra energy in the emission of photons, the wavelength of which can be predicted by quantum theory. Any given phosphor has several different quantum levels to which electrons can be excited, each of which results in the emission of light of a different wavelength.

Electrons in some excited levels of the phosphor are less stable and therefore return to the unexcited state more rapidly than others. *Fluorescence* is the light emitted as these very unstable electrons lose their excess energy while the phosphor is being stimulated by the electron beam. *Phosphorescence* is the light given out after the stimulation has ceased and the relatively more stable electrons return to their ground state. Figure 7.6 illustrates both the delay in the onset of light emission from the phosphor after the pulse of excitation by the electron beam and also the exponential decay of the

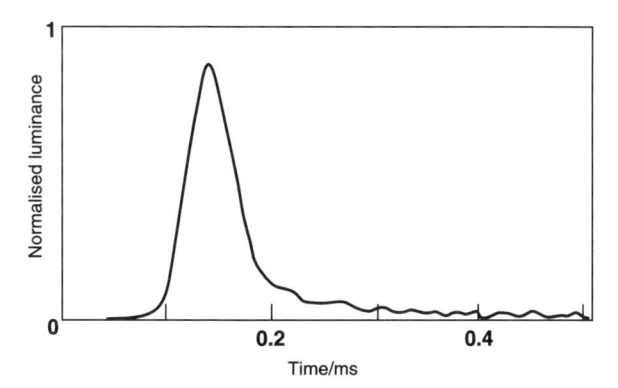

Figure 7.6 Decay of luminescence from a P40 phosphor (from ref. 3, reproduced by permission of Macmillan Press)

subsequent luminescence. The radiant intensity I_e of the emitted light can be described by Eqn 7.3:

$$I_e = I_p N_0 \exp[-\alpha(t - t_0)] \tag{7.3}$$

where I_p = peak radiant flux of the emitted photons per phosphor centre
N_0 = number of phosphor centres excited by the electron beam (a function of beam current and spot size)
α = time constant = $2.3/t_p$
t_p = persistence time period of phosphor decay
t_0 = time corresponding to moment of peak radiant output
t = time.

The *persistence* of a phosphor is defined to be the time from the removal of excitation to the moment when the luminescence has decayed to 10% of the initial peak light emission. For phosphor types such as the P22 used in most television and graphics displays, the persistence is in the range 1 to 10 ms, but long-persistence phosphors such as P7 (bluish-white) and P39 (green) are available with persistence times of around half a second. By smoothing the display's light output these long-persistence phosphors reduce flicker and can improve the viewing characteristics of static displays; with moving images, however, they produce a smeared after-image which is generally undesirable except in some target-tracking applications such as radar screens.

The ability of the phosphor to convert the energy of incident electrons into output radiant energy is termed its *radiant efficiency* and is typically in the range 10% to 20%, i.e. 100–200 mW of radiated energy per watt of electron beam energy. For visible light output this figure is converted to *luminous efficiency* by weighting the spectral radiant power distribution by the V_λ function (section 1.6). Typical values for the luminous efficiency of P22 colour phosphors are 12 (red), 50 (green) and 5 (blue) lm W^{-1} [6].

Ageing of phosphors affects all CRT displays and manifests itself as a gradual reduction in phosphor efficiency. It is a function of the total amount of charge deposited on the phosphor and the ability of the combination of substrate, phosphor and aluminium backing layer to dissipate that charge. Typically phosphor efficiency will fall rapidly at first and then decline more gradually, finally starting to fall rapidly again near the end of the CRT's usable life. Ageing is cumulative, and a given reduction in phosphor efficiency may be related to the total charge per unit area deposited over the lifetime of the phosphor screen [1]. Figure 7.7 shows test results for a typical P22 phosphor set, in which the efficiency of the blue phosphor after 15 000 h (about 21 months of continuous usage) has dropped to 70% of its initial value. This indicates the importance of screen-saver software in the host computer, which will automatically reduce the luminance of the display to a low average value when left inactive for a few minutes.

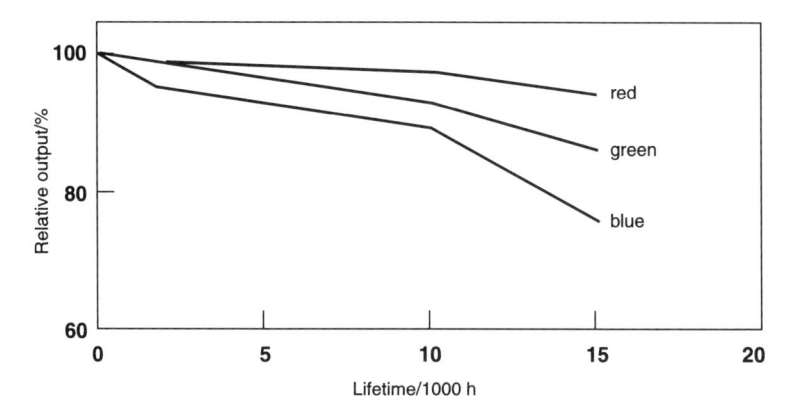

Figure 7.7 Ageing of P22 phosphors in a 20-inch high-resolution CRT display (reproduced by permission of Barco Graphic Displays)

7.2.5 Spectral characteristics of phosphors

The older types of zinc sulphide phosphors for CRT displays were relatively broadband, with a normal spectral power distribution around the dominant wavelength. These gave good colour reproduction for blue and green but suffered from reds that were decidedly orange. Improved red phosphors resulted in the mid-1960s from the introduction of rare-earth activators, such as europium, into a vanadate or oxysulphide

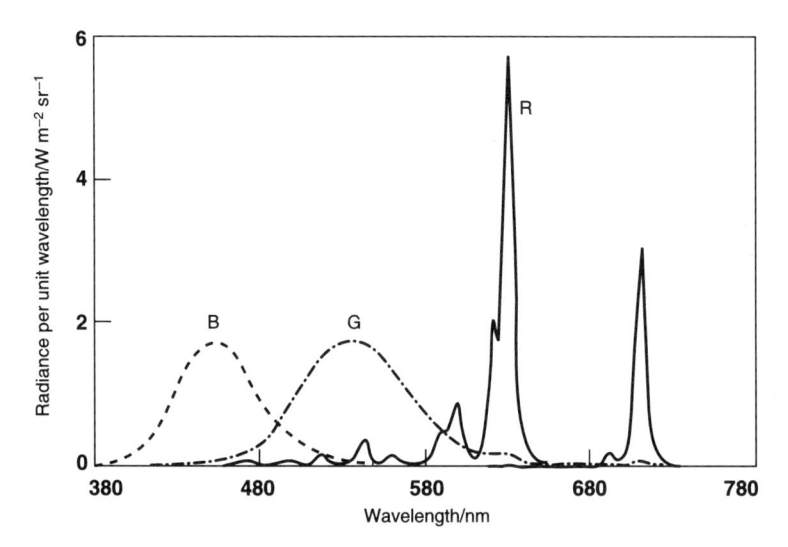

Figure 7.8 Spectral radiant power distribution of a Barco CRT with P22 phosphor set; measurements taken over the range 380–780 nm at 2 nm intervals (reproduced by permission of L D Silverstein)

host material. These gave a better red with longer dominant wavelength, improved brightness and higher colour purity. In contrast to the smooth spectral curves of the silicate and sulphide phosphors, however, these rare-earth phosphors have most of their energy concentrated in a few narrow spikes (Figure 7.8). This can cause problems with the colour measurement of displays, as described later in this chapter.

Most modern phosphors include added inorganic pigments, which absorb all wavelengths outside the primary spectral band of emission. This provides a method of contrast enhancement because the unwanted wavelengths within the visible spectrum are thereby prevented from reducing the purity of the overall colour and the unwanted

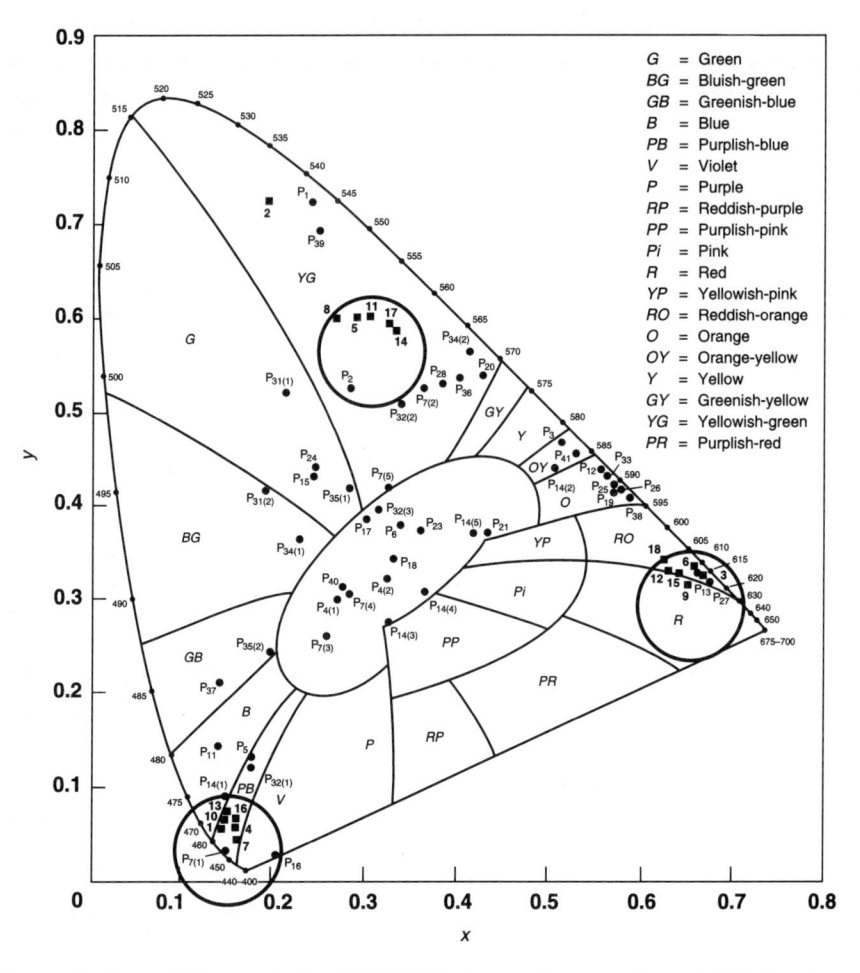

Figure 7.9 Chromaticity coordinates (CIE 1931) for the most commonly used CRT phosphors; circled phosphors belong to the family designated P22 (reproduced by permission of L D Silverstein)

wavelengths outside the visible spectrum are prevented from causing secondary emissions from the other phosphors. Sophisticated methods are now available for modelling the light generation and light propagation in thin phosphor layers [7].

For colorimetric purposes the colour of a phosphor is usually specified in terms of its chromaticity coordinates x and y. A wide variety of phosphors is available for different applications and the range of colours obtainable is clear from Figure 7.9. There is a certain ambiguity about the JEDEC nomenclature, which is based on usage and approximate colour rather than on spectral radiance and chemical composition. The classification P22, for example, which is the most common designation for three-colour television and computer graphic displays, covers six different phosphor systems, each with different chemical composition, different chromaticity coordinates and different emission spectra [2]. Add to this the substantial variations among different batches of phosphor material, even from the same manufacturer, and the need for reliable means of display measurement and calibration for colour-critical applications becomes evident.

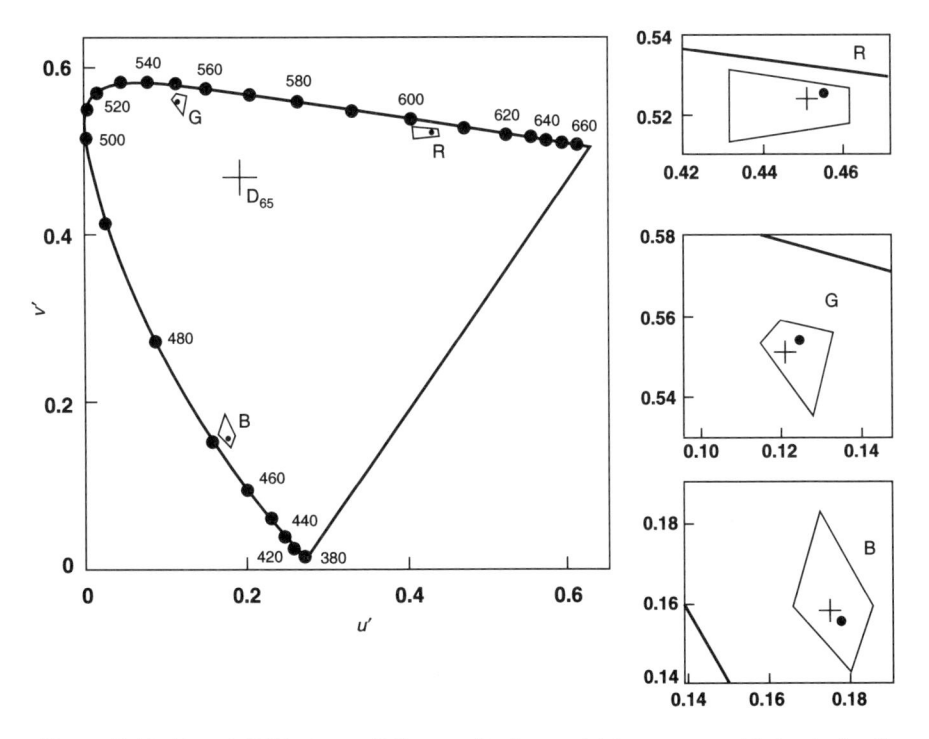

Figure 7.10 Target RGB chromaticity coordinates and tolerance quadrilaterals for the standardised EBU primaries for studio monitors; crosses are target coordinates and dots are typical achieved coordinates (reproduced by permission of the European Broadcasting Union)

In order to circumvent these problems of phosphor colour variability, the European Broadcasting Union (EBU) standard for television studio monitors specifies tight tolerances on the $u'v'$ chromaticity coordinates of the three RGB phosphors [8]. A quadrilateral region is defined around the target chromaticity coordinates for each primary, and in order to comply with the standard the coordinates of an actual phosphor must fall within the region (Figure 7.10).

7.2.6 Raster scan and flicker

Because the electron beam is focused to a small spot, it can stimulate only a very small area of the screen phosphors at any given instant. In order to produce an image over the whole display area, therefore, the electron beam is scanned in a regular pattern known as a *raster*, effectively a series of parallel lines, left to right and top to bottom of the screen (Figure 7.11). At the end of each line the beam is rapidly retraced back to the start of the next line, and similarly at the bottom of the screen the beam is rapidly retraced to the top. Because the light output from the phosphors at each point is only a very short pulse (Figure 7.6), the whole scan has to be repeated many times per second in order to sustain the visual illusion of a continuous image. In television systems the screen refresh rate is equal to the mains frequency, namely 50 Hz in Europe (PAL and SECAM) and 60 Hz in the USA (NTSC). Television scan patterns are also interlaced, in order to conserve transmission bandwidth, so that the scan of the odd lines alternates with the scan of the even lines. Computer graphic displays initially used the same scan rates and interlace patterns, but because they are not tied to national transmission standards in the same way as television receivers they have developed to the point where non-interlaced scan in excess of 70 Hz is now the norm. Much of the impetus

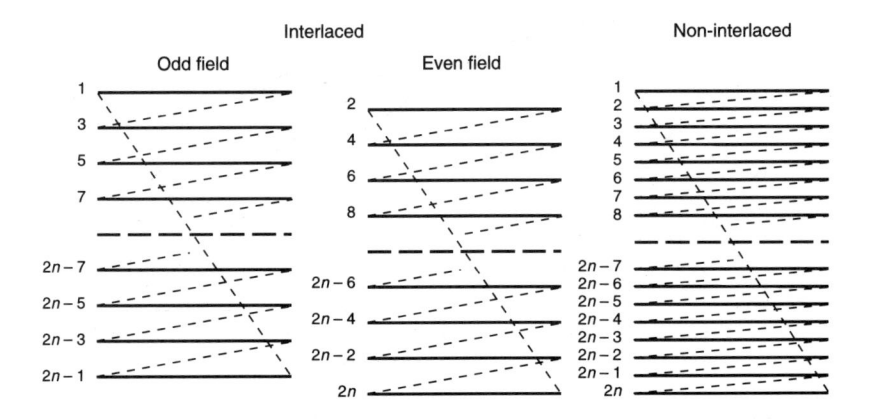

Figure 7.11 Raster scan patterns in a CRT display

for this development has been the desire to eliminate the jitter due to interlaced scanning (especially in fine detail and near-horizontal lines) and to minimise display flicker.

Flicker is a perceptual phenomenon, related to the persistence of vision in the retinal photoreceptors and the cortical integration of successive images received from the visual scan pattern of the human eye (section 8.8.4). As the repetition rate of a flashing light increases, there comes a point when an observer will no longer see the individual flashes but a steady source of light. This frequency, known as the *critical flicker fusion frequency*, is dependent on several variables and cannot easily be predicted. Flicker is undesirable in displays because it can produce distractions and fatigue, as well as biases in apparent brightness and colour perception.

The main display parameters that affect flicker are refresh frequency, luminance and phosphor persistence. These can be traded off to reduce the probability that users will be troubled by flicker [3]. Figure 7.12 shows how the critical flicker fusion frequency (CFF) decreases with longer phosphor persistence but increases with display luminance. This is unfortunate because the current trend is toward displays with ever higher levels of luminance in positive presentation (black text on a white page) which can be used in normal well-lit office environments. Changes in chrominance (hue) have virtually no effect because flicker sensitivity is independent of wavelength at photopic levels of retinal illuminance.

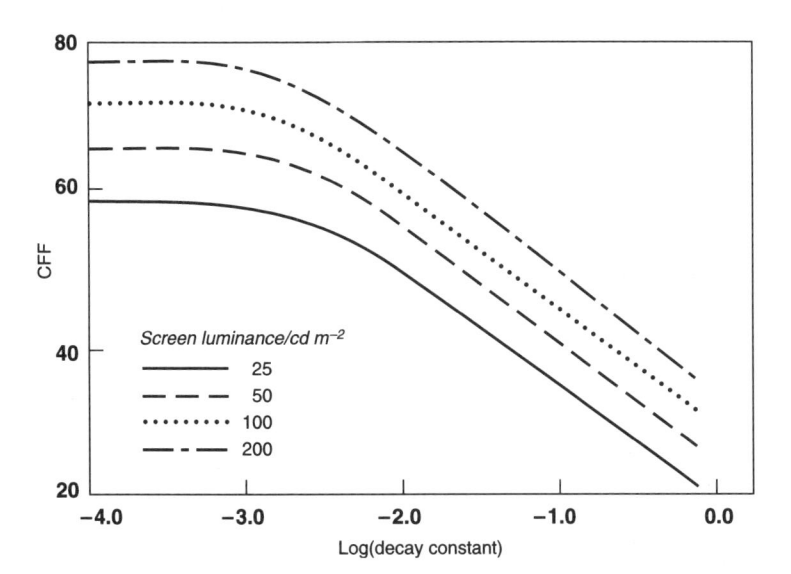

Figure 7.12 Critical flicker frequency as a function of phosphor persistence with display luminance as a parameter (from ref. 3, reproduced by permission of Macmillan Press)

Other factors that influence flicker perception are the display field size, the level of ambient illumination, the nature of the task and even the age of the viewer. Flicker is more obvious in the peripheral visual field, which is better equipped to detect rapid motion, than in the central foveal field and hence it is advisable to design the display screen layout, if possible, to avoid white or bright colours around the edges of the display. High levels of ambient light reduce flicker in two ways: by raising the black level and so reducing the contrast of the display, and also by causing the pupils of the eyes to contract, thereby reducing the amount of light entering the eyes, especially from peripheral areas. Task requirements affect flicker because of the different types of eye movement they demand of the viewer: flicker is less obvious when the eyes are darting all around the screen, searching for a visual cue or tracking a target, than when fixated upon fine detail. Age affects flicker perception as it does everything else, steadily reducing one's sensitivity as the years advance [10].

One useful way to determine the necessary refresh rate of a display is to plot the percentage of users observing flicker against the refresh rate (Figure 7.13), from which it is clear that for a given refresh rate the probability of perceiving flicker is much higher in the peripheral visual field. In order to achieve a situation where only a small proportion of users are aware of flicker under common viewing conditions, the manufacturers of raster-scanned CRT displays are moving toward a refresh rate of 84 Hz; it is probable that by early next century refresh rates of around 100 Hz will be standard.

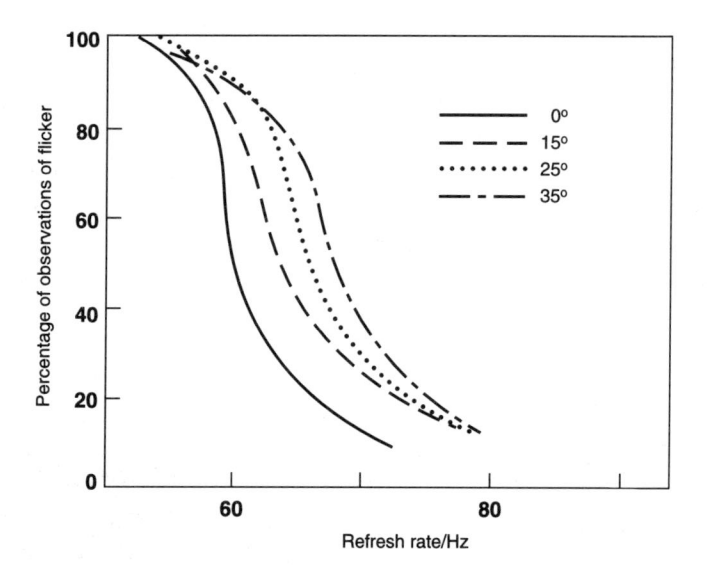

Figure 7.13 Perception of flicker as a function of display refresh rate with retinal angle as a parameter; the experiments used a CRT with a P4 phosphor, luminance 100 cd m^{-2} and viewing distance 50 cm (reproduced by permission of G Murch)

7.3 LIQUID-CRYSTAL DISPLAYS

Over the past twenty years a formidable rival to the CRT has steadily been developed: the liquid-crystal display (LCD). Initially limited to small alphanumeric displays, such as watches and calculators, it has increased in resolution and versatility to the point where it is now a major competitor in applications such as public information systems and portable television sets, and has opened up new markets for products such as laptop computers, in-flight entertainment and overhead projection panels. The main features that make LCDs so attractive include their flatness and lightness, low power consumption, low operating voltage, option of transmissive or reflective mode of operation, long lifetimes, high reliability and competitive manufacturing costs.

7.3.1 Properties of liquid crystals

The term 'liquid crystal' was first used in 1890 by Lehmann to describe a substance that flows in the manner of normal liquids but with optical behaviour similar to that of an anisotropic crystal. It designates a state of matter intermediate between a solid and a liquid, having some of the properties of each. Liquid crystals (LCs) can be defined as complex anisomeric organic molecules which, under certain temperature conditions, exhibit the fluid characteristics of a liquid and the molecular orientation characteristics of a solid [9].

The molecular ordering in liquid crystals falls into three categories: smectic (layer-like), nematic (thread-like) and cholesteric (helical, as encountered in derivatives of cholesterol) (Figure 7.14). In the smectic phase the molecules tend to form into layers, though there is no long-range positional order within the layers. In the nematic phase all positional order is lost, with only the orientation order remaining. In the cholesteric phase the asymmetry of the molecules causes a small angular twist between molecules, which results in a macroscopic helical structure. Any LC material may pass through several of these mesophases as its temperature changes, but usually the more highly ordered smectic phase is found at lower temperatures than the less-ordered nematic and cholesteric phases.

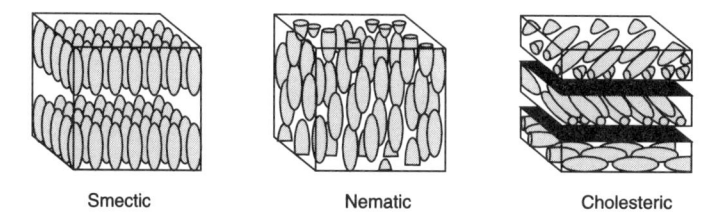

Figure 7.14 Three common phases of liquid-crystal materials

Because liquid-crystal materials are indisputably liquid and flow with quite low viscosity, their orientation order is normally preserved only over short distances (typically a few tenths of a millimetre) and not over indefinite distances as it would be in a solid crystal. The molecular orientation within an ordered region is defined by the *director*, a unit vector representing the alignment direction of the long axes of the molecules. With most LC materials it is possible to align the molecules by mechanical abrasion (rubbing) of the surface in one direction, allowing very large areas of uniform orientation and texture to be produced, which is essential for the uniform appearance of displays [10].

Liquid crystals are highly anisotropic in most of their physical properties when measured along axes parallel and perpendicular to the director. In particular, the anisotropy of the dielectric constant means that electric fields can be used to control the orientation of the director; the refractive index of the material also differs markedly along the two axes. These properties are used to advantage in the TN and STN displays discussed below.

7.3.2 The twisted-nematic (TN) cell

The most common LC configuration for display applications employs nematic materials, arranged with a 90° twist (TN cell). These are frequently used in watches, calculators and instrument displays. A thin (5–20 µm) layer of nematic material is placed between glass plates, with the top surface rubbed so that the director of its top layer is rotated by 90° to the director of its bottom layer, with intermediate layers forming a continuous twist between them (Figure 7.15). Incoming light is linearly polarised by passing through a layer of polarising material, then twisted by the LC material and finally escapes through a second polarising layer on the bottom of the cell. The twisting of the light is a consequence of the birefringence of the medium, that is, the existence of different indices of refraction for the horizontally and vertically polarised light waves, so that in each LC layer the light polarisation preferentially takes the angle of the local director.

When an electric field of a few volts (3–5 V) is applied to the cell all the layers in the LC nematic material line up in the same direction, so that the light is not twisted and hence is blocked by the exit polariser. This crossed-polariser configuration gives a normally white mode, i.e. the LC cell turns dark only when the voltage is applied. In some applications a parallel polariser configuration is employed, in which the cell is dark *except* when a voltage is applied. Figure 7.16 shows the transmittance of each type of TN cell as a function of voltage.

To construct an LC display from an array of TN cells, the inner surfaces of the glass plates are covered by electrode patterns which define the active areas of the display. In

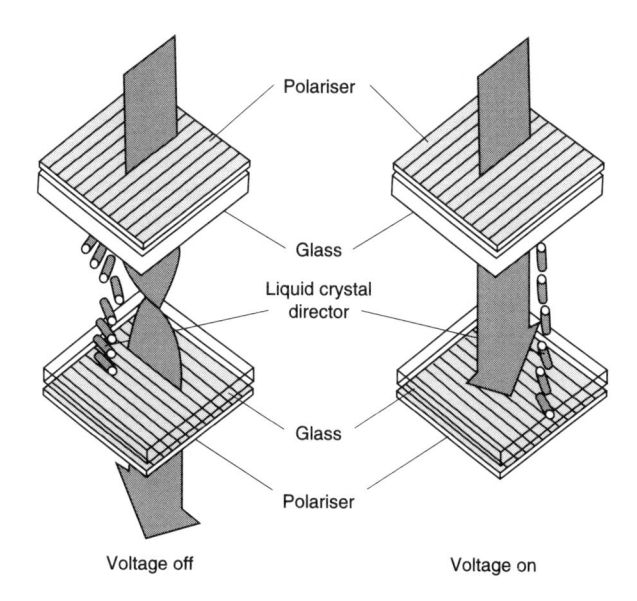

Figure 7.15 Principle of operation of a twisted-nematic LC cell (reproduced by permission of L D Silverstein)

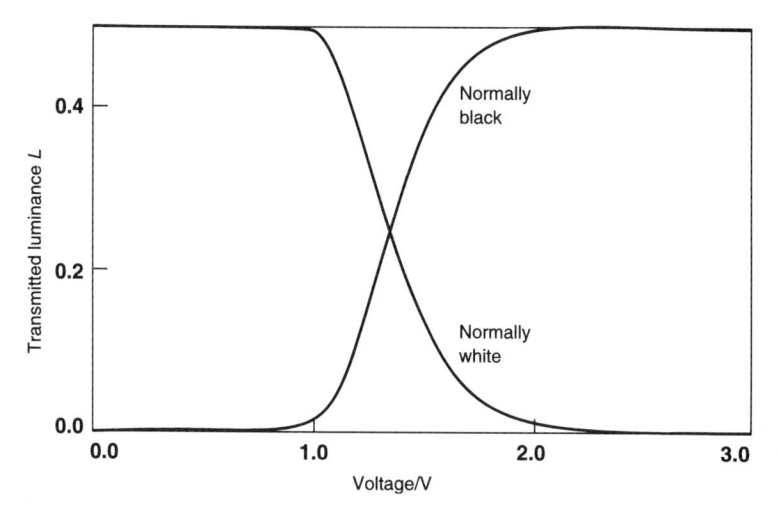

Figure 7.16 Transmittance of a TN cell in the normally white and normally black modes as a function of voltage (after ref. 11)

transmissive cells, as used with backlighting, both electrodes are made of a transparent conductor such as an indium tin oxide (ITO) mixture, whereas in reflective cells for use with ambient lighting the rear electrode plane may be metallic, allowing reflection of the incident light back through the cell.

The TN cell works well for simple display applications, especially those requiring low power consumption, but has certain limitations that prevent its use in displays requiring a high contrast range and a large number of picture elements. First, the contrast ratio between the light and dark states is rather poor because although the crossed polarisers are effective in blocking light in the dark state they tend to have low transmittances, typically only 20–30%, in the nonblocking state. Secondly, the switching time of the cells is relatively long, of the order of 100 ms rise or fall time, and becomes even longer at low temperatures, when the LC material gets sluggish.

The third limitation of TN cells is the temperature dependence of the switching threshold voltage, which becomes critical when attempting to address a display with a large number of cells. The simplest method of addressing, known as *multiplexing*, is to have two sets of conductive electrodes, one set connecting the cells in rows (X direction) and the other connecting the cells in columns (Y direction). Multiplexed addressing allows a display of $M \times N$ pixels to be addressed with only $M + N$ drivers, but suffers from crosstalk because the voltages applied to select the desired cell can also have an effect on other interconnected cells. In a display with 100 multiplexed lines, a selected 'on' pixel receives only about 10% more voltage than a nonselected 'off' pixel. This property makes TN displays unsuitable for highly multiplexed operation, especially in variable temperature environments. One solution has been to employ active matrix addressing by introducing a suitable thin-film switching device, either a transistor or a diode network, into each TN cell but the increased manufacturing costs and lower production yields of this approach combine to make such displays much more expensive.

The most obvious limitation of TN displays is the dependence of contrast ratio on viewing angle. Figure 7.17 compares the viewing-angle characteristics of the normally black (NB) and normally white (NW) modes of a TN cell. Curves of equal contrast ratio (100 : 1 and 20 : 1) are presented in polar form, where the radial coordinate represents the angular deviation from the normal to the plane of the display. It is clear that the two have substantially different viewing characteristics and that neither approaches the full hemispherical viewability of the CRT.

Many of the limitations of the TN cell have been overcome since the mid-1980s by the development of the supertwist (STN) and double-supertwist (D-STN) versions, in which the angle of twist in the LC material is increased to 180° or 270°. The high twist angle is combined with a high pre-tilt angle at the LCD's alignment layer and the result is a substantial improvement in both contrast ratio and range of viewing angle. Greater control over the electro-optic switching curve has also made it possible to achieve intermediate grey levels between the selected and nonselected states. The addressing limitations and long LC switching times have both been overcome by Scheffer and his colleagues through a technique called *active addressing*, which makes it

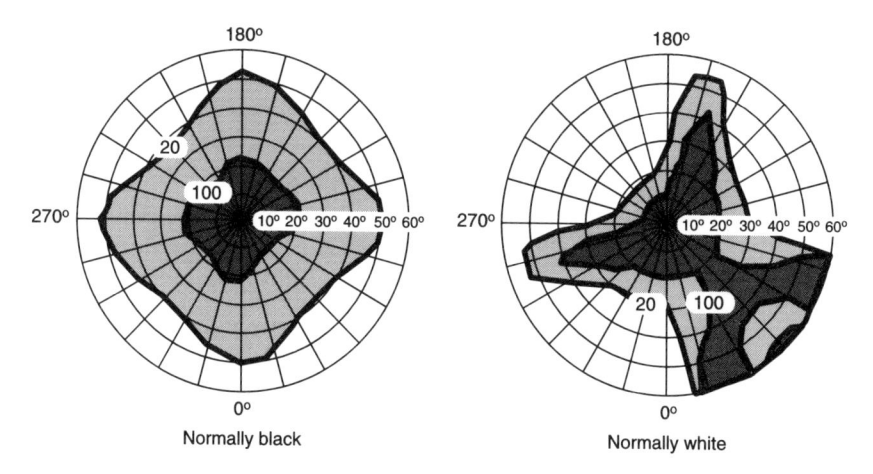

Figure 7.17 Iso-contrast viewing diagrams of two modes of a TN display; the centre represents on-axis viewing and the periphery 60° off-axis (from ref. 11, reproduced by permission of Scheffer & Nehring 1993)

possible now to achieve high-contrast, full-colour, full-motion video images in STN displays [11].

7.3.3 Coloured liquid-crystal displays

The variable birefringence of liquid-crystal material makes LC cells sensitive to both the applied voltage and the wavelength of the transmitted light. Thus it is possible to achieve a colour change through the cell if white light is used and the applied voltage altered appropriately. Despite this desirable characteristic, however, it has not yet proved possible to employ this technique to control colour in a commercially viable display because of the limited gamut of colour generated, restricted angle of view, lack of grey scale control and temperature sensitivity [12].

To obtain a multicolour STN display using conventional colour filter technology it is essential to avoid the colour shifts produced by variable birefringence in order to achieve a high-contrast, achromatic 'light valve'. The double-layer supertwist display (D-STN) is now the standard, in which a second STN layer is placed beneath the first, identical in thickness but without electrodes and twisted in the opposite sense [13]. This passive layer optically compensates the active STN layer in its dark (nonselect) state by undoing whatever the active layer does to the transmitted light (Figure 7.18). The advantages are that the nonselect state thereby remains dark for all wavelengths, and the cell remains compensated over the entire operating temperature range independent of the device and material parameters.

The standard approach to colour LCD design is to place a mosaic of colour filters

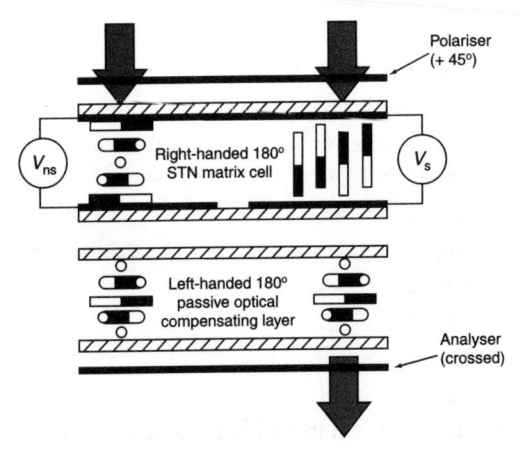

Figure 7.18 Construction of a double-layer supertwist (D-STN) display; light is transmitted normally in the select state (right) but in the nonselect state (left) is compensated by the second layer at all wavelengths (from ref. 11, reproduced by permission of Scheffer & Nehring 1993)

over the LC cells, together with a fluorescent backlight (Figure 7.19). In this form the individual cells act as 'light valves' controlling the passage of light through the colour filters, and each pixel of the image is a composite group of (usually three) coloured subpixels. From the observer's point of view the display then acts as a trichromatic emissive device, with each group of cells as one picture element. The optical performance of the whole display can be modelled by computing the transmission, reflection and polarisation state of the light as it passes through each layer of the structure.

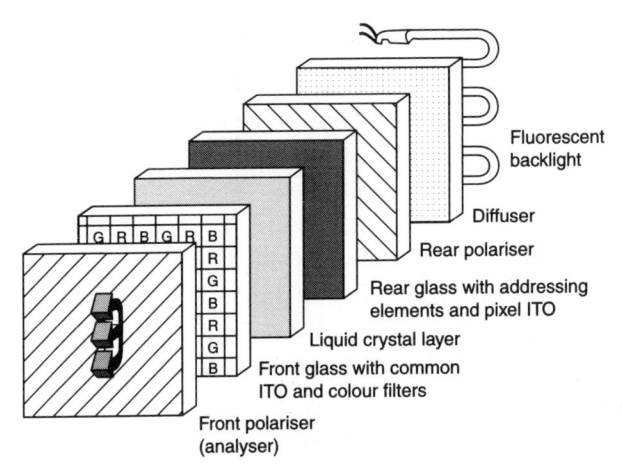

Figure 7.19 Multilayer structure of a colour LC display (reproduced by permission of L D Silverstein)

Non-emissive full-colour displays are unacceptably dark when used in reflected ambient light, because of the absorption of the polarisers. Full-colour LCD displays are thus almost invariably designed for transmission of a backlight, the usual form of which is a serpentine fluorescent lamp of either the hot-cathode or the cold-cathode type, with a rear reflector to direct the light forward. The hot-cathode lamp requires a lower starting voltage and has higher efficacy but suffers from a shorter lifetime (5000 to 15 000 h) and poorer impact resistance than the cold-cathode lamp. Since the total luminous energy transmitted through the multilayer sandwich of LC material, polarisers and RGB filters is usually no more than about 5%, quite a powerful backlight is required: the luminance is generally in excess of 25 000 cd m^{-2}.

Tri-band phosphor mixtures, which have three distinct peaks in their spectral power distribution, are typically employed in backlights to improve colour rendering with a correlated colour temperature of around 4600 K. There is much interest in the development of fluorescent phosphors that have spectral peaks close to the peak spectral transmittance wavelengths of the colour filters, because by this means considerable improvement of efficiency may be achieved [14]. Figure 7.20 shows the spectral power distribution of a typical lamp containing a mixture of three phosphor types.

The colour behaviour of the LC layer is complex to model and an exact solution requires a (4 × 4) propagation matrix approach, but this is computationally very intensive and simpler methods such as the (2 × 2) Jones matrix are often employed [15]. The LC cell gap (thickness of the LC material) is a critical parameter for tuning its colour

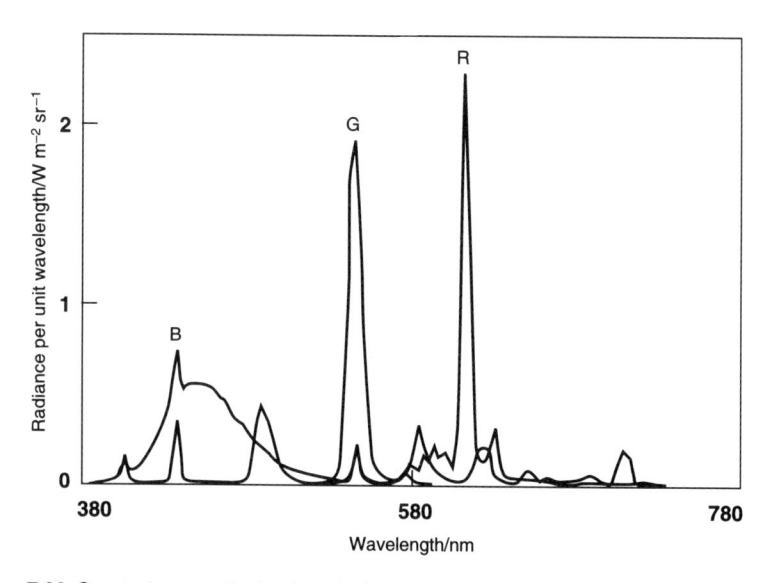

Figure 7.20 Spectral power distribution of a hot-cathode fluorescent lamp for LCD backlighting, containing a mixture of three phosphor types (reproduced by permission of L D Silverstein)

behaviour. For given values of cell gap and birefringence of LC material, the transmittance of the cell is a function of wavelength (Figure 7.21). In design the cell gap is selected to maximise some combination of display contrast and colour gamut.

The colour filter layer is normally laminated to the front glass of the LC array, beneath the front polariser. Various processes have been used to fabricate the thin-film absorption filters, including printing, evaporation, electrodeposition and photolithography. In one process, for example, yellow (isoindolinone), magenta (quinacridone) and cyan (metal phthalocyanine) organic pigments are evaporated through a metal stencil mask [16]. The variety of dyes and pigments compatible with LC materials and a given manufacturing process is limited, but the filter thickness and dye concentration can both be adjusted within a moderate range [17]. Leakage of light between the cell boundaries is prevented by adding a black matrix, normally a thin opaque metallic layer. With careful design the light leakage area can be reduced to less than 1% of the total viewing area, allowing the display's contrast ratio to be maximised.

The triplets of colour filters may be arranged in various patterns. Recommendations are that for data graphic displays the RGB triplets should be arranged in vertical stripes, in order to minimise aliasing effects (the 'jaggies') in vertical graphic lines, whereas for video displays they should be in triangles (delta formation), which are less obtrusive in imagery [18]. These arrangements are now customary in commercial products. Diagonal arrangements should be avoided because they lead to an asymmetry of spatial resolution in the diagonal direction [19]. Other designs place the cells in 2 × 2

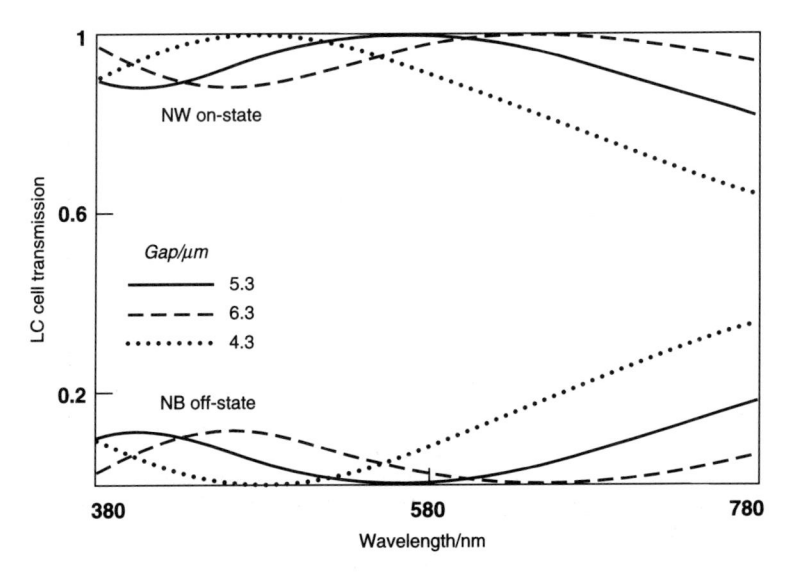

Figure 7.21 Spectral transmittance of an LC cell for three cell gap thicknesses (reproduced by permission of L D Silverstein)

square tiles, adding either a white or second green cell in the fourth position. These improve the overall display luminance at the expense of resolution and may noticeably worsen interference patterns for some types of graphics.

The filters convert the white backlight into the additive primary colours of red, green and blue. Whilst this is a necessary step in generating an emissive trichromatic display, it has the drawback of immediately reducing the overall light output to one-third, because the transmissive waveband of each of the three filters spans only about one-third of the visible spectrum. Figure 7.22 shows the spectral transmittance curves of a typical set of RGB filters; it is clear that there is considerable loss of light, especially from the blue and green filters. In order to achieve a primary green, for example, only the green pixels can be illuminated so the maximum possible display luminance is reduced to one-third of the white level. Because the peak transmittance of the green filter is only just over 50%, moreover, the peak radiant output is less than 20% of the level that would be obtained with the same backlight from an unfiltered white display.

A new development uses thin-film dichroic optical interference filters, manufactured by vacuum deposition on to oxide materials. These filters can be tailored to have spectral characteristics nearer to the ideal 'block dye' forms, and have higher transmittance and greater purity and greater stability over time than conventional dyes [20]. The drawback is that they only work properly when the light passes through at angles within 15° of the normal to the plane of the filter and at temperatures below 20 °C. At higher temperatures and greater angles of incidence the spectral transmission band

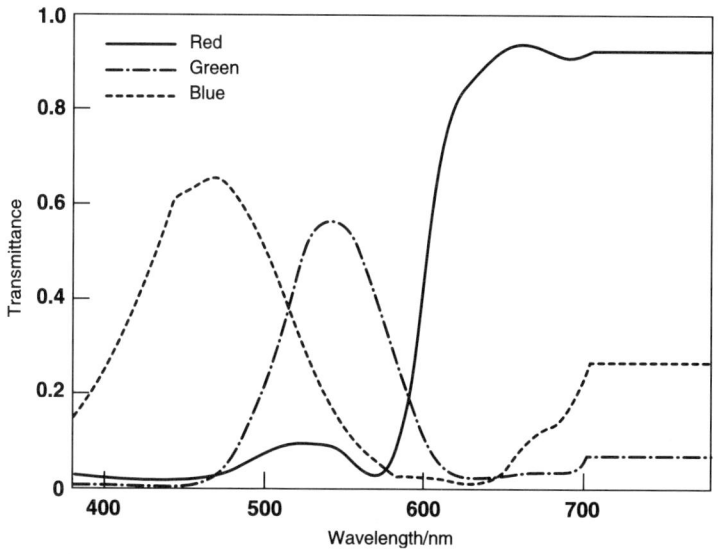

Figure 7.22 Spectral transmittance functions of RGB filters in a colour LCD (reproduced by permission of L D Silverstein)

shifts to shorter wavelengths. Figure 7.23 compares the spectral transmittance curves for a magenta (green-absorbing) filter made from conventional dye and one made using dichroic multilayer interference technology.

Holography is now being considered as a third approach to making filters for LCDs, because these have the sharp spectral cutoff characteristics of interference filters and are easier to manufacture than by means of a dye process. Such filters can be made with a pass-band transmittance of over 95% and a blocking-band transmittance less than 0.001%, with an edge slope in transmittance from 90% down to 10% over only 10 nm change in wavelength. Problems remain in controlling the angle of exposure to change the holographic reconstruction wavelength whilst avoiding spurious interferences, but this colour filter technology looks promising for the future [21].

Recently an ingenious new type of colour STN projection display has been devised using LC cells with subtractive primary CMY filters stacked in three layers [22], similar in principle to the three dye layers in colour transparency film. By switching the cell in each layer on or off, eight different colour combinations can be achieved (Figure 7.24). By using the grey scale capability of the cells, a full gamut of colours is possible. Because each pixel transmits some combination of filtered light, not just one out of three filtered cells in a triad, this stacked subtractive method produces an image about three times brighter than the conventional mosaic filter approach. The LCDs are only used at normal angle of incidence because the light must travel through all three cells to

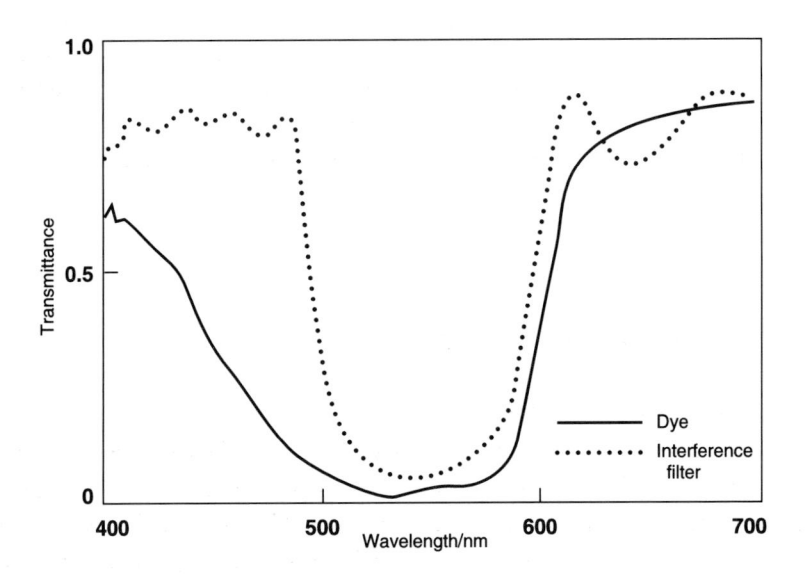

Figure 7.23 Spectral transmittance for dye and interference magenta filters (from ref. 20, reproduced by permission of Conner et al. 1993)

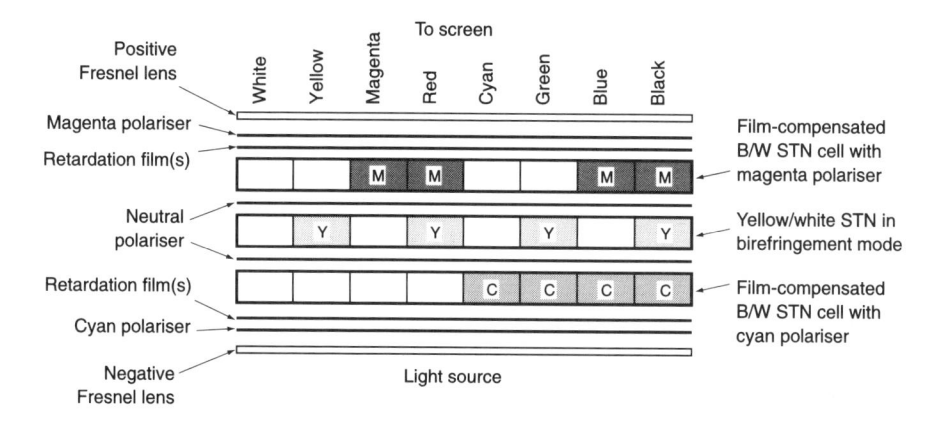

Figure 7.24 Subtractive colour scheme for an overhead projector using CMY cell stack (from ref. 11, reproduced by permission of Scheffer & Nehring 1993)

emerge with the proper colour, and parallax is avoided in the projector design by using Fresnel lenses to ensure a parallel light path through the stack.

The LC cell is not a binary device. Between the light and dark states there is a range of grey levels that can be obtained by applying suitable voltages between the limiting values. In contrast to the gamma function of CRTs, no general theoretical voltage–luminance transfer function can easily be defined for LCDs, as the behaviour of a given device is strongly dependent on the LC material and the cell construction parameters. In practice it is satisfactory to fit an interpolation function, either piecewise linear or (more accurately) cubic spline, to a set of measured points. Figure 7.25 shows a typical transfer function, similar but not identical for each of the three channels.

7.4 COLORIMETRY OF DISPLAYS

7.4.1 Chromaticity of primaries

Both CRT and backlit LCD colour displays are emissive devices in which the light emitted by each of the three primary sources (phosphors or filters) is defined in terms of its spectral power distribution (SPD). Once this is known, the CIE system can be applied to calculate the tristimulus values of the corresponding colour, by taking the product at each wavelength of the SPD and the three CIE colour-matching functions (section 3.5). Full colorimetric characterisation of the display requires the calculation

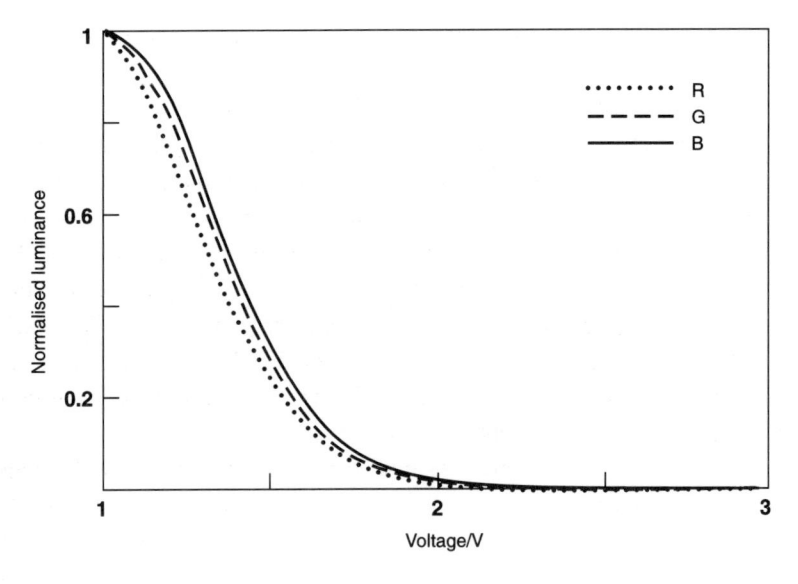

Figure 7.25 Normalised voltage–luminance transfer function for an LCD (reproduced by permission of L D Silverstein)

to be performed for each of the three primary SPDs, yielding nine tristimulus values. The summation for the red display primary is shown in standard notation in Eqn 7.4:

$$X_r = k \int_a^b S_{r\lambda}\, \bar{x}_\lambda\, d\lambda$$
$$Y_r = k \int_a^b S_{r\lambda}\, \bar{y}_\lambda\, d\lambda \qquad (7.4)$$
$$Z_r = k \int_a^b S_{r\lambda}\, \bar{z}_\lambda\, d\lambda$$
$$L_r = 683 Y_r$$

where X_r, Y_r, Z_r = tristimulus values for red display primary
S_r = spectral power distribution of red primary (radiance in W m^{-2} sr^{-1} at each unit of wavelength)
a, b = lower and upper limits of the visible spectrum, typically 380 and 780 nm
$\bar{x}, \bar{y}, \bar{z}$ = CIE colour-matching functions
k = normalising factor
$d\lambda$ = wavelength sampling interval, typically 5 nm
L_r = luminance of red primary, in cd m^{-2} (see section 7.5.1 for explanation of the factor 683).

Similar calculations are required to obtain the tristimulus values for the green and blue primaries.

The areas of colours considered in display applications usually have quite small angular subtense and the CIE 1931 standard colorimetric observer (2° field) provides the appropriate set of colour-matching functions. But this may not always be the case, depending on the viewing distance, the size of the screen and the proportion of the screen covered by a particular colour patch. When viewing a television screen at 3 m distance each 1 cm on the screen subtends 12 min of arc at the eye, and the full width of the screen (say 45 cm for a 23-inch diagonal screen) subtends about 9°. For a desktop computer graphic display at arm's length (nominally 60 cm), each 1 cm on the screen subtends approximately 1° at the eye, and the full width of the screen (say 30 cm for a 16-inch diagonal screen) subtends about 28°.

The CIE 2° colour-matching functions apply for an angular subtense in the range 0.5–4°. In the example above this range is equivalent to a patch diameter between 2.5 and 20 cm for the television display or between 0.5 and 4 cm for the desktop display. Angles smaller than about 30 min of arc result in colour assimilation by the retina, reducing the eye's ability to discriminate one colour from another and causing the colour-matching relationships to break down, especially for images with a great deal of edge content such as text and linework. The dominant factor affecting colour matching and colour perception for such small visual angles is small-field tritanopia, a term used to describe the eye's inability to see fine detail at short wavelengths because of the virtual absence of 'blue' photoreceptors in the fovea (section 8.4.3). This is the reason for the graphic design guideline never to present fine text or graphics on a display where the only colour difference between foreground and background is in the blue channel: blue text on a black ground, for example, or yellow text on a white ground.

For angles larger than 4° the CIE 1964 supplementary 10° colour-matching functions (the so-called 10° observer) are recommended (section 3.9.2), because of the eye's increased short-wavelength (blue) sensitivity for large colour fields. For precise colorimetric analysis or visual research studies, the Judd modification to the CIE 1931 colour-matching functions should be used [23], on account of the retina's greater photopic sensitivity at wavelengths less than 460 nm. This modification was endorsed by the CIE in 1988 [24]; without it the 1931 CMFs can underpredict the luminance of the blue primary in a display by up to 5%, depending on its SPD.

When a display is viewed at low luminance levels the colour-matching functions, which are defined for photopic conditions (good illumination, in which vision depends only on the cone cells of the retina), may also begin to break down because of the transition from cone vision to rod vision (mesopic vision). As the luminance of the white-adapting field falls below about 20 cd m^{-2} the rod contribution starts to become significant (greater than about 10% of the achromatic visual signal) and leads to

402 COLOUR IN VISUAL DISPLAYS

relatively increased sensitivity at shorter wavelengths (the Purkinje shift, see section 8.7.3). Thus when viewing a low-luminance display in a dim room the red and orange hues will look darker than normal and the blues and cyan hues will look lighter. Under such conditions various correcting factors need to be introduced into the colorimetric calculations; alternatively, and preferably, resort should be made to a more comprehensive model of colour appearance [25].

Once the tristimulus values for each primary are known, their chromaticity values can be calculated by the standard formulae for either xy (1931 system, section 3.8) or $u'v'$ (1976 system, section 3.14). Figure 7.26 shows the chromaticity coordinates for the three phosphors of a high-quality CRT plotted on both the 1931 and 1976 diagrams.

What the eye sees is the additive sum of the luminance contributions of each of the three display primaries. By adjusting the light output of each primary between zero and its maximum value, the colour of the sum can be set to any chromaticity coordinates within the triangle joining the three primary coordinate positions. Because the CIE system is also based on additive primaries the process of calculating tristimulus values is relatively straightforward, but there are two significant difficulties to be overcome before CIE values for displays can be specified [26].

The first difficulty is that the CIE system is not absolute but relative, with the usual convention for surface colour that the Y tristimulus value of the illuminant is scaled to 100. The usual solution to this difficulty is to take the white reference for the display as the colour obtained with all three colour channels set to maximum output, and then to normalise the Y value of this white to 100.

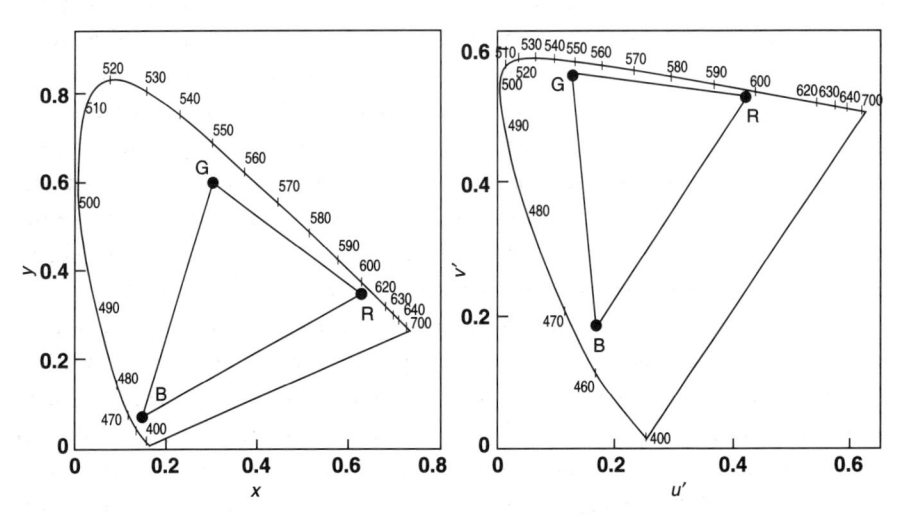

Figure 7.26 Colour gamut of a CRT plotted on the CIE 1931 and 1976 chromaticity diagrams; the display was a Barco Calibrator with P22 phosphor set, the SPDs of which are shown in Figure 7.8 (reproduced by permission of L D Silverstein)

The second difficulty is how to define the 'amounts' of the three primaries. This is dealt with by assuming that the luminance contribution of any of the three channels to the total luminance is directly proportional to the luminance of that channel measured in isolation, i.e. that the system is additive. (This is true only for fields of less than 4° angular subtense, where the 2° CMFs apply [24].) Thus we can write (Eqn 7.5):

$$X = X_r + X_g + X_b$$
$$Y = Y_r + Y_g + Y_b \quad (7.5)$$
$$Z = Z_r + Z_g + Z_b$$

Given this additivity, the colours generated by any specified amounts of the three primaries can be calculated in terms of their XYZ tristimulus values from a knowledge of the chromaticity coordinates of the three primaries ($x_r y_r$; $x_g y_g$; $x_b y_b$), and the output luminances (Y_r, Y_g, Y_b) for each channel. Since it follows from the definition of chromaticity coordinates that $X_r = (x_r/y_r)Y_r$ etc., Eqn 7.5 can be rewritten in terms of a (3 × 3) matrix multiplication as Eqn 7.6 [27]:

$$\begin{bmatrix} X \\ Y \\ Z \end{bmatrix} = \begin{bmatrix} x_r/y_r & x_g/y_g & x_b/y_b \\ 1 & 1 & 1 \\ z_r/y_r & z_g/y_g & z_b/y_b \end{bmatrix} \begin{bmatrix} Y_r \\ Y_g \\ Y_b \end{bmatrix} \quad (7.6)$$

The triangle formed by joining the three phosphor coordinate points by straight lines represents the gamut of all colours obtainable by a particular display and is well inside the spectral locus. The colours outside the boundary of the triangle are known as out-of-gamut colours, i.e. saturated colours that could be seen by the human visual system but cannot be produced by the display. It is tempting but misleading to regard the interior area of the two-dimensional chromaticity triangle as representing the full range of colours that can be produced on a display, because each of the primary colours actually occurs at a different luminance level (Y). In the three dimensions of xyY space the display gamut is seen to be akin to a truncated triangular pyramid, where at high luminance levels the colour range becomes very restricted (Figures 7.27 and 7.28).

The gamut of colours achievable by a particular display can readily be visualised as a solid volume in any three-dimensional colour space, in the same way that the Munsell Color Tree shows the gamut boundaries and colour relationships for a particular set of surface colorants (Plate 2). Most useful for visualisation and intuitive manipulation of colours are perceptually uniform colour spaces such as CIELAB 1976 and CIELUV, in which equal Euclidean distances between any two points in the colour coordinate space should ideally represent equal perceived differences between the corresponding colours (section 3.14). The $L^*u^*v^*$ colour space is preferred for displays because the

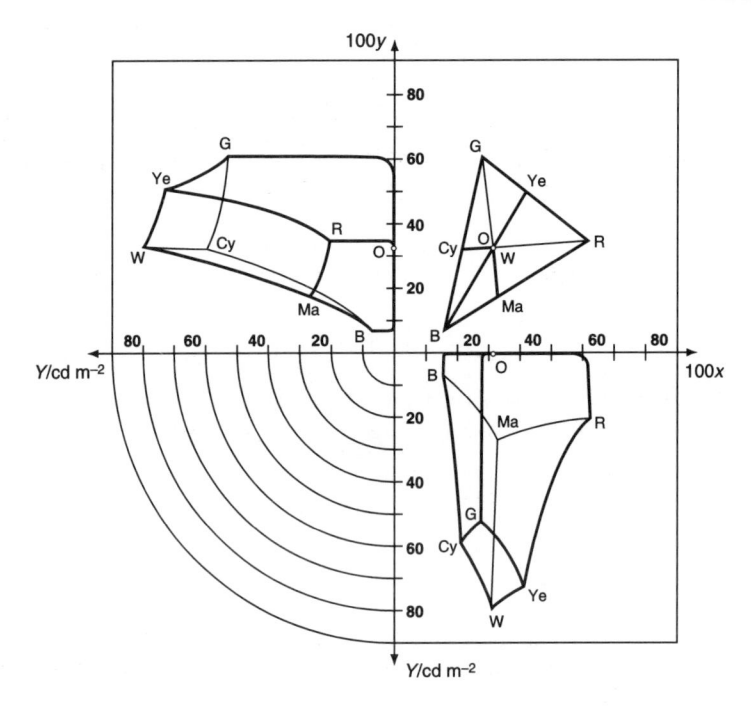

Figure 7.27 The range of colours obtainable from CRT phosphors depends on the luminance level; showing the colour solid in xyY space makes it clear how the chromaticity diagram shrinks from the full triangle to a single point at white (reproduced by permission of Barco Graphic Displays)

u^*v^* axes are linear transformations of the 1931 xy coordinates and therefore preserve the linear additivity properties of the chromaticity diagram. Two-dimensional cross-sections of uniform colour spaces, particularly planes of constant lightness or constant hue, are commonly employed to show the extent of a display's colour gamut (Figure 7.29) [28,29].

7.4.2 Setting the white point

The white point in a scene is of crucial importance because it sets the reference by which all other colours are judged. The human visual system scales the response from the eye's photoreceptors according to the illuminant colour [30], and hence a fundamental part of the CIE system is the definition of a reference white, normally a perfectly reflecting (or transmitting) diffuser. For displays the reference white is the colour produced by maximum output of all three channels. The correlated colour temperature of this white depends on the relative luminances of the three channels, as given by Eqn 7.6, from which the white chromaticity coordinates can be calculated. In practice the

COLORIMETRY OF DISPLAYS 405

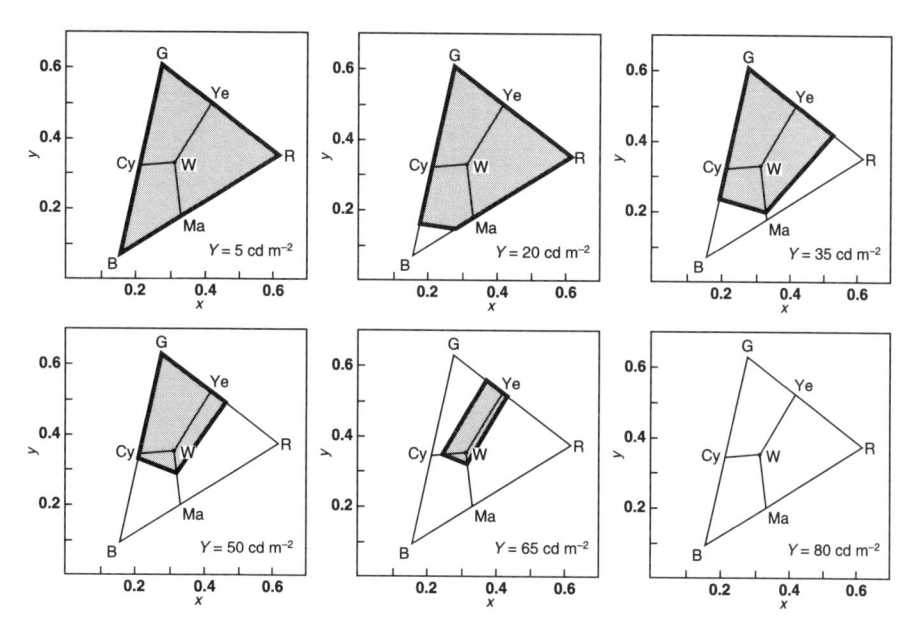

Figure 7.28 Six cross-sections of the solid in Figure 7.27 on the CIE 1931 chromaticity plane; the range of colours obtainable depends on the luminance level Y (reproduced by permission of Barco Graphic Displays)

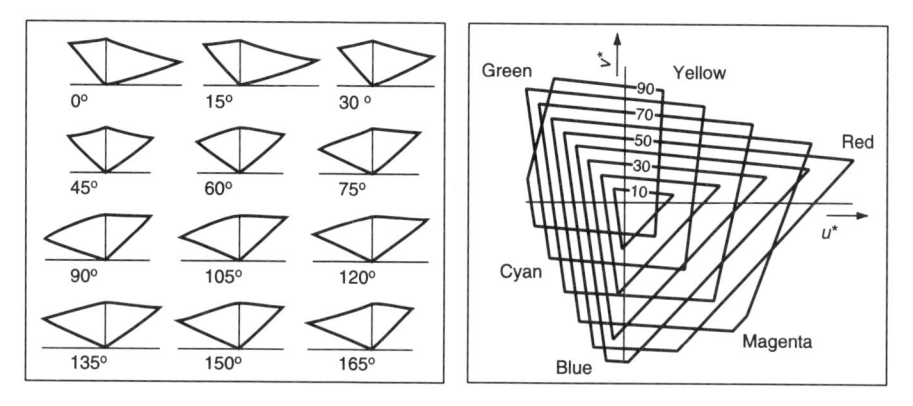

Figure 7.29 Two-dimensional cross-sections through the CIELUV gamut of a colour display (from ref. 28, reproduced by permission of Robertson 1988)

white point can vary from below 4000 K to above 10 000 K, following the Planckian locus in the CIE chromaticity diagram (Figure 7.30).

The most common standards in use for setting the display white point are the CIE daylight (D series) illuminants D_{50}, D_{65}, D_{75} and D_{93}, correlated to 5000, 6500, 7500 and 9300 K respectively, whose chromaticity coordinates are given in Table 7.1. D_{75} is

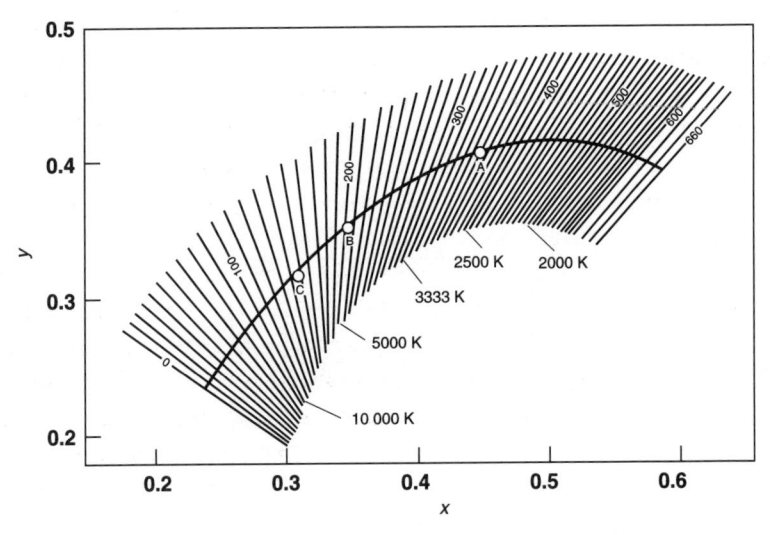

Figure 7.30 Lines of constant correlated colour temperature in the CIE 1931 chromaticity diagram; isotherms can be specified in either kelvin or mirek (= μK^{-1}) (from ref. 31, reproduced by permission of Miller 1987)

used as a white standard by some US-based manufacturers of television CRTs, although the NTSC white point is CIE illuminant C, correlated to 6774 K. D_{65} was adopted instead by the EBU as a white point for PAL broadcast television, because of both technical limitations and perceptual considerations. In the graphic arts D_{50} has been used for many years as the standard illuminant for proof viewing of prints [32]. Commercial cinema projectors use either carbon arc lamps of approximately 5000 K or xenon arc lamps at 6500 K. Experiments in which the colour temperature of a cinema projection lamp is varied by means of filters have shown that observers prefer a mean colour temperature of about 5400 K as being neither too cool nor too warm [27]. Given that this colour temperature approximates to daylight on a clear sunny day, the experimental result is hardly surprising.

Table 7.1 Chromaticity coordinates of the four most common display white point settings

Colour temp./K	x	y	u'	v'	Usage
5000 (D_{50})	0.3457	0.3587	0.2091	0.4882	Graphic arts print standard
6500 (D_{65})	0.3128	0.3292	0.1978	0.4684	Textiles and broadcast TV
7500 (D_{75})	0.2991	0.3150	0.1935	0.4586	US television
9300 (D_{93})	0.2832	0.2972	0.1888	0.4457	Computer graphics

Computer graphics CRT monitors are usually factory-set at 9300 K, not for reasons of colour balance but in order to utilise more of the radiance from the blue phosphor and hence to achieve the higher luminance levels required for satisfactory operation in the ambient light levels of a normal office working environment. Visual chromatic adaptation leads to partial compensation for the very blue colour cast, but such a high colour temperature is unsatisfactory for applications requiring critical colour judgements to be made from the display.

There are two approaches to setting the white point of a computer graphic display to a desired colour temperature: either true setting within the monitor itself or simulated setting by control of the digital signal values from the host computer. For true setting the gain and bias of the amplifiers in the three channels of the monitor drive electronics are adjusted so that the correct relative luminance is attained when maximum input signal is applied to all three channels. Calibration adjustments are generally provided in a display monitor for this purpose, although it may be necessary to get access to the internal electronics to locate them.

For simulated white point setting the digital values supplied to the graphics controller in the host computer are scaled by the software driver so that the display appears to have a white of the correct colour temperature, even though the true white point may be a different colour. This approach is commonly used in desktop publishing systems, in which a standard monitor with a factory-set colour temperature of 9300 K can be driven to produce a white of D_{65} or D_{50} without needing internal adjustments to the monitor itself. The drawback is that in order to achieve the correct balance of the RGB channels only one or at most two of them can be kept at maximum signal level while the others need to be reduced, resulting in lower overall luminance level. Tests have shown that the luminance of the white point on a CRT display adjusted to true 6500 K falls to 95.6% with a simulated D_{50} white point; the luminance of the same CRT display adjusted to true 9300 K falls substantially to 74.5% with a simulated D_{50} white point [33].

7.4.3 Display characterisation

An essential step in knowing how to drive a display to produce the desired colours is determining the exact relationship between the input signals and the output visual colour stimuli. In a computer graphic display the input signals are typically the digital-to-analogue converter (DAC) pixel values, while the output stimuli are measured red, green and blue XYZ tristimulus values or one of their CIE derivatives. The procedure for establishing this relationship is known as *device characterisation* and is generally performed by a combination of both analytical (modelling) and empirical (measurement) methods.

RGB DAC values → [Signal linearisation] → $Y_r Y_g Y_b$ Luminance levels → [Colorimetric transform] → XYZ Tristimulus values

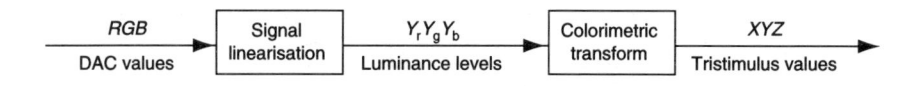

Figure 7.31 Two-stage transformation from *RGB* voltage signals to *XYZ* tristimulus values

Various mathematical models have been proposed [34–36] for display characterisation. These models usually comprise two stages (Figure 7.31). Stage one is a one-dimensional nonlinear transformation for each of the three channels from the DAC values to the output luminance levels, which is known as signal linearisation or gamma correction. Stage two is a linear three-dimensional transformation of these luminances to the tristimulus values of the colour displayed.

Before commencing the modelling, certain colorimetric information needs to be determined, namely the chromaticities and luminances of the three primaries, plus the white and black points. Each primary's chromaticity coordinates are obtained by measuring the colour at maximum DAC signal value (normally 255 for an 8-bit per channel device) with the other two channels set to zero. The desired luminances of the white and black points can be achieved by interactively adjusting the display's brightness and contrast controls while measuring the display luminance from suitable patches on the screen with a photometer.

The importance of setting the brightness and contrast controls to a known and optimal state can scarcely be overemphasised. Their names are actually rather misleading, because 'brightness' adjusts the offset (light output at zero drive voltage), whereas 'contrast' adjusts the gain of the amplifiers and hence the slope of the voltage–luminance transfer function. These two controls are provided for the operator to adjust freely on most monitors, yet they can have a dramatic effect on the colour rendering of the display. To make matters worse they are somewhat interdependent in their action, and it is possible by setting both controls too high to drive the monitor beyond its normal operating limits. Figure 7.32 shows their effect on the luminance of white on a typical CRT monitor and the extent of the interaction between them. The maximum luminance in this example is limited to about 80 cd m^{-2}, above which the transfer function becomes distorted [37].

Once the display's operating state has been stabilised to a known and repeatable state, the first stage of modelling (signal linearisation) can begin. This requires measurement of the light output for a series of DAC values applied to each of the RGB channels in turn. Then by plotting the logarithm of the measured output luminances against the logarithm of the normalised DAC values, a simple linear relationship (the gamma of the display) can be derived. If the 'gamma law' of Eqn 7.2 held perfectly the graph of log(luminance) versus log(signal) would be a straight line as in Figure 7.3. In

COLORIMETRY OF DISPLAYS 409

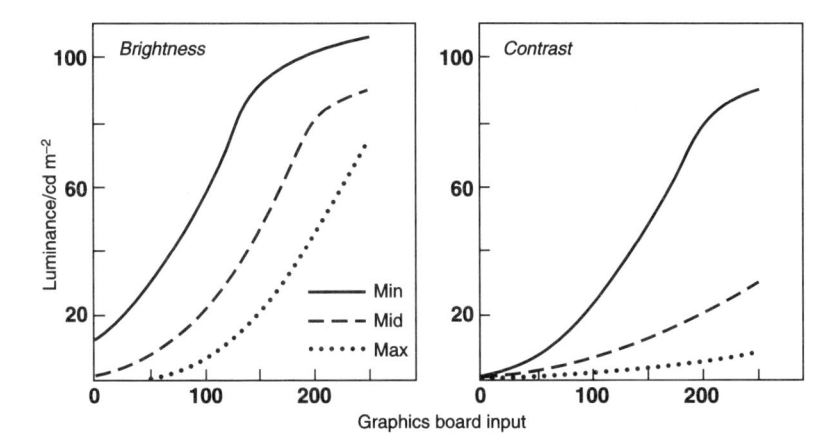

Figure 7.32 Gamma functions obtained from a Sony monitor: (left) three settings of 'brightness' with 'contrast' at maximum; (right) three settings of 'contrast' with 'brightness' at maximum (reproduced by permission of J D Mollon)

practice, departures from the straight line sometimes occur at low signal levels (DAC values less than about 50 out of 255) due to the effects of ambient light and nonlinear behaviour of the electron gun (Figure 7.33). The measured data can usually be more accurately modelled by a second- or third-order polynomial in log–log space, using a standard regression or least-squares fitting algorithm [34]. Even better results can be

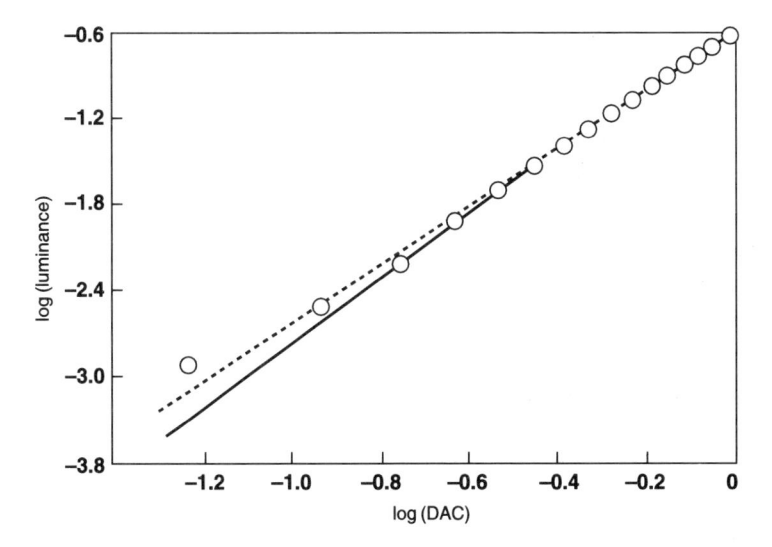

Figure 7.33 Normalised log(luminance) versus log(DAC) data measured on a Barco monitor, plotted against best-fit linear (dashed) and second-order (solid) polynomials in log–log space (reproduced by permission of M R Luo)

obtained by piecewise linear or cubic spline interpolation in log–log space of the measured data from a suitable set of DAC values, spaced more closely at low signal levels where the behaviour departs from linearity.

However the signal linearisation is performed, whether by simple gamma function or by polynomial curve fitting or by piecewise interpolation, it is usually implemented by means of a one-dimensional look-up table (LUT) for each channel in order to reduce the computational load. The second stage of the transformation of Figure 7.31, converting luminance to tristimulus values, is then performed accurately by the (3×3) matrix multiplication of Eqn 7.6.

The inverse transformation is required in order to determine what RGB signals are needed to drive the display in order to produce a specified (known XYZ) on the display. This can be performed by applying the inverse (3×3) matrix to the XYZ values and the inverse interpolation of the measured luminance versus signal data. The latter can always be determined uniquely because of the monotonic and continuous relationship between input signal and output luminance for both CRT and LCD devices, but note that their characteristic curves are quite different (compare Figures 7.25 and 7.32).

The model-based approach described above requires relatively few measurements to achieve a complete characterisation of a display, but its accuracy depends on the degree to which the theoretical model represents the actual behaviour of the display. An alternative approach, which makes no *a priori* assumptions about the display's behaviour, is to measure a large number of colours generated by systematic combinations of input signal values. A three-dimensional interpolation procedure can then be applied to the measured data set in order to determine all intermediate values [35,38]. In practice, this approach is relatively easy to implement in the forward direction (signal to colour) but requires very complex interpolation procedures in the reverse direction (colour to signal). While it has become the standard approach for other classes of imaging devices that are difficult to model, such as colour printers, it is generally not worth the effort for well-behaved displays.

Four important assumptions are normally applied in characterising a display [39]:
(a) temporal stability (the same colorimetric properties persist over time)
(b) spatial uniformity (the same colorimetric parameters are preserved at different positions on the screen)
(c) channel independence (the exitance of one channel at a given pixel location is not dependent on the other channels at that pixel)
(d) phosphor constancy (the chromaticities of each phosphor are independent of the voltage applied to the channels).

Each of these, if untrue, would affect the predictive accuracy of the characterisation models, and various methods have been proposed to measure such discrepancies [36,40].

7.4.4 Colour appearance

The CIE system was originally defined to measure the visual equivalence of coloured lights, and was later applied to the measurement of surface colours and assessment of colour differences between two colorant mixtures on similar media. It has proved very successful for comparing, for example, two dyes applied to a similar fabric or two paints applied to a similar substrate, when viewed under a specified illuminant with given viewing geometry and surround. Provided that similar constraints are imposed, the CIE system can be equally effective as a metric for assessing the accuracy of a colour match in image reproduction on displays.

When considering how to achieve identical colours between two media, such as between a display and a reflection print, there are four aspects of applying colorimetry that can cause problems [41]:

(a) white point equivalence
(b) media differences
(c) measurement geometry
(d) gamut mapping.

A reproduction will look exactly the same as the original image only if both have the same XYZ values for the white point, and if the two media have similar surface characteristics and are observed under similar viewing conditions, and if the reproduction medium can produce all the colours present in the original. In practice these parameters vary considerably from one medium to another, as Table 7.2 indicates. The result is that two stimuli presented on different media under distinct viewing conditions, even though they are measured to have identical XYZ values, will usually not match visually.

Table 7.2 Characteristics of different media

Parameter	Real scene	CRT display	Print in office
Type of medium	Mixed	Self-luminous	Reflective
Dimensions	Dynamic 3-D	Dynamic 2-D	Static 2-D
Illuminant type	Sun + lights	Self + ambient	Lamps + daylight
Luminance/cd m^{-2}	1000–5000	50–120	150–500
White point/K	2000–10 000	5000–9300	2700–4500
Surround	Light	Dark–light	Light
Dynamic range	1000 : 1	100 : 1	50 : 1
Spatial resolution	Ideal	Low	Very low
Colorants	Mixed	Phosphors	Ink pigments
Colour gamut	Very large	Large	Medium

Surround plays an important part in the appearance of colour. The simultaneous contrast effect is well known for a single coloured patch on a coloured background [42], but applies also for a coloured border around a complex image. On a CRT display there are four zones surrounding the central image area on the screen (Figure 7.34). The most important is the image background, i.e. the addressable area of the screen outside the image itself, over which the application designer has full control but which is often neglected. The effect of a dark image background is to make the image appear lighter, requiring an increase in the tonal contrast gradient of the image to compensate. Conversely, a light background makes the image appear darker, requiring a decrease in image contrast. A light background also tends to make the image appear less colourful. The lightness and colour of the monitor faceplate, being further away, affect the appearance of the image to a lesser degree but still cannot be ignored. Ideally they should be unobtrusive, neutral in colour and graded in lightness from the image [43]. Outermost is the surrounding environment in which the display is viewed, which also impinges on the peripheral visual field of the display user.

Several other important parameters affect the colour appearance of an image, which relate to the nature of the medium or environment in which the image is viewed. For an image presented on a display monitor, in particular, the factors described in the following text must be considered.

The brightness, or luminance level, of the display affects the apparent colourfulness

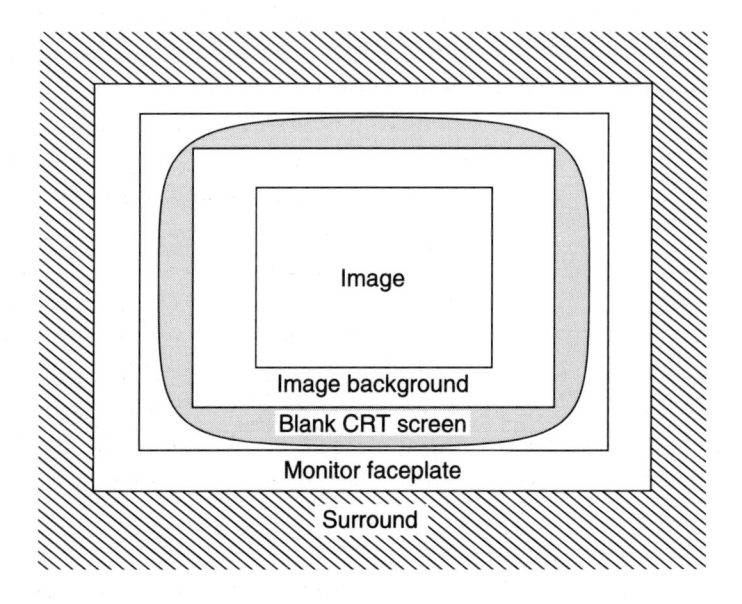

Figure 7.34 Surround regions of a typical display monitor

of colours in the image: the brighter the image, the more colourful it will appear. In general, an image will look more like the original scene if the intensity of the display is adjusted to be closer to the original.

The spectral power distribution of the illuminant determines the colour of the reference white in a scene or in a reflection print. On a display, the reference white is determined by the relative outputs of the red, green and blue channels (section 7.4.2). The human visual system is very good at compensating for both the colour and level of the illumination. As a result of this adaptation, objects tend to be recognised as having nearly the same colour in a wide range of conditions, a phenomenon known as *colour constancy*. But the adaptation is not perfect so that when, for example, transparencies are projected in a dark room with tungsten light the colour of the screen may still appear yellowish rather than pure white, especially in the light areas of the image. In general, adaptation to illuminants of different colours becomes less complete as the purity of the colour increases and more complete as the intensity increases [44].

The level and directional distribution of ambient illumination can affect appearance in various ways. For monitors it obviously affects the brightness of the faceplate and background, but also its reflection from the surface of the screen has a similar effect to reflection from the gloss on a photographic print in reducing image colourfulness and contrast. The effect is more marked when lighting is diffuse rather than directional. To compensate for flare, the image contrast needs to be increased for dark tones more than for light ones, i.e. more for shadow regions than highlights. Close specification over the room lighting and geometry is also desirable, although it is not always achievable in an office working environment [45]. The display colour gamut, especially for dark colours, is strongly affected by the ambient illumination [46].

Over the past ten years great progress has been made in the development of mathematical models of colour appearance, which predict from a knowledge of both the visual stimulus (input XYZ) and the viewing condition parameters (such as white reference, luminance level, surround colour and medium) the colour that an observer actually sees in terms of the perceptual attributes of lightness, colourfulness and hue (LCH). Research by Luo and coworkers at Loughborough University has shown conclusively that the Hunt model [25] gives accurate predictions for a wide range of viewing conditions and different media, including CRT displays [47]. The CIE Technical Committee TC1-27 has issued guidelines for coordinated research on the evaluation of colour appearance models for comparison of images on reflection prints and self-luminous displays [48]. The availability of such colour appearance models will enable the mapping of images between media in a way that faithfully preserves the appearance of image colours when viewed under different conditions. Such transformations will be essential for the future application of displays as 'soft proof' devices for previewing the colour appearance of printed images [49].

7.5 MEASUREMENT AND CALIBRATION

7.5.1 Units

The primary determinant of display performance is the nature of the radiant energy emitted by the display. The science of measuring radiant energy is called radiometry, and its principal terms and units are set out in Table 7.3 [50]. The range of wavelengths covered is in general much broader than the visible spectrum.

When the measurement range is confined to visible wavelengths (380–780 nm) the science is instead called photometry, and it has a corresponding set of terms and units (Table 7.4) which are derived from the radiometric units by multiplying at each wavelength by the spectral luminous efficiency function V_λ (section 1.6.1) and normalising by the factor 683 as, for example, in Eqn 7.7:

$$\Phi = 683 \int_{380}^{780} \Phi_e \, V_\lambda \, d\lambda \qquad (7.7)$$

The factor 683 in Eqn 7.7 for luminous flux converts the watts of radiant flux into lumens of light and has the units of lm W^{-1}. It is the luminous efficacy of monochromatic radiation at 555 nm, which is the wavelength of the eye's maximum response and the peak of the V_λ function. It is needed because by definition the V_λ function is the relative spectral response of the eye, normalised to 1.0 at its peak value [50].

In the US the usual unit for luminance is the foot lambert (fL), defined as 1 candela per square foot. The conversion factors are 1 foot lambert = 3.426 cd m^{-2} and 1 cd m^{-2} = 0.2919 fL.

An additional set of units exists for radiation falling on to a surface (irradiance, illuminance and so on), used in the characterisation of ambient light and print viewing

Table 7.3 Principal terms and units used in the measurement of radiant energy from displays

Name	Symbol	Unit	Definition
Radiant energy	Q	joule (J)	Energy emitted or received as radiation
Radiant flux	Φ_e	watt (W)	Radiant energy per unit time (= radiant power)
Radiant exitance	M_e	W m^{-2}	Radiant flux leaving surface per unit area
Radiant intensity	I_e	W sr^{-1}	Radiant flux per unit solid angle (in steradians)
Radiance	L_e	W sr^{-1} m^{-2}	Radiant intensity per unit projected area
Radiance factor	β_e	–	Ratio of radiance to that of perfect diffuser
Radiant efficiency	η_e	–	Ratio of radiant flux emitted to power consumed

Table 7.4 Principal terms and units used in the measurement of luminous energy

Name	Symbol	Unit	Definition
Luminous flux	Φ	lumen (lm)	Radiant flux weighted by 683 times V_λ function
Luminous exitance	M	lux (= lm m^{-2})	Luminous flux leaving surface per unit area
Luminous intensity	I	candela (cd)	Luminous flux per unit solid angle
Luminance	L	cd m^{-2}	Luminous intensity per unit projected area
Luminance factor	β	–	Ratio of luminance to that of perfect diffuser
Luminous efficiency	η	–	Ratio of luminous flux to radiant flux
Luminous efficacy	K	lm W^{-1}	Ratio of luminous flux emitted to power consumed

conditions. Any of the terms in Tables 7.3 and 7.4 may be prefixed by 'Spectral ...', the revised term denoting the quantity per unit wavelength, or its value for monochromatic light of a given wavelength. All these units are physical and can be measured instrumentally. They should not be confused with the perceptual units of brightness and lightness, which are attributes of visual sensation in a human observer and must be determined through psychophysical experiment.

7.5.2 Spectroradiometers

The instrument used for precise measurement of the distribution of radiant flux from a display as a function of wavelength is the telespectroradiometer (TSR). A block diagram of a typical 'slow-scan' TSR is shown in Figure 7.35. Light from a small area of the display is collected through an aperture by the telescope and conveyed by a fibre optic link to the monochromator, which disperses the incoming light by means of a prism or diffraction grating and samples a small wavelength interval (typically 5 nm) through a narrow slit. The radiant flux in this interval is detected (i.e. converted into electrical energy) by a photomultiplier tube, then amplified as an analogue voltage, converted into a digital value and stored. Software in the host computer controls the process, stepping the monochromator sequentially through the full range of wavelengths. Adjustments can be made to the lens and sampling aperture in the telescope (10 min to 2° arc) and to the wavelength interval (1 to 10 nm). Such instruments can achieve high levels of accuracy, but may take several minutes to complete one scan across all wavelengths.

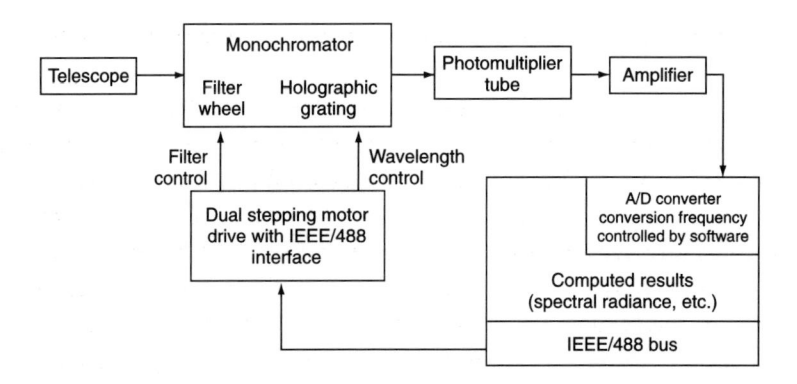

Figure 7.35 Schematic diagram of a scanning telespectroradiometer (reproduced by permission of Bentham Instruments)

Another type of spectroradiometer is the 'fast-scan' variety in which, instead of a monochromator, the spectral dispersion is performed by a polychromator which simultaneously measures the flux at all wavelength intervals. This can be achieved by directing the light output of the dispersive element (prism or diffraction grating) on to a silicon detector array. An array of 256 elements spread uniformly over the visible spectrum, for example, can give a wavelength resolution of about 1.5 nm. Such instruments can complete a measurement in a fraction of a second, but have lower inherent accuracy because of the more limited dynamic range (and hence poorer signal-to-noise ratio) of the silicon detectors compared with photomultiplier tubes. Also, because all the wavelengths are measured simultaneously, there is no possibility of adjusting the sensitivity to suit different wavelengths, which again limits dynamic range when spectral peaks are present.

In order to calibrate the spectroradiometer to a known radiance standard, a reference source is always required. This typically takes the form of a tungsten lamp inside an integrating sphere coated on the inside with a uniform white reflective material such as barium sulphate, with a small aperture through which the diffuse radiation is emitted. Such lamps are available as transfer standards from national agencies such as the National Physical Laboratory (NPL), where they are calibrated against the national primary standards, and are supplied with data for the radiant flux at each wavelength throughout the operating range. By measuring the standard lamp and comparing the results with the reference data, the control software for the TSR can determine a correction factor at each wavelength to convert the measured signals into accurate radiance values. The procedure should be repeated from time to time to compensate for drift or changes in the sensitivity of the detector and amplifier. Wavelength calibration can be checked by scanning a light source that has a line emission spectrum, such as a mercury discharge lamp, and comparing the measured wavelengths of the lines against their known reference values.

For effective use of a spectroradiometer in measuring display radiance, several factors must be taken into account [51]. The overall wavelength range must be sufficient to include all radiant energy from the display primaries visible to the eye. The sampling bandwidth must be narrow enough to resolve any spikes in the SPD of the display (a particular problem for red phosphors, as evident from Figure 7.8), but wide enough to maintain a satisfactory signal-to-noise ratio. The dynamic range of the detector must be sufficient to handle the variations in light level, which can be up to 1000 : 1 in a raster-scanned CRT over one refresh cycle. Stray light leakage through the monochromator must be eliminated if possible. Where the light from the display is polarised, as is the case with most backlit LCDs, the average of two perpendicular measurements should be taken [52].

7.5.3 Tricolorimeters

Instead of scanning a large number of narrow wavelength intervals, it is often cheaper and more convenient to measure the radiant flux from a display through three broadband filters. If a radiometer could be equipped with filters having exactly the spectral transmittances of the CIE tristimulus functions $\bar{x}_\lambda$, $\bar{y}_\lambda$ and $\bar{z}_\lambda$ then the tristimulus values X, Y and Z of the displayed colour could be measured directly. In practice it is difficult (and therefore expensive) to match the CIE functions exactly, especially the double lobe of the $\bar{x}_\lambda$ function. Some high-performance instruments, such as those manufactured by LMT, come close to the ideal by using a mosaic of up to eight filters for each of four detectors. Others obtain good results by having two detectors with separate filters to measure $\bar{x}_\lambda$, one in the 380–500 nm portion of the spectrum and the other in the 500–780 nm portion. Cheaper instruments use just three filters with reasonably broad overlapping spectral sensitivity distributions and peaks in the red, green and blue regions of the spectrum (at around 610, 550 and 450 nm respectively) and add a fraction of $\bar{z}_\lambda$ to approximate the short-wavelength lobe of $\bar{x}_\lambda$ (Figure 7.36). The measured R, G and B values are then converted by (3 × 3) matrix multiplication to obtain the X, Y and Z values. A different matrix may be used for each white point to optimise the accuracy of the results.

Tricolorimeters give good results when measuring light from sources that have smooth broad-band spectra, such as daylight, and they can also give good results for any source when the filter spectral characteristics are a close match to the colour-matching functions [53]. Because the filter spectral sensitivities generally differ from true colour-matching functions ('instrumental metamerism'), best results are obtained only when the calibrating source has an SPD very similar to that of the source being

418 COLOUR IN VISUAL DISPLAYS

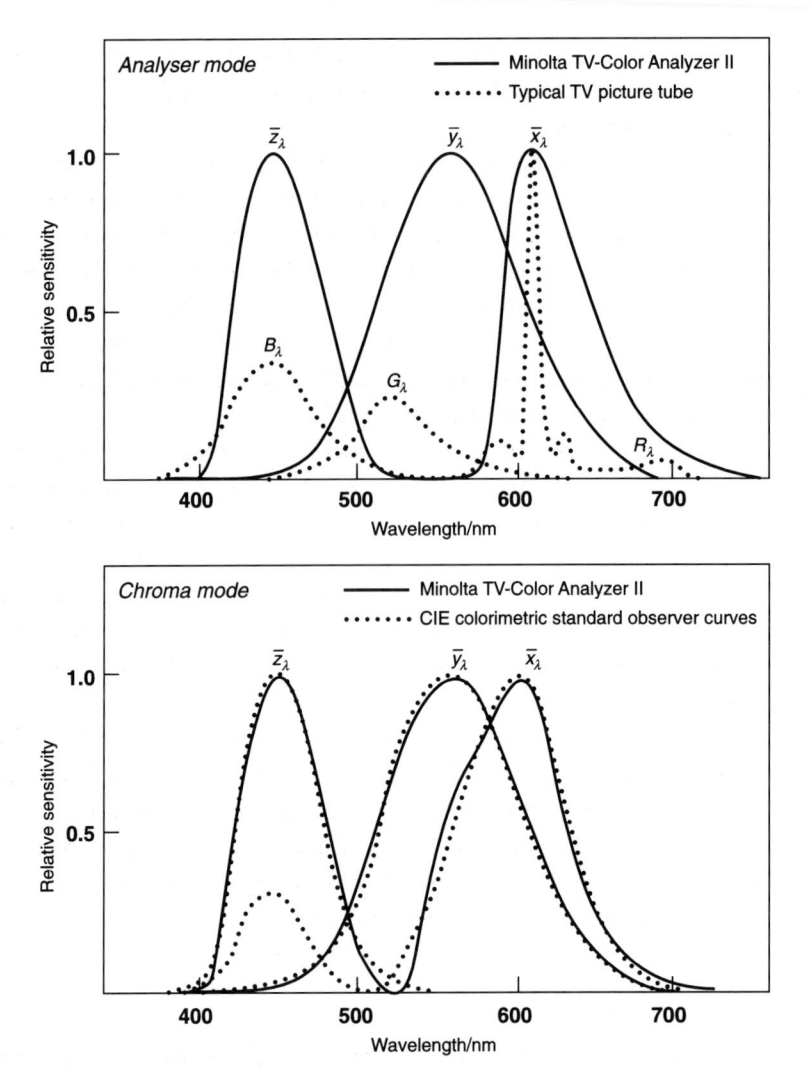

Figure 7.36 Typical RGB set of filters in a tristimulus colorimeter; the instrument can be used in analyser mode (optimised for phosphor SPDs) or chroma mode (matched to the tristimulus functions) (reproduced by permission of Minolta (UK) Ltd)

measured. Unfortunately, the emission spectra of displays are not ideal candidates for such measurement, as both CRT phosphors and LCD backlights exhibit narrow high-energy spikes. The value of tricolorimeters in measuring displays, therefore, lies not in their absolute accuracy, which is rarely better than ±0.01 in xy chromaticity (and may be much worse for a typical P22 red phosphor), but in their stability, portability and relatively low cost. Such instruments can be very useful for examining small colour differences or for verifying a display's stability, and they are commonly used in production environments to check whether displays of near-identical SPD differ significantly from the reference standard.

For both spectroradiometers and tricolorimeters the following factors need to be considered to ensure the validity and accuracy of the display colour measurements [51].

(a) The spatial properties of displays, such as variation of luminance over the surface of a CRT and the fine line-and-dot structure of the light-emitting pixels, require careful consideration. The measurement area needs to be large enough to average out the light over a substantial number of neighbouring pixels, typically at least 10 mm diameter on the screen surface, and the displayed colour field should completely fill this area and be uniform within it.

(b) Variations in colour with viewing angle, which occur with most LCD displays and also with CRT screens coated with wavelength-dependent interference filters, mean that the measurement geometry must be considered. A complete characterisation may involve measuring colour at a large number of angles throughout the front-of-screen hemisphere, either by using a goniospectroradiometer, which mechanically scans all the angular viewpoints, or by means of a conoscopic device, which maps all angles within a viewing cone on to the image plane of a CCD sensor [54].

(c) The temporal properties of displays, such as the repeated short pulses of light emitted by a raster-scanned CRT, are crucial but often overlooked. The integration time of the measurement must be long enough to collect and average a large number of pulses, normally at least 1 s, and also should be synchronised to the refresh rate (vertical sweep frequency) so that every measurement starts at the same point and counts the same number of pulses.

7.5.4 Display calibration

As displays are being used in new and more exacting applications, the need for accurate set-up and control of colour is becoming increasingly important. Whereas in a television display large changes in colour balance and tonal gradation are usually tolerable, in 'soft proof' preview of images for reproduction in print every nuance of tone and colour variation can affect the fidelity of the display. Precise calibration is a system problem affecting not only the colour and uniformity of the display screen itself, but also the accuracy of measuring equipment, the colour temperature selected for the white point, the design of the video drive electronics and the type of calibration method employed.

The first step in calibrating a display, before commencing colorimetric correction, is to set its operating state to some predetermined standard. For a CRT this can involve degaussing (or demagnetising) and adjusting the colour purity, convergence, focus, DC-offset and gain controls [55]. Sometimes such adjustments may interact with one another (convergence and purity, for example), requiring lengthy iterative tweaking in order to achieve the best compromise.

For CRT displays the single most critical factor affecting colour is the effect of the phosphor chromaticity coordinates on the white point. The typical tolerance band for phosphors is ±0.02 xy with the primary luminance levels held constant. With the combined tolerance variations of all three phosphors, the chromaticity coordinates of white can vary by as much as ±0.05 [56].

For verifying the white point setting of a display, manufacturers may use a white colour standard CRT as a reference device. Various methods are available for fabricating such a device, including the mixing of phosphors in a monochrome CRT to produce the required SPD, or adjusting the relative weights of the three standard RGB phosphors to achieve the correct white balance. In operation the special white CRT can be employed either as a calibration standard for colorimeters used on the production line or as a visual reference for side-by-side comparison. For two adjacent white displays of the same luminance, colour differences can be perceived with a change of as little as 0.003 in either of the xy chromaticity coordinates [57], equivalent to about 3 units in CIELUV space (i.e. $3\Delta E^*_{uv}$).

One way to reduce the white point variation from one display to another is to measure the primary coordinates and primary luminances of each display *in situ* and to adjust their relative luminances to accommodate the differences from specification of the phosphor. This approach, which can accurately achieve a simulated white point setting, is the basis of the calibration devices that are now being offered as accessories for high-quality computer graphic displays. In some cases, such as the Barco Calibrator, the calibration process is controlled by a microprocessor within the monitor itself (Figure 7.37). An optical sensing head containing a photodiode is attached by a rubber suction ring to the screen surface and the internal processor initiates a test sequence in which it displays a series of colour patches and reads back the measured luminance of each from the sensor. The data is used to optimise the parameters (corrector signals) of the display drive electronics in order to achieve the specified luminance and colour temperature for the white point.

This design has the advantage of being self-contained, so that no special control software is required in the host computer. The monitor can therefore retain its individual characteristics, such as gamma, gain and offset values, within its own memory; it can compensate very effectively for phosphor ageing (Figure 7.7) and can also achieve good reproducibility of a specified white. In a study the reproducibility performance of three Barco Calibrator monitors using a test set of 729 colours was found to be better than 0.005 $\Delta x \Delta y$ (mean) or less than $4\Delta E^*_{uv}$ [38]. Being self-contained is also a

MEASUREMENT AND CALIBRATION   421

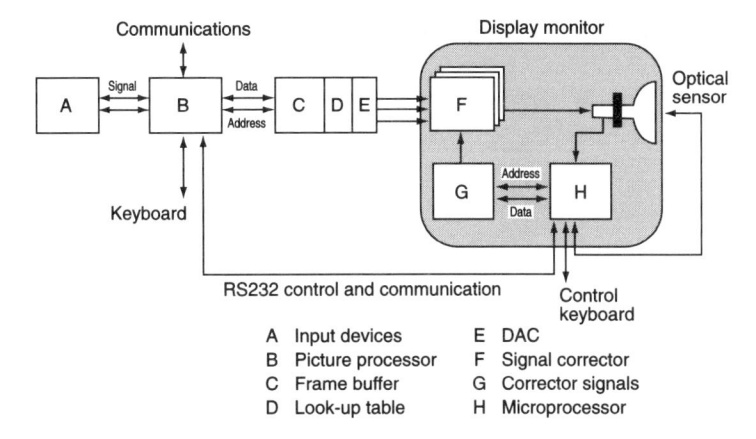

Figure 7.37 Schematic diagram of the Barco Calibrator monitor showing self-calibrating display (reproduced by permission of Barco Graphics)

A Input devices   E DAC
B Picture processor   F Signal corrector
C Frame buffer   G Corrector signals
D Look-up table   H Microprocessor

limitation of the design, however, because it is a closed system and cannot take into account the behaviour of the graphics controller card, which in practice may have significant errors in gain, offset and linearity. Another limitation is that the optical sensor contains only one photodiode, and is therefore 'colour blind'. Its luminance measurements can only be interpreted with knowledge of the SPD of the phosphors in the CRT, and the device is not easily transferable to any other type of display.

An alternative approach to CRT self-calibration is to monitor regularly the beam current of each of the three electron guns, based on the premise that the light emitted by each phosphor primary is linearly proportional to its beam current (Eqn 7.1). By feeding back digital measurements to the host computer, the software driver can automatically set white and black points and compensate for inaccuracies in the signals from the analogue video drivers. This degree of control of the display by the host computer provides desktop computer manufacturers with an opportunity to 'close the loop' around many of the variables that contribute to uncertainty of the white point. Because it is measuring only beam current and not the actual light output, however, this method cannot compensate for beam landing errors, phosphor chromaticity errors or ageing of the phosphors and glass [58].

Improved calibration performance and versatility can be achieved by direct luminance measurement, by including the graphics controller within the loop and driving the calibration process from software executing within the host computer. This reduces the costs of added components within the monitor, provides a more powerful processing environment for the calibration algorithms and allows the user to control the process through an interactive graphic user interface. A further advantage is that the workspace parameters seen by the application software storing an RGB image in the

422 COLOUR IN VISUAL DISPLAYS

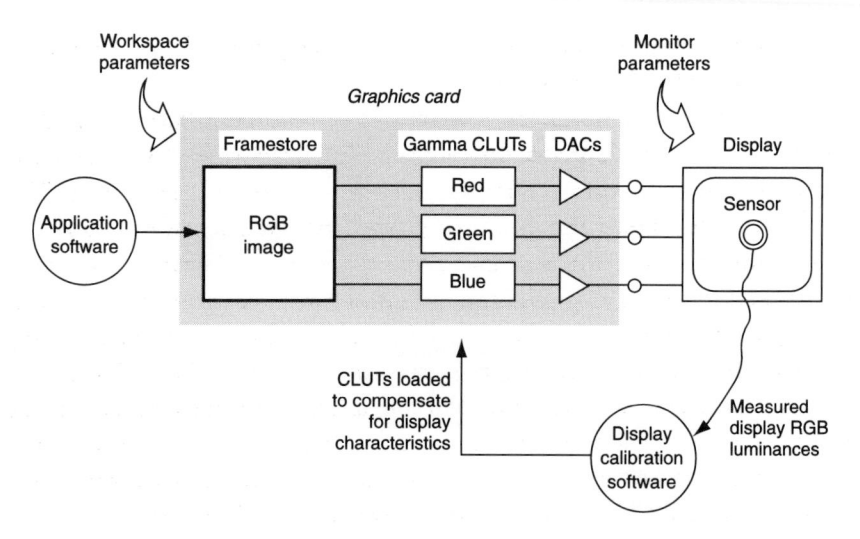

Figure 7.38 Display calibration: a display calibrator compensates the actual monitor parameters to provide idealised workspace parameters to the application software

frame store can be separated from the physical monitor parameters of the monitor itself (Figure 7.38). Thus a computer graphics application could always assume an idealised workspace gamma of 1.80 even though the actual monitor gamma (which varies from one monitor to another) might be about 2.3. The compensation is effected by suitable loading of the three colour look-up tables (CLUTs) in the graphics controller, based on analysis of measurements of a series of displayed colour patches.

The manufacturers of high-quality CRT monitors for computer imaging and desktop publishing are now beginning to provide display calibrator devices, which are in effect low-cost tricolorimeters, incorporating three photodiodes covered by red, green and blue filters. Such devices measure the red, green and blue components for each colour patch displayed and relay the data back to the control program in the host computer, which can calculate the chromaticities and hence correct for variations in the colours of the phosphor primaries. By measuring a number (typically 10–16) of achromatic grey levels the software can ensure that the colour temperature specified by the user is maintained throughout the full tonal range from white to black, rather than just at the white point. Recent studies have shown that a well-designed tricolorimeter can provide more accurate and consistent calibration of CRT displays than is obtained with an arbitrary spectroradiometer [59]. By using the device also as a photometer for measuring illuminance in a print viewing cabinet, the luminance level of the display can be set to an equal value to permit optimum side-by-side comparison of image colour. By using the device as a photometer for ambient illuminance, moreover, the effects of ambient light on the displayed image can be compensated by increasing contrast at the dark end of the tonal range.

This trend should continue over the next few years, with manufacturers embedding more intelligence within displays and also supporting standards that allow the accurate characterisation of the display and transfer of colorimetric image data. The colorimetric and photometric signature or 'personality' of the display will be held in memory within the display 'head' or associated controller, including the chromaticity coordinates and the voltage-to-luminance transfer functions and maximum luminance for each primary channel, plus the white reference. In-built calibration functions, based on measurements from an integral sensor, will permit closed-loop calibration of white point, maximum luminance and voltage-to-luminance transfer functions, as well as compensation for variations in luminance and colour purity across the full screen area. The colorimetric display signature will be provided to colour-management systems implemented within the operating system software of the host computer (such as Apple's ColorSync), for storage in data structures known as 'device profiles'. These will then be used automatically by the colour-management algorithms in reconstructing colour images to render them with correct colour appearance on the display [60].

REFERENCES

1. G H Hunt in *Display engineering – conditioning, technologies, applications*, Ed. D Bosman (Amsterdam: Elsevier, 1989) 93.
2. T R H Wheeler and M G Clark in *Color in electronic displays*, Ed. H Widdel and D L Post (New York: Plenum Press, 1992) 221.
3. J E Farrell in *Vision and visual dysfunction*, Vol. 15. The man–machine interface, Ed. J A J Roufs (Basingstoke, UK: Macmillan, 1991).
4. R M Merrifield in *Color and the computer*, Ed. H J Durrett (Orlando, FL: Academic Press, 1986).
5. L D Silverstein and J S Lepkowski, *SID Conf. Digest*, **17** (1986) 416.
6. S Sherr, *Electronic displays*, 2nd Edn (New York: John Wiley, 1993).
7. J Schug, I Koehler and R Loschek, Proc. 13th Int. Display Res. Conf. 'EuroDisplay 93' (1993) 345.
8. European Broadcasting Union Tech. Report 3273-E (Oct 1993).
9. L D Silverstein, 'Color in CRT and LCD displays', SID Conf. seminar notes (1994).
10. A J Hughes in *Display engineering – conditioning, technologies, applications*, Ed. D Bosman (Amsterdam: Elsevier, 1989) 141.
11. T Scheffer and J Nehring, 'Supertwisted nematic (STN) LCDs', SID Conf. seminar M-7 notes (1993).
12. M G Clark in *Color in electronic displays*, Ed. H Widdel and D L Post (New York: Plenum Press, 1992) 257.
13. N Kimura *et al.*, *SID Conf. Digest*, **19** (1988) 49.
14. I Lewin, 'Principles of LCD backlighting', SID Conf. seminar M-9 notes (1993).
15. H L Ong, Proc. 11th Int. Display Res. Conf. (1991) 1.
16. T Ueno *et al.*, *Proc. SID*, **29** (1988) 105.

17. K Tsuda, *Displays*, **14** (2) (Apr 1993) 115.
18. F E Gomer et al., *SID Conf. Digest*, **19** (1988) 435.
19. K D Ruelberg and S Zander, *Displays*, **14** (3) (July 1993) 166.
20. A Conner et al., *SID Conf. Digest*, **24** (1993) 577.
21. J Biles, *SID Conf. Digest*, **25** (1994) 403.
22. A Conner, Proc. 10th Int. Display Res. Conf. 'EuroDisplay 90' (1993) 362.
23. J J Vos, *Col. Res. Appl.*, **3** (1978) 125.
24. P K Kaiser, CIE Publication No. 86 (1990).
25. R W G Hunt, *Col. Res. Appl.*, **19** (1994) 23.
26. C J Hawkyard, *Rev. Prog. Col.*, **21** (1991) 43.
27. W N Sproson, *Colour science in television and display systems* (Bristol: Adam Hilger, 1983) 27.
28. P K Robertson, *IEEE Comp. Graphics Appl.*, 8 (5) (Sept 1988) 50.
29. G Derefeldt and C Hedin, *Displays*, 10 (3) (July 1989) 134.
30. I Overington, *Computer vision* (Amsterdam: Elsevier, 1992) 285.
31. K A Miller, 'Display measurement technology', SID Conf. seminar 6 notes (1987).
32. A J Johnson and M Scott-Taggart, *Guidelines for choosing the correct viewing conditions for colour publishing* (Leatherhead, UK: PIRA Press, 1993).
33. E Jennings, Proc. TAGA Conf. (1993).
34. W B Cowan, *Comp. Graphics*, **17** (1983) 315.
35. D L Post and C S Calhoun, *Col. Res. Appl.*, **14** (1989) 172.
36. R S Berns, R J Motta and M E Gorzynski, *Col. Res. Appl.*, **18** (1993) 299.
37. J D Mollon and M R Baker, Colour Vision Deficiencies XII (1995) 423.
38. M R Luo et al., *Proc. CIE*, **22** (1991) 97.
39. W B Cowan and N Rowell, *Col. Res. Appl.* (Supplement), **11** (1986) S35.
40. D H Brainard, *Col. Res. Appl.*, **14** (1989) 23.
41. A J Johnson, Proc. TAGA Conf. (1992)
42. J Albers, *Interaction of color* (New Haven: Yale University Press, 1963) 18.
43. L W MacDonald, M R Luo and S A R Scrivener, *J. Phot. Sci.*, **38** (4) (1990) 177.
44. R W G Hunt, *Measuring colour*, 2nd Edn (Chichester, UK: Ellis Horwood, 1991) 57.
45. D Travis, *Effective color displays: theory and practice* (London: Academic Press, 1991) 89.
46. D L Post and C J Lloyd, *Displays*, **15** (1) (Jan 1994) 39.
47. M R Luo, *Col. Res. Appl.*, **18** (1993) 98.
48. P J Alessi, *Col. Res. Appl.*, **19** (1994) 48.
49. L W MacDonald, Proc. IS&T/SID Color Imaging Conf. Transforms & Transportability of Color (1993) 193.
50. R McCluney, *Introduction to radiometry and photometry* (Norwood, MA: Artech House, 1994).
51. A R Hanson, *The colorimetry of visual displays* (Teddington, UK: National Physical Laboratory, 1994).
52. S Pefferkorn et al., Proc. 13th Int. Display Res. Conf. 'EuroDisplay 93' (1993) 443.
53. R S Berns, M E Gorzynski and R J Motta, *Col. Res. Appl.*, **18** (1993) 315.
54. T Leroux, Proc. 13th Int. Display Res. Conf. 'EuroDisplay 93' (1993) 447.
55. D L Post in *Color in electronic displays*, Ed. H Widdel and D L Post (New York: Plenum Press, 1992) 299.
56. L Virgin et al., *SID Conf. Digest*, **17** (1986) 334.
57. R L Donofrio, *Displays*, **14** (4) (Oct 1993) 216.
58. R Cappels, *SID Conf. Digest*, **25** (1994) 15.
59. M D Fairchild, *SID Conf. Digest*, **25** (1994) 865.
60. L W McDonald, *Displays*, **16** (4) (May 1996) 203.

RECOMMENDED FURTHER READING

Color and the computer, Ed. H J Durrett (Boston: Academic Press, 1987).

R W G Hunt, *The reproduction of colour*, 5th Edn (Kingston-upon-Thames UK: Fountain Press, 1995).

R Jackson, L MacDonald and K Freeman, *Computer generated colour* (Chichester, UK: John Wiley, 1994).

Color in electronic displays, Ed. H Widdel and D L Post, Defense Research Series Vol. 3 (New York: Plenum Press, 1992).

Display systems: technology and applications, Ed. L MacDonald and A Lowe (Chichester UK: John Wiley, 1996).

EQUIPMENT SUPPLIERS

Barco UK Ltd, 50 Suttons Park Avenue, Reading, Berkshire RG6 1AZ, UK.

Bentham Instruments Ltd, 2 Boulton Road, Reading, Berkshire RG2 0NH, UK.

Minolta (UK) Ltd, Rooksley Park, Precedent Drive, Milton Keynes, Buckinghamshire MK13 8HF, UK.

CHAPTER 8

How we see colour

Adrian R Hill

8.1 THE EYE AND THE BRAIN

The act of seeing is a perceptual process and takes place in the brain. Therefore any theory of colour vision, if it is to be an adequate description of how we see colour, must account for the way in which nerve impulses originating in the retina of the eye are subsequently analysed by the brain. The role of the eye in the visual process is that of a transducer; it provides the means of converting electromagnetic energy in the form of light into nerve impulses. Following an initial coding within the layers of neurones (nerve cells) in the retina, the nerve impulses are transmitted along the optic nerve and relayed, via nerve fibres situated near the base of the brain, to terminate in a part of the brain at the back of the head known as the occipital cortex. Somewhere between the photoreceptors in the retina and the termination of the nerve signal in the higher centres of the brain, information about the shape, depth, movement, lightness and colour of the external world are coded and decoded to provide the sense of sight.

From a series of elegant experiments on the mixing of coloured lights, James Clerk Maxwell made the profound deduction in 1872 that 'the science of colour must ... be regarded as essentially a mental science'. Since those early and almost forgotten experiments of Maxwell much has been discovered about the way in which information concerning the stimulus of light is coded by the visual system, but the synthesis and interpretation of light stimuli into a meaningful perception still remains something of an enigma. In the last 40 years or so considerable advances have been made in our understanding of how the nerve signal is processed between the retina and the brain. These studies have embraced the fields of psychology, photochemistry, electrophysiology, anatomy and the clinical vision sciences. Of necessity, most of the experiments involving an investigation of the electrical activity in the neurones of the visual system have been conducted on animals. But there is now sufficient knowledge from studies of comparative anatomy and physiology on primates to be able to make inferences, with reasonable confidence, about the human visual process. By drawing on this research, as well as on the results of experiments with human observers, it is possible to trace the route taken by nerve signals from the retina to the brain. This is the approach taken in this chapter. In order to provide a framework for the currently accepted model of the colour vision process we start with a description of the structure and function of the eye as an optical instrument.

8.2 STRUCTURE AND FUNCTION OF THE EYE

The essential feature of the eye of vertebrates is its ability to form an image; by contrast, the eyes of many lower animals are able only to detect the presence of light. A horizontal section of the human eye is shown in Figure 8.1. In common with all vertebrate eyes, its major optical component is the cornea. This tough, transparent membrane is only about 0.5 mm thick. It forms part of the outer coat of the globe of the eye and provides approximately two-thirds of the eye's total optical power. (The total power of the resting eye, when relaxed for distant vision, is about +60 dioptres. The *dioptric power* of a lens is the reciprocal of its focal length when that focal length is expressed in metres; the dioptric power of a converging lens is conventionally given a

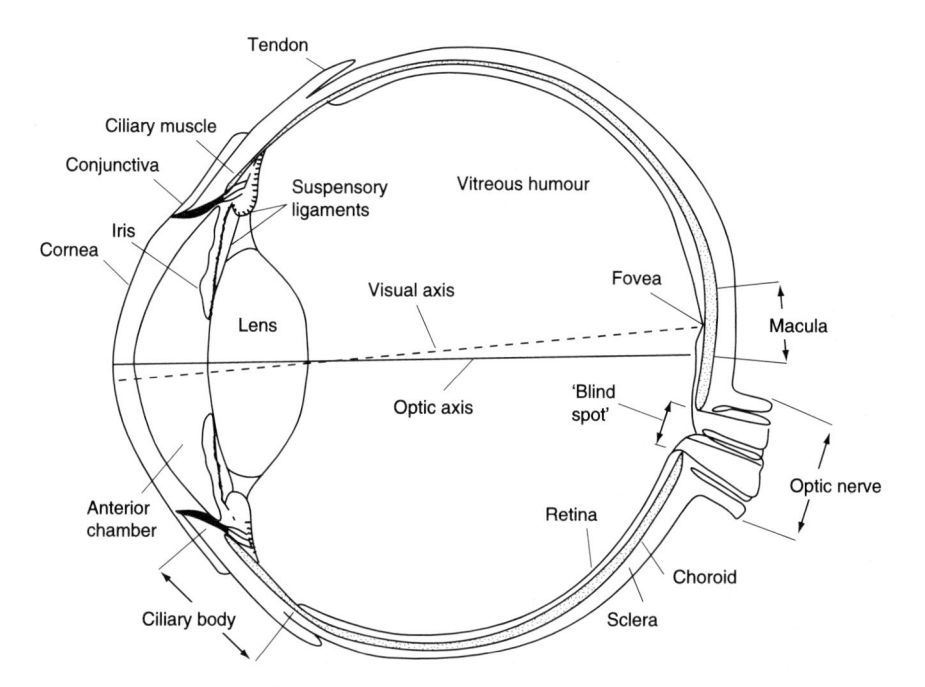

Figure 8.1 Horizontal section of the human eye

positive sign and that of a diverging lens a negative sign.) Situated just behind the cornea is a filled chamber containing a clear fluid called the aqueous humour, within which is suspended the crystalline lens and iris. The crystalline lens accounts for almost all the remaining one-third of the eye's optical power. In the young adult eye, constriction of a circular sphincter muscle just behind the iris (the ciliary muscle) can make the lens become more bulbous, thereby increasing its power by up to 50%; this ability to change its power is known as *accommodation,* and is the process by which objects at varying distances from the eye can be focused on the retina. With advancing age the lens increases in thickness and becomes less able to change its shape, with the consequence that its ability to accommodate for the correct focus of near objects decreases; this is why many people require spectacles to aid near vision from the fourth or fifth decade of life.

The amount of light entering the eye is controlled by the iris, a membrane which has a circular aperture (the pupil) and is positioned just in front of the lens. In darkness the pupil expands to its greatest diameter of about 8 mm, while in very bright light it contracts to about 1.5 mm. Since the focal length of the eye is about 17 mm the effective aperture of the pupil can vary between f2.5 and f13, giving a range of nearly 30 : 1 in the amount of light admitted to the retina. (The *f-number* of a lens, often encountered in photography, is the ratio of its focal length to its diameter; for example, f13 means the focal length is 13 times the lens diameter.)

Immediately behind the crystalline lens is the vitreous humour, a clear gel-like mass that occupies about 60% of the internal volume of the eye. While the aqueous humour is the source of nutrition for the cornea and lens, the function of the vitreous gel appears to be primarily to keep the optical path clear of floating cellular debris and to minimise the likelihood of a retinal detachment.

All these pre-retinal structures are designed to provide a clearly focused image on the retina. This is a layer of neural elements about 0.1 mm thick, lining the inside of the globe of the eye and posterior to the lens. The outermost layer of the retina (i.e. the layer closest to the outside of the globe) contains the photosensitive receptors known as 'rods' and 'cones', so called because of their morphological appearance under the microscope. In functional terms, rods are responsible for *scotopic vision,* that is, vision at low levels of illumination (less than 0.1 lux). At these illuminance levels the cones do not function. At higher illuminances (above about 500 lux) the rods saturate, leaving the cones to provide us with both colour vision and the ability to see fine detail; this is referred to as *photopic vision.* Between these two illuminances vision involves a mixture of rod and cone activity and is known as *mesopic vision.*

The rods and cones vary in their density distribution across the retina (Figure 8.2) and there is no firm evidence that they are neurologically interconnected except perhaps between the rods and short-wavelength-sensitive cones. While the density of

STRUCTURE AND FUNCTION OF THE EYE   429

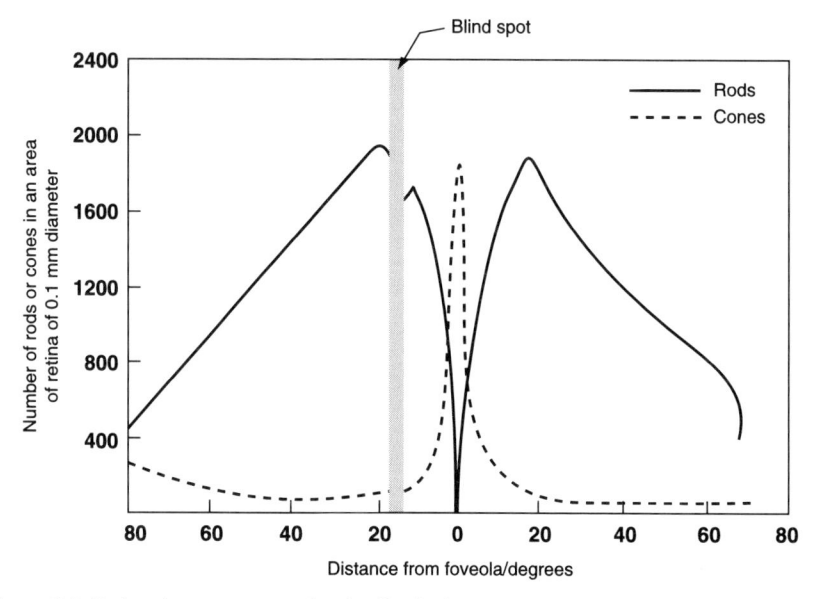

Figure 8.2 Rod and cone receptor density distribution across the retina

the receptor types varies across the retina, they are not arranged in a systematic pattern. Only cones are present at the point where the primary visual axis of the eye intersects the retina, and here they occur at their maximum density. This is a small approximately circular area subtending about 1° in diameter and is called the foveola. It corresponds to the point of maximum resolution in the retina where it is possible to resolve detail of the order of 1 min of arc (i.e. 1/60th of a degree). This resolving power falls off rapidly with increasing distance from the foveola (Figure 8.3), an effect that is paralleled by the presence of an increasing number of neural interconnections between receptors, giving a grouped response in the more peripheral parts of the retina. Because such changes in function can be loosely identified with underlying retinal structures, a series of central retinal zones have acquired special significance in fundamental studies of vision, especially colour perception. These zones are concentrically centred on the intersection of the visual axis with the retina (Figure 8.1). They are the foveola, fovea and macula, which respectively have an angular subtense of 1°, 5° and 20°. Although these dimensions are widely accepted, the boundaries between the zones are not abrupt but are rather smooth transitions of underlying neural structure and function. A typical size of the field of view for many colour-matching or colour-discrimination experiments is 2°; this is contained well within the foveal zone and includes a very high density of cones.

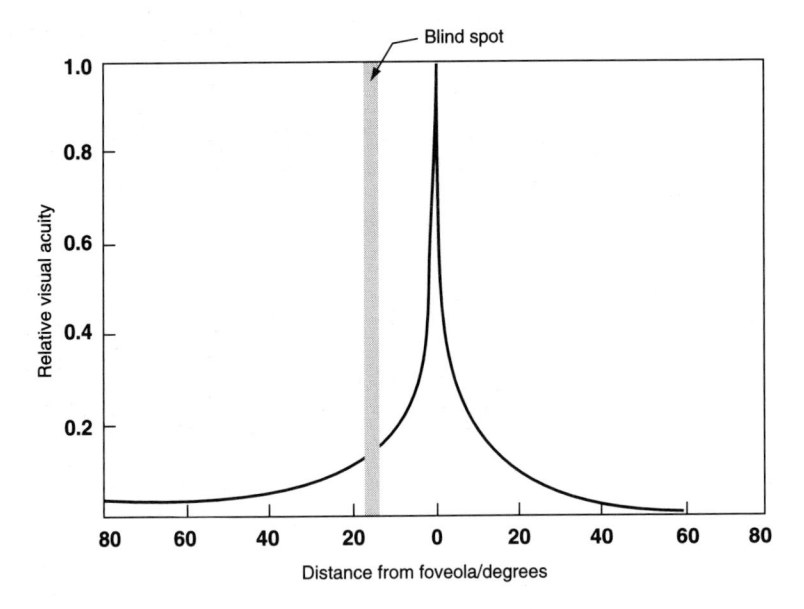

Figure 8.3 Resolving power of the eye as a function of eccentricity from the foveola

Figure 8.4 shows a simplified schematic diagram of a transverse section through the retina illustrating its various structural elements. Although the retina is a multilayered tissue, its layers can be divided into two major parts. The outer section, the neuro-epithelium, contains the sensory receptors (rods and cones) upon which an image needs to be focused. The innermost layer of the retina, through which light must travel to reach the photoreceptors, comprises an extremely complex arrangement of nervous tissue. This tissue is similar in both structure and function to the central nervous system, of which it is in fact an outlying part.

Metabolism of the retina is maintained by the supply of oxygen and nutrients from the highly vascular underlying choroid. Enclosing all these tissues is the outer coat of the eye, the sclera. It is a thin, but strong, layer of collagen-fibre connective tissue which maintains the globe in a roughly spherical shape of approximately 25 mm in diameter.

8.3 THE MECHANISM OF VISION

8.3.1 Receptor photopigments

The photosensitive pigments in the retina are contained within the outer segments of the rods and cones, and appear as a series of vertically stacked platelets when viewed

THE MECHANISM OF VISION 431

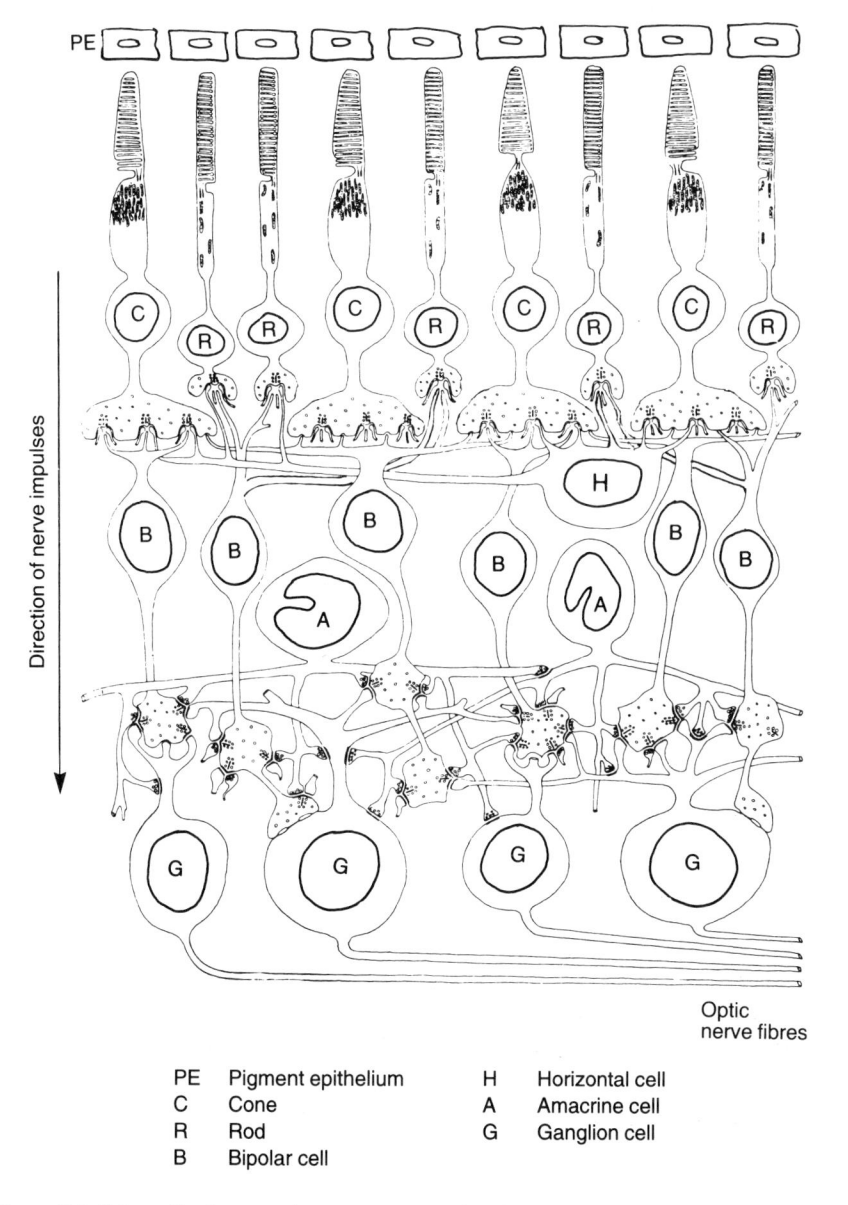

PE	Pigment epithelium	H	Horizontal cell
C	Cone	A	Amacrine cell
R	Rod	G	Ganglion cell
B	Bipolar cell		

Figure 8.4 Schematic diagram of a transverse section through the retina showing the different nerve cells and their interconnections; light must pass through all the retinal layers, except the pigment epithelium, to reach and stimulate the rods and cones

under the microscope (Figure 8.5). Four different photopigments are present in the normal human retina, one contained within the rods and three others distributed amongst the population of cones. All four photopigments are believed to consist of a

432 HOW WE SEE COLOUR

Cone **Rod**

Direction of nerve impulse ↓

Outer Segment

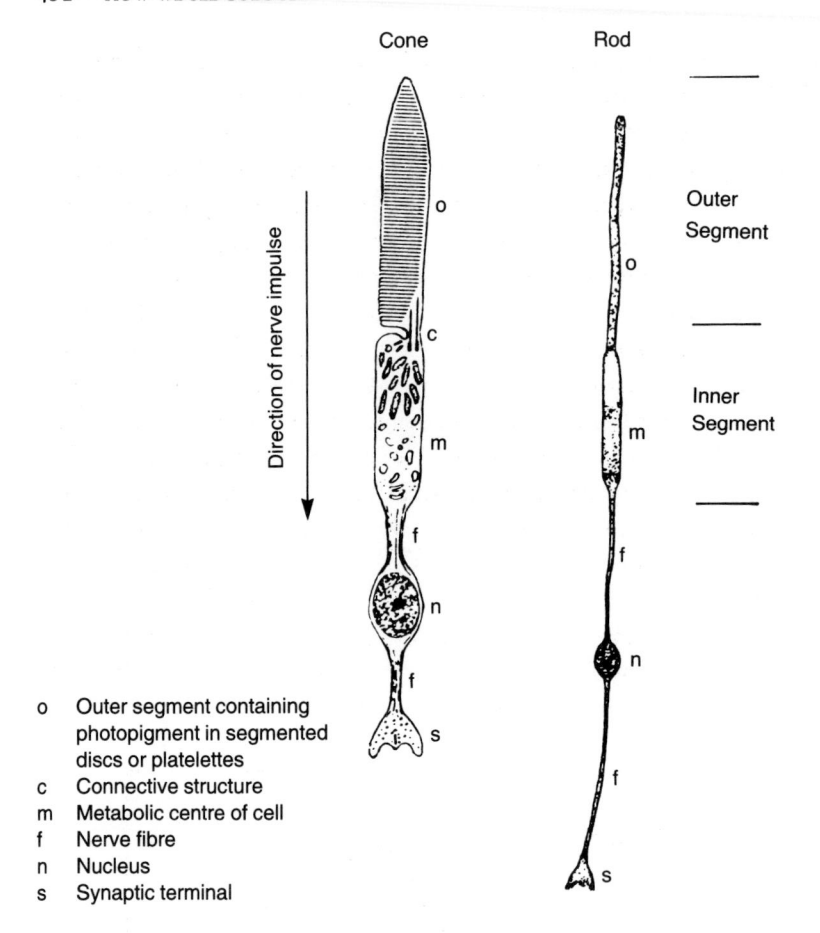

Inner Segment

o   Outer segment containing photopigment in segmented discs or platelettes
c   Connective structure
m   Metabolic centre of cell
f   Nerve fibre
n   Nucleus
s   Synaptic terminal

Figure 8.5 Schematic diagram of the human retinal rod and cone receptors

protein molecule (the opsin) to which is bound a derivative of vitamin A_1, the isomer 11-*cis*-retinal (called the chromophoric group). This chromophoric group is common to all four photopigments, but its spectral absorption properties differ according to the protein to which it is bound.

The visual pigment found in the outer segments of the rods is known as rhodopsin. It absorbs light mostly in the middle wavelengths of the visible spectrum (at a wavelength of about 496 nm) and transmits light at both the short- and long-wavelength ends of the spectrum. When light is absorbed by rhodopsin the bonds in the retinal group are rearranged. The molecule changes its shape, untwisting and straightening out in a procedure called a *cis–trans* isomerisation (the terms *cis* and *trans* refer to the two shapes of the molecule). The untwisting and straightening of the isomerising molecule cause it to break loose from the protein opsin. This is the process by which rhodopsin is bleached; a similar process is believed to occur on isomerisation of the visual pigments in the outer segments of the cones. Before being bleached, rhodopsin appears purple by transmitted light and hence is often called 'visual purple'.

The first unequivocal evidence that there are three different photopigments among the population of cones in the human retina was obtained in 1964 [1,2]. These early studies involved microspectrophotometric measurements through the full thickness of an intact retina (i.e. along the length of the cone). More recently, in 1983, absorption spectra were recorded transversely across the outer segments of isolated human cones from the eyes of seven individuals by using a dual-beam Liebman microspectrophotometer [3]. (The eyes had been removed surgically for clinical reasons.) This approach provided absorption spectra of the cone pigments independently of the absorption properties of the surrounding retinal tissues. Together these studies demonstrated conclusively that the three cone pigments and that from the rods all have maximum absorption in different parts of the visible spectrum, with a considerable spectral overlap (Figure 8.6).

The peak sensitivities of the cones lie in the 'blue' (c. 420 nm), 'green' (c. 530 nm) and 'yellow-green' (c. 560 nm) parts of the spectrum, while that for the rods lies at 496 nm. When the absorption spectra are plotted as a function of wavenumber (the

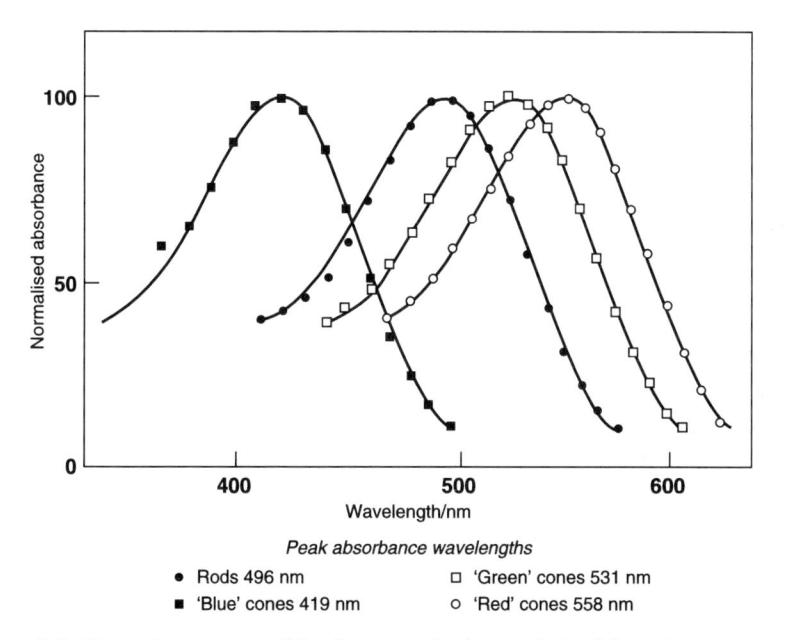

Figure 8.6 Absorption spectra of the four receptor types, derived from transverse microspectrophotometric measurements across the outer segments of the receptors [3]

reciprocal of wavelength) they take on similar shapes and bandwidths. Although the three cone pigments are most sensitive in the short-, middle- and long-wavelength ranges of the spectrum, it is traditional to speak of 'blue', 'green' and 'red' cones respectively. This nomenclature is inappropriate, however, because the cone photopigments themselves are not coloured; rather, their absorption spectra represent the probability that an individual photon will be absorbed at certain wavelengths. Moreover, because of the overlapping absorption spectra of the three cone pigments, there is a unique combination (i.e. triplet) of absorption probabilities for each wavelength in the visible spectrum. This means that the visual system is able to discriminate wavelength by comparing the amounts of absorption in different classes of cone. Furthermore, the presence of just three types of cone photoreceptor in the normal retina provides the basis of trichromacy (section 8.4).

The three cone types do not occur with equal frequency or density throughout the retina. The ratio of their relative abundances is approximately 40 'red' : 20 'green' : 1 'blue'; in the fovea itself the 'blue' cones are almost entirely absent. The probability of absorption of light across the retina therefore varies not only according to wavelength but also on the relative distribution of cone types, and is least in the short-wavelength region of the spectrum. Another result of the varying density of the different types of cone is that colour vision in the periphery of the visual field is different from that from the fovea in its centre. It is for this reason that colour-vision standards are based on judgements made within the central few degrees of vision.

Under constant light stimulation, a steady state exists in which the rates of photopigment bleaching and regeneration are equal; this is achieved by the continual supply of new vitamin A derivative molecules from the adjacent layer of retinal pigment epithelium. By comparison, when the eye is kept in complete darkness for an extended length of time (about an hour) every protein molecule in the rods is bonded with a molecule of retinal and in this state the retina is at its most sensitive for 'catching' or absorbing light quanta. (It will be recalled that rods are adapted for operating at low illumination.) In such a state the eye is said to be fully dark-adapted and vision is scotopic. Because scotopic illumination levels are too low to stimulate the cones, the eye in its dark-adapted state has poor spatial vision (i.e. poor visual acuity) but it has extraordinary powers of light detection. A series of extremely elegant experiments conducted in Oxford during the late 1940s led to the conclusion that, under the conditions of full dark adaptation, a single quantum of light is sufficient to cause isomerisation of a molecule of rhodopsin, but that about nine quanta (presumably stimulating nine separate rods) are required before a light flash could be detected and seen with 60% certainty [4]. This extreme sensitivity of the visual system surpasses that of any electro-optical device yet developed.

The method by which isomerisation of the visual pigments transmits information to the adjacent neural elements within the retina is still unknown. While the opsin remains unchanged under the influence of light, it appears that the process by which the retinal molecule eventually breaks loose from the opsin somehow gives rise to a change in electrical potential of approximately 2 µV across the cell membrane in the outer segment of the rod or cone. This electrical potential change is detectable at the foot or base of the receptor about 2 ms after the arrival of a light stimulus on the retina. In accordance with the so-called principle of univariance, which individual receptors are thought to obey, the cell's electrical polarisation (and hence its output) increases with the rate at which photons are absorbed by the receptor outer segment. Distinguishing between the wavelength and the intensity of different stimuli must, therefore, be a postreceptoral activity.

8.3.2 Postreceptoral retinal pathways

The photochemical activity in the rods and cones is the first of many stages in the transmission of information about light and colour from the retina to the brain. By the release at nerve endings of chemicals known as neurotransmitters, the nerve signal jumps the gap (synapse) between adjacent neurones within the retina. Only a few neurotransmitters have been identified in the retina; most are peptides. The number of candidates is bewildering but glutamate is believed to be the principal chemical messenger released by rods and cones following stimulation, and it is probably also present at the synapse junctions of most bipolar cells (see Figure 8.4).

The passage of the nerve impulse through the thin layer of retinal tissue is not straightforward. It results in a series of complex interactions between adjacent neurones which have an effect similar to that of a small-gain amplifier for improving the signal-to-noise ratio and thereby heightening contrast boundaries. This very elaborate coding of the signal occurs for information about not only brightness (i.e. luminance) but also colour, and it is likely that different retinal neurotransmitter chemicals subserve specific coding functions by acting at selected synapses. In the course of this retinal coding some neurones are actually made less receptive to the passage of a signal, while others are excited and transmit the signal to the next stage in the pathway. Those cells which at any instant have a reduced sensitivity are said to be in a state of neural inhibition; this is the result of a form of 'cross-talk' from adjacent cells which act laterally within the retina and release neurotransmitters that block the passage of a signal across a synapse. The cells that perform this initial coding of the signal from the rods and cones are the horizontal cells (Figure 8.4) and they appear to behave differently from most other neurones.

The characteristic response of most neurones to stimulation is a fixed-magnitude change in the electrical potential difference across its membrane. This is brought about by a double movement of ions across the permeable neurone membrane, causing the interior of the cell to become electrically negative with respect to its inactive surround. For most neurones in the body the ions concerned are potassium and sodium, and the difference in their concentrations across the membrane provides for a *resting potential* of the cell. When the neurone is stimulated, the signal is transmitted along the fibre in the form of a wave of change in the sodium/potassium ionic balance across the cell membrane. The typical output from most neurones is a brief reduction in electrical potential: the cell is said to become depolarised. Typically this depolarisation is of a fixed magnitude lasting only a few milliseconds before the cell returns to its former resting potential. This brief electrical activity, known as an *action potential* or *spike discharge*, is characteristic of most neurones. In their performance they are said to obey the 'all or none' principle of neurone response. In other words, a neurone responds to stimulation with either a fixed discharge or none at all, regardless of the type of stimulation. Information about stimulus intensity in these instances is conveyed by the rate, or frequency, with which the spike discharges travel down the nerve fibre and the rate at which they are propagated along adjacent neurones.

Although these changes in cell activity can be recorded as changes in electrical potential, a nerve impulse is not an electric current flowing down the length of a nerve fibre in the way that electricity passes down a wire. It is a progression of ionic changes whose electrical signs constitute the action potential. The propagation of a nerve impulse along a fibre is therefore characterised by a wave of electrochemical activity, and the energy for that transmission comes not from the stimulus but from the nerve itself. The speed of conductance of any nerve impulse depends on the cross-sectional area of a fibre and the amount of the surrounding membrane's insulating sheathing (a white fatty substance called myelin). Furthermore, there are slight delays in the propagation of an impulse as it crosses the junction (i.e. synapse) between successive neurones. In the passage of a nerve impulse from the retinal photoreceptors to the visual cortex of the brain, there are only three synapses; these occur at the junctions between the retinal receptor, the bipolar cell and the retinal ganglion cell and at the lateral geniculate nucleus in the thalamus (part of the brain, see section 8.3.3). Neither the horizontal nor the amacrine cells in the retina (Figure 8.4) are part of the direct chain of neurones between the retina and the brain; their role is to modify the signal or even cancel it at an early stage.

The retinal horizontal cells respond to output from the photoreceptors not by a spike discharge as found in other neurones in the body, but by a graded potential change comprising either an increase (hyperpolarisation) or a decrease (depolarisation) of the normal resting potential. These properties of the neurone correspond to two forms of the cell's response to stimulation; they are respectively referred to as inhibition and excitation. The graded response is such that the change of electrical potential in a horizontal cell increases with increasing stimulus intensity, unlike the fixed magnitude of a spike discharge found in other neurones which is 'all-or-none' in character. This graded response and duality of polarisation (i.e. hyperpolarisation or depolarisation) is an essential feature of neural coding in the visual system, and is found first at the base of the photoreceptor due to its interaction with the horizontal and bipolar cells. The occurrence of hyperpolarisation in a neurone is generally associated with an inhibitory action of nerve conductance and is believed to be a consequence of neural feedback to earlier sites of neighbouring cells. (Paradoxically, inhibitory activity is often referred to as a negative response, even though it is the result of hyperpolarisation.)

These graded potentials have been demonstrated electrophysiologically in primates by placing a fine microelectrode into the inner nuclear layer of the retina. They are named S-potentials, after Gunnar Svaetichin who first demonstrated their existence in 1956 [5]. Depending on the depth of recording in the retina, there are two main types of S-potential response. These are believed to originate in two types of horizontal cell:

(a) L-type cells, which respond uniformly to light of any wavelength and are thought to be responsible for coding information about luminosity
(b) C-type cells, which change their polarity depending on the part of the spectrum from which the stimulating light originates, thereby providing the basis of a wavelength-coded signal.

It appears that the L-type horizontal cells receive their input from all three types of cone photoreceptor, whereas the C-type cells exhibit response characteristics that originate either from 'red' and 'green' cones, or from 'red', 'green' and 'blue' cones. Examples of the change in polarity of C-type cells as a function of the cones' stimulating wavelength are shown in Figure 8.7. There is either an excitatory (positive) response when the cell is activated, or an inhibitory (negative) response where the receptiveness of the cell to signals is reduced. These two features of horizontal cells correspond respectively to their increased (i.e. depolarisation) or decreased (i.e. hyperpolarisation) firing rate according to the wavelength of the stimulus. Although C-type horizontal cells have not been recorded in primates, the two types of antagonistic or opponent wavelength responses are certainly present in the retinal ganglion cells, and this is the initial stage in the neural processing of a response of colour.

Before reaching the ganglion cell layer, a nerve impulse travels along the bipolar cell neurone and undergoes further coding from the lateral interaction effects of amacrine cells. These are cells that integrate information over a relatively large area of the retina (i.e. up to 0.5°) within the macular region and an even larger area in the peripheral retina; they link input from groups of bipolars to ganglion cells and other amacrines. As a result of these complex overlapping synaptic fields of the amacrine cells, several

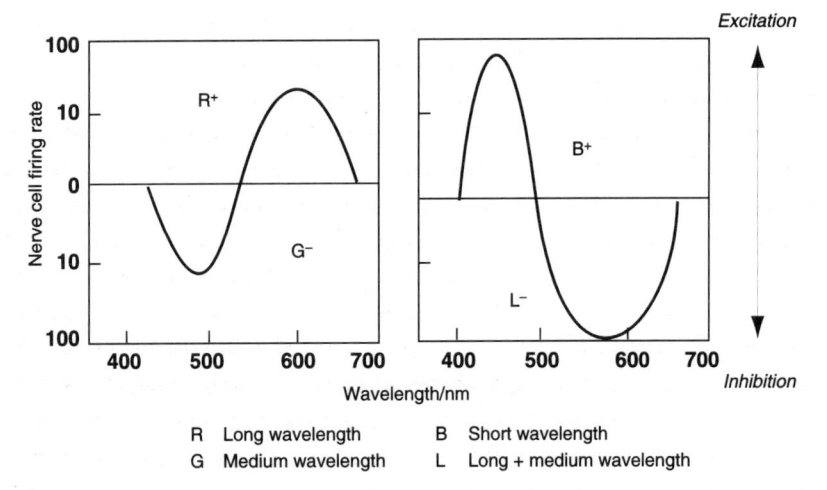

Figure 8.7 Spectrally opponent response characteristics found in the horizontal cells of the retina

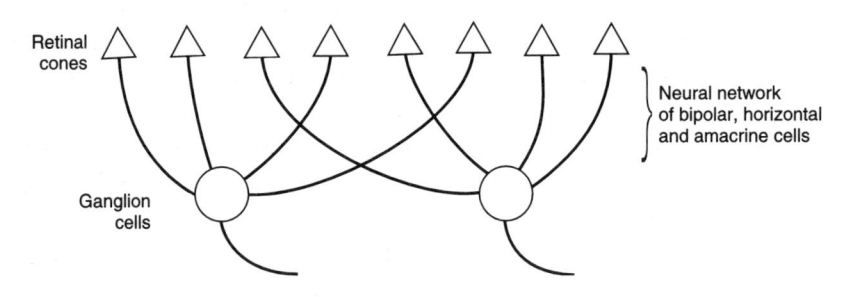

R Long wavelength
G Medium wavelength
B Short wavelength
L Long + medium wavelength

Figure 8.8 Nerve signals from several photoreceptors converge on to a single retinal ganglion cell; this neural convergence in the retina defines the receptive field of a ganglion cell

ganglion cells could in theory share input from the same 'red' and 'green' cones, but it is believed that separate sets of bipolars and amacrines maintain distinction between the luminosity and colour signals. The rod system is believed to have its own separate set of bipolar and amacrine cells, thereby achieving further functional separation for responses under low levels of illumination (scotopic vision).

There are about a million ganglion cells in the adult eye (each having its own unique optic nerve fibre), while there are about 150 million rod and cone receptors. A single retinal ganglion cell thus has to receive input signals from several bipolar cells. Output signals from the photoreceptors must therefore be combined within the retinal layers to provide a common input to specified ganglion cells. Neurophysiologists term this the process of convergence (Figure 8.8). The presence of this convergence is not a sign of receptor redundancy, but an essential feature of the coding mechanism within the retina. It is equally important to the detection and discrimination of luminance as to the detection and discrimination of form, movement and colour.

Convergence of receptor output on to a ganglion cell occurs over well-defined areas of the retina. These areas are known as receptive fields, and they have characteristic spatial distributions across the retina for gathering information about both luminance and colour. The ganglion cell receptive field is delimited by the lateral distribution of many fine nerve fibre branches, or dendrites, which conduct impulses towards the neurone body. Most receptive fields are circular in their organisation, but they are not totally determined by fixed anatomical properties of the ganglion cell dendrites, nor do they occupy mutually exclusive spatial areas of retina. Individual photoreceptors have output neural pathways that potentially lead to more than one ganglion cell. The preferred route is influenced by the surrounding stimulus conditions and hence is governed to a large extent by the characteristics of the retinal image. These conditions determine the amount of inhibitory feedback between adjacent sets of receptors via the lateral interconnections of horizontal cells. A typical ganglion cell's response is characterised by a single receptive field, for which the central receptors provide an excitatory input while the surround receptors form an annulus of inhibition in which the firing rate is decreased. Figure 8.9 illustrates an *on-centre* field, showing how the excitatory (+) response of the on-centre receptors is transmitted to bipolar cells. Signals along the horizontal cells cause inhibitory neurotransmitter chemicals to be released on adjacent bipolar cells. The normal excitatory response of the surround receptors is therefore suppressed and this has an inhibitory (–) influence on the ganglion cell. (The amacrine cells may also be responsible for converting excitatory activity in bipolar cells to inhibitory action on the ganglion cell.)

Despite its sophistication in nerve signal modification, a single ganglion cell receptive field cannot provide the coding for brightness, shape and size of the retinal image. This is achieved by a comparison of signals in optic nerve fibres receiving input from many adjacent overlapping receptive fields. Some of these fields are of the on-centre kind described above. About an equal number are *off-centre* receptive fields, with the opposite response characteristics: light falling in the centre inhibits firing of the ganglion cell and illumination of the surrounding annulus has an excitatory effect. The opponent characteristics of the receptive fields, which can be recorded as adjacent spatial areas of excitatory and inhibitory activity, are fundamental to the way the retina encodes much of its information about the visual world. Opponency in the retina is particularly well developed for the processing of information about spatial luminance contrast, and is the reason why we perceive a heightened contrast at a luminance boundary. Uniform illumination of the entire receptive field leads to mutual antagonism between the cell's centre and surround in which excitatory input is

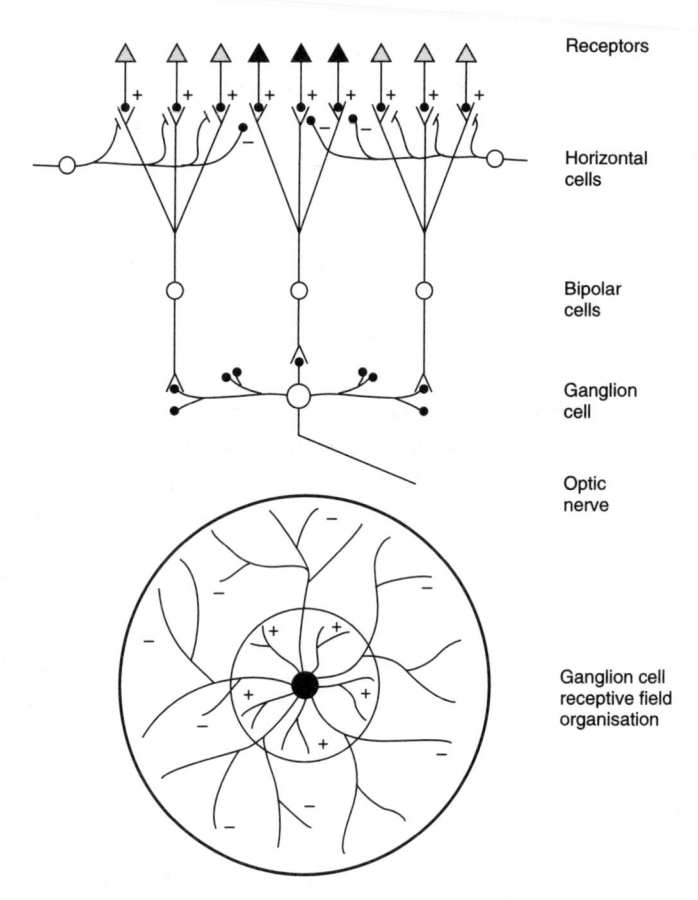

Figure 8.9 Simplified functional diagram of retinal nerve connections and retinal structure of an on-centre/off-surround ganglion cell receptive field

balanced by an opposing influence of inhibition, thus producing no effect or response in the ganglion cell. Consequently the opponent spatial organisation of the receptive fields discriminates against diffuse illumination in favour of luminance gradients or luminance contrast in the retinal image.

The size of the receptive fields is determined by the intensity and size of the stimulus. Under high levels of illumination the inhibitory effects of the lateral interaction along horizontal cells are at their strongest for all neighbouring receptive fields, with the result that the effective field size is reduced. Under high levels of illumination, therefore, the number of functional receptive fields per unit area of retina is increased. Increasing illumination level thus leads to an increase in the fineness of the functional neural analysing elements within the retina. It is as though illumination level determines whether the retinal neural units will be operating as fine-grain or coarse-grain detectors (analogous to a fine-grain or coarse-grain photographic film). By this means the resolving power of the retina increases with increasing illumination level, thus making it possible to discriminate small detail better and thereby achieve a higher visual acuity.

So far this discussion has focused on the principle of opponency in the context of luminance contrast. Information about colour is also coded through an opponent mechanism. Mention has already been made about the possible existence of spectrally opponent C-type horizontal cells. But there are also retinal ganglion cells that are spectrally opponent, whose receptive fields are such that they are connected with at least two (sometimes three) sets of cones containing different photopigments and thus having different spectral sensitivities. As with the receptive fields which are specifically organised for coding luminance contrast, many of those coded for wavelength are also organised concentrically, and the lateral synaptic connections are such that one type of cone excites the ganglion cell and another type inhibits it. These spectrally opponent characteristics mean that some ganglion cells are organised such that their rate of firing will increase following photoreceptor stimulation by one set of wavelengths, but decrease to another set. These ganglion cells are said to have the properties of wavelength (or spectral) opponency, and they provide the means by which the visual system discriminates against white light. There are several different kinds of such cell in the retina, each of which exhibits characteristically different wavelength-specific responses. As with the organisation of ganglion receptive fields for the coding of luminance and luminance contrast, so also are there specific ganglion cell receptive fields for coding information about wavelength. In both instances the opponent properties of the receptive fields are organised spatially by the convergence of signals from several photoreceptors on to a single ganglion cell. Indeed, there is a suggestion that some of the receptive fields may be shared for conveying information about luminance, colour and also retinal position.

As with the receptor organisation for coding luminance contrast, the receptive fields for having spectral opponency do not occupy mutually exclusive topographic areas of the retina. They are functionally rather than anatomically defined units of neural activity and their size depends on the absolute intensity and relative spectral intensity distribution in the stimulus, both of which govern the amount of lateral inhibition present among the amacrine cells. Output from a single cone, for example, has the potential of feeding into more than one ganglion cell and the favoured route is determined by the spatial arrangements of the luminance and chromatic distributions within the stimulus. Thus, these wavelength-opponent cells enhance the boundaries between regions of different chromaticity in a similar way to the heightening of lightness contrast that takes place in other ganglion cells.

There are primarily three classes of retinal ganglion cell coded for wavelength: two show spectrally antagonistic responses and one is characterised by a positive response that is independent of wavelength (Figure 8.10). The antagonistic cells (type I and type II) exhibit an increased firing (positive) response to some wavelengths and a decreased firing (inhibitory) response to others, whereas the type III cells give a positive response to a broad band of wavelengths across most of the visible spectrum. The spectral sensitivity of the type III cells is very similar to the photopic luminosity function (Figure 1.19), which represents the relative spectral sensitivity to the combined output from all three cone types at high illuminance.

But such a simple classification comprising just three cell types belies the subtleties of spectral responsiveness that seem to be present among retinal ganglion cells. Recent neurophysiological studies, for example, have shown a wide range of differing spectral sensitivities (*action spectra*) for cell types I and II, which together comprise about 60% of all the retinal ganglion cells [6]. Most have an antagonistic response to medium-wavelength (M) and long-wavelength (L) ranges, the rest exhibiting an antagonism between the short (S) and medium-plus-long (M + L) wavelengths. All spectrally opponent ganglion cells receive input from specific cone types characterised by the two opponent groups of C-potentials in the horizontal cells. The input signals to the ganglion cells are combined in well-defined spatially arranged approximately circular receptive fields, and are such that paired receptor types (such as 'red' and 'green' cones) oppose each other in their neural coding. For example, the centre of a receptive field coded for long (L) and medium (M) wavelengths may be excited by signals originating in 'red' cones (i.e. responding preferentially to long wavelengths), whereas the

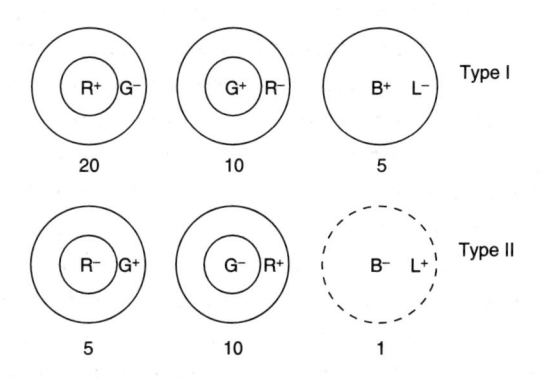

Figure 8.10 Centre/surround ganglion receptive fields which are selective for wavelength; characteristic spectral responses of these opponent receptive fields are shown in Figure 8.7

adjacent surround to the same receptive field may be inhibited by signals arising from the 'green' cones (i.e. medium wavelengths).

The output from such an (L+/M−) antagonistic ganglion cell depends on the relative amount of stimulation it receives from either the 'red' or the 'green' cones. If signals from the red cones dominate, indicating a greater intensity of light of longer wavelengths in the stimulus, then the ganglion cell will be activated. By comparison, an excess of medium wavelengths in the stimulus will have a dominant inhibitory influence on the same ganglion cell, effectively switching it off. If the incident light from the stimulus includes equal amounts of long and medium wavelengths, then the output from the 'red' and 'green' cones in ganglion cells with (L+/M−) centre/surround receptive fields will be mutually antagonistic and the ganglion cell will not fire. Such a null effect would occur if the receptive field of an (L+/M−) ganglion cell were stimulated by light of middle-range wavelengths.

Not all (L/M) ganglion cell receptive fields are affected in the same way by light of middle-range wavelengths; ganglion cells vary widely in wavelength sensitivity (or action spectrum) depending on the relative dominance in their receptive fields of input of either the 'red'- or the 'green'-sensitive cone mechanism. As a result, not only are there variations in the spectral sensitivity maxima of excitation and inhibition among these classes of ganglion cell, but also their neutral points can be found in almost any part of the visible spectrum. (These *neutral points* are so called because they represent the stimulus wavelength which neither inhibits nor activates the ganglion cell.) The action spectrum of such a cell is therefore largely influenced by the predominant wavelengths of the stimulating light. The consequent range of spectral sensitivities among the two main groups of ganglion cell (i.e. L+/M− and L−/M+) gives a heterogeneous array of spectral opponency throughout the retina. Although the exact role of this feature is not known with certainty, it appears to be essential to fine hue and wavelength discrimination, and possibly also to maintaining the sense of constancy of perceived colours (section 8.3.3). Because of the centre/surround spatial arrangements of the 'red' and 'green' cone inputs to such a receptive field, these ganglion cells are particularly well tuned to detecting chromatic boundaries in a stimulus. They give a somewhat heightened response output when their activity is averaged over several adjacent receptive fields. Similar arguments also hold for the manner in which the S versus (L + M) antagonistic receptive fields respond to input from the blue cones versus the combined signals from red and green cones. The common (though erroneous) practice of giving colour names to the three cone types has resulted in these same names being assigned to the spectrally selective ganglion cells. Consequently an L/M antagonistic ganglion cell has become known as a 'red'/'green' cell, and those with an S/(L + M) antagonistic response as 'blue'/'yellow' cells.

Although these colour names are widely used when describing retinal structures having wavelength-specific response characteristics, strictly speaking it is inappropriate to refer to spectrally antagonistic ganglion cells as 'colour-opponent cells'. They are more correctly termed 'wavelength-opponent cells' because they are only responding to intensity changes from two types of receptor having different spectral sensitivities. This is not just a semantic distinction, but one that is fundamental to understanding the way in which we see colour. Almost 300 years ago Isaac Newton emphasised that rays of light themselves are not coloured; it is also true that different retinal cones do not give a response of colour. Their response is purely a function of the probability of the absorption of light of different wavelengths. Their individual output signals are therefore related to the different wavelength intensities of a stimulus and bear no relation to its perceived colour. While the neural elements in the retina play a key role in the initial coding of the photoreceptor output signals by organising them in such a way that the information is represented in terms of response ratios, no single cell can distinguish the colour of a stimulus. Information about colour is contained in the neural network between the retina and brain, and the process of decoding this information requires the simultaneous examination of signals from several adjacent cells having slightly different spatial or wavelength response characteristics. It is the brain and not the retina that has such a decoding system, and it is the comparison and combination of the multitude of signals that enables us to perceive colour. Let us therefore now consider what happens to the neural signals once they leave the retina and travel into the brain.

8.3.3 Visual processing beyond the retina

About 50 ms after a photon is absorbed in the outer segment of a receptor cell, the resultant nerve signals leave the retina and pass along the optic nerve, away from the eye towards the base of the brain. Soon after leaving the eye, fibres in the optic nerve segregate into two bundles; those from the left half of each retina of each eye (i.e. to the left of the fovea) continue to the left half of the brain (left hemisphere) and those from the right halves of the retinas are routed to the right hemisphere. This involves some anatomical crossing-over of certain nerve fibres (known as a process of *semi-decussation*), which occurs at the optic chiasma (Figure 8.11); there is, however, no mixing of the nerve signals from the two eyes at this point. This crossing-over of half the optic nerve fibres from each eye has the consequence that nerve signals arising from objects positioned in the visual field to the right of the point of fixation, and imaged on the left halves of both retinas, are processed in the left hemisphere of the brain, and objects in the left field of view, which are imaged on the right half of the retinas, are processed in the right cerebral hemisphere. Nerve fibres from the macular region of the retinas appear to have some shared representation in both cerebral hemispheres. One of the products of having two eyes arranged in this way, with each viewing the world from a different vantage point, is that the right and left eyes receive slightly different images. These small geometric disparities in the retinal images of the eyes are responsible for our stereoscopic sense by which we are able to say that one object is nearer or further away than another.

A short distance behind the optic chiasma, close to the base of the brain, neural signals from both eyes meet in a complex cellular body known as the *lateral geniculate nucleus* (Figure 8.11). Here the retinal ganglion cell fibres terminate, forming synapses with other nerve fibres leading to another area of the brain concerned with vision. In each cerebral hemisphere the signals from right and left eyes are represented in anatomically separate laminae. Although the receptive field characteristics of some of the cells in the lateral geniculate nucleus are slightly more complex than those of the retinal ganglion cells, their spatial arrangements for luminance and wavelength-selective antagonistic coding appear to be basically unchanged. It would appear that the principal function of this relay nucleus is to group signals from related areas of the retinas in the two eyes. From here the signals travel along the long fibres of the

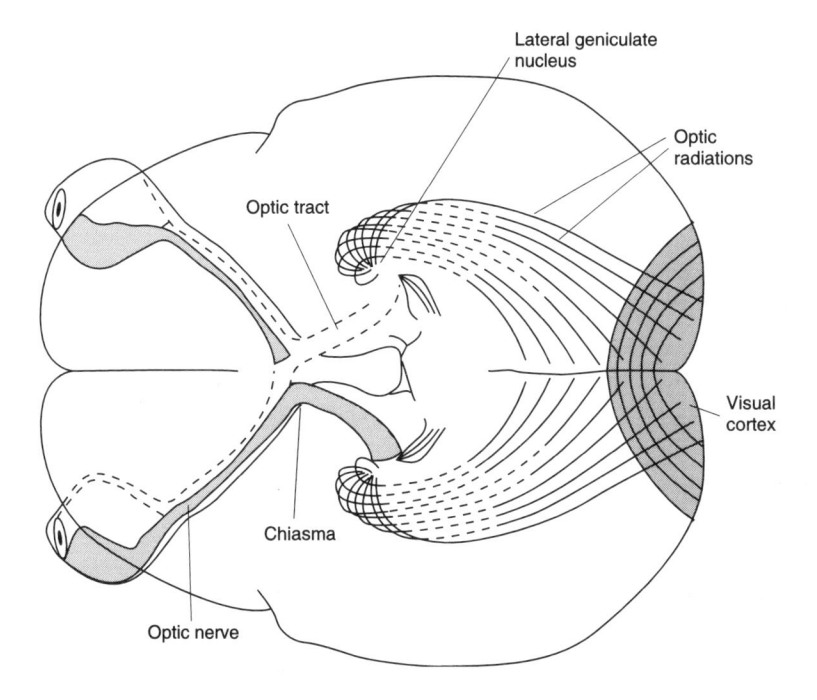

Figure 8.11 Path taken by nerve signals from the eye to the visual cortex at the back of the brain, showing anatomical relationships of the different neural features in the visual process as viewed from the top of the head

geniculate nucleus known as the *optic radiations*, through the lower part of the midbrain to a region of the cerebral hemispheres at the back of the head called the *visual cortex*.

The visual cortex is aptly named because it is the centre of our visual perception. But it does not contain geometrically representational images of the world in the way that these are formed optically on the retinas. Cells in the visual cortex appear to be arranged in areas, each of which is responsive to different basic attributes of a stimulus. Thus, for example, there is a region in which cells respond primarily to the movement of an object, another in which cells are specific to object orientation and yet another that responds primarily to those properties of a stimulus associated with its perceived colour. It is only in recent years that this functional organisation of the visual cortex has been demonstrated. As a consequence, it has now become possible to identify areas in the visual cortex, each of which is characterised by cells that are predominantly responsible for analysing different properties of the retinal image. The first electrophysiological evidence of this functional specialisation of cells in the visual cortex was produced in 1962 by two Harvard physiologists, Hubel and Wiesel [7], both of whom received Nobel prizes for their work in this field.

On entering the visual cortex via the optic radiations, nerve impulses from both eyes terminate close to each other near the surface of the brain in an area known as the *primary visual cortex*, or V1. Like the rest of the cerebral cortex, the visual cortex has a considerable surface area due to the extensively folded topography of the brain. When these folds are taken into account, the spatial distribution of nerve fibres terminating in the primary visual cortex can be mapped topographically on the surface of the brain in a way that corresponds reasonably with well-defined positions on the retina. This mapping demonstrates that signals arising from objects in different parts of the visual scene have been kept segregated along the neural pathways as far as area V1 of the visual cortex. The topographic distribution of nerve fibre terminals in the cortex is, however, distorted in such a way that fibres from the macular region of the retina cover a proportionately larger area of the brain than do nerve fibres from the peripheral retina. It is as though area V1 in the visual cortex has differential scaling or magnification factors with respect to ganglion cell distributions in the retinas.

This is best illustrated by looking at the cortical representation of the macular zone in the retinas. The macular area of the retina has an exaggerated area of representation in the visual cortex which can be likened to a nonlinear map projection of the retina on to the brain. This projection of a small area of the retina on to a large surface area of the brain provides for the greatest detail in vision; the macula also has the greatest concentration of cones, which are responsible for colour vision. In fact, the exaggerated cortical representation of the central area of the visual field roughly corresponds to the highest density of the retinal ganglion cells. So it would seem that the more complex are the visual functions, the greater is the area of the brain devoted to information processing. But rather than providing for a total analysis of the visual scene like some distorted image on the surface of the brain, the primary visual cortex acts as a segregator of signals by parcelling out different kinds of information to other adjacent areas of the brain for further processing.

These other visual areas of the brain are in a belt of cortex surrounding V1. They appear to be organised in a multiplicity of distinct areas and each must receive its own separate input of signals from each part of V1. Much of our increased understanding of this part of the brain has come from the meticulous neurophysiological researches of Semir Zeki at University College, London [8,9]; in particular, Zeki identified those parts of the brain that detect colour and movement. The accumulation of findings, from both neurophysiological and psychophysical studies, has led to a theory of functional specialisation of the brain which states that different attributes of the visual scene (such as movement, form and colour) are processed in different regions of the visual cortex. These areas have been labelled V2 to V5 (although there may well be more than four). With the exception of area V2, they are believed to be largely functionally specific. The different functions associated with each area are still somewhat speculative, although V4 appears to be primarily responsible for processing information about colour. Experimental work on monkeys shows that cells in V3 respond primarily to the orientation of an object; V5 seems to be concerned with the analysis of movement of an object and changes in its direction, whereas cells in V2 respond to most stimulus properties and are thought to provide further segregation of nerve impulses arriving from V1 before they proceed to the other areas of V3 to V5.

In V1 the network of interconnections in the few millimetres of nerve tissue near the surface of the brain is complex and is found in several well-defined layers where fibres from right and left eyes terminate very close to each other. Cells responsive to the same stimulus property (orientation or wavelength, for example) are grouped together in parallel columns perpendicular to the surface of the brain. In this area of the visual cortex there are three different types of cell responsible for the further coding of information about colour. Only one of these three has similar response characteristics to cells found in the lateral geniculate nucleus and also in the retinal ganglion cells. This type of cortical cell has response characteristics that are wavelength-opponent within a spatially defined receptive field. Unlike their retinal ganglion cell counterparts they have no antagonistic surround to their receptive fields, but they do possess wavelength opponency. For example, some cells are activated by long wavelengths and inhibited by medium wavelengths, and are designated as (L+/M–) or sometimes as 'red-on' and 'green-off' cells. Cells giving the opposite response (L–/M+) are also found in this area of the visual cortex; so too are cells exhibiting opponency for short versus medium-plus-long wavelengths, sometimes referred to as 'blue'/'yellow' opponent cells. An important feature of all these opponent cells in the primary visual cortex is that their response is a function of the relative intensities of different wavelengths present in the stimulus and not of perceived colour.

To appreciate the distinction between wavelength and perceived colour it is necessary to realise that the perceived colour of an object cannot always be predicted from the composition of short, medium or long wavelengths of light reflected from its surface. It is the wavelength composition of light coming from the object *and* from its surrounding areas that is important for the discrimination of colour. For example, when we view a green apple in a bowl of mixed fruit out of doors under natural daylight, the reflected light reaching our eyes will be mostly of medium wavelengths, and an opponent ganglion cell (L–/M+) will respond positively if the image of the apple is in its receptive field. But if we take that same bowl of fruit indoors and view it under tungsten light in which long wavelengths predominate, then the light reflected from the apple may now contain more long than medium or short wavelengths: the (L–/M+) opponent cortical cell would be inhibited, yet we would still perceive the apple as green. This phenomenon, where objects that are part of a complex multicoloured scene look the same colour under different illuminants that have radically different spectral properties, is known as *colour constancy*. Colour constancy is a property of vision that is essential to the stability of colour perception in the real world, but it can present problems to the colour technologist whose task is to discriminate small differences in colour. The very existence of the phenomenon of colour constancy means that the behaviour of opponent cells in V1 of the visual cortex are not the basis of colours as we see them; the cells are merely antagonistically responsive to different wavelengths. To explain the phenomenon of colour constancy we need to find cortical cells whose responses will correlate with perceived colour and are independent of the relative amounts of long, medium or short wavelengths present in the reflected light from the stimulus.

Other cells in V1 have very specific sensitivities to narrow ranges of wavelengths of 50–100 nm bandwidth, behaving in a manner similar to a bandpass filter (Figure 8.12). It is believed that these narrow-bandpass spectrally selective cortical cells behave like 'zero detectors' by responding maximally to the wavelength of the neutral point of retinal opponent ganglion cells. A variety of such cells have been found in the retinas of monkeys, covering a wide range of neutral points at almost all wavelengths between 420 nm and 650 nm, and it is likely that others occur across the entire visible spectrum. Furthermore, the sensitivity maxima of excitation and inhibition of retinal opponent ganglion cells can differ by almost two log units. It has been proposed that these two response characteristics of retinal ganglion cells provide the means by which the visual system can make fine discriminations of wavelength or chromaticity [10]. The fact that there are many different 'zero detectors' in area V1 of the cortex, all finely tuned spectrally, enables wavelength differences of only a few nanometres to be

THE MECHANISM OF VISION 449

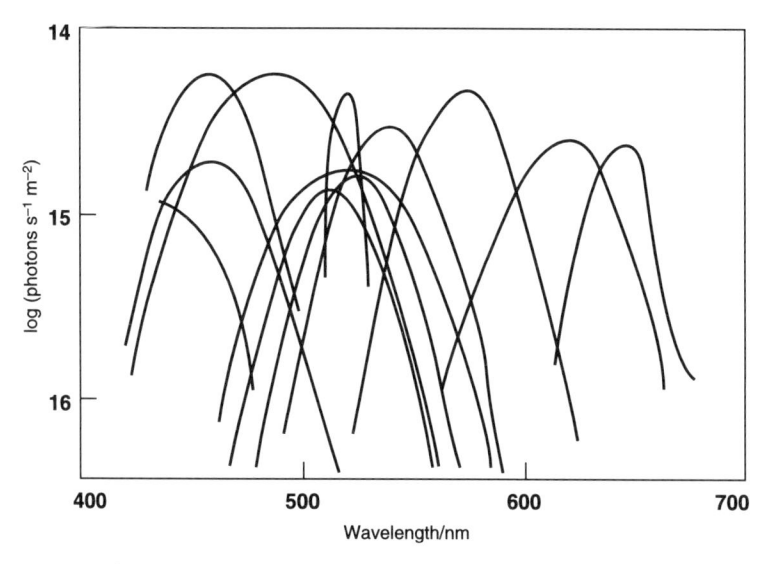

Figure 8.12 Some nerve cells in the visual cortex have narrow-waveband response characteristics, and the action spectra of these cells range throughout the visible spectrum

detected. For example, consider the task of discriminating a chromatic boundary separating two equiluminant juxtaposed coloured papers having different spectral reflective properties. Underlying each side of the retinal image of the chromatic boundary will be opponent ganglion cells having a wide array of neutral points. Whether an opponent cell is activated or inhibited will depend upon the wavelength of its neutral point, and this will be different on the two sides of the chromatic boundary. Signals from clustered groups of ganglion cells, all having the same neutral point, underlying each side of the retinal image of the chromatic boundary, will be transmitted to finely tuned wavelength-specific cells in area V1. Certain groups of these wavelength-selective cortical cells will then fire according to the relative dominance of different wavelength distributions in the retinal image on either side of the chromatic boundary of the stimulus.

The closer together the wavelengths of ganglion cell neutral points, the closer together will be the 'preferred' wavelengths of the finely tuned cortical cells, thus giving finer wavelength discrimination. This is at its best over the middle range of wavelengths in the visible spectrum. Based on this proposition, variations in wavelength discrimination across the spectrum can be accounted for by an unequal distribution of ganglion cell neutral points.

Some finely tuned cortical cells show a positive response to a stimulus within their waveband selectivity, while others are inhibited on stimulation. But, as with the opponent cortical cells in V1, they are not colour-coded cells in the sense of being responsive to a particular perceived colour. They are simply another set of wavelength-selective cells, responding only when the stimulus contains sufficient light of their preferred waveband. They cannot therefore explain the phenomenon of colour constancy. The spectral response characteristics of these cells is somewhat analogous to that of the cone photoreceptors in the retina, except that their wavelength tuning is narrower. While a single cell cannot distinguish between different wavelengths, it can signal changes in the amount of light of its preferred wavelengths relative to light of other wavelengths. This sensitivity to differences in intensity, which is independent of colour, represents information about brightness distribution in the visual scene.

A common characteristic of both the above types of cortical cell is that their receptive fields have no antagonistic surrounds, so that they are unable to compare the wavelength composition of light reflected from surrounding regions in the visual field. Cells that can do this are concentrated in a particular zone of area V1 and have what may be called *double-antagonistic* or *double-opponent centre/surround* receptive fields [11]. Thus, for example, the central receptive field of one such cell may be activated by long wavelengths (i.e. L+/M−), while its surround would have the converse response characteristics (L−/M+). The properties of these cells appear to be the consequence of interactions of overlapping pairs of different single-opponent centre/surround receptive fields through interconnecting neurones.

These double-opponent cortical cells can make wavelength comparisons, which other cells in the cortex cannot, and since colour vision is based on a comparison of light energies of all wavelengths reflected from different surfaces, they must play an important role. At present that role remains obscure, but it does seem that double-opponent cells may be the initial step in generating responses of 'lightness' from signal strengths at their preferred wavelengths, which are then compared elsewhere in the brain to construct colours. Where that comparison occurs is also uncertain, but the best candidate appears to be in the part of the brain known as area V4, surrounding V1. This is the first location in the visual cortex where cell responses seem to correlate with perceived colours. The cells in V4 have similarly narrow-waveband tuning to those found in V1, but their response is considerably influenced by the activity of cells whose receptive fields are immediately adjacent to their own. It is as though each cell has its own receptive field lying in the centre of an extensive surround receptive field covering several degrees. It is possible that the double-opponent cells found in V1 contribute to this extended surround via their relay and interactions in the intermediate cortical area V2. The colour-coded cells in V4 have very special response characteristics. While they respond to light of specific wavelengths, as do the wavelength-selective cells in V1, their response also depends on the spectral composition of the surrounding visual field. For example, such cells do not respond when the entire visual scene is illuminated by monochromatic light, even if that light is within the waveband selectivity of the cell. Under such extreme lighting conditions, of course, the perception of different colours does not exist and boundaries between objects in the visual scene are encoded merely in terms of lightness contrast. In order to respond to a particular colour these wavelength-selective cells need an area of a visual scene to be illuminated by light comprising a wide array of different wavelengths. White light is typical of such illumination, although its wavelength composition can be very varied, ranging from a bias of short wavelengths (as in daylight) to a bias of long wavelengths (as in tungsten light) or even a three-band fluorescent lamp (section 1.5.4). Whatever the spectral bias of the light source, provided it contains a sufficient range of widely spaced wavelengths, we will perceive colours in a complex visual scene. This is because the cells in V4 make a comparison of the relative light energies of different wavelengths reflected from the surfaces of adjacent objects. So whether a green apple in a bowl of fruit is viewed under tungsten light or daylight it will still be perceived as green, even though the composition of the light reflected from its surface may change dramatically in the two viewing situations from predominately long to predominately short wavelengths. This constancy in our perception arises because all other objects in the bowl of fruit are also illuminated with the same light. In order to construct the perception of colour associated with each fruit in the bowl, it seems that the cortical cells respond by making ratio comparisons of neural activity which corresponds with the composition of the reflected wavelengths from all objects in the visual scene. Furthermore, these wavelength-related ratios will remain unchanged, not only under considerable variations in the spectral composition of the light source, but also for extensive changes in the amount of light illuminating the visual scene. Consequently, the perceived colours of objects viewed under such varied conditions of illumination will remain unchanged. What is required for achieving this phenomenon is that the observer's visual field contains objects with a richly varied array of spectral properties; without this, there can be no perception of colour. Fortunately such an environment does not exist in practice. Imagine, however, that an unfortunate person was placed inside a large integrating sphere where the illumination level and spectral properties of the light were uniform over the entire internal surface of the sphere. Even if the surface were illuminated with monochromatic light, the individual would not perceive it as coloured. This is because there would be nothing against which the ganglion or cortical cells could make a comparison. The existence of colour constancy in our perception of the visual world emphasises the fact that all judgements about colour are based upon a comparison of signals between cells that possess different wavelength sensitivities.

8.3.4 Overview

The elegant coding principles in the retina are directed primarily towards one end: to discriminate between stimuli having different spectral properties. To determine the colour associated with each part of our field of view, the brain has to analyse the wavelength composition of light from all areas and then compare them with one another. The special ability of the visual system to construct a sense of constancy of colour appearance under a wide range of lighting conditions is a consequence of the simultaneous comparison of the spectral reflectance of one region of the field of view with that of an adjacent region. The spatial properties of the receptive fields of cells involved in the coding of information for wavelength are thus an essential part of how we see colour.

It should now be evident that colours are not a property of the real world, but are rather a construct of the brain. Caution should therefore be exercised when describing regions of the spectrum or even unique wavelengths by colour names, because light energy itself possesses no colour. Only the evocation of responses in different sets of cortical cells by light energy according to its wavelength composition enables us to distinguish objects in terms of their spectral reflectances and eventually perceive them as coloured.

8.4 MODELLING THE COLOUR VISION PROCESS

8.4.1 Historical

Since the mid-18th century the field of colour science has produced an abundance of colour-vision theories. Foremost among these is a proposition put forward by the scientist Thomas Young in 1802. He argued that there must be three types of photoreceptor in the eye because, simply as a result of the possible permutations, this number could account for the limited number of principal hues we can identify. All combinations of three primary hue variables give seven principal hues, and in the 18th and early 19th centuries the mysticism associated with these numerical values almost certainly encouraged a wide acceptance of the idea. Although Young was not the first person to suggest that human colour vision depended on just three variables, he received the credit for stating it clearly at the right place and right time (in a lecture to the Royal Society in London). The time was right because interest had been stirred by a paper published four years earlier by John Dalton, the chemical scientist from Manchester, in which he had described his own defective colour vision.

Consistent with Newton's proposition that light itself is not coloured, Young believed that the three variables responsible for our colour vision were in the eye and not in the light. But it was not until 1855 that experimental evidence in favour of this hypothesis was provided. This came from the work of James Clerk Maxwell who demonstrated, by means of spinning sectored discs of different coloured papers, that any nonspectral colour could be matched by the appropriate mixture of just three suitable colours, which he called primaries. (Some colours produced by monochromatic light or light of high spectral purity are exceptions to this rule, but they can be matched by a special technique using just three primaries.) A few years later, in 1866, Herman von Helmholtz gave respectability to this experimental data by proposing a physiologically based hypothesis of three channels with different, but somewhat overlapping, spectral sensitivities. The combined theoretical propositions were eventually formulated into one theory, which became known as the Young–Helmholtz trichromatic theory of colour vision. Nevertheless, while the names of Young and Helmholtz are permanently associated with this theory, it would perhaps be more appropriate to honour Newton and Maxwell for providing the early experimental evidence upon which the theory is based.

That the three hypothetical spectrally sensitive channels were not simply additive in their response to light was shown by colour-mixing experiments in which certain hues appeared to cancel each other. For example, appropriate mixtures of 'red' and 'green' can form a yellow that appears to be totally devoid of any 'red' or 'green'. Similarly mixtures of 'yellow' and 'blue' can produce 'white' that is devoid of any trace of 'yellow'. On the other hand, 'blue' or 'yellow' can mix with 'red' or 'green' without total loss of the hue properties of the original components. Ewald Hering pointed out that this specific antagonism between certain hues required a totally different explanation from that suggested by Young and Helmholtz. In 1878 Hering put forward his opponent colour theory, which proposed that there were three separate antagonistic channels in the retina: two for processing information about colour and one for lightness. The two colour channels were 'red' versus 'green' and 'yellow' versus 'blue', and the third opponent channel was to account for responses of 'white' versus 'black'.

Neither the hypothesis put forward by Young and Helmholtz nor that of Hering can explain all the phenomena of colour vision. Nevertheless, they have survived many other relatively short-lived alternative propositions and are now considered as the basis of modern colour-vision theory. In 1959, however, a paper by Edwin Land caused a stir among colour scientists [12,13]. It reported some amazingly simple yet apparently startling experiments involving two slide projectors, two monochromatic photographs and two different broad-band spectral filters. In attempting to explain his observations, Land put forward his so-called retinex theory in which he claimed that our experience of colour of the real world needs only three separate and independent channels for comparing 'lightnesses'. What he had in fact demonstrated was the importance of a visual system which could analyse contrasts between the reflected wavelength and brightness properties of adjacent objects in a complex visual scene. Land had shown that these analyses of contrast were essential for preserving the phenomenon of colour constancy. Young, Helmholtz and Hering have survived the attack, and our understanding of colour vision has become enriched in the process.

8.4.2 Three apparently conflicting theories

The three major colour-vision theories that have attracted the attention of colour scientists are the trichromatic theory of Young and Helmholtz, Hering's opponent colour theory and the retinex theory put forward by Land.

The *trichromatic theory* is firmly rooted in the most fundamental property of colour vision, that of trichromacy: that is, the ability to match any colour with an appropriate mixture of just three suitably chosen primary wavelengths. Although the three fixed wavelengths used to provide the mixture are called primaries, trichromacy does not depend upon any particular set of primaries. There are many primary wavelength triplets that will fulfil the requirements of trichromacy. The choice of a set of primaries is dictated by the rule that no one of them must be able to be matched by mixing the other two. This means that the actual set of primary wavelengths used in colour-mixing experiments is selected somewhat arbitrarily. In order to minimise experimental error, however, primaries are conventionally chosen from wavelengths that are widely spaced in the visible spectrum. The selection generally includes a long, a medium and a short wavelength, which typically evoke the colour names of red, green and blue respectively, and these colour names have become associated with the additive primaries. Experiments involving colour mixing of different monochromatic lights show the relative luminosities of the three primary stimuli required to provide a colour match with each wavelength in the visible spectrum (section 3.6). What is important about all the possible sets of primaries is that any three spectral functions able to satisfy human colour-mixing properties of trichromacy can be linearly transformed into any other set. It was from the results of such colour-mixing experiments, pioneered by Maxwell, that attempts were made to derive speculative spectral response functions for the (hypothesised) three cone types in the retina. Extensive independent colour-matching investigations in London during the 1920s [14–17] eventually led the Commission Internationale de L'Éclairage (CIE) to publish standardised values in 1931 (Figure 3.6). Averaged values from these studies became known as the *1931 standard observer*; as averaged values, however, they give no indication of the variations in colour-matching functions that exist between different people.

While the principles of colorimetry evolved out of the experimental fact of trichromacy, its validity and computational convenience as a method of colour measurement does not depend on the correctness of any theory of colour vision. The only connection between colour vision theory and trichromacy is that, in attempting to explain the latter, the Young–Helmholtz *trichromatic theory* is built around the assumption that there are three types of retinal receptor.

The most convincing early evidence for defining the cone sensitivity functions came from two sets of experiments. The first group consisted of colour-matching experiments using people with defective colour vision, in whom one cone type was assumed to be missing, and the second group comprised experiments of hue detection where the sensitivity of one cone type was reduced considerably, although only temporarily, by exposing the eye to high-intensity lights of specific wavelengths (these latter experiments are known as selective adaptation studies, involving the determination of incremental colour thresholds using techniques developed by W S Stiles in the late 1930s) [18,19]. Direct confirmation that there were three types of human retinal cone with different spectral absorption properties was not achieved, however, until 1964. Then it came from two independent groups of investigators, one in the United States of America [1] and one in Britain [2], both of whom carried out microspectrophotometry on single cone receptors through the full thickness of retinal tissue. More recently, microspectrophotometry results have been obtained by transverse measurements across the outer segments of isolated human cones [3]. There is a remarkably close similarity between the original hypothesised response functions of the receptors and the objectively determined cone absorption spectra. This was the evidence that supporters of the trichromatic theory were looking for, but it came at a time when considerable advances were being made on other aspects of retinal and visual physiology, and these were more in accordance with an opponent mechanism of colour vision.

Hering's *opponent colour theory* was proposed as an attempt to explain some of the facts about colour that cannot be predicted from the trichromatic theory. For example, the result of mixing light of certain pairs of wavelengths cannot be predicted intuitively from people's everyday experiences with pigments; although mixtures of 'red' and 'blue' light give a reddish-blue perception, a mixture of 'red' and 'green' light gives an appearance of yellow, not a reddish-green. The last-named combination is an example of colours that Hering considered to oppose each other. His theory was based on the proposition that separate opponent colour channels exist in the retina for conducting information about colours that seemingly oppose each other. Hering suggested that there are three such channels which he described in terms of perceptual responses of colour: one for red versus green, another for yellow versus blue and the third for the apparent opponency between black and white. He formulated the notion of opponency in the proposition that channels of neurones within the retina respond with different polarity depending upon the prevailing stimulus wavelength. For example, he argued that one channel signalled redness when it responded in one direction and greenness when it responded in the opposite direction, and there was no output when neither red nor green was seen.

In its original form, the theory seemed able to explain the existence of five apparent psychological primaries, namely red, green, yellow, blue and white. It also provided a satisfactory explanation of complementary colours seen in negative after-images (section 8.8.4). For example, the after-image of a bright 'red' spot of light when seen against a white background is perceived as green, and the after-image of a bright 'yellow' spot of light viewed on a white background is perceived as blue.

Experimental evidence in favour of the principle of opponency derives from two sources. Firstly, there is electrophysiological evidence of spectral antagonism from neurone responses both in the retina and in the visual pathways of the brain. Secondly, results of special colour-mixing studies using human observers also support the phenomenon of opponency. The physiological evidence has been described in section 8.3. The evidence from colour-mixing studies comes largely from investigations using a hue-cancellation technique. In this method, pairs of wavelengths that elicit opposite hue responses (red and green, for example, or yellow and blue) are mixed by superimposition and their relative energies are varied until a ratio is determined at which neither of the opposite hues can be detected in the mixture. The relative energies required for the hue cancellation serve as a measure of the amplitude of the relevant colour-opponent function at each wavelength.

While most current models of colour vision incorporate the two colour-related opponent channels of Hering's theory, very few accept the idea of a black versus white opponency – the third of Hering's opponent channels, where mixtures are not so clearly antagonistic. Mixtures of black and white cannot eliminate the original components of the mixture as can the colour pairs of red/green and yellow/blue. The perception of whiteness and blackness depends considerably on spatial contrast and the distribution of reflectances among adjacent objects, aspects of the retinal image which are also critical in colour vision.

The *retinex theory* of colour perception was advanced in 1959 by Edwin Land [12] to explain the phenomenon of colour constancy (section 8.3.3). The retinex theory (the word is a contraction of the words <u>retin</u>a and cort<u>ex</u>) suggests that, somewhere in the visual pathway between the retina and the cortex, the nervous system assigns a colour to an area of the field of view. The theory assumes that this is achieved by comparisons of the lightnesses of surfaces having different wavelength-reflecting properties. The only prerequisite for such wavelength comparisons to be made is that the illuminating light comprises a wide range of wavelengths.

In support of this theory, Land described several strikingly simple experiments with colour, the results of which could be explained neither by the Young–Helmholtz trichromatic theory nor by the Hering opponent colour theory. His most graphic demonstrations involved viewing a multicoloured scene of abstract shapes, as in a Mondrian painting. The merit in choosing an abstract painting was that none of the shapes had any visual meaning. When viewed under white light (such as daylight) the colour names associated with each area of the scene were noted along with the wavelength composition of the light coming from each area. Typically, for example, the wavelength composition of the light coming from each area seen as red would be rich in long and sparse in medium and short wavelengths. When the spectral composition of the incident light was changed so that the area previously seen as red now reflected a much greater amount of light of medium and short wavelengths rather than long wavelengths, the same area still looked red. If, however, all the surrounding areas were masked with black card to permit the viewing of a single coloured shape, the area that was previously seen as red now appeared green, and this was more consistent with the greater amounts of light of medium and short wavelengths coming from its surface. This was a beautiful and simple demonstration of the dependence of colour on the availability of comparative judgements in the field of view. Clearly, for the perceived colour of a designated area of the Mondrian display to be independent of the wavelength composition of the light reflected from its surface, the perceptual response must be a function of the light of different wavelengths reflected from all other adjacent areas in the visual scene.

The phenomenon of colour constancy, therefore, can only be explained if we assume that somehow the visual system's analysis of colour does not depend upon absolute energies. The corollary of this is that colour vision is essentially a process of comparative judgements. Consider the example of a multicoloured scene illuminated with monochromatic light, or alternatively viewed through a narrow-bandpass interference filter. When the scene is illuminated with monochromatic light of long wavelengths, all the objects and areas of that scene will be perceived to differ only in terms of lightness, depending on the amount of long-wavelength light reflected from each surface. Objects that reflect mostly short wavelengths, and are typically seen as blue in white light, will look very dark under monochromatic long-wavelength light, while objects with high reflectance properties for long wavelengths will look very light. When objects with other spectral reflective properties are viewed under monochromatic light they will appear to differ in lightness and, if special viewing conditions prevail, they will be perceived as being of various shades of grey. In other words, they will not be distinguishable in terms of hue. Furthermore, the lightness within the scene will not change even if the total amount of energy of the long-wavelength light illuminating the scene increases or decreases. If the same scene were next to be illuminated with monochromatic light of medium wavelength and finally with short-wavelength monochromatic light, the same objects would generate two further, but different, sets of lightnesses depending on their selective wavelength-dependent reflectances. In each instance, the visual system is assigning lightnesses to the different objects by comparing them with each other. Now, since the long-wavelength light will be absorbed preferentially by the 'red' cones in the retina, and the medium and short wavelengths by the 'green' and 'blue' cones respectively, we now have a system for coding lightnesses by each set of photoreceptors. This, of course, is the only way in which the receptors can code information, for the cones themselves cannot see colour.

Land's demonstrations show too that the precise amounts of each of the long, medium and short wavelengths reflected from a surface are unimportant, since the lightness produced by each wavelength individually is independent of its physical intensity. Lightness discrimination, therefore, is the basis of colour vision; since the perception of different lightnesses is a comparative phenomenon, it follows that colour vision itself is a comparative phenomenon. Consequently, in order to maintain constancy of the colours in a scene, it is necessary for the analysis of colour to proceed not only without reference to the absolute energies of the light incident on the retina, but also without reference to the precise wavelength composition of the light. Our constructs of the real world clearly approximate more closely to the true constants of surfaces, namely their relative reflectances.

While the retinex theory is elegant in its explanation of experimental observations, its major shortcoming is that it is inconsistent with the physiological facts of visual processing in three significant respects. First, it supposes (against considerable evidence to the contrary) that there are three independent neural channels, identical to the three cone response characteristics, which are entirely segregated between the retina and cortex. Secondly, it has no place for the principles of opponency and argues that all colours can be generated by comparing responses to lightnesses in each of the three hypothetical receptor-based channels. Finally, it supposes that there are just three neural channels providing direct input for comparisons in the visual cortex, yet current physiological evidence points to the presence of multiple cortical channels having narrow-band wavelength sensitivities that are distributed widely throughout the visible spectrum. It is instructive, therefore, to see how we can construct a framework to explain both the physiological facts and the observable phenomena of colour vision. That framework will inevitably be a composite model, incorporating aspects of trichromacy, opponency and lightness comparisons.

8.4.3 A composite model of colour vision

Of the three major contending theories of colour vision, none alone provides an adequate explanation of how we see colour. Most of the modern approaches to the modelling of colour vision have involved theories that combine elements of both the trichromatic and opponent processes. These models have been termed 'zone theories' of colour vision because they propose that the different stages of trichromacy and opponency occur within different layers of the retina. Hence, in a strictly neurological sense, the operations on the nerve signals are zoned according to the coding taking place in different retinal neurones. For example, trichromacy is represented at the receptor level by the three cone types, and opponency occurs in the neural processing of the lateral interconnections among the horizontal and amacrine cells of the retina. But a shortcoming of all these models is that they offer no scheme to account for the processing of signals in the cortex. Furthermore, none of them provides an adequate explanation of colour constancy. In this latter respect there is considerable divergence of the major theories. To explain colour constancy the Young–Helmholtz hypothesis attributes to the visual system the ability to discriminate between the physical spectral properties of surface reflectances and illumination. Since the photopigments in the visual receptors respond only to the spectral energy distribution of the light entering the eye, there is no suggestion in the trichromatic model as to how that discrimination is achieved. In the opponent theory it is assumed vaguely that colour constancy can be explained through antagonistic response ratios. Land, on the other hand, proposed in the retinex theory that a particular colour sensation is derived from the comparison of three lightness values, the colour constancy of an object being related to the colours of adjacent objects in the visual scene. Land's theory accords more closely with empirical observations since colour constancy (and also lightness constancy) is not maintained when adjacent areas with different spectral reflectances are masked from view. In such instances the visual response is mediated purely in terms of the spectral composition of the light entering the eye and not in terms of the surface spectral reflectances of the object. Explanations for the propositions of Land are centred on the visual cortex, where there are complex spatially organised colour-opponent cells which would be capable of comparing lightness among selective waveband sets derived from adjacent areas in the field of view.

The features of receptoral trichromacy and neural opponency are now widely accepted by most colour scientists as fundamental properties of the colour-vision process. The inclusion of waveband-specific lightness detectors in the model has still to receive a general acceptance. Substantial evidence is now accruing in its favour, however, and it is therefore instructive to incorporate this feature in a composite model of colour vision. The following descriptive model is based upon this premise.

The first stage of the coding of information for colour is via three types of retinal photoreceptors (cones) having overlapping absorption spectra with peaks at approximately 560, 530 and 425 nm. The spectral sensitivities of different types of cone receptor are functions of the relative rates of absorption of photons by the photopigments, but the output signal of an individual receptor preserves no information about the wavelength of the light absorbed. By spatial integration of signals across the neural network in the retina, input to the ganglion cells consists of excitation or inhibition from different cone types. This includes pairing of signals in an opponent manner from long- versus medium-wavelength cones and also short- versus medium-plus-long-wavelength cones. These are sometimes referred to as 'red'/'green' and 'blue'/'yellow' opponent signals, where the word yellow represents the combined input from medium- and long-wavelength receptors, although at this level the cells contain information only about lightness and not about colour. The spatial arrangement of the receptive fields of these opponent ganglion cells is particularly well suited for detecting contrast boundaries. Information about hue and colour saturation is, however, encoded by opponent ganglion cells that have different spectral sensitivities (action spectra).

Within a given class of 'red'/'green' opponent cells (R+/G– or R–/G+) there exists a wide variety of cells responding positively or negatively to different wavelengths. The wavelength to which a cell is unresponsive is its neutral point. The distribution of different neutral points throughout the retina means that very small differences in wavelength can be signalled via the ganglion cells. For example, groups of colour-opponent ganglion cells, whose neutral point lies between the spectral loci (or principal wavelength compositions) of two adjacent surfaces that give rise to a chromatic border, will respond maximally either by excitation (positively) or inhibition (negatively) and thereby signal the presence of that border. The minimal chromatic difference detectable depends on the relative frequency distributions of ganglion cells with different neutral points. Since this distribution is somewhat heterogeneous, colour discrimination ability varies across the visible spectrum. Furthermore, since it is only the red/green opponent cells that exhibit this feature and since the neutral points of most of these cells are at medium wavelengths, it is not surprising that colour discrimination is best in the middle of the visible spectrum. On the other hand, the 'blue'/'yellow'-sensitive cells all appear to have the same neutral points. Consequently, wavelength discrimination is particularly poor at the short-wavelength end of the spectrum.

The visual system also has a poorer spatial resolving power at the short-wavelength end of the spectrum. This is because the 'blue' cone receptors are scarce throughout the retina; additionally, the receptive fields of the 'blue'/'yellow' opponent ganglion cells are much larger than those of the 'red'/'green' mechanism. The short-wavelength receptors, therefore, contribute very little either to the resolution of detail or to the detection of luminance contrast. Since the fovea is very sparsely populated with 'blue' cones and contains a large number of 'red'/'green' opponent ganglion cells with small receptive fields, both resolution and chromatic discrimination of medium and long wavelengths are best in this region of the retina. The 'green' cone mechanism appears to predominate among the ganglion cell activity in the fovea, while the 'red' cone mechanism becomes increasingly dominant towards the retinal periphery. This variation in dominance of the different types of opponent ganglion cell at different distances on the retina from the foveola is the principal reason why there are differences in colour vision between central and peripheral areas of the field of view.

The 'red'/'green' opponent cells contribute to the sensations of both brightness and colour contrast because of their multiplicity of wavelength neutral points and their centre/surround receptive field spatial characteristics. The 'blue' cones, however, appear to play no part in the processing of information about brightness. On the other hand, the perception of whiteness requires involvement of signals from all three receptor types and is, therefore, a truly trichromatic response achieved by means of opponent processing.

The interactions between the opponent retinal ganglion cells result in a fine spectral tuning of colour-coded cells in the visual cortex. In addition to these wavelength-selective cells in the cortex, which have extensive and complex surround receptive fields, this part of the brain also has wavelength-opponent and wavelength-differencing double-opponent cells. Between them, these cells encode lightness in the stimulus in accordance with an algorithm having the following three properties:

(a) the ability to gauge wavelengths
(b) the ability to compare wavelengths between adjacent boundaries, thus providing for contrast detection, and
(c) the ability to assign colours to a stimulus while largely discounting the wavelength composition of the light entering the eye, thereby preserving colour constancy.

The presence of luminance and chromatic contrast boundaries in the physical stimulus is essential for this complex processing of visual signals. In a visual field devoid of contrasts there is no perception. It is almost as though the eye is used as a hand that rubs across the contours of a scene in order to enhance their contrast, and constancy is preserved where the *ratio* of the responses to adjacent surface areas is conserved.

Consequently, as so often in scientific endeavour, we have the framework of a model that combines the fundamental elements of several theories. In this composite framework for a model of colour vision the work and different hypotheses of Newton, Maxwell, Helmholtz, Hering and Land can now be reconciled. What is missing from the model, however, is the final link in the process by which a mere sensation is converted into a meaningful perception. There is as yet no satisfactory model to explain how we synthesise the mass of information about lightness and colour in the visual scene, for example, to enable a car-owner to recognise his own little red Citroën 2CV among many other vehicles in a car park. Perceptual synthesis of this nature clearly occurs at another level in the brain and seems to be based on some form of hypothesis testing that is influenced by selective attention. In other words, our ability to recognise a visual object not only depends upon our familiarity with its form, brightness and colour, but also on what we expect to see. In this respect we can certainly agree with Maxwell that the recognition of colour is a mental activity, largely independent of the physical properties of the objects in the real world; in fact the whole act of perception may be regarded as a mental construct by which our sensory systems attempt to place meaning on the ever-changing stimuli in the world around us.

8.5 ATTRIBUTES OF COLOUR

The fundamental principles behind our understanding of the attributes of colour were succinctly summarised by Isaac Newton when writing about his classical experiments with a prism conducted in the second half of the 17th century. In describing the properties of a spectrum produced by prismatic dispersion he wrote with profound insight about the distinction between subjective and objective attributes of colour: 'For the Rays to speak properly are not coloured. In them there is nothing else than a certain Power and Disposition to stir up a Sensation of this or that Colour.'

By distinguishing between light and colour, Newton showed he was aware that the perception of colour is dependent upon a particular relationship between physical stimulus properties and the subjective interpretation of a sensation. In this respect he correctly used the term 'colour' to indicate a subjective response. Since that time, the word colour has been widely misapplied to refer both to the wavelength of light and also to the subjective response. When we now commonly speak of 'red light', for example, we really mean light of long wavelengths which is *perceived* as being red. This dual use of the word can sometimes be confusing, particularly since a colour perception can result from a variety of different stimulus conditions. Hence in colour science it is necessary to distinguish the electromagnetic and photometric properties of the stimulus from the psychological terms used to describe an observer's perception. It is simply a matter of convenience that we are able to define three physical properties of light that are loosely related to our perception of colour. But, because of phenomena such as colour constancy, there can never be a one-to-one match between the physical properties of a stimulus and the resulting perception.

Whereas colour perception may be described in terms of just three properties, it is dependent upon many more than three stimulus parameters. The three properties or psychological attributes of colour are hue, saturation and luminosity (i.e. brightness or lightness). These are principally, but by no means wholly, related to the stimulus dimensions of dominant wavelength, spectral purity and luminance respectively (though there are many exceptions to this generalisation). Changes in one stimulus variable may affect all three psychological attributes of colour. For example, a change in the luminance of a stimulus will not only affect its luminosity but also its hue and saturation. Similarly, a change in dominant wavelength will produce not only a change in hue but also changes in luminosity and saturation; likewise, changes in spectral purity will affect saturation, hue and luminosity. Furthermore, the lack of a simple relationship between the physical properties of a stimulus and perception (such as colour constancy or lightness constancy) is demonstrated by the observation that the perceived hue of a stimulus will depend on the context or set of circumstances in which it is viewed. This is an attribute of colour that presents considerable problems to the colour technologist whose task may be to assess the physical characteristics of an object from a knowledge of its colour appearance. The vision scientist, however, can use a knowledge of these complex interrelationships of physical properties and psychological attributes to provide a better understanding of the visual processes involved.

8.6 METHODS OF INVESTIGATING THE PERCEPTION OF COLOUR

8.6.1 Principles

The quest for a satisfactory explanation of how we see colour embraces several disciplines. These range from the physics and chemistry of the photoreceptor process to the physiology of nerve conduction from the retina to the visual cortex, finally including the psychology of perception. Although this may sound rather daunting to people new to the field, they should be encouraged by the knowledge that some of the greatest contributions in science have come from studies that cross the boundaries of the major disciplines of physics, chemistry, physiology and psychology. *Psychophysics* is the name given to studies which examine the relationship between stimulus (physics) and response (psychology), whereas studies of the relationship between sensory mechanisms (physiology) and what is perceived (psychology) come under the group heading of *psychophysiology*. For technical and ethical reasons, there are very few studies of the latter type in the investigation of human colour vision. Most such studies are based upon animal experiments from which only tentative inferences about human colour perception can be drawn. Therefore the main body of research into human colour perception has been in the field of visual psychophysics. Where animal behavioural data match the results of human psychophysical experiments, however, and where the anatomical organisation of the visual system is known to be similar in both animals and humans, inferences about the physiology of the human colour vision process are both possible and widely accepted.

In common with all biological processes, data on human colour vision is characterised by rather large variations between individuals in their responses to a constant set of stimuli. In psychophysical studies these variations are due, in part, to the measurement technique used for quantifying the perceptual response. There are two major types of psychophysical experiment. One depends on an observer reporting what is seen or perceived, such as by the use of colour names. The other approach involves the quantification of visual performance thresholds on tasks of detection, discrimination or colour-matching ability.

There are limitations to the use of colour naming as a scientific method for investigating the properties of colour vision. Colour names depend upon both a common understanding of language and a common use or application of colour words. To say that something is 'red' implies that we are also able to say when something is 'not red', and it is not surprising that there are large individual differences in such judgements. The task can become even more problematic when a colour name is used in conjunction with a known object such as a flower or fruit. Consider, for example, use of the colour word green. What I recall as green 'in my mind's eye' is highly unlikely to be the same as your mental image of the colour name. This will depend largely on how important it is to each of us to make discriminations of green in everyday life situations. The task also reveals the inadequacy of the English language in the existence of just the single word 'green' to describe the myriad of colour variations present in tree foliage and vegetation. In other languages, it is possible to find either a more sparse or even a richer vocabulary for describing colours. For example, some Australian aboriginal tribes appear to have a very limited repertoire of colour terms, whereas Inuits (Eskimos) have many verbal descriptions for the subtle variations in the colour of snow. It appears that colour names develop where they are most needed, and the colour names associated with different regions of the spectrum relate neither to equal wavelength intervals nor to equal perceptible intervals. Even Helmholtz, in his classical 19th-century treatise on physiological optics, commented on the somewhat capricious nature of colour names by observing that: 'borders between colours in the spectrum do not exist in reality but are drawn arbitrarily by us out of our love of nomenclature'.

For example, we would find considerable ambiguities in the words used to describe particular regions of the spectrum if we were to examine among people with normal colour vision the pattern of colour names associated with monochromatic light of different wavelengths. The regions of the spectrum that elicit different colour names represent the arbitrary colour-name boundaries to which Helmholtz refers. Such ambiguities are present not only when using the principal colour names (red, orange, yellow, green, blue and violet) but to an even greater extent when less commonly understood words such as aquamarine, turquoise or beige are used. Language, therefore, is a very imprecise clue to what we see, and our understanding of the perceptual properties of colour vision has been acquired largely from experimental psychophysical studies of visual thresholds.

8.6.2 Determining thresholds

Most experimental studies into the psychophysical investigation of human colour vision involve measuring performance thresholds. There are three principal psychophysical procedures for determining these thresholds and the precision of the threshold measurement is largely influenced by which of these procedures is adopted. The three principal methods are:
(a) method of adjustment or average error
(b) method of limits
(c) method of constant stimuli.

All three methods derive their names from the manner in which the stimulus presentation is controlled. The only prerequisites for their successful application are that the physical stimulus properties (wavelength, spectral purity and luminance, for example) be amenable to systematic quantifiable change and that the subjective attributes under investigation (such as hue, saturation and luminosity) are clearly understood dimensions of perceptual experience.

The *method of adjustment* is a simple procedure involving a continual adjustment of the stimulus properties by bracketing on either side of the visual threshold being investigated until a satisfactory endpoint is reached. By this means the error is averaged. In this procedure, the variation in stimulus property is usually controlled by the person who is making the judgements (i.e. the observer).

The *method of limits* involves a smooth progressive change in the magnitude of a defined stimulus property up to the visual threshold of detection or discrimination. The direction of change of the stimulus property (i.e. increase or decrease in stimulus magnitude) almost always influences the determined threshold value. Where this is likely, a better estimate of the threshold under investigation may be obtained from independent assessments of both ascending and descending methods for changing stimulus magnitude.

The *method of constant stimuli* is usually considered as the psychophysical method that provides the best estimate of a visual threshold; unfortunately, it is the most time-consuming procedure for obtaining the relevant information. It involves the observer making judgements on many presentations of the stimulus, at each of which the magnitude of the stimulus property is fixed (constant). Judgements from multiple presentations of the stimulus are made over a predetermined range of the stimulus magnitude. The task of the observer is usually to make either an absolute or a comparative judgement on each stimulus presentation; the essential feature of this method is that the stimulus properties are not varied during an observation. Finally, the threshold is determined statistically from a series of repeated judgements made at each discrete value of stimulus magnitude.

There is an element of imprecision in each of these psychometric methods, as is the case with any measurement procedure. So it is unrealistic to measure a threshold of detection or discrimination by a single observation. The quantum nature of light and the random fluctuations in neural 'noise', as well as variations in the criterion of judgement being used by the observer, all dictate the probabilistic nature of any threshold. A threshold, therefore, must be viewed as a statistical construct rather than being represented as a single value; ideally, the investigator should establish its confidence limits from repeated measurements. Frequently, for convenience, the results of psychophysical studies are presented in terms of vision sensitivity – as, for example, in expressing the relative sensitivity of the visual system to differing wavelengths. The choice of sensitivity rather than threshold measures is often made to normalise data from judgements made by different people. Since sensitivity measures may be expressed on a relative scale, their use makes it is possible to minimise (or eliminate) the effects of individual differences in absolute threshold values. In mathematical terms, sensitivity is simply expressed as the reciprocal of the threshold.

The choice of psychophysical procedure is an important factor influencing the precision with which the threshold can be specified. Of the three principal techniques, the method of adjustment is typically associated with having the greatest measurement error and the method of constant stimuli gives the best estimate of measurement error. Unfortunately, the latter method requires many more observations than the former in order to obtain an estimate of the threshold. In a simplified form, however, the method of constant stimuli is often used in commercial or industrial situations where colour samples are compared visually with a reference standard. In these instances, the sample and reference are usually viewed simultaneously side by side and the observer has to make a judgement of whether the sample looks the same as or different from the reference. While this paired-comparison application of the method of constant stimuli is satisfactory for judgements of 'same' or 'different', it is a poor means of determining the perceived magnitude of a difference between two stimuli.

Judgements of whether a sample and reference colour look the same are simplified estimates of a colour-difference threshold, which can also be expressed statistically in probabilistic terms. Such threshold values are sometimes called *just-noticeable differences* (JNDs). Consider, for example, a range of fabric samples all of which differ slightly in colour from a reference sample, and which are paired with the reference on many repeated occasions in a random order. Observers are asked to make judgements of 'same' or 'different' for each pair. Because of the variability in sensory function and human judgement, some errors in the judgement decisions are inevitable. Thus, decisions of 'same' would be given on some occasions for certain sample/reference pairs, while on other occasions the opposite judgement of 'different' would be given for the same pairs. If a sufficiently large number of paired comparisons are made between the reference and sample colours, the data can be represented graphically as a plot of the frequency of pairs judged as different versus a physical measure of the colour difference between each of the sample pairs. The resultant graph would be an S-shaped curve. This curve can easily be converted to show the probability of the physical parameter of colour difference being associated with judgements that the paired stimuli visually appear as different or the same. When expressed as a probability function this curve obeys the binomial distribution, from which it is possible to determine the likelihood that samples of a stated colour difference will be judged as different from the reference. An understanding of the statistical nature of a colour-difference threshold is crucial in deciding what differences can be tolerated for a satisfactory match, and it clearly has commercial implications.

Both the principles of psychometric method and the consequent statistical inferences are as important to the colour technologist as they are to the colour scientist. This is because they are major determinants of the confidence that can be placed on the results of studies involving subjective judgements. For similar reasons, it is often instructive to view the results of standard clinical or occupational tests of colour vision in terms of the psychometric method which the tests employ. The common feature of all these measurement problems is that they involve determining thresholds. These measurement limitations are an inevitable feature of all psychophysical investigations of vision, and the colour technologist must have an understanding of how they vary throughout the gamut of colour experience. Section 8.7, therefore, describes the sensitivity limits for discrimination of the three stimulus properties of wavelength, spectral purity and luminance, since it is these three properties that primarily determine colour quality.

8.7 DISCRIMINATION OF COLOUR ATTRIBUTES

8.7.1 Wavelength discrimination

It has been estimated that a person with normal colour vision can discriminate something in the region of seven million different colours. Such powerful sensory discrimination is, of course, not solely related to the spectral properties of light; indeed, there are fewer than 400 unique wavelengths in the visible spectrum. (The limits of the visible spectrum are generally considered to lie at around 380 nm and 770 nm although, inevitably, there are small differences between people.) The vast number of different colours we can perceive are the consequence of the very large number of combinations of just-noticeable differences (JNDs) in wavelength, purity and luminance. For example, a change from high to low luminance of a surface pigment whose dominant wavelength is 580 nm and which has a purity of about 50% would result in a perceived change in colour from yellow to brown. Yet this change in colour appearance occurs without any change in the dominant wavelength of reflected light from the stimulus, which emphasises the complexity of the relationship between physiological properties of the visual system and the resultant perception. Clearly, colour naming is not an appropriate means of describing our ability to discriminate differences in wavelength.

Most investigations of wavelength discrimination employ either the psychophysical method of limits or the method of constant stimuli. A typical method of making such measurements is to arrange two samples of monochromatic light in a circular field divided horizontally into two halves (i.e. a bipartite field). These can then be viewed and compared simultaneously. For example, if the top half of the bipartite field contains a reference light of wavelength λ, small incremental changes in wavelength (i.e. $\lambda + \Delta\lambda$) are then made to the bottom half of the comparison field until the two halves are no longer judged to be the same colour. (In making such judgements, adjustments must be made, for each change in wavelength, to maintain a constant luminosity of both halves of the bipartite field by weighting the luminance of each wavelength according to the relative spectral sensitivity response of the observer's visual system.) The value of $\Delta\lambda$ is then said to be the JND in wavelength; quantitatively it represents the wavelength-difference threshold. When such judgements are made using all wavelengths throughout the visible spectrum as a reference, the resultant difference thresholds provide a wavelength-discrimination curve (Figure 8.13).

The essential characteristics of wavelength discrimination for a person with normal colour vision are the presence of three threshold minima in the long-, medium- and short-wavelength parts of the spectrum. These are the values at which sensitivity to changes in wavelength are greatest and are of the order of 1 nm. Their presence is due, in part, to the existence of the three types of retinal cone receptors that have different spectral absorption characteristics (section 8.3.1) and the opponent retinal ganglion cells' varying action spectra and neutral points. In fact, it is possible to obtain a very close approximation to the shape and magnitude of the wavelength discrimination curve from theoretical consideration of the principles of trichromacy. The approach

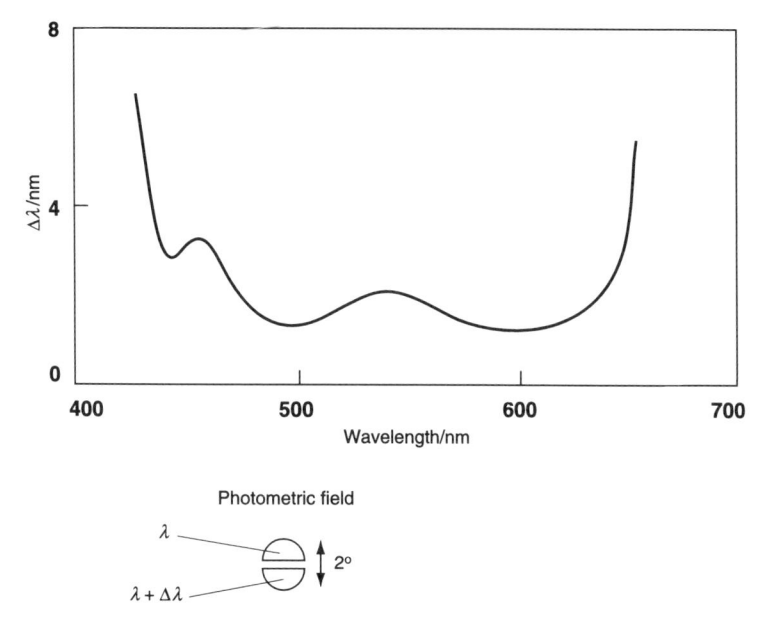

Figure 8.13 Wavelength-discrimination curve

uses the mathematical notion of the line element (i.e. distance element) by which the trichromatic properties of the visual system are represented as weighted linear combinations of the three empirically determined relative spectral sensitivities of the retinal response systems (sections 3.6 and 3.7) [20]. Where the rate of change in the response ratio of the three cone systems is greatest for a given wavelength difference, the sensitivity to a change in wavelength is at a maximum. The principle of trichromacy is not the whole story here, however. Electrophysiological studies on monkeys (whose wavelength discrimination is similar to that of humans) have shown that good wavelength discrimination is related to a high change in the firing rate ratio of the major opponent neurone types (i.e. 'red'/'green' and 'blue'/'yellow' cells) at the level of the lateral geniculate nucleus in the brain [21], and presumably also in the retinal ganglion cells. It would seem that short-wavelength discrimination is governed largely by the 'blue'/'yellow' opponent cells and long-wavelength discrimination by the 'red'/'green' opponent cells.

Since wavelength discrimination is a function of the strength of opponent signals, it is not surprising to find that certain viewing conditions influence the size of the JNDs in wavelength. These include luminance level, field size (i.e. angular subtense of the stimulus), retinal eccentricity and the chromaticity of the surround visual field. Decreasing luminance and reducing field size to less than 1° both show progressive losses in wavelength discrimination, which are greater in the short-wavelength range of the spectrum. For these same viewing parameters there is also a shift in short-wavelength discrimination minima towards shorter wavelengths. These effects occur because reduced luminance and field size cause a greater reduction in the number of responsive spectrally opponent cells of the 'blue'/'yellow' variety than of the 'red'/'green' type. Again, it is the size of the *ratio* of signals in the two types of opponent cell that seems to correlate best with discrimination. Changes in wavelength discrimination with increasing distance from the fovea are also the result of systematic variations in the distribution of the different types of spectrally opponent ganglion cells across the retina. While the fovea is dominated by 'red'/'green' opponent ganglion cells, there is a gradual shift in dominance towards 'blue'/'yellow' opponency with increasing eccentricity towards the peripheral retina. Thus, with increasing retinal eccentricity, wavelength discrimination shows a rapid fall-off in the midspectral region, until eventually (at about 40° away from the foveola) it ceases to be trichromatic.

The nature and content of the surrounding visual field in which we view an object significantly influence the way in which we perceive that object. This is as important for all the parameters of vision (such as size, shape and movement) as it is for judgements about colour. To investigate the effects on colour of variations in the properties of the surround field, a bipartite stimulus field can be used surrounded by a large background annulus. The two surround field properties of particular interest are its luminance and chromaticity. Where the luminance ratio between test field and surround exceeds 10 : 1, wavelength discrimination rapidly deteriorates. Surround fields of different chromaticity, on the other hand, produce systematic shifts in the positions of the minima of the just-noticeable wavelength-difference thresholds. In both these instances lateral interaction effects within the neural organisation of the retina modify the dominance of the receptive field characteristics. This, in turn, alters the signal ratios among the wide range of spectrally different opponent ganglion cells.

To people whose work requires critical judgements of colour difference, the practical consequences of the many ways in which wavelength discrimination is influenced by the viewing context should be self-evident. Clearly, to obtain repeatable judgements of discrimination, the viewing conditions should not change and should be standardised. For maximum sensitivity the test field should be no smaller than 2° angular subtense, viewed foveally at photopic levels of luminance against a neutral surround of a luminance level similar to that of the target.

8.7.2 Discrimination of spectral purity

Deciding when an object no longer appears white, but contains just a hint of colour, is one of the more difficult perceptual judgements to make. This is known as the *saturation threshold* of a colour and it may be determined psychophysically. For example, the saturation of the colour green depends on the relative quantities of perceived greenness and perceived whiteness. As a psychological attribute of a colour, saturation has no meaningful units. Hence it is expressed by reference to the physical stimulus property of spectral (or colorimetric) purity, which is defined as the percentage ratio of the luminance of monochromatic light to the total luminance.

A typical method of determining saturation thresholds is to present pairs of stimuli in a bipartite field, using the method of paired comparisons. For absolute thresholds, one half of the bipartite field (the reference) is kept as neutral and the other half is a mixture of a monochromatic light and a neutral light. As the luminance of the monochromatic component is increased, so the luminance of the neutral component is decreased in such a way as to keep the sum of the luminances constant. For determining difference thresholds, the proportion of monochromatic to neutral components is varied between the two halves of the bipartite field.

Threshold measures of JNDs in spectral purity from white show marked variations throughout the spectrum (Figure 8.14). Thresholds are smallest, and hence sensitivity is greatest, for both short and long wavelengths. As might be expected, the smaller the absolute threshold, the greater is the number of JND steps from white to a fully saturated colour. For example, we can perceive and discriminate many more steps of degrees of redness, from white through pink to deep red, than we can with yellowness. Purity discrimination is therefore poorest at those wavelengths where absolute

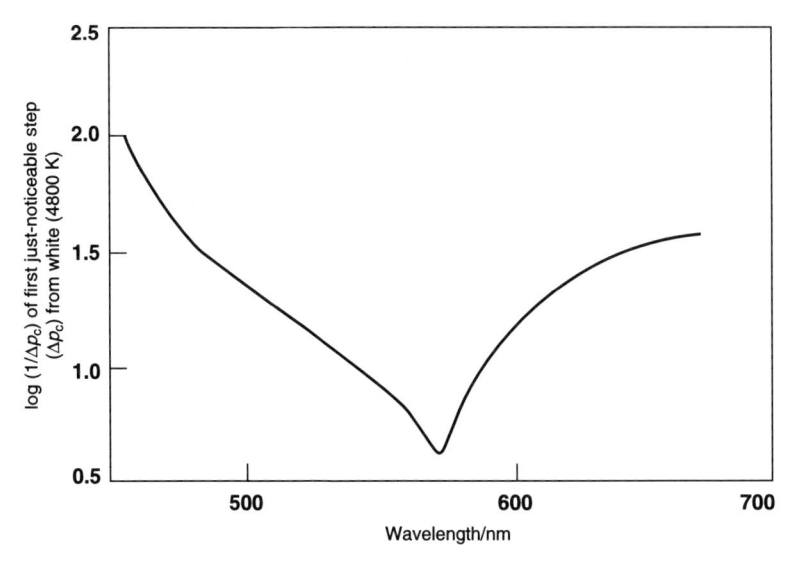

Figure 8.14 Log ($1/\Delta p_c$) of first just-noticeable step (Δp_c) from white

sensitivity is also poor. In people with normal colour vision these wavelengths are in the region of 570 nm. Such marked variations in saturation thresholds have clear implications for the colour technologist when establishing commercial tolerance limits for the quality control of colour.

Since perceived saturation is a function of the wavelength and luminance components of a stimulus, it is reasonable to seek an explanation for the phenomenon in the combined activity of both chromatic and luminosity (i.e. achromatic) channels in the visual system. If the opponent cells code hue and the non-opponent cells code achromatic luminosity, then it might be supposed that the spectral purity of the stimulus would be coded by the ratio between opponent and non-opponent cell activities. Recent studies have shown that the function given by this ratio at different wavelengths satisfactorily describes the purity-discrimination curve [22–24]. It will be recalled that the term 'non-opponent' refers only to the non-opponent characteristics of such cells with respect to variations in wavelength. Cells of this sort do indeed show opponent response characteristics with respect to the spatial properties of a stimulus within their receptive fields for changes in luminance.

8.7.3 Discrimination of luminance

For achromatic stimuli, a change in luminance is perceived as a change in luminosity (i.e. the lightness of an object or the brightness of a light source). Discrimination in this domain is considerably influenced by the prevailing stimulus intensity; as luminance increases we need progressively larger incremental changes in luminance to produce a just-noticeable difference (Figure 8.15). The visual system is much more sensitive to a change in luminance at lower rather than higher intensities, and it is at low intensities that the incremental luminance steps are smaller for equal changes in luminosity.

This variation in discrimination as a function of stimulus intensity is a property of most sensory modalities. It was first investigated in detail by Weber, a 19th-century German physiologist, in a series of delightful experiments on the sense of touch and weight discrimination. These original studies on perception of stimulus magnitude provided the foundation of present-day psychophysics. The first detailed investigation into luminance discrimination was carried out by Fechner, who was Weber's successor as professor of physiology at Leipzig. Fechner showed that, over a fairly large luminance range, the JND in luminance is a constant fraction of the stimulus intensity: in other words, the ratio $\Delta L/L$ is a constant (where L represents the stimulus luminance and ΔL the JND in luminance). This relationship is so fundamental in visual psychophysics that it is now known as the *Weber–Fechner law*, and it holds true over most of the photopic range. Under optimum viewing conditions the Weber–Fechner constant is of

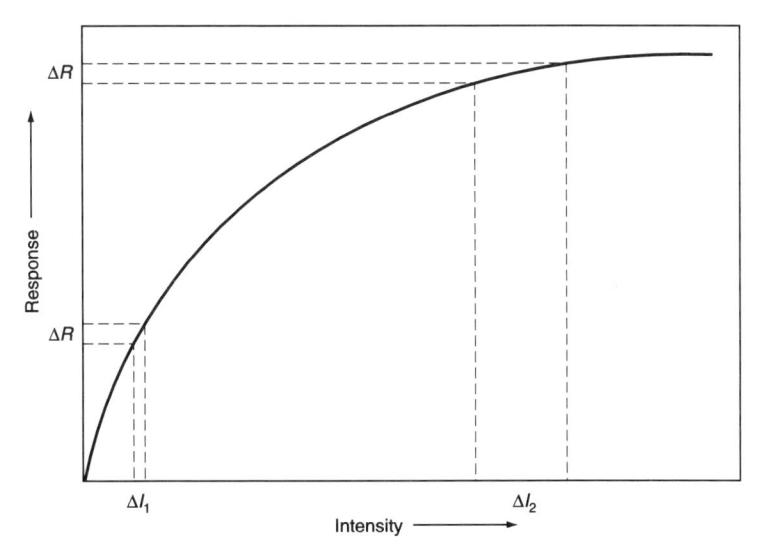

Figure 8.15 Relationship between response and stimulus intensity

the order of 0.01; this means that, at best, we can detect or discriminate a 1% change in luminance. The relationship suggests that, over a wide photopic luminance range, the discrimination of luminance is based upon the ratio of nerve impulse frequencies from retinal ganglion cells with adjacent receptive fields. Departures from the Weber–Fechner law occur at high luminances, because of saturation in cone receptor activity, and also at low luminances, where the lack of linearity in the relationship between stimulus intensity and subjective response occurs principally because under these conditions vision is mediated by rods (cones and rods operate over different luminance ranges and have different receptive field properties).

The averaged amount of light incident across the retina determines whether vision is mediated as a photopic, a mesopic or a scotopic response. But there is no sharp switch-over from one vision state to another. In the process of adapting to a new average light intensity the visual system changes very quickly when moving from a low to a high light intensity (i.e. light adaptation), but the sensitivity change is very slow when moving from a high to low light intensity (i.e. dark adaptation). If, after a period of high-intensity light adaptation, a person's sensitivity to light is measured as a function of time, a discontinuity in the rate of adaptation is observed. This occurs where there is a change-over in the predominance of rod and cone activity in the retina and is found for absolute light-threshold judgements as well as for luminance-difference thresholds. These adaptation characteristics are just one of several properties of vision that provide evidence for the existence of two fundamentally different detection mechanisms (i.e. the rod and cone receptor systems). Further examination of the luminance-difference thresholds across the visible spectrum reveals that, when the luminance-difference threshold is expressed as a function of wavelength, it closely mirrors the absorption spectrum of the rod photopigments at low scotopic luminances. This is clear evidence that the wavelength-related response of rods is simply a function of the probability of the absorption of light by rhodopsin and, because only one receptor type is activated, perception is achromatic. There is one exception to this relationship, however. For wavelengths greater than 600 nm both rod and cone thresholds are similar and, even at absolute light thresholds, a colour of dull red is perceived.

At high photopic luminances, the luminance-difference thresholds vary as a function of wavelength (Figure 8.16) but there are systematic differences between the scotopic and photopic response functions. The luminance discrimination for cones is poorer than that of rods for all wavelengths less than 600 nm, and the wavelength that corresponds to the minimum JND (i.e. maximum sensitivity) at photopic luminances is shifted to a longer wavelength than its counterpart under scotopic conditions. This photopic curve of relative spectral sensitivity represents the combined response functions of the three cone types and is largely a result of the absorption spectra of the three cone photopigments. At photopic luminances, the intensity of each wavelength that corresponds with the absolute threshold for detection will be perceived as coloured. The interval between the scotopic (achromatic) and photopic (chromatic) luminance-difference thresholds is, therefore, aptly known as the *photochromatic interval*.

Another distinction between the scotopic and photopic spectral luminance-difference thresholds is the shift in wavelength at which the smallest JND occurs. The change in wavelength of peak sensitivity from 507 nm to 555 nm is known as the Purkinje shift, after the Czechoslovakian physiologist who first reported it in 1825. He observed that in daylight the eye is most sensitive to light perceived as a yellow-green colour, whereas at night the maximum sensitivity is to light perceived as blue-green. Observations like this are part of everyday experience and can be noticed, for example, in the change in the relative brightness of 'yellow' and 'blue' flowers between daylight and dusk; at midday 'yellow' flowers appear brighter than 'blue' flowers but the reverse is the case at dusk as the adaptation level moves into a low mesopic state.

In colorimetry, the absolute differences in these wavelength-dependent luminance-discrimination thresholds are ignored. What is of primary interest is the relative sensitivity of the eye to different wavelengths. Consequently it is customary to express the data in terms of sensitivity scale, which is the reciprocal of luminance-difference threshold measures. When the data for each function are normalised to provide unitary sensitivity at their most discriminable wavelengths (i.e. 555 nm and 507 nm), the

DISCRIMINATION OF COLOUR ATTRIBUTES 475

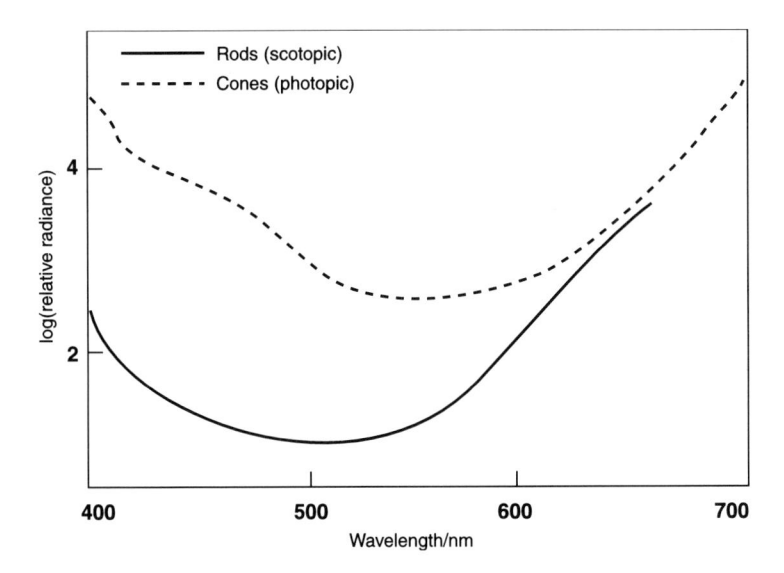

Figure 8.16 Luminance difference threshold of wavelength under photopic and scotopic viewing conditions

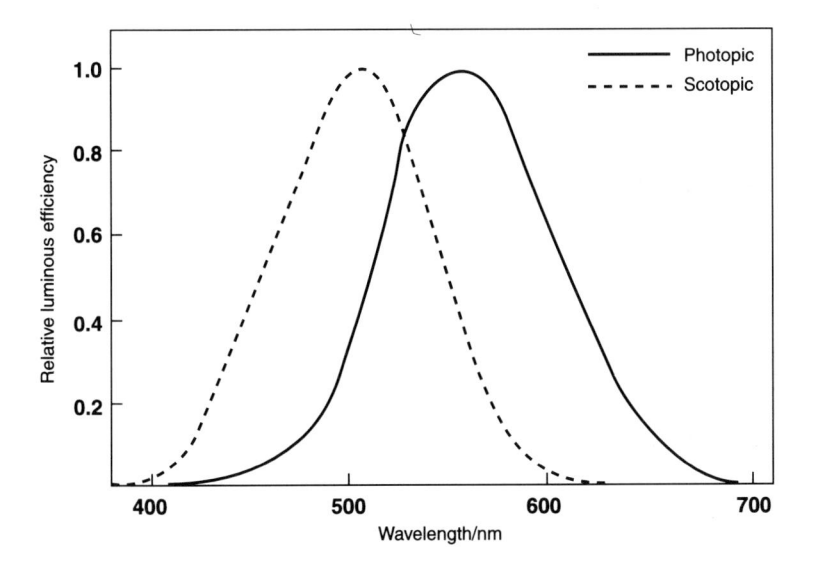

Figure 8.17 Photopic and scotopic relative spectral luminosity functions, obtained by normalising the threshold curves in Figure 8.16 with respect to the minimum luminance difference thresholds

resultant curves are the photopic (V_λ) and scotopic (V'_λ) relative spectral luminosity functions (Figure 8.17). They are so called because they represent the relative amount of light required from each wavelength of an equal-energy spectrum to give a threshold response such as detection. The shape of these wavelength response functions is the same, however, whether the criterion of judgement is one of light detection or of equal luminosity (brightness). Helmholtz, in the mid-19th century, remarked on the perceptual consequences of this property of vision when he stated that: 'the most luminous of the prismatic colours are the yellow and orange. These affect the senses more strongly than all the rest together, and the next to these in strength are the red and green. The blue compared with these is a faint and dark colour and the indigo and violet are much darker and fainter, so that these compared with the stronger colours are little to be regarded.'

Intermediate relative spectral sensitivity curves may also be determined for light of intensities ranging between photopic and scotopic luminance levels. These are known as mesopic relative spectral luminosity functions. Their shape, and the wavelength of maximum sensitivity, change progressively with luminance level from full photopic to full scotopic levels and they reflect the relative dominance of rod and cone activity. It is unnecessary to postulate the existence either of inhibition of the rods by the cones at high luminance levels, or of inhibition of the cones by the rods at low luminance levels, to explain the shapes of the many mesopic luminosity functions that exist between full photopic and full scotopic adaptation levels. Nevertheless, there does appear to be increasing evidence for some interaction between rods and the 'blue' cone mechanism, and it is believed that this is the reason for the greater variation in short-wavelength threshold values between individuals. The argument is that this rod and 'blue' cone interaction can explain why there is so much variability in mesopic spectral sensitivities, and this feature of vision is the principal reason why no agreement has yet been reached on producing standardised mesopic relative spectral luminosity functions. Although it is known that these variations and anomalies exist for mesopic vision, in practice most commercial photometers use correction filters to weight their response either in terms of the CIE photopic or scotopic luminosity functions depending on whether high or low illuminances (i.e. illumination levels) are being measured.

The number of just-noticeably different steps of luminance we can discriminate varies not only with the wavelength of light but also with the surrounding luminance. The area immediately surrounding an object is sometimes called the inducing or adapting field. As the luminance of the inducing field increases, the range of experienced luminosity or brightness also increases. This means that, for a given luminance range, more just-noticeable luminance-difference steps become discernible. The effects can largely be explained by reference to lateral (i.e. spatial) interactions within the nerve fibres of the retina. Nerve signals between adjacent receptive fields, receiving light of different intensities, are modified by inhibition, which increases the perceived luminosity difference and thus heightens the contrast boundary. These characteristics of the perception of luminosity emphasise the marked influence of viewing context both on luminance discrimination and also of the perceptual scaling of luminosity. Such factors clearly need to be appreciated by colour scientists or technologists if accuracy and reliability is to be achieved in those colour aspects of commercial quality control, which are based upon visual judgements of luminance difference.

8.7.4 Chromaticity discrimination

While information about discrimination thresholds and the perceptual scaling properties of the independent physical variables of dominant wavelength, spectral purity and luminance are essential to an understanding of the visual process, it has limited practical significance to the colour technologist who is interested primarily in chromaticities. *Chromaticity* is the term used to describe some property of a stimulus in terms of its colour quality. It is an intermediate term that is neither a measure of perceived colour (i.e. hue) nor a measure of radiant energy. Chromaticity is derived from the proportions of the three standardised primary reference wavelengths (primary colour-matching stimuli) required to provide a colour match with the stimulus being described. In specific colorimetric terms, the chromaticity of a stimulus is defined by its chromaticity coordinates, which are the ratios of each tristimulus value of a colour to their sum.

For the person involved in colour-quality control who has to decide whether two colour samples would be accepted visually as the same, it is usually more relevant to know what chromaticity difference can be tolerated before two samples are judged as being different. For this purpose it is instructive to derive JND thresholds for chromaticity discrimination throughout the relevant area of colour space. The first serious attempt at such a study was undertaken by MacAdam at the Eastman Kodak Research Laboratories in New York [25,26]. To do this, he designed a special colorimeter having a bipartite comparison field of view that allowed very small variations in chromaticity to be made while automatically maintaining constant luminance. Chromaticity discrimination thresholds were determined for 25 'colour' samples and the results plotted as small areas in the chromaticity diagram. The centre of each of these small areas was the reference colour from which chromaticity difference judgements were made. The result was a large number of small areas over a large part of colour space, which described the extreme limits within which differences in chromaticity centred on the reference colour could not be discriminated.

Initial inspection of MacAdam's results suggested that these small discrimination areas could be described by ellipses, differing both in size and in orientation of major axis, throughout the CIE *xy* chromaticity diagram (Figure 4.6). These are now known as MacAdam discrimination ellipses. Since the chromaticity diagram is nothing more than a geometric model for describing the physical properties of a stimulus in terms of the ratios of tristimulus values, there is no reason to suppose that equal linear distances in the diagram should represent equal perceptible differences. MacAdam was the first to quantify the lack of perceptual uniformity in the CIE *xy* colour space. His work provided the basis for attempts to find a mathematical transformation that would produce a perceptually uniform geometry of colour space. This would be achieved if all the chromaticity discrimination ellipses could be represented as circles. A first approximation to perceptual uniformity of colour space used a very simple linear transformation of *xy* space. Such a procedure is the basis of the 1960 CIE *uv* chromaticity diagram. Under this transformation, however, the chromaticity discrimination data of MacAdam still depart from circles sufficiently that equal linear distances in the *uv* diagram cannot be assumed to represent equal perceptible differences in colour.

Many other, more complex, nonlinear transformations have been attempted, one of the earliest being that by Farnsworth (Figure 8.18), but none have yet achieved the goal of perceptually uniform colour space. More recently it has been shown that the MacAdam ellipses are not true ellipses; they seem to exhibit repeatable irregularities of contour and are essentially asymmetric. These are further illustrations of the inappropriateness of colour space for accurately representing the perceptual attributes of colour, and the complexities of this relationship are such that the problem cannot be solved in terms of simple Euclidean geometry. It seems that a three-dimensional Riemannian space is likely to be needed for an adequate description of a perceptually uniform colour space. In general, however, in order to map a Riemannian space of three-dimensions a six-dimension Euclidean space may be required. Unfortunately, most people find such a geometry almost impossible to visualise and it is therefore unlikely to gain practical acceptance in the fields of colour science and technology. To achieve a simple solution to this vexed problem has for decades been the 'holy grail' of colour physics and the quest remains unsolved. Even the more recent CIELAB system fails to provide complete perceptual uniformity in chromaticity space [27].

Furthermore, for a complete representation of chromaticity difference thresholds, the MacAdam ellipses need to be extended into discrimination 'solids' to include the third dimension of luminance. This would mean that, instead of using two-dimensional ellipses, we would need to use three-dimensional irregular ellipsoids throughout the entire colour solid. On account of these difficulties, several of the more recent successful attempts to define a perceptually uniform colour space have used

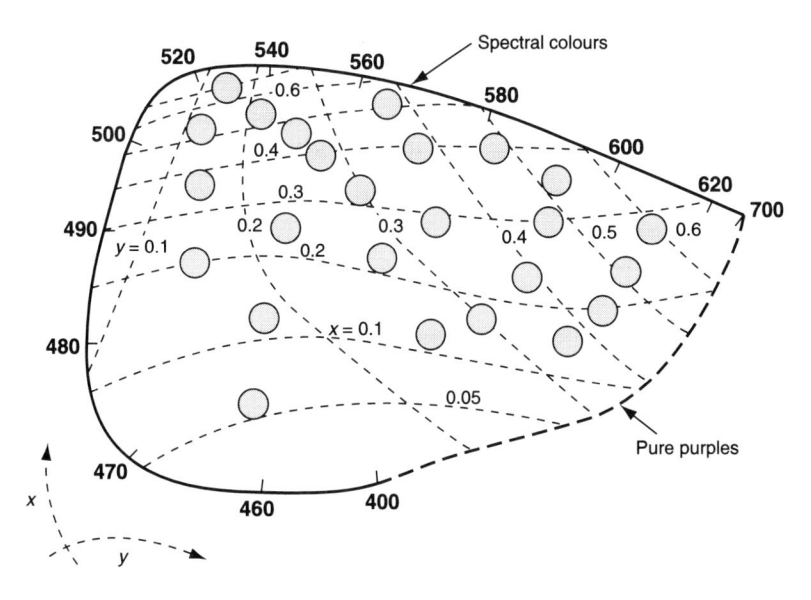

Figure 8.18 Nonlinear transformation of a CIE chromaticity diagram suggested by Farnsworth in an attempt to produce a geometry in which chromaticity difference thresholds are represented as uniform differences in colour space; the MacAdam ellipses approximate very closely to circles

different mathematical transformations for discretely defined areas of the chromaticity chart. Using such transformations, small colour differences within limited areas of the chromaticity diagram can be shown to be reasonably well correlated with perceived discriminable steps in chromaticity.

8.8 SOME COLOUR APPEARANCE PHENOMENA

8.8.1 Colour constancy

One of the most striking features of visual experience is that under everyday conditions the colour appearance of most objects is largely independent of changes in the spectral composition of the illumination. This is the phenomenon of *colour constancy*. There is a comparable term for the phenomenon by which judgements of lightness are independent of the absolute light energy levels but are based upon a comparison of reflectances. This is called *lightness constancy* and it is the means by which we are able to judge a piece of coal as black and a piece of chalk as white under very different illumination levels. Lightness constancy will only hold if direct comparisons are possible between the coal and chalk, even if under the low illumination levels the absolute amount of reflected light from the piece of chalk is *less* than that reflected from the coal at high illumination levels. It is these properties of vision that enables us to see a world of stable colours and lightness rather than the capricious physical world in which the wavelength and intensity of light reflected from objects in a visual scene are constantly changing. (The physiological basis for colour constancy is described in detail in section 8.3.3.)

The term *chromatic adaptation* is sometimes used to describe the way in which the visual system successfully accounts for variations in the spectral composition of illumination while still maintaining as constant the perceived colour of objects in the visual scene. But other conditions can also result in a failure of colour constancy. For example, constancy is particularly well maintained when a textured object is viewed in a complex visual scene that contains several surrounding objects of relatively high reflectivity. Under these conditions, provided that a fairly wide range of wavelengths are present in the illuminant, many response comparisons will be possible among the double-opponent wavelength-selective cortical cells. Even when the illuminant has discontinuous spectra or missing wavelength bands (a triphosphor fluorescent lamp, for example), colour constancy is preserved. On the other hand, where the illuminant is primarily monochromatic (such as a low-pressure sodium street lamp), the cortical cell wavelength-related response ratios become distorted. The consequent changes in colour appearance are referred to as *colour constancy failure*. Similar colour distortions also occur if there are marked variations in the spectral composition of the illumination across the visual scene, or if there are localised highlights: for example, where shafts of sunlight filter through the leaves in a shady wooded area, or where spotlights are used in a domestic or theatrical lighting arrangement. In these instances, the ratios of reflected light intensities of different wavelengths will not be maintained across the scene and there will be either partial or complete failure of colour constancy. The fewer the wavelength reflectance comparisons available in the field of view, the less able will be the visual system to conserve constancy. Similarly, colour constancy and lightness constancy will usually fail when isolated objects are viewed against a dark background because there is nothing with which the wavelength of the light reflected from the object may be compared. This is particularly so for small objects that have minimal surface variations of spectral reflectance and texture. Furthermore, there is a progressive failure of constancy as the visual field is decreased to a very small angular subtense to an eventual state of 'tunnel vision'. Again, this occurs because, as the field of view decreases, there become progressively fewer features upon which to base lightness and wavelength comparisons.

The visual system is thus particularly well suited to the analysis of complex scenes such as those encountered in everyday life. Under these circumstances ordinary changes in illumination and viewing conditions do not result in any disturbing changes of object colour perceptions. If this were not the case, our view of the world would be an ever-changing kaleidoscopic array of colours, lightness and shapes; the scene would be so confusing that we could not make rational or reliable inferences about what we see.

8.8.2 Colour phenomena associated with spatial relationships

While it is generally true that many objects of different spectral reflectances are necessary for preserving colour constancy, the spatial relationship between objects within the field of view can also alter the perception of hue, saturation and luminosity. These effects are known as *simultaneous contrast phenomena*. They are due to lateral (spatial) interaction effects between adjacent wavelength-coded receptive fields within the neural network of the retina.

Simultaneous colour contrast phenomena are, not surprisingly, much more varied than simultaneous contrast effects of luminosity. Nevertheless, in all instances colour contrast gives rise to an increase in the perceived colour difference across a boundary. This heightening of colour contrast results from selective interactions between the two types of colour-opponent cell. Whether for chromatic or achromatic stimuli, simultaneous contrast effects are the visual system's method of enhancing contrast boundaries, with the retina behaving like a small differential amplifier.

When the chromaticity differences between adjacent stimuli are large the colour distortions that occur in simultaneous colour contrast phenomena are also large, and they are reasonably predictable. Under these conditions the lateral inhibitory interaction effects are maximised. Thus where two adjacent stimuli are perceived as colours of the same hue but one having a high saturation and the other a low saturation, they will generally appear more and less saturated respectively than they would if each were viewed in isolation. If the illumination level is high the neural inhibitory effects in the retina will be particularly strong, thereby enhancing the perceived saturation differences; the stimulus perceived as having the lower saturation may actually appear either achromatic or, possibly, as a complementary hue to the other colour. Stimuli of complementary hue generally appear more saturated when placed adjacent to each other, and the juxtaposition of stimuli perceived as noncomplementary hues makes them appear more different in hue. In general, when objects are viewed adjacent to each other they appear to be more different than when viewed in isolation. By this means the visual system is acting in such a way as to enhance our ability to detect differences, a feature which makes life particularly difficult for the colour technologist.

There is one major exception to these simultaneous colour contrast effects. When a chromatic object of small angular subtense (<1°) is viewed against a larger area of uniform but different chromaticity, both object and surround appear more alike rather than more different in colour contrast. For example, fine white threads woven into a fabric generally cause the entire area of fabric to appear lighter in colour. Conversely, fine interwoven black threads produce a overall darkening effect. This is known as the *spreading* or *assimilation* effect. Explanations of this phenomenon in terms of scattered light within the eye are not satisfactory because these spreading effects can also occur in fine patterns of almost any chromaticity, even where the luminosities are the same. Although the spreading effect can sometimes be a nuisance to the colour technologist, it is often used to good purpose in the design of complex patterns for carpets or polychrome patterned brick façades of buildings.

8.8.3 Luminance-related colour phenomena

The relationship between the perception of hue and the level of luminance has been known since the second half of the 19th century. Although the effects are small, rarely exceeding 30 nm in perceived dominant wavelength, they are systematic. For example, when viewing conditions change from low to high luminances within the photopic range, greenish-blue hues change in appearance to look more blue while both greenish-yellow hues and reds look more yellow. There are three exceptions: the colour responses typically evoked by three wavelengths (approximately 475 nm, 500 nm and 575 nm), which remain essentially unchanged by changes in illuminance. The phenomenon is known as the *Bezold–Brucke hue shift*, after the two investigators who independently reported it in the 1870s.

Although the colour-appearance changes in the Bezold–Brucke effect are somewhat complex, they may be accounted for in terms of neural coding in the retina. It is here that the independent colour-opponent systems exhibit different degrees of responsiveness, the 'blue'/'yellow' system being more sensitive than the 'red'/'green'. Consequently there are differential rates of response of the paired chromatic systems as functions of luminance for a fixed spectral distribution of light. As the relative dominance of the opponent cells changes from 'red'/'green' at the fovea to 'blue'/'yellow' in the mid- to peripheral retina, the colour changes of the Bezold–Brucke effect also reverse. Even so, the three wavelengths mentioned above evoke colour-appearance responses that remain essentially the same. Such observations point to the possibility that, at least in part, the physiological feature underlying this phenomenon is due to a change in the spectral tuning (i.e. a shift in the neutral or null point) within each colour-opponent system as luminance increases.

Another colour phenomenon related to changes in luminance is the variation that occurs in the perceived saturation of all colours. Very dim colours appear relatively desaturated, maximum saturations are perceived at intermediate luminance levels, and at extremely high luminances all colours again become relatively desaturated. The explanation for this effect is similar to that given for the Bezold–Brucke hue shift, except that in this instance both the chromatic (opponent) and the achromatic systems are involved. As luminance increases, the achromatic neural response rate (i.e. nerve conduction firing frequency) increases much more rapidly than do those of the two opponent chromatic systems. Hence the response ratios between the achromatic and chromatic systems vary with luminance in such a way that with increasing luminance levels the more sensitive achromatic response becomes progressively predominant.

In addition to the effect of luminance on perceived saturation, as reflected light increases in spectral purity it also appears to increase in luminosity, even when the luminance remains constant. This phenomenon was originally described by Helmholtz in the mid-19th century and first investigated by Kohlrausch in the 1920s. It is now known as the Helmholtz–Kohlrausch effect. In order to describe the consequences of this effect to the science of colorimetry we need to consider the fundamental dimension of luminance more closely.

A particular feature of the stimulus property of luminance is that higher intensities are assumed to result in a greater perceived luminosity. This is not an unrealistic expectation, because the dimension of luminance is a construct that depends both on the radiometric properties of the stimulus energy and on the relative spectral luminosity function of the visual system. According to this mathematical construct, it is assumed that the total luminance of any mixture of lights is equal to the sum of the luminances of the components. This is known as *Grassman's fourth rule of colorimetry*; when it relates specifically to the luminance additivity of spectrally different lights the relationship is known as *Abney's law of luminance additivity*.

While the properties of luminance additivity may be demonstrated with a photoelectric photometer, the Helmholtz–Kohlrausch effect means that they do not always extend to the perception of luminosity. Consider, for example, matching a light perceived as saturated yellow with a mixture of two spectrally different lights that independently are perceived as saturated red and saturated green. Let us call the luminance of these lights L_y, L_r and L_g respectively. If we arrange (by means of a photoelectric photometer) for these three luminances to match each other, then the law of luminance additivity would imply that a mixture of L_r and L_g, each at half their luminance intensity, would equate with the value of L_y (i.e. luminance additivity, according to Abney's law, would mean that $L_y = L_r + L_g$). While this relationship will hold for a photocell, however, it will not do so for a subjective luminosity match. To the visual system, the mixture of lights appearing as red and green appears much darker than a reference yellow. Removal of one component of the mixture results in a slight increase in luminosity for the remaining component, although this will not be sufficient to produce a luminosity match with L_y.

The phenomenon of apparently increasing luminosity when light is taken away can be demonstrated by observing a scene through a photographer's 'minus blue' filter. This is a yellow-appearing filter that absorbs all wavelengths shorter than about 450 nm and transmits all wavelengths longer than 500 nm. (Because of the sharp transition between the wavelengths of maximum and minumum transmission, this is known as a cutoff filter.) If this filter is held before one eye and a scene viewed using each eye alternately, the scene will look brighter (have higher luminosity) when the observer looks through the filter than without it – even though less light reaches the eye behind the filter.

In the 'minus blue' filter demonstration, elimination of the shorter wavelengths results in an enhanced luminosity. The change in luminosity that occurs for a spectral mix depends on both wavelength and spectral purity, however, and is particularly marked when mixing lights perceived as complementary colours. These phenomena present considerable problems in photometry since, when measuring spectrally complex light sources with a photoelectric cell, no simple relationship between luminance and luminosity can be assumed. Furthermore, although luminance additivity failure has for some time eluded a satisfactory physiological explanation, two associated observations indicate a possible basis for resolving this problem. Firstly, luminance additivity failure does not occur in low scotopic luminances where the rod achromatic system predominates and cone activity is negligible. Secondly, it does not occur in heterochromatic photometry where luminosity matches of two alternately flickering lights of different wavelength are based primarily on the achromatic system: this is because the temporal sensitivity of the visual achromatic system is higher than that of the chromatic system. Failure of luminance additivity only occurs at photopic luminances and under conditions of simultaneous luminosity comparisons when both the chromatic and achromatic systems are operational. It is as though the two colour-opponent (chromatic) systems have a dampening or cancellation effect on the luminosity signal derived from the long-, medium- and short-wavelength cone receptors. Thus in these circumstances there would be an altered ratio of output signals from the chromatic and achromatic systems.

For the colour technologist, a practical consequence of the Helmholtz–Kohlrausch effect is that it may have some influence on judgements of two samples looking the 'same' when simultaneously comparing object samples that are highly metameric.

8.8.4 Temporal colour phenomena

Most people will have observed that following the prolonged viewing of an object of high intensity and high saturation, there is an after-effect on images that fall on the same part of the retina as the adapting object: objects in the centre of the field of view appear tinged with the complementary hue. The effect is known as an *after-image*. If the adapting object appeared red, for example, then the after-image would be the complementary hue of green. Indeed, selective adaptation of the visual system to any spectrally pure light of high intensity results in an after-effect lasting a few seconds in which the colour of the visual scene appears tinged with a hue complementary to that of the initial adapting light. The effect may be simply demonstrated by steadily looking at a high-intensity circular red patch on a white surround for about half a minute. Upon removal of the red patch, a circular after-image is observed in its place that has the appearance of a desaturated green. The after-image disappears after several seconds, during which time it may change slightly in colour. This effect is the result of a selective desensitisation of one of the cone receptor types; in this example it is the cones responsive to long wavelengths that are selectively desensitised. Under normal circumstances, when viewing an achromatic or neutral field, the signals from the three cone receptor types are 'balanced' and give rise to the perception of white. Once the adapting stimulus has been removed, subsequent signals from the three cone receptor types in the presence of an achromatic field are no longer balanced. The ratios of neural impulse frequencies in the bipolar cells of the retina are therefore distorted in favour of the unadapted wavelengths. As the photopigments in the selectively adapted cones regenerate to the former balance of sensitivity with other cones, the after-image gradually fades and disappears.

After-images have practical consequences for the colour technologist, who must take care to avoid making judgements of hue after viewing areas that emit spectrally selective light of high intensity. In order to minimise the effects of after-images on critical colour judgements, observations should be made after viewing a large evenly illuminated achromatic surface.

A further temporal property of colour, referred to above, is that associated with viewing brief stimuli. For viewing times of less than 0.1 s, most hues appear progressively desaturated. For very brief stimuli (about 3 ms duration), monochromatic lights in the region of the spectrum between 490 nm and 520 nm are perceived as achromatic (i.e. white or grey). It is believed that this effect occurs because the achromatic system has a much shorter response latency than the chromatic system. This property of vision provides the basis of heterochromatic flicker photometry, in which the luminosity of lights having small wavelength differences may be compared independently of their chromatic differences.

Interestingly, colours may be evoked even when a variety of black and white stimulus patterns are illuminated intermittently with 'white' light at temporal frequencies much less than those required for producing the state of critical flicker fusion (i.e. the temporal frequency at which a light no longer appears as flickering). These are known as *Fechner's colours*, and are best appreciated when viewing a rotating patterned disc. Even when such a disc contains just black line patterns on a neutral background, different colours appear along the radius of the rotating disc, depending on the pattern configuration and the speed of rotation. The phenomenon is the basis of a 19th-century children's toy known as Benham's top. The effect is believed to be due to differences in the response rise and decay periods of the different chromatic systems in the eye which, under these special viewing conditions, allow different colours to be perceived even without changes in any of the wavelength properties of the illuminating light.

8.9 INDIVIDUAL DIFFERENCES IN COLOUR VISION

8.9.1 Receptor properties

Our perception of colour is the product of a complex sequence of physiological activities. Like all physiological systems, the characteristics of the visual system vary between individuals. It is not surprising therefore that physiological differences, however small they may be, result in individual differences of perceived colour. One of the principal ways in which these individual differences could occur is in variations of the spectral absorption properties of the retinal cone photopigments. Recent microspectrophotometric studies on the human visual pigments from seven eyes of individuals with normal colour vision have identified variations of almost 20 nm in the wavelength of peak absorption for cones that are responsive to long and medium wavelengths [3]. The variations in peak absorbance are thought to be not wholly attributable to microspectrophotometric measurement error due to a suggested bimodality in the distribution of peak absorbance wavelengths. It is hypothesised that each major type of cone includes discrete subpopulations of cones, characterised by different peak absorptions, and that individuals with normal colour vision may possess more or fewer receptors from each of the subpopulations. The cause of these variations in spectral absorption within a major class of cones has not been identified, but they could result from differential effects of certain chemical substances in the retina on the vitamin A components of the photopigments during development. Whatever their origin, the consequence is that quite large individual differences in spectral sensitivity can be expected, even among individuals with 'normal' colour vision.

8.9.2 Pre-receptoral properties

The optical media of the eye that lie in front of the retina do not transmit light equally at all wavelengths. There is therefore a wavelength-selective filtering of the light reaching the visual photopigments. Both the spectral nature and the amount of the filtered light differs between individuals with normal colour vision.

In particular, the crystalline lens absorbs strongly at short wavelengths; at 400 nm its absorption is about ten times that for wavelengths of 600 nm or more. In this respect the crystalline lens acts something like a photographer's 'minus blue' filter. The absorption of the lens increases with age; in some people this presents as a cataract later in life. This spectrally selective, age-related increase in absorption has practical implications for the colour scientist and technologist. These are particularly relevant to judgements of fine colour matching involving 'blue' materials, or to absolute judgements of the blueness of materials (for example, in diamond grading).

A second pre-receptoral pigment is found in the macular area of the retina, just in front of the photoreceptors. This macular pigment is yellow in appearance with an absorption spectrum very similar to that of xanthophyll (a carotenoid pigment), having a peak absorption at about 450 nm. Although the presence of this inert macular pigment does not appear to change with age, there are marked individual differences in its absorption properties.

The variations in both the macular pigment and the lens contribute to colour vision differences between people even though the individuals concerned may be classified as having normal colour vision.

8.9.3 'Normal' colour vision

Clearly, individual differences in pre-receptoral and receptoral absorption properties result in individual variations of relative spectral luminosity, chromatic discrimination, colour matching and colour perception. Some idea of the considerable variations in the photopic luminosity function may be derived from Figure 8.19, which shows the spread of measurements obtained from several observers with normal colour vision. Not only are there variations in peak sensitivity of almost 20 nm; there are also quite large differences in the relative sensitivity at different wavelengths. Since the photopic luminosity curve for an individual is essentially a linear combination of that person's colour-matching functions, considerable individual differences also exist in colour-matching coefficients. Consequently, CIE standardised values for the photopic luminosity function and related colour-matching coefficients are appropriate only to a hypothetical average normal observer. For a full description of normal trichromacy it would be necessary to specify statistical confidence limits associated with each of these standardised averaged relative spectral luminosity functions, but such data would clearly be cumbersome to use in both photometry and colorimetry. In the CIE system normality is defined in terms of an average, in contrast to most other fields of human performance where normality includes a specification of variance (usually 95% confidence limits) derived from individual performance differences. The colour technologist must appreciate the practical significance of the distinction between

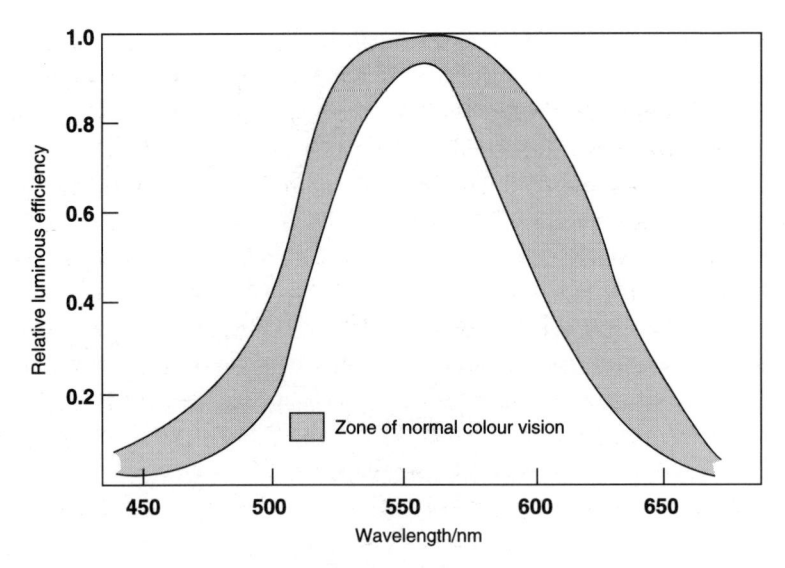

Figure 8.19 Individual differences in the photopic relative spectral luminosity function

these two definitions of normality, because colorimetric calculations based upon an average norm cannot be expected to be judged similarly by 95% of normal trichromats.

At photopic luminances, individual variations in normal colour vision can have far-reaching practical consequences for the colour technologist. Over the years, many attempts have been made to derive a perceptually uniform colour space, with little regard for the significance of individual differences. For example, one of the earliest attempts at transforming the MacAdam chromaticity discrimination ellipses into circles was based upon the data of a single observer. Despite its obvious limitations, it was the data of this observer that provided the transformation coefficients eventually adopted by the CIE as the standard uniform chromaticity space diagram (known as the 1960 CIE UCS diagram). Although this transformation is an internationally recognised standard, this does not imply that it always has practical relevance. Particular care should be exercised when drawing inferences about population limits based on the data of a single observer and, where these decisions are likely to be crucial, it is preferable to determine the required data from a relevant sample population of people with normal colour vision. These issues are of particular significance to those industrial processes that set fine tolerance limits of colour difference for quality control acceptability standards.

8.9.4 Observer metamerism

It is possible for a pair of objects to appear matched in colour even if their spectral reflectances differ. This happens if the combined parameters of the spectral properties of the illuminant, the spectral reflectance characteristics of the objects and the colour-matching coefficients of the observer's eye match for the two objects being viewed. Such object pairs are described as having metameric colours (section 3.15). The colour match will fail if there is a difference in the spectral characteristics of any one of the three components (illuminant, surface reflectance or observer response). Since, as we have seen, no two individuals are likely to have exactly the same wavelength-dependent response characteristics, a pair of objects of slightly different spectral reflectance characteristics that appear matched in colour for one observer will not necessarily appear matched for another. This state of affairs is frequently encountered in commercial or industrial contexts and is known as *observer metamerism*. It is to be distinguished from a breakdown of colour match of two objects for a single observer that occurs with a change in the spectral composition of the illuminant (daylight to tungsten light, for instance), which is known as *illuminant metamerism*, or that arises from different surface reflectance characteristics due to textural differences in the objects which become manifest with a change in viewing direction (*geometric metamerism*).

Metameric colours are, theoretically, defined as colour stimuli of identical tristimulus values but different spectral energy distributions. Such a definition is satisfactory only when standardised colour-matching functions are used. It is under these conditions that the terms 'illuminant metamerism' and 'geometric metamerism' are relevant. The existence of observer metamerism, however, is a consequence of individual variations in colour vision. These variations imply the existence of different colour-matching functions and hence different tristimulus values between two or more observers. They are the cause of many customer complaints to the colour industry, where colour matching is an important feature of product components. In order to take account of the practical consequences of observer metamerism, commercial decisions of the acceptability of a colour match should preferably be based on a consensus of judgements from a small group of observers, all with normal colour vision.

8.10 DEFECTIVE COLOUR VISION

8.10.1 Anomalous trichromacy

One of the fundamental properties of normal colour vision is the experimental fact of trichromacy (see section 8.4). Evidence of this dates back to the mid-19th century when Maxwell and Helmholtz independently demonstrated that light of any spectral composition could be a component of a colour-matching equation that included appropriate amounts of light intensity from just three primary wavelengths. The essential feature of this trivariant property of vision is the algebraic equivalence of colour-matching equations. An exact visual match of any light, including the spectrally pure colours, is however not achievable in practice by an additive mixture of three primary stimuli, even for observers with normal colour vision. For example, in attempts to match a pure spectral colour a mixture of the three primaries always appears a little too desaturated. In practice this can be overcome by adding one of the three primaries to the sample light stimulus and this combination may then be matched by an appropriate mixture of the other two primaries. This is the reason why some terms are negative in the algebraic expression of colour-matching equations. When expressed in this way, it is the triplet proportions of the three primary stimulus intensities that provide the basis of the tristimulus colour-matching coefficients.

The usual method for determining the colour-matching coefficients is to present an observer with a split circular field (i.e. a bipartite viewing field), one half of which contains the sample stimulus plus one of the primaries and the other half a mixture of the remaining two primaries. When both halves of the bipartite field visually match yet have very different spectral energy distributions, the pair of stimuli are said to be metameric. In other words, they are equivalent in physiological and perceptual terms but not physically.

If a large unselected population of observers is used to perform such a colour-matching experiment, it is found that most people achieve an acceptable match within a very small distribution of colour-mixture proportions. For some colour-mixture equations, however, there will be a few widely divergent values: these are extreme cases of observer metamerism. Not only do such deviations in colour-mixture ratios occur systematically; they are also very repeatable for a given individual. In fact, some colour mixtures produce a match so deviant yet so stable that their existence is clearly not the result of random physiological or perceptual variations in a single population group. Statistically, they comprise a quite distinct but small group of individuals who are defined as having defective colour vision. Such people, who have trichromatic vision but use extremely deviant proportions of the three primaries in their colour-matching equations, are known as *anomalous trichromats*.

Normal and anomalous trichromats may be distinguished by the use of tests that make use of only two primaries for colour-matching performance. These colour matches may be performed on special visual colorimeters known as anomaloscopes. Using special colour-matching equations, analyses of the colour-mixture proportions used by anomalous trichromats shows that three distinct population groups may be identified. There are those people who have a weakened sensitivity to long wavelengths and require much more of the 'red' primary than normal trichromats to obtain a colour match. Such people are known as *protanomalous trichromats*. Those with a weakened sensitivity to midband wavelengths are known as *deuteranomalous trichromats* and those with weakened short-wavelength sensitivity (if they can be found) as *tritanomalous trichromats*. Not surprisingly, these three distinct groups of colour vision deficiencies are more commonly known as 'red-defective', 'green-defective' and 'blue-defective', although the use of such colour names in this context is not particularly appropriate.

The special terms used for defective colour vision are derived from the Greek words *protos*, *deuteros* and *tritos*, meaning first, second and third respectively. This is a purely nominal classification based upon the order in which the defects were originally described in the late 19th century, and it indicates neither the severity nor prevalence of the different types of defective colour vision. But while the small individual differences in normal trichromacy may be explained largely by random physiological variations both in optical medial transmittance and photoreceptor absorption, certain very specific photoreceptor (i.e. cone) characteristics account for the anomalous trichromacies. It has long been postulated that an anomalous photopigment may be involved and not until the 1980s was experimental evidence obtained to support this hypothesis. This came from studies undertaken by two independent groups of investigators [28,29]. One team carried out a lengthy series of experiments on squirrel monkeys involving both behavioural colour-matching and wavelength-discrimination measurements followed by single-receptor microspectrophotometry on the excised retinas of the same animals [29]. In the behavioural experiments, several of the monkeys were known to have a colour-vision deficiency because they were shown to have a colour-matching performance resembling either the protanomalous or deuteranomalous variants of human colour vision. From all the anomalous trichromatic animals the microspectrophotometric results revealed a cone photopigment that was different from the three cone photopigments found among normal trichromatic animals. The absorption of this anomalous photopigment had a bandwidth and shape similar to those of the other cone photopigments, and a peak roughly midway between those of normal middle- and long-wavelength cones. (The absorption characteristics and wavelengths of peak absorption of the cone photopigments in squirrel monkeys with normal trichromatic vision are similar to those found in normal human trichromats.)

Only three cone types were found in the retina of each anomalous trichromatic monkey. In the behaviourally protanomalous trichromats, cones with the anomalous photopigment replaced the 'normal' long-wavelength cones, while in the deuteranomalous trichromats they replaced the middle-wavelength cones. Consequently in these two types of anomalous trichromacy there is a reduced wavelength difference of peak absorption between the remaining normal 'green' cone pigment and the anomalous pigment. This has the effect of reducing wavelength discrimination over selected regions of the spectrum. Although such studies have so far been confined to monkeys, the analogy with human colour vision would appear to be valid because of the close similarity between the 'normal' cone absorption spectra and the comparable behavioural/psychophysical colour-matching performance.

Even if the anomalous photopigment hypothesis is the basis of anomalous trichromacy in humans, however, it is still not clear how the putative pigment is derived. There are two main contenders for its method of production. Firstly, variations in absorption spectra are thought to arise from small changes in the protein portion of the photopigment molecule (the opsin), the chromophore remaining unchanged. Secondly, it has been suggested that the cone receptors containing the anomalous pigment include a mixture of the photopigment molecules from both the middle- and long-wavelength receptors. The simplicity of the latter hypothesis is very attractive because different mixture proportions of the two 'normal' middle- and long-wavelength photopigments could easily explain the wide variations that occur in both colour-matching performance and wavelength discrimination among protanomalous and deuteranomalous trichromats. For example, the smaller the wavelength difference between the absorption spectra of two adjacent photopigments (one 'normal', the other anomalous), the poorer is the wavelength discrimination in that region of the spectrum and the more deviant the colour-matching performance.

It is believed that an altered short-wavelength photopigment could also be the basis of tritanomalous vision if it existed, but no cases of congenital colour-vision deficiencies of this type have been reported. While an abnormality of the 'blue' cone system occurs in several pathologies of the eye, in the only form of inherited congenital 'blue' colour-vision deficiency in humans the 'blue' cone receptors are either missing or functioning in a way which is similar to the 'green' cones.

8.10.2 Dichromacy

About 2–3% of people with defective colour vision can match any light, including all spectral colours, by a suitable mixture of only two appropriately selected primaries. Such people are said to have dichromatic vision and they are called *dichromats*. This condition arises when only two of the three 'normal' cone photopigments are present in the retina, but there are characteristic differences in colour vision depending upon which photopigment is absent. If the long-wavelength-sensitive cones are absent the resulting colour vision is described as *protanopic*, and if the middle-wavelength cones are absent vision is described as *deuteranopic*. Finally, and more rarely, absence of the short-wavelength cones results in a colour-vision deficiency defined as *tritanopic* vision. These three states of inherited defective colour vision are commonly, but inappropriately, termed 'red-blind', 'green-blind' and 'blue-blind' respectively. The labels are inappropriate because dichromats still perceive reds, greens and blues, although their perceptions are different from those of people with normal colour vision. For both anomalous trichromats and dichromats the general descriptive term 'defective colour vision' is, therefore, to be preferred. The nature of the differences between the different types of dichromacy and also between dichromatic and trichromatic vision may be characterised in terms of differences in wavelength-discrimination and colour-matching performance.

8.10.3 Monochromacy

When any light or spectral colour can be matched by adjusting the intensity of a single primary stimulus, a condition of monochromacy exists and vision is said to be *monochromatic*. It would be more appropriate to describe this kind of vision as *achromatic* (without colour), since such people perceive the visual world merely in terms of shades of grey. People with vision of this nature are therefore correctly called 'colour-blind', and their colour-matching equation contains only two algebraic components because they can match lights of any two wavelengths merely by adjusting the radiance of one.

There are two principal causes, and therefore two types, of monochromacy. Both are extremely rare, but the less rare is that in which the cones are either absent or nonfunctional. These people are called *rod monochromats*. Their response to light of different wavelengths is entirely dependent upon the absorption spectra of the rods. The spectral sensitivity of rod monochromats therefore corresponds to the scotopic relative luminous efficiency curve of normal trichromats (Figure 8.17). Rod monochromats also have poor visual acuity, because the population density of rods at the fovea is low and their corresponding receptive fields are large. The visual performance of people with this kind of vision is, moreover, better at low rather than high levels of illumination. It is not surprising, therefore, that rod monochromacy is a particularly disabling eye condition.

In the rarer form of inherited monochromacy there is evidence of some cone function. This defect is known as *cone monochromacy*. The spectral sensitivities of people with this condition vary quite widely because they may differ in the relative predominance of the short-, middle- or long-wavelength-sensitive cones, but they have normal visual acuity. While the major types of defective colour vision are receptoral in nature, it is generally considered that the pathological site of the defect in cone monochromats is postreceptoral, probably in the neural layers of the retina.

8.10.4 Principal characteristics of defective colour vision

There are many ways in which the perceptual characteristics of colour vision show systematic differences between the several types of colour-vision deficiency. Only four such properties are discussed here: spectral sensitivity, wavelength discrimination, purity discrimination and chromatic discrimination. Differences in terms of colour-matching ability are discussed in section 8.11.2.

Any change in the absorption spectra of the cone photopigments will result in a change of the photopic spectral sensitivity of the visual system. In protanopia the absence of the long-wavelength-sensitive cones means that the photopic spectral sensitivity response is based upon the absorption characteristics of just the short- and medium-wavelength cones. By comparison with the normal trichromat, the protanope shows considerably reduced sensitivity to the longer wavelengths: stimuli typically perceived as red to the normal trichromat appear dull or dark to the protanope. Furthermore, because the photopic luminosity curve is a function of the combined absorption spectra of the cones, the wavelength of peak sensitivity is shifted in protanopes towards shorter wavelengths (Figure 8.20). On the other hand, the photopic relative spectral luminosity curve of deuteranopes is derived from the absorption spectra of the short- and long-wavelength cones. Its peak sensitivity and overall shape are therefore not very different from those of the normal trichromat. In deuteranomalous eyes, however, where it is believed there is an anomalous medium-wavelength-sensitive cone pigment whose peak absorption is shifted towards the longer wavelengths, there is an associated slight increase in long-wavelength sensitivity. By a similar argument, people with tritanopic defects are assumed to show a reduced sensitivity to short wavelengths with a consequent slight shift towards longer wavelengths in the peak of their photopic relative spectral luminosity curve. Colours perceived as bright blue to the normal trichromat therefore appear much darker to the tritanope. Insufficient data have been collected to provide a representative description of the relative spectral luminosity curve of people who are tritanopic.

The discrimination of wavelength is a function of the different spectral absorption rates, and hence response ratios, of the three cone receptor types. In normal trichromats the average just-noticeable wavelength difference throughout the visual spectrum is in the range 1–5 nm, suggesting that there are unique responses that can be expressed as triplet ratios from the three cone systems. These differences occur at approximately 1–5 nm intervals across most of the visible spectrum. In anomalous trichromacy, where the anomalous photopigments have absorption spectra intermediate between those of the adjacent photopigments of normal colour vision, there are identifiable parts of the spectrum in which the same triplet response ratios cover a relatively wide spectral band. Within these spectral bands, wavelength discrimination is either non-existent or very poor. The mean wavelength separation of peak absorption between the adjacent anomalous and normal photopigments is about 15 nm in both protanomaly and deuteranomaly. This compares with a peak absorption

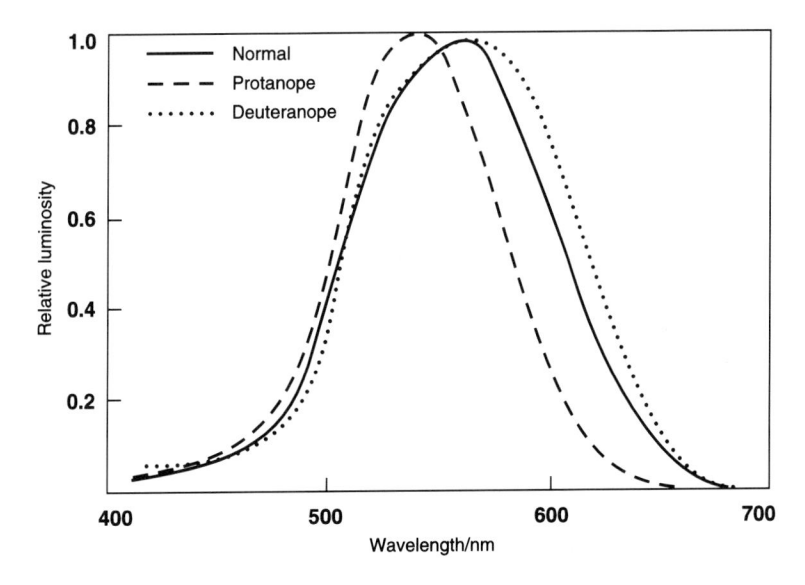

Figure 8.20 Cone absorption spectra in the anomalous trichromacies and dichromacies compared with those for people with normal colour vision

wavelength difference between the long- and medium-wavelength cones of about 30 nm in the normal trichromat. The smaller the difference in absorption spectra between two spectrally adjacent cone photopigments, the poorer is the wavelength discrimination in that region of the spectrum. The anomalous photopigment hypothesis for anomalous trichromacy would also explain the existence of slight differences in maximum wavelength discrimination that exists between protanomalous, deuteranomalous and normal trichromats. In dichromats, there are regions of the spectrum where only a single cone photopigment predominates. In these spectral regions, which differ for the different types of dichromacy, there are no comparative response functions and therefore no wavelength discrimination.

The purity-discrimination function describes the way in which the observer perceives differences in the saturation of a colour, ranging from white through to its full spectral purity (for example, from white through pink to red). Wavelengths that give rise to colours perceived as having the greatest saturation are those for which only a small increment of the spectral component is necessary to make a white light appear tinged with colour. For the normal trichromat, this occurs for colours perceived as blue and red. Also, there are more discriminable steps between white and full spectral purity for spectral hues that appear highly saturated than for those spectral hues that appear less saturated. Since purity discrimination appears to be dependent upon both the chromatic (i.e. opponent) and luminance (non-opponent) channels within the visual system, it is not surprising that characteristic differences are observed among the various colour-vision deficiencies. In general, purity discrimination is poorest where wavelength discrimination is best. It is almost as though there is a trade-off between the luminance information transmitted to the brain along the non-opponent channels and the colour information transmitted along the wavelength-opponent channels. To the dichromat, certain spectral wavelengths appear completely colourless. These are known as the achromatic points or neutral points in the spectrum, and they are defined as the spectrally pure wavelength that matches white. (For their precise specification, it is necessary to know the colour temperature of the white light used for achieving the match.) For the three types of dichromacy, the neutral points are centred around the following wavelengths: 490 nm in protanopia, 500 nm in deuteranopia and 570 nm in tritanopia. These values represent the midpoints of a small range of wavelengths that also appear achromatic. The range of spectral wavelengths that can be exactly matched to white is known as the *neutral zone*. Its existence and associated wavelength characteristics are diagnostic of the different types of dichromacy. A spread of the neutral zone over a wide range of wavelengths is indicative of a severe colour-vision defect. Individual differences in the neutral zone wavelengths are inevitable and they are due, in part, to individual variations in the selective light-filtering effects of the macular pigment.

With systematic differences occurring in the relative spectral luminosity response, wavelength discrimination and purity discrimination, it is to be expected that the different types of colour-vision deficiency will show systematic differences in chromaticity discrimination. When the chromaticity-discrimination thresholds for anomalous trichromats are depicted in the CIE *xy* diagram, the discrimination ellipses (MacAdam ellipses) become markedly elongated and somewhat asymmetric. There are three distinct populations of these altered discrimination ellipses and they correspond to the three types of inherited colour-vision deficiency. In fact, the orientation and distribution of these different populations of chromaticity-discrimination ellipse are so characteristically different from each other that they are actually diagnostic of the type of colour-vision defect. Since each ellipse represents the boundary of chromaticity-discrimination thresholds, all chromaticities within an ellipse will be perceived as having the same colour. Reference to these chromaticity thresholds in the CIE *xy* colour space will therefore give some indication of the colour confusions experienced by people having the principal types of colour-vision deficiency. The length of the major axis of the discrimination ellipse is an indication of the severity of any colour-vision defect. In dichromacy, the major axis of each discrimination ellipse extends the full length or width of the chromaticity diagram and is typically depicted as a line. These lines are known as *isoconfusion lines* because they describe the locus of chromaticities which all appear the same colour (Figure 8.21). Reference to the colour names given by normal trichromats to chromaticities falling along any isoconfusion line will provide an indication of the colour confusions typically experienced by each type of dichromat. A useful feature of illustrating the properties of colour-vision deficiency in this way is that the neutral point (the

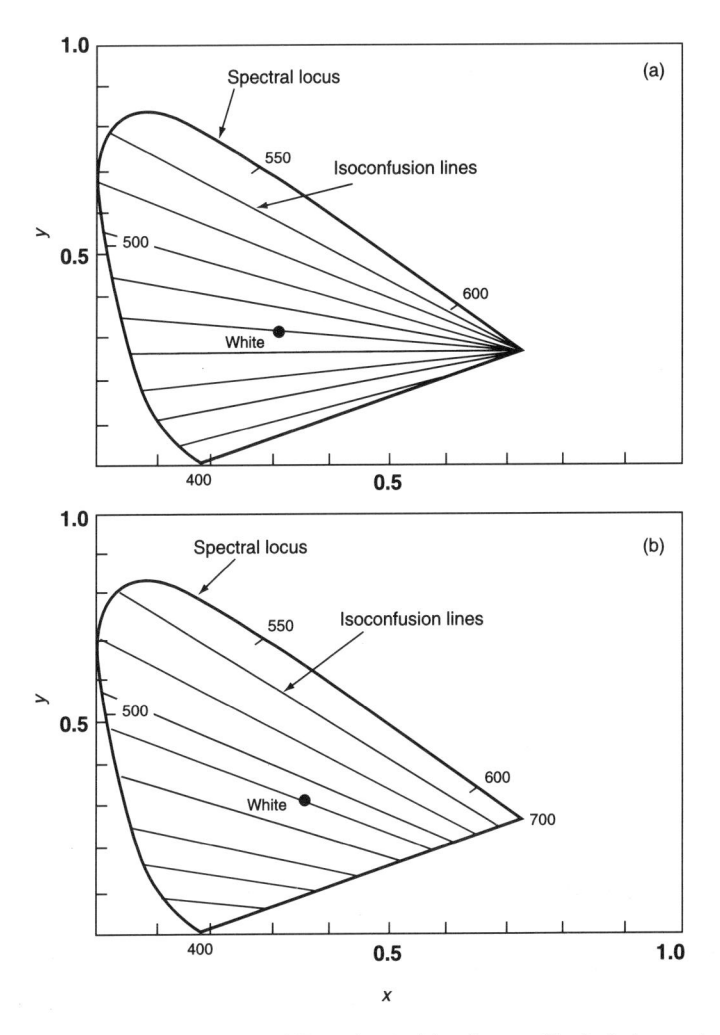

Figure 8.21 Isoconfusion lines in the CIE *xy* chromaticity diagram (the loci of zero chromaticity discrimination) for (a) protanopia, (b) deuteranopia; the point on the spectral locus which joins an isoconfusion line with the white point in the chromaticity diagram indicates the neutral point for that type of colour-vision deficiency; in the anomalous trichromacies, chromaticity discrimination thresholds are represented by elongated MacAdam ellipses with their major axes corresponding to the orientation of the relevant isoconfusion lines

wavelength which is confused with grey or white) can easily be determined: it is that point on the spectral locus of the CIE xy chromaticity diagram that falls on the same isoconfusion line that passes through the white point. All chromaticities along this particular isoconfusion line are perceived as achromatic and will be confused with white (or grey).

8.10.5 Origin and incidence of defective colour vision

Defective colour vision may be either inherited as a genetic trait or acquired as a consequence of pathology. The inherited colour-vision deficiencies are much more common in men than in women. This is because they are associated with a defective gene on the X-chromosome, and this same chromosome is also one of the determinants of the sex of an offspring. Males have just one X-chromosome (plus a Y-chromosome) while females have two X-chromosomes.

If a man's single X-chromosome carries the colour-defective gene, then he will have defective colour vision. In women, however, both X-chromosomes need to be affected before a colour-vision deficiency becomes manifest. If a woman has only one affected X-chromosome her colour vision is normal but she is a *carrier* of the defective gene. Since any offspring will receive only one X-chromosome from their mother, there is a 50% chance that any of her sons or daughters may inherit the colour-defective gene. This means that there is an equal (50%) chance that a woman who is a carrier of a defective colour-vision gene will produce a son with a colour-vision deficiency or a daughter who is a carrier of the deficiency. Since sons do not inherit the X-chromosome from their father, defective colour vision cannot be passed from father to son. In the relatively rare instance where both parents have defective colour vision, all their X-chromosomes must carry the defective gene and all their sons and daughters will inherit the defect. Recent gene-mapping studies have shown that the genes for protanopic or protanomalous and deuteranopic or deuteranomalous characteristics are located at different but fairly close sites on the X-chromosome [30,31]. These two defective genes will therefore be inherited independently of each other. Tritanopic deficiencies, however, are inherited by a completely different mode of inheritance, which is not sex-linked. These, and the state of monochromacy, are extremely rare conditions that occur with approximately equal frequency in males and females. There are slight variations in the incidence of the different types of defective colour vision according to the geographical area or ethnic origin of a population; incidence values for European races are given in Table 8.1.

In all inherited colour-vision deficiencies, both eyes are equally affected and the conditions remain unchanged throughout life. There is no method for their treatment or correction, although a person with a colour-vision deficiency may be able to pass some of

Table 8.1 Incidence of defective colour vision among Europeans

Type	Variant	Incidence/% Males	Females
Anomalous trichromacy	Protanomaly	1.0	0.03
	Deuteranomaly	5.0	0.35
	Tritanomaly	?	?
Dichromacy	Protanopia	1.0	0.01
	Deuteranopia	1.0	0.01
	Tritanopia	0.005	0.005
Monochromacy	Rod monochromacy	0.003	0.003
	Cone monochromacy	0.000001	0.000001

the conventional colour-vision tests by viewing through a 'coloured' filter. This use of a spectrally selective filter simply has the effect of changing the orientation of isoconfusion lines and therefore shifts the neutral point upon which the design of most colour-vision tests are based: all that it can do is to change the nature of the colour confusions that a colour-defective person typically makes. It cannot give a colour-defective person normal colour vision, nor can it improve chromaticity discrimination.

Colour-vision deficiencies can be acquired from any systemic or ocular pathology affecting the retina or the optic nerve [32]. Such conditions include most chronic vascular disorders (such as diabetes) and chronic disorders of the nervous system (such as multiple sclerosis). The most common of the acquired colour-vision deficiencies, however, are due to pathologies of the eye associated with ageing (glaucoma, for example) and certain pathologies of the macula (such as age-related maculopathy). Several drugs, including nicotine from excessive smoking, may act as neurotoxic agents and will result in damage to the retina and/or optic nerve. The macula is particularly susceptible to damage of this kind.

Although the colour confusions in acquired colour-vision deficiencies are essentially similar to those of the different types of inherited defects, they are much more variable in expression; they are often unequal between the two eyes and they are much less stable over time. In acquired deficiencies the most prevalent defect is tritanlike, and is referred to as a 'yellow'/'blue' deficiency. (The shortened terms *protan*, *deutan* and *tritan* are used when reference is made to each type of defective colour vision; they are collective nouns which include both the anomalous trichromacy and dichromacy variants.) It would appear that the particular vulnerability of shortwavelength cones to damage is primarily due to their paucity. In general, acquired colour-vision defects become progressively more marked as the disease progresses and usually end eventually as an acquired monochromacy known as *achromatopsia* ('vision without colour'). The incidence of these vision defects is difficult to estimate but is likely to be about 2% of the general population and is increasingly more prevalent in the older age groups.

8.11 TESTS FOR DEFECTIVE COLOUR VISION

8.11.1 Principles

The design principles of most tests for defective colour vision are based upon the properties of inherited colour-vision defects. These include the known systematic variations in colour-matching ability, as well as in chromatic discrimination thresholds. The design principles of most colour-vision tests can therefore be ilustrated by reference to the characteristic isoconfusion lines of colour-vision deficiencies as represented in the CIE xy chromaticity diagram. In these diagrams it is possible to select pairs of stimuli whose chromaticities will be discriminated by people with normal colour vision yet appear indistinguishable to either a protan, a deutan or a tritan colour-defective person.

In laboratory studies, a detailed examination of defective colour vision usually involves measuring an individual's wavelength discrimination and also determining his or her relative spectral luminosity function. Both these measures require sophisticated optical instruments and are time-consuming. For practical purposes, simpler test procedures are used. A wide variety of these simple practical tests is available; they differ in both the terms of their design principles and also their method of administration. Classification of the tests is not altogether straightforward, but for simplicity they may be grouped into four types:

(a) tests based on the principles of matching a reference stimulus by a mixture of light intensities from two spectral primaries; this group of tests involves the use of the special visual colorimeters known as anomaloscopes, and they are widely accepted as the reference standard against which the efficiency of all other colour-vision tests should be compared

(b) tests based largely on the principle of colour confusion; most of these are known as *pseudo-isochromatic tests* because they employ stimuli that may be described as falsely appearing of the same colour (usually, these pseudo-isochromatic tests are presented in book form, on each page of which is printed a geometric pattern or design of specified chromaticity that is viewed against a background of different but carefully selected chromaticity; the choice of chromaticities is such that both the symbol and background lie along an isoconfusion line)

(c) tests designed principally to assess colour discrimination, the most widely known of which is the Farnsworth–Munsell 100 hue test

(d) specialised tests designed to measure specific aspects of colour-vision performance, such as judgements of colour difference and colour memory.

8.11.2 Colour-matching tests

The first person to describe a quantitative investigation of the variations in colour-matching ability among normal and colour-defective people was Lord Rayleigh, in 1881; one of his early colour-deficient subjects was his brother-in-law, Arthur Balfour, who eventually served as prime minister of Great Britain from 1902 to 1905. Rayleigh, who conducted his experiments at home, used a spinning disc comprising two different sectors which appeared as red and green to an observer with normal colour vision. The proportions of the red and green sectors were adjusted to achieve a match with a reference disc appearing as yellow. Colour-mixing experiments, which are based on varying the proportion or intensities of 'red' and 'green' to match a reference 'yellow', are now described as being based on the Rayleigh colour-mixing equation. A particular feature of the Rayleigh equation is that it corresponds with the isoconfusion line in the CIE xy chromaticity diagram that is common to both protan and deutan colour-vision deficiencies. This feature makes the equation particularly amenable to quantifying colour-vision deficiencies, as well as providing a simple means of distinguishing between protan and deutan defects. To achieve a match with 'yellow', for example, people with protan deficiencies select more 'red' in a 'red'/'green' mix, while deutans require more 'green' in their mixture. The anomaloscopes referred to in sections 8.10.1 and 8.11.1 are devices that use the Rayleigh equation to specify and quantify anomalous colour vision in this way; they may also be used to measure the subtle variations in colour vision found between people with normal colour vision. All these instruments use light of known spectral composition (achieved either by filters or by prismatic dispersion). The mixture of long- and middle-wavelength light is presented in one half of a bipartite viewing field and the other half contains the reference stimulus, usually light of an intermediate wavelength. When a colour match is achieved the two halves of the bipartite viewing field are metameric.

Other colour-mixing equations using different pairs of primaries are also possible. These correspond with the tritan isoconfusion lines for investigating the tritan colour vision deficiency (Engelking–Trendelenburg equation) and an equation for measuring the effects of ageing on colour vision due to an increase in the inert short-wavelength-absorbing pigment in the crystalline lens of the eye (Pickford–Lakowski equation).

Several anomaloscope instruments are available but not all use the same method of producing the primaries. The two most widely used instruments are the Nagel anomaloscope (a narrow-band spectroscope) and the Pickford–Nicolson anomaloscope (employing broad-band glass filters). Slight differences in colour-matching measurements may occur between these two instruments because the primaries they use for their Rayleigh equation are not identical. Additionally, there are slight differences between the instruments in the angular subtense and luminance of the viewing field. The advantage of the Pickford–Nicolson anomaloscope is that it provides a direct, rather than telescopic, view of the stimuli; unlike the Nagel, its disadvantage is that small batch-to-batch variations in the spectral transmission characteristics of the glass filters mean that each instrument needs individual calibration with observers from a sample population of normal trichromats.

In administration, a method of constant stimuli or method of limits is used for finding the range of the two primaries (670 nm and 545 nm in the Nagel anomaloscope) with the reference stimulus (589 nm). For each mixture comparison the luminance of the 589 nm reference stimulus must be adjusted to account for differences in the relative spectral luminosity functions of protans and deutans. For example, the 670 nm primary appears very dim to the protanope, and therefore only a low luminance of the 589 nm reference stimulus will be needed to achieve a visual match.

Several different types of result may be obtained with an anomaloscope, by which different groups of normal colour-vision defects may be identified. This is achieved by reference to two statistics derived from use of the instrument: the matching range and the midpoint of the colour matches. The former is an indication of the observer's wavelength discrimination, whereas the latter is a demonstration of the direction of shift in the peak spectral sensitivity of the putative pigment in the anomalous trichromats. The matching range for a given observer includes all the light-mixture ratios of 670 and 545 nm that are judged as a match to the reference 589 nm. The midpoint of the colour match is determined as the middle of the matching range of different proportions of mixtures of light of 670 and 545 nm. If the midpoint of the colour matches is shifted towards the longer wavelength (i.e. contains a greater proportion of the 670 nm primary in the Rayleigh equation) and is more than three standard deviations away from the range of match points for the normal, then a diagnosis of protanomalous trichromacy is made. Conversely, if the midpoint is shifted towards the shorter wavelength of the two primaries by a similar statistical distance from the norm, the diagnosis is one of deuteranomalous trichromacy. If colour matches are achieved for the full range of the Rayleigh equation (i.e. all possible proportions of mixtures of light of 670 and 545 nm), then the observer would be classified as having dichromacy. In this state of dichromacy, the differentiation between protanopia and deuteranopia is made on the basis of the changes in luminance required in the 589 nm reference stimulus to achieve a match with the light mixtures of 670 and 545 nm.

Protanomalous and protanopic people, because of their reduced sensitivity to long wavelengths, perceive any light mixture which contains a predominance of long wavelengths (670 nm, for example) as being very dark compared to those containing a predominance of middle-range wavelengths (such as 545 nm). Deuteranopes, on the other hand, perceive light of differing proportions of long and middle-range wavelengths as all having approximately the same brightness.

For most normal trichromats there is a unique proportion of 670 and 545 nm light intensities that will match to 589 nm, and the distribution of match midpoints for a large number of observers may be described by a Gaussian curve. Within this group of normal trichromats it is possible to identify statistically two subtypes whose colour vision may be unsuitable for jobs demanding either representative colour vision or fine colour discrimination. A *deviant colour normal* is a normal trichromat whose Rayleigh equation is of the normal range but whose midpoint is displaced more than two standard deviations from the mean of average observers. By statistical definition, these people will comprise 4% of the colour-normal population. On the other hand, a *colour-weak* person is a normal trichromat who has a Rayleigh equation matching range more than twice that of the most frequent range for the colour normal population, but whose colour-matching midpoint is within the normal range. Such people comprise approximately 20% of the population of normal trichromats; in addition to their atypical colour-matching performance, they also have a slightly reduced wavelength-discrimination ability.

The statistical distinction of two subtypes of anomalous trichromat in terms of the size of the Rayleigh equation matching range (i.e. simple and extreme) has a practical as well as a theoretical significance. From the practical viewpoint it is a useful distinction because some occupational activities may involve relatively coarse colour-discrimination tasks that can be performed adequately by simple anomalous trichromats without any special risk to their own or public safety.

8.11.3 Colour-confusion tests

The most commonly used colour-vision tests are based primarily (but not entirely) upon the principle of colour confusion; most are pseudo-isochromatic tests, described in section 8.11.1. Pseudo-isochromatic tests were first developed in 1878 by the Swedish ophthalmologist Stilling, who constructed his own test empirically by comparing the performance of two colour-defective colleagues with that of someone having normal colour vision. There are now very many variants of the pseudo-isochromatic test, of which the most widely employed is the Ishihara test. In this test, the symbol and background chromaticities on most of the pages (plates) are close to the isoconfusion line used as the Rayleigh equation. It is in this region of colour space that the protan and deutan isoconfusion lines have a closely similar orientation. This feature, together with the range of chromaticities for the spotted symbol and background, means that each plate of the test permits the detection of both protan and deutan deficiencies. For testing adults the symbol is an Arabic numeral; on a few plates, intended for testing children, it is a wavy line that can be traced. Three different design principles are used in construction of the designs. The symbol may be visible to the colour-normal and not to the colour-defective ('vanishing' type), or vice versa ('hidden symbol'), or it may appear as one digit to the colour-normal and as a different one to the colour-defective ('transformation type'). In certain types of colour-vision deficiency some of the symbols may appear to become confused with the background; these serve as qualitative diagnostic tests of the vanishing type. In all instances, the choice of chromaticities of the symbol and the background is based upon the direction of the average dichromatic isoconfusion lines in the CIE xy diagram, and the average chromaticity difference between the symbol and its background is roughly the same for all plates. Since the Ishihara test is designed to detect only 'red' or 'green' colour-vision deficiencies, only the protan and deutan isoconfusion lines are relevant to its construction.

In the 'vanishing' type of plate the chromaticities have been selected so that the perceived colour difference between symbol and background is greater for people with normal colour vision than for those with defective vision. This is achieved by choosing the chromaticities of the symbol and background to fall on, or close to, a known isoconfusion line, and to be separated by large colour differences as seen by the normal observer.

The 'hidden symbol' plate uses principally three different hues with small variations in saturation. The colour differences between adjacent sample spots are so large that, to the colour-normal observer, the outline of the symbol is concealed in the background random dot pattern. To protan or deutan observers, however, the 'hidden symbol' plate has two distinct colour groupings which occur along the separate mean isoconfusion lines for protan and deutan colour-vision deficiencies. To such people the sample spots within these two groupings are confused with each other or appear so similar that they provide a sense of figural unity to the symbol which may be discriminated against the background. The principles for the design of this type of plate are exactly those employed for colour camouflage used in warfare, and its application in the Ishihara test shows that people with defective colour vision can be used to break such camouflage codes.

The 'transformation' plate is a particularly ingenious design. It involves the use of four chromaticity clusters for the sample spots, with large colour differences between the clusters. Two pairs of these clusters fall on different isoconfusion lines in such a way that, to the colour-defective, one pair is perceived as symbol and the other as background. A colour-normal observer, however, whose reference point for colour is the chromaticity position of illuminant C (the illuminant under which the test should be viewed), perceives a different grouping of chromaticity clusters that alters what he or she can recognise as a figure, compared with the background. In a typical example of this type of plate a colour-normal observer would see the digit 5, which would become transformed into the digit 2 for the colour-defective.

The fourth type of design is the qualitative diagnostic plate. It is similar in principle to the 'vanishing' type, except that only two chromaticity clusters are used to print the different symbols; one is on the protan isoconfusion line and the other on the deutan isoconfusion line with respect to a common background. From tests using these plates it is possible to make a tentative classification of the type of colour-vision deficiency, depending on which symbol is not detected.

Although the mean colour difference between symbol and background is the same on each plate in the Ishihara test, the test is not designed to grade the severity of a colour-vision defect. Some pseudo-isochromatic tests can, however, be used to do this. Typical of these grading tests is the Tokyo Medical College test which has a series of 'vanishing' plates, each having one of three magnitudes of mean colour difference between the symbol and background chromaticities. The test can be used to classify colour-defective responses as mild, moderate or severe. The function of the Ishihara test, by contrast, is purely to provide a pass/fail decision between colour-normal and colour-defective. The ability of a colour-defective person to read a set of plates depends not only on his or her chromatic discrimination ability and the colour difference between the symbol and background, but also on how appropriate the selected confusion chromaticities in the test are for that individual. Just as a person with normal colour vision may make colour matches that differ from the group average and do not correspond with those expected of a CIE standard observer, so also will the isoconfusion lines of a single colour-defective person differ from their population group averages. Since the design of any pseudo-isochromatic test is based upon averaged isoconfusion lines, it is not surprising to find that some people with defective colour vision can 'pass' certain plates while some with normal colour vision 'fail' others. These inevitable misclassification errors lead to a slight reduction in the confidence that can be placed on decisions from performance results. Coupled with this inherent feature of test design, it is necessary to realise that, when screening for defective colour vision in an unselected population, the evidence is usually heavily weighted in favour of normal colour vision. This is because normal colour vision is a much more prevalent state than defective colour vision in an unselected healthy population (92% of males and 99.5% of females have normal colour vision). Consequently, for the recommended pass/fail cutoff criterion on the test, decisions that a person has normal colour vision given that they 'pass' can always be made with a much higher confidence than decisions of colour-defective given that the person 'failed' the test. For example, when using the Ishihara test to screen young male adults, the probability that a person has normal colour vision given that they 'pass' the test (i.e. make no more than the recommended three page errors) is 0.99, whereas the probability that a person is colour-defective given that they 'fail' the test with more than three errors is only 0.82. The Ishihara test cannot indicate the severity of a colour-vision deficiency, and was not designed to do so. Different pseudo-isochromatic tests (because of small differences in plate design and construction) have different performance error rates and, therefore, different confidence levels associated with decisions of colour-normal or colour-defective. For screening and selection purposes, however, the Ishihara test is one of the more efficient available.

Not all colour-vision tests based on the design principle of colour confusion are of the pseudo-isochromatic type. Another colour-confusion test widely used for personnel selection in industry is the Farnsworth dichotomous D15 test. This particular test was designed to distinguish between individuals with a severe colour-vision deficiency and those who are either colour-normal or have a mild colour-vision deficiency. As such, the test fails only about 60% of colour-defectives, including the dichromats and extreme anomalous trichromats. In design it comprises 15 circular discs (plus one reference disc) of different hues but having the same value and saturation in Munsell notation. The disc samples are chosen to represent approximately equal perceived hue steps in colour space and are centred on the white point of the CIE xy chromaticity diagram. All the discs are presented to the candidate simultaneously. His or her task is to arrange the discs in order according to colour, starting with the disc closest in colour appearance to the reference disc. Normal trichromats and simple anomalous trichromats arrange the discs in order without error. Extreme anomalous trichromats and dichromats confuse discs that lie on their isoconfusion lines, and their sequencing of the discs reflects these confusions. Because the isoconfusion lines have characteristically different orientations for protan, deutan and tritan defects, different pairs of discs in the test are confused by people with the different types of colour-vision defect. The test makes no assumption about the orientation of individual isoconfusion lines; it therefore also has qualitative diagnostic value. Because of the relatively close spacing of the disc samples along any chromaticity line, however, its diagnostic accuracy is much less than that of the anomaloscope. Other versions of this test have been produced with reduced saturation of the discs, thereby increasing its sensitivity for detecting mild colour-vision deficiencies but at the same time decreasing its differential diagnostic accuracy.

The City University colour-vision test has gained popularity in recent years [33]. It uses the same Munsell colours as the Farnsworth D15 test, but they are arranged in book form so that on each page four discs are placed at four points of the compass and a fifth disc, the reference sample, at the intersection of the two axes. The four peripheral discs have been chosen such that three fall on each of the protan, deutan and tritan isoconfusion lines and the fourth on none of these. The candidate's task is to say which one of the peripheral discs is the closest in colour appearance to the disc in the centre. Different colour comparisons are presented on ten pages. Although the test is intended to be a confusion test, it is actually one of colour-difference estimation because – especially for the person with normal colour vision – none of the paired samples actually provides a visual match. The candidate's task is to decide which of the pairs of coloured discs has the smallest perceived colour difference; this is particularly difficult when the paired coloured samples vary in hue. Since the colour differences employed in the test are all roughly the same, it cannot be used to grade the severity of a colour-vision defect, contrary to the designer's claims.

8.11.4 Colour-discrimination tests

The differences in wavelength discrimination between the various types of colour-vision deficiency are large, systematic and well documented. The measurement of just-noticeable wavelength difference thresholds, however, is both time-consuming and requires sophisticated apparatus. A simplified test that is relatively quick to use and produces results that are closely correlated with the wavelength discrimination function is the Farnsworth–Munsell 100 hue test (FM 100 hue). The test consists of 85 discs (originally 100) made from Munsell papers of different hues but constant saturation and value. Each disc subtends approximately 2° at a viewing distance of about 40 cm and the samples have been chosen to represent perceptually equal steps of hue that form a natural hue circle. For people with normal colour vision, the colorimetric spacing between adjacent discs is roughly equivalent to an average of ten just-noticeable hue steps. The task of a candidate is to arrange the discs in colour order so that discs placed next to each other have a minimal perceived colour difference. Where the discs lie on the isoconfusion line of the candidate, hue discrimination is poor and the perceived ordering of the discs is confused. These colour confusions are represented as performance errors in the test by comparing the selected disc sequence with the theoretically ideal sequence. The errors are then plotted on a polar graph which can be used to distinguish four colour-vision types by reference to the performance characteristics of each group: the three types of colour-vision deficiency (protan, deutan and tritan) and also normal colour vision. The errors on this polar plot are bipolar. They occur where discs lie along isoconfusion lines and are represented on the polar plot by an error that is perpendicular to those isoconfusion lines. In order to avoid colour confusions occurring across the colour circle (as in the D15 test), the discs in the FM 100 hue test are grouped into four boxes which are presented one at a time.

Performance on the FM 100 hue test is not dependent on any standardised average observer, whether or not they have normal or defective colour vision. This test has therefore proved popular for the assessment of acquired colour-vision deficiencies, in which the direction of the isoconfusion lines can be predicted far less reliably than in congenital colour deficiencies. For any person with defective colour vision, the error axis of the FM 100 hue polar plot, when represented in the chromaticity diagram, provides an approximate indication of the orientation of that person's own isoconfusion lines. Furthermore, the total number of errors made on the test (i.e. disc order inversions) is an indirect measure of wavelength-discrimination ability and also correlates with the matching range on the anomaloscope. Testing and manual scoring on this test can take as much as 30 minutes, and this is often given as the reason for it not being chosen as an aptitude test in the colour industry. An automated scoring version of the test is now available (Huematic) which reduces the scoring time to less than a minute.

Since performance on the FM 100 hue test is a function of hue discrimination, it is not surprising that it is more readily affected by illumination level than most other colour-vision tests. Errors on the FM 100 hue decrease with increasing illuminance in parallel with the improvement with wavelength-difference discrimination. Errors on the test also vary systematically with age, with the best performance being achieved between the ages of 20 and 25 years. This largely parallels the decline in sensory function which occurs with increasing age; age-related population norms have been published for the test. While the departure from these norms by two standard deviations in excess of the mean have been used for several years for decisions about clinically significant hue-discrimination losses, nothing has yet been published on maximal error scores that are acceptable for jobs that involve colour quality-control decisions. Moreover, small but significant improvements in performance on the test can be made by familiarity through frequent repeated use. Standardised administrative procedures are therefore needed to achieve meaningful test results.

8.11.5 Some special tests

In the selection of potential employees for certain occupations, particularly those involving judgements or decisions based upon the perception of colour, it may not be sufficient just to distinguish between people whose colour vision is defective and those in whom it is normal. Frequently it is more appropriate to ask whether the colour-vision characteristics of the individual are suitable for the job, or whether the candidate has an aptitude for specific skills required by the nature of the proposed work.

Practical questions of this type resulted in the development of some of the earliest colour-vision tests. For example, the Holmgren wool test was developed by a Swedish ophthalmologist following a major rail accident in that country in 1876. This was a crude colour-confusion test, and only those who performed the test without error were permitted to work on the railways. At about the same time the merchant navy also recognised the need for marines to have good colour vision, and by 1912 the UK Board of Trade had adopted a standardised colour-vision test designed by a British ophthalmologist, the Edridge–Green lantern test. The test was designed to simulate real signals in terms of their chromaticity, angular subtense and intensity, and the observer's task was to name the colours (either red, green or white); it was intended to be used in the dark where colour discrimination is not at its best. Since only signal colours used by the Board of Trade were included in the test, in general only dichromats and extreme anomalous trichromats would fail. In about 1980 a new test, the Holmes–Wright lantern test, was adopted by the merchant navy, the armed services and the civil aviation authorities. It employs standardised filters to produce signal light chromaticities which are just inside the boundaries recommended by the British Standard for coloured signals and codes (BS 1376: 1974).

In other occupations, the principal problem of personnel selection may not be one of distinguishing between individuals in terms of 'safe' and 'unsafe' colour-vision performance. If the nature of the task involves fine and accurate colour-quality control, it may be more important for economic reasons to select personnel who have good colour discrimination and a colour-matching ability that is representative of a population norm. Industries that need to consider such factors include the textile, chemical, paint and electronics industries. Clearly, not all these applications have the same colour-vision requirements; moreover, it is frequently difficult to predict job performance on the basis of standard clinical colour-vision tests. For this reason, certain special colour-vision tests have been designed to achieve a close approximation with specific occupational tasks. In selecting personnel for the electronics industry, for example, it may be more relevant to determine the accuracy with which a wide and representative range of resistor and capacitor colour codes can be identified under typical working conditions, rather than to screen vision on a pseudo-isochromatic test such as the Ishihara or to assess work performance on the basis of the FM 100 hue error score. With careful design and choice of samples, these task-simulation tests can provide a very useful guide to work performance. But there may be practical reasons against the economical construction of such tests. In the printing and dyeing trades, for example, a full range of printed or dyed material is not always available or even appropriate for use in a colour-vision test. To overcome this problem two special tests of colour-vision aptitudes have been developed: the colour-aptitude test and the Burnham–Clark–Munsell colour-memory test [33]. Although some attempts have been made to provide standardised scores for these two tests, it would be more appropriate for users to determine pass/fail cutoff scores relevant to their own particular requirements by examining the relationship between test score and work performance.

The colour-aptitude test of the Inter-Society Colour Council was designed to help select personnel directly involved in colour-matching and -sorting tasks. It is essentially a saturation-discrimination test and is intended to be used only by people with normal colour vision. It will not distinguish between normal and defective colour vision. The test consists of twelve 'reference' plastic tiles in each of four different hues mounted in random arrays on a display board. The chromaticity coordinates of the plastic tiles are approximately along four different hue lines in CIE xy colour space. A duplicate set of the same 48 tiles is available for matching against the reference set. The matching tiles are presented to the observer one at a time in a random sequence and the task is to identify the corresponding matching reference tile. After a match has been made, the matching tile is removed from the display board so that each successive tile is compared against all twelve reference tiles of the same hue. Within each hue set of twelve tiles, the colour difference between the tiles is very small, about only five JNDs. Even people with normal colour vision are therefore likely to make errors on the test, which means that it is particularly good for distinguishing between individuals with different colour-matching performance. Scoring on the test reflects the closeness and accuracy with which colour matches are made, and performance is unrelated to FM 100 hue or anomaloscope results. The colour-aptitude test therefore measures some ability which the other two tests do not.

The Burnham–Clark–Munsell colour-memory test was designed to assess how well a person can memorise a colour. A not uncommon feature of many tasks in industry and commerce is the comparison of a sample in one place with a standard reference elsewhere. In everyday life, too, we frequently rely on colour memory – for example, to judge the freshness of food by its appearance. The colour-memory test provides a quantitative measure of that memory ability. The test consists of 43 comparison discs of different hue mounted in a hue circle. The set is made up of every alternate disc from the FM 100 hue test. Duplicates of 20 of these hues are used as test samples. The observer is presented with one test sample disc for 5 s and after an interval of 5 s (or any other specified time interval) is asked to select from the hue circle of 43 discs the one that resembles the first hue shown. Performance is measured in terms of accuracy. This is neither a colour-confusion nor a chromaticity-discrimination test, and the results do not correlate with performance on either the FM 100 hue test, the anomaloscope or the colour-aptitude test. The Burnham–Clark–Munsell colour-memory test can, therefore, be said to be measuring a unique property of colour-vision ability.

8.11.6 Some practical points

Since performance on different colour-vision tests is invariably poorly correlated, it would seem that different colour-vision tests measure different attributes of colour perception. Any attempt to provide a detailed description of a person's colour-vision abilities will therefore require the use of several different tests. For the greatest efficiency of information, the selected tests should be those in which the performance is known to be uncorrelated or poorly related. Such a collection of tests is often called a 'test battery', and its specific compilation will depend upon the application for which the information is to be used. For example, a test battery designed to identify inherited colour-vision deficiencies would be inefficient and inappropriate for the detection of acquired colour-vision deficiencies. Likewise, tests designed for the rapid differentiation between normal and defective colour vision will provide at best only minimal information on the severity or type of a colour-vision defect. In all testing of human abilities, there are no sharp boundaries between normal and defective performance; in the realm of colour-vision testing, this means that some people with normal colour vision fail certain tests, while some people with defective colour vision may pass the same tests. It is the responsibility of the test user to familiarise him/herself with the benefits and limitations of each test and to be aware of the likelihoods associated with decision errors of pass/fail or acceptable/unacceptable. Although some errors are inevitable in psychophysical tests, the ambiguities and misclassifications that occur in certain colour-vision results are due largely to individual differences in colour-vision characteristics – particularly among mild anomalous trichromats – of which many colour-vision tests take no account. Of course, the variability in individual performance on any test is minimised when the viewing conditions and psychophysical method of stimulus presentation are standardised and carefully controlled. For example, most colour-vision tests that comprise surface reflecting material as stimuli were designed for use under CIE standard illuminant C, or its near equivalent. It is of especial importance that the spectrum of the light source should be continuous and not discontinuous, as with most fluorescent lamps. A stable illumination level is also important; ideally, this should be at least 500 lux if the visual response is to be primarily in the photopic range. Furthermore, if complex colour-vision tests such as the anomaloscope are used, the user must not only be particularly skilled in the testing procedure but also knowledgeable about interpretation of the results.

For practical colour-related tasks in science, industry or commerce, colour vision is only one of a range of personal attributes affecting an individual's job performance. It is not surprising, therefore, that the results from single colour-vision tests, or even colour-vision test batteries, are not good predictors of job performance. If tests or test batteries are to be used in this way, then ideally their predictive value in terms of job performance should be determined empirically for individual task-specific requirements. Nevertheless, people who are familiar with the principles of colour-vision test design and have a knowledge of the properties of normal and defective colour vision will find that it is possible to develop relevant aptitude tests using task-related material. Of course, the threshold levels of acceptability for such aptitude test performance will depend on what is required, either for commercial or safety reasons, in terms of acceptability levels of the job performance or product quality. In certain industrial processes (excluding those involving personal or public safety), the acceptability criteria used for setting the standards of product quality may change as consumer demands change. If, therefore, personnel selection procedures for colour-specific tasks are to be made on a rational basis, the colour-vision requirements must accord with the tightest production standards likely to be encountered in the performance of that task. Establishing the demands of the job should always precede setting the aptitude selection criteria. In this way, it may even be possible to determine the type of job or occupational task for which an individual is best suited, even if they have slightly atypical colour vision, rather than simply selecting those individuals deemed fit for the job.

REFERENCES

1. W B Marks, W H Dobelle and E F MacNichol, *Science*, **143** (1964) 1181.
2. P K Brown and G Wald, *Science*, **144** (1964) 45.
3. H J A Dartnell, J K Bowmaker and J Mollon, *Proc. Royal Soc.*, **B220** (1983) 115.
4. S Hecht, S Schlaer and M H Pirenne, *J. Gen. Physiol.*, **25** (1942) 819.
5. G Svaetichin, *Acta Physiol. Scand.*, Suppl. 134, **39** (156) 17.
6. E Zrenner, *Neurophysiological aspects of colour vision in primates* (Berlin: Springer, 1983).
7. D H Hubel and T N Wiesel, *J. Physiol.*, **195** (1968) 215.
8. S Zeki, *J. Physiol.*, **277** (1978) 278.
9. S Zeki, *A vision of the brain* (London: Blackwell Scientific, 1993).
10. E Zrenner in *Central and peripheral mechanisms of colour vision*, Ed. D Ottoson and S Zeki (London: Macmillan, 1985) 165.
11. C R Michael in *Central and peripheral mechanisms of colour vision*, Ed. D Ottoson and S Zeki (London: Macmillan, 1985) 199.
12. E H Land, *Proc. Nat. Acad. Sci.*, **45** (1959) 115, 636.
13. E H Land, *Vision Res.*, **26** (1986) 7.
14. J Guild, *Phil. Trans. Royal Soc.*, **A230** (1931) 149.
15. W D Wright, *Trans. Opt. Soc. London*, **30** (1928) 141.
16. W D Wright, *Researches on normal and defective colour vision* (London: Kimpton, 1946).
17. W D Wright in *Handbook of sensory physiology*, Vol 7, Part 4, Ed. D Jameson and L V Hurvich (Berlin: Springer 1972) 434.
18. W S Stiles, *Doc. Ophthalmol.*, **3** (1939) 141.
19. W S Stiles, *Mechanisms of colour vision* (London: Academic Press, 1978).
20. W S Stiles, *Proc. Phys. Soc. London*, **58** (1946) 41.
21. R L De Valois, *Cold Spring Harbour Symp. Quant. Biol.*, **30** (1965) 567.

22. D A Palmer, *Vision Res.*, **7** (1967) 619.
23. D A Palmer, *Nature*, **262** (1976) 601.
24. Y Le Grand in *Handbook of sensory physiology*, Vol. **7**, Part 4, Ed. D Jameson and L M Hurvich (Berlin: Springer, 1972) 413.
25. D L MacAdam, *J. Opt. Soc. Amer.*, **32** (1942) 247.
26. D L MacAdam, *J. Opt. Soc. Amer.*, **33** (1943) 18.
27. M R Luo and B Rigg, *J.S.D.C.*, **102** (1986) 164.
28. E F MacNichol et al. in *Colour vision: physiology and psychophysics*, Ed. J D Mollon and L T Sharpe (London: Academic Press, 1983) 13.
29. G H Jacobs, J K Bowmaker and J D Mollon, *Doc. Ophthalmol. Proc. Series*, **33** (1982) 269.
30. J Nathans, D Thomas and D S Hogness, *Science*, **232** (1986) 193.
31. J Nathans, T P Piantanida, R L Eddy and T B Shows, *Science*, **232** (1986) 203.
32. *Vision and visual dysfunction*, Vol. 7 – Inherited and acquired colour vision deficiencies, Ed. D H Foster (London: Macmillan Press, 1991).
33. J Voke, *Colour vision testing in specific industries and professions* (Windsor: Keeler, 1980).
34. R Lakowski, *Br. J. Indust. Medicine*, **26** (1969) 265.

FURTHER READING

M Longair in *Colour: art and science*, Ed. T Lamb and J Bourriau (Cambridge: CUP, 1995).

APPENDICES

Values of colour-matching functions and illuminant spectral energy distributions are defined by the CIE. Numerous slightly different tables of tristimulus weighting factors have been published, however. To promote uniformity of practice, the Colour Measurement Committee of the Society of Dyers and Colourists has recommended that all such values should be taken from the American standard ASTM E 308–95. All the values in these appendices have been taken from that standard, with permission. Further details and additional tables (for example, weighting factors at 10 nm intervals) are given in the standard, which can be obtained from:

American Society for Testing and Materials
1916 Race Street
Philadelphia
PA 19103
USA

Appendix 1 Relative spectral energy distributions of CIE standard illuminants A, C and D_{65}

λ/nm	E_λ A	C	D_{65}	λ/nm	E_λ A	C	D_{65}
300	0.93		0.03	395	13.35	55.17	68.70
305	1.13		1.7	400	14.71	63.30	82.75
310	1.36		3.3	405	16.15	71.81	87.12
315	1.62		11.8	410	17.68	80.60	91.49
320	1.93	0.01	20.2	415	19.29	89.53	92.46
325	2.27	0.20	28.6	420	20.99	98.10	93.43
330	2.66	0.40	37.1	425	22.79	105.80	90.06
335	3.10	1.55	38.5	430	24.67	112.40	86.68
340	3.59	2.70	39.9	435	26.64	117.75	95.77
345	4.14	4.85	42.4	440	28.70	121.50	104.86
350	4.74	7.00	44.9	445	30.85	123.45	110.94
355	5.41	9.95	45.8	450	33.09	124.00	117.01
360	6.14	12.90	46.6	455	35.41	123.60	117.41
365	6.95	17.20	49.4	460	37.81	123.10	117.81
370	7.82	21.40	52.1	465	40.30	123.30	116.34
375	8.77	27.50	51.0	470	42.87	123.80	114.86
380	9.80	33.00	49.98	475	45.52	124.09	115.39
385	10.90	39.92	52.31	480	48.24	123.90	115.92
390	12.09	47.40	54.65	485	51.04	122.92	112.37

Appendix 1 – continued

490	53.91	120.70	108.81	640	157.98	87.80	83.70
495	56.85	116.90	109.08	645	161.52	87.99	81.86
500	59.86	112.10	109.35	650	165.03	88.20	80.03
505	62.93	106.98	108.58	655	168.51	88.20	80.12
510	66.06	102.30	107.80	660	171.96	87.90	80.21
515	69.25	98.81	106.30	665	175.38	87.22	81.25
520	72.50	96.90	104.79	670	178.77	86.30	82.28
525	75.79	96.78	106.24	675	182.12	85.30	80.28
530	79.13	98.00	107.69	680	185.43	84.00	78.28
535	82.52	99.94	106.05	685	188.70	82.21	74.00
540	85.95	102.10	104.41	690	191.93	80.20	69.72
545	89.41	103.95	104.23	695	195.12	78.24	70.67
550	92.91	105.20	104.05	700	198.26	76.30	71.61
555	96.44	105.67	102.02	705	201.36	74.36	72.98
560	100.00	105.30	100.00	710	204.41	72.40	74.35
565	103.58	104.11	98.17	715	207.41	70.40	67.98
570	107.18	102.30	96.33	720	210.36	68.30	61.60
575	110.80	100.15	96.06	725	213.27	66.30	65.74
580	114.44	97.80	95.79	730	216.12	64.40	69.89
585	118.08	95.43	92.24	735	218.92	62.80	72.49
590	121.73	93.20	88.69	740	221.67	61.50	75.09
595	125.39	91.22	89.35	745	224.36	60.20	69.34
600	129.04	89.70	90.01	750	227.00	59.20	63.59
605	132.70	88.83	89.80	755	229.59	58.50	55.01
610	136.35	88.40	89.60	760	232.12	58.10	46.42
615	139.99	88.19	88.65	765	234.59	58.00	56.61
620	143.62	88.10	87.70	770	237.01	58.20	66.81
625	147.24	88.06	85.49	775	239.37	58.50	65.09
630	150.84	88.00	83.29	780	241.68	59.10	63.38
635	154.42	87.86	83.49				

Appendix 2 Spectral tristimulus values (colour-matching functions) for the CIE 1931 standard (2°) observer

	Tristimulus values				Tristimulus values		
λ/nm	$\bar{x}_\lambda$	$\bar{y}_\lambda$	$\bar{z}_\lambda$	λ/nm	$\bar{x}_\lambda$	$\bar{y}_\lambda$	$\bar{z}_\lambda$
380	0.0014	0.0000	0.0065	420	0.1344	0.0040	0.6456
385	0.0022	0.0001	0.0105	425	0.2148	0.0073	1.0391
390	0.0042	0.0001	0.0201	430	0.2839	0.0116	1.3856
395	0.0076	0.0002	0.0362	435	0.3285	0.0168	1.6230
400	0.0143	0.0004	0.0679	440	0.3483	0.0230	1.7471
405	0.0232	0.0005	0.1102	445	0.3481	0.0298	1.7826
410	0.0435	0.0012	0.2074	450	0.3362	0.0380	1.7721
415	0.0776	0.0022	0.3713	455	0.3187	0.0480	1.7441

Appendix 2 – *continued*

460	0.2908	0.0600	1.6692	630	0.6424	0.2650	0.0000
465	0.2511	0.0739	1.5281	635	0.5419	0.2170	
470	0.1954	0.0910	1.2876	640	0.4479	0.1750	
475	0.1421	0.1126	1.0419	645	0.3608	0.1382	
480	0.0956	0.1390	0.8130	650	0.2835	0.1070	
485	0.0580	0.1693	0.6162	655	0.2187	0.0816	
490	0.0320	0.2080	0.4652	660	0.1649	0.0610	
495	0.0147	0.2586	0.3533	665	0.1212	0.0446	
500	0.0049	0.3230	0.2720	670	0.0874	0.0320	
505	0.0024	0.4073	0.2123	675	0.0636	0.0232	
510	0.0093	0.5030	0.1582	680	0.0468	0.0170	
515	0.0291	0.6082	0.1117	685	0.0329	0.0119	
520	0.0633	0.7100	0.0782	690	0.0227	0.0082	
525	0.1096	0.7932	0.0573	695	0.0158	0.0057	
530	0.1655	0.8620	0.0422	700	0.0114	0.0041	
535	0.2257	0.9149	0.0298	705	0.0081	0.0029	
540	0.2904	0.9540	0.0203	710	0.0058	0.0021	
545	0.3597	0.9803	0.0134	715	0.0041	0.0015	
550	0.4334	0.9950	0.0087	720	0.0029	0.0010	
555	0.5121	1.0000	0.0057	725	0.0020	0.0007	
560	0.5945	0.9950	0.0039	730	0.0014	0.0005	
565	0.6784	0.9786	0.0027	735	0.0010	0.0004	
570	0.7621	0.9520	0.0021	740	0.0007	0.0002	
575	0.8425	0.9154	0.0018	745	0.0005	0.0002	
580	0.9163	0.8700	0.0017	750	0.0003	0.0001	
585	0.9786	0.8163	0.0014	755	0.0002	0.0001	
590	1.0263	0.7570	0.0011	760	0.0002	0.0001	
595	1.0567	0.6949	0.0010	765	0.0001	0.0000	
600	1.0622	0.6310	0.0008	770	0.0001		
605	1.0456	0.5668	0.0006	775	0.0001		
610	1.0026	0.5030	0.0003	780	0.0000		
615	0.9384	0.4412	0.0002				
620	0.8544	0.3810	0.0002				
625	0.7514	0.3210	0.0001	Total	21.3714	21.3711	21.3715

Appendix 3 Tristimulus values and chromaticity coordinates for a sample reflecting 100% of the incident light (based on 1931 (2°) observer and weighting factors at 20 nm intervals)

Illuminant	X	Y	Z	x	y
A	109.850	100.000	35.585	0.4476	0.4074
C	98.074	100.000	118.232	0.3101	0.3161
D_{65}	95.047	100.000	108.883	0.3127	0.3290

Appendix 4 Spectral tristimulus values (colour-matching functions) for the CIE 1964 standard (10°) observer

λ/nm	Colour-matching function $\bar{x}_{10,\lambda}$	$\bar{y}_{10,\lambda}$	$\bar{z}_{10,\lambda}$	λ/nm	Colour-matching function $\bar{x}_{10,\lambda}$	$\bar{y}_{10,\lambda}$	$\bar{z}_{10,\lambda}$
380	0.0002	0.0000	0.0007	590	1.1185	0.7774	
385	0.0007	0.0001	0.0029	595	1.1343	0.7204	
390	0.0024	0.0003	0.0105	600	1.1240	0.6583	
395	0.0072	0.0008	0.0323	605	1.0891	0.5939	
400	0.0191	0.0020	0.0860	610	1.0305	0.5280	
405	0.0434	0.0045	0.1971	615	0.9507	0.4618	
410	0.0847	0.0088	0.3894	620	0.8563	0.3981	
415	0.1406	0.0145	0.6568	625	0.7549	0.3396	
420	0.2045	0.0214	0.9725	630	0.6475	0.2835	
425	0.2647	0.0295	1.2825	635	0.5351	0.2283	
430	0.3147	0.0387	1.5535	640	0.4316	0.1798	
435	0.3577	0.0496	1.7985	645	0.3437	0.1402	
440	0.3837	0.0621	1.9673	650	0.2683	0.1076	
445	0.3867	0.0747	2.0273	655	0.2043	0.0812	
450	0.3707	0.0895	1.9948	660	0.1526	0.0603	
455	0.3430	0.1063	1.9007	665	0.1122	0.0441	
460	0.3023	0.1282	1.7454	670	0.0813	0.0318	
465	0.2541	0.1528	1.5549	675	0.0579	0.0226	
470	0.1956	0.1852	1.3176	680	0.0409	0.0159	
475	0.1323	0.2199	1.0302	685	0.0286	0.0111	
480	0.0805	0.2536	0.7721	690	0.0199	0.0077	
485	0.0411	0.2977	0.5701	695	0.0138	0.0054	
490	0.0162	0.3391	0.4153	700	0.0096	0.0037	
495	0.0051	0.3954	0.3024	705	0.0066	0.0026	
500	0.0038	0.4608	0.2185	710	0.0046	0.0018	
505	0.0154	0.5314	0.1592	715	0.0031	0.0012	
510	0.0375	0.6067	0.1120	720	0.0022	0.0008	
515	0.0714	0.6857	0.0822	725	0.0015	0.0006	
520	0.1177	0.7618	0.0607	730	0.0010	0.0004	
525	0.1730	0.8233	0.0431	735	0.0007	0.0003	
530	0.2365	0.8752	0.0305	740	0.0005	0.0002	
535	0.3042	0.9238	0.0206	745	0.0004	0.0001	
540	0.3768	0.9620	0.0137	750	0.0003	0.0001	
545	0.4516	0.9822	0.0079	755	0.0002	0.0001	
550	0.5298	0.9918	0.0040	760	0.0001	0.0000	
555	0.6161	0.9991	0.0011	765	0.0001		
560	0.7052	0.9973	0.0000	775	0.0000		
565	0.7938	0.9824		780			
570	0.8787	0.9555					
575	0.9512	0.9152					
580	1.0142	0.8689					
585	1.0743	0.8256		Total	23.3294	23.3324	23.3343

518 APPENDICES

Appendix 5 Tristimulus weighting factors for calculation of tristimulus values from spectral data obtained at 20 nm intervals and uncorrected for bandpass dependence

Note: In some cases in the tables below, small negative values appear. These are correct. The white point values do not correspond exactly to the sums of the values in the tables, because of round-off errors. The white point values given should be used in the calculation of, for example, CIE L^*, a^* and b^*.

λ/nm	$E_\lambda \bar{x}_\lambda$	$E_\lambda \bar{y}_\lambda$	$E_\lambda \bar{z}_\lambda$
Illuminant A, 1931 (2°) observer			
360	0.000	0.000	0.000
380	0.013	0.000	0.060
400	−0.026	0.000	−0.123
420	0.483	0.009	2.306
440	1.955	0.106	9.637
460	2.145	0.385	12.257
480	0.848	1.119	7.301
500	−0.112	3.247	2.727
520	0.611	9.517	1.035
540	4.407	15.434	0.274
560	10.804	18.703	0.055
580	19.601	18.746	0.034
600	26.256	15.233	0.018
620	23.295	10.105	0.003
640	12.853	4.939	0.000
660	4.863	1.784	0.000
680	1.363	0.495	0.000
700	0.359	0.129	0.000
720	0.100	0.036	0.000
740	0.023	0.008	0.000
760	0.006	0.002	0.000
780	0.002	0.001	0.000
White point	109.850	100.000	35.585
Illuminant A, 1964 (10°) observer			
360	0.000	0.000	0.000
380	0.007	0.000	0.037
400	−0.016	0.000	−0.088
420	0.691	0.066	3.226
440	2.025	0.285	10.278
460	2.158	0.796	12.345
480	0.642	2.043	6.555
500	−0.160	4.630	1.966
520	1.284	9.668	0.721
540	5.445	14.621	0.171
560	12.238	17.766	−0.013
580	20.755	17.800	0.004
600	26.325	15.129	−0.001
620	22.187	10.097	0.000

Appendix 5 – *continued*

640	11.816	4.858	0.000
660	4.221	1.643	0.000
680	1.154	0.452	0.000
700	0.282	0.109	0.000
720	0.068	0.026	0.000
740	0.017	0.007	0.000
760	0.004	0.002	0.000
780	0.001	0.000	0.000
White point	111.144	100.000	35.200

Illuminant C, 1931 (2°) observer

360	0.000	0.000	0.000
380	0.066	0.000	0.311
400	−0.164	0.001	−0.777
420	2.373	0.044	11.296
440	8.595	0.491	42.561
460	6.939	1.308	39.899
480	2.045	3.062	18.451
500	−0.217	6.596	4.728
520	0.881	12.925	1.341
540	5.406	18.650	0.319
560	11.842	20.143	0.059
580	17.169	16.095	0.028
600	18.383	10.537	0.013
620	14.348	6.211	0.002
640	7.148	2.743	0.000
660	2.484	0.911	0.000
680	0.600	0.218	0.000
700	0.136	0.049	0.000
720	0.031	0.011	0.000
740	0.006	0.002	0.000
760	0.002	0.001	0.000
780	0.000	0.000	0.000
White point	98.074	100.000	118.232

Illuminant C, 1964 (10°) observer

360	0.000	0.000	0.000
380	0.043	0.002	0.213
400	−0.122	−0.004	−0.622
420	3.216	0.301	15.025
440	8.476	1.239	43.144
460	6.668	2.577	38.431
480	1.430	5.320	15.661
500	−0.249	8.742	3.219
520	1.734	12.466	0.897
540	6.364	16.891	0.187
560	12.790	18.284	−0.014
580	17.338	14.617	0.004
600	17.597	10.019	−0.001

Appendix 5 – *continued*

620	13.045	5.925	0.000
640	6.283	2.581	0.000
660	2.055	0.800	0.000
680	0.488	0.191	0.000
700	0.100	0.039	0.000
720	0.021	0.008	0.000
740	0.004	0.002	0.000
760	0.001	0.000	0.000
780	0.000	0.000	0.000
White point	97.285	100.000	116.145

Illuminant D_{65}, 1931 (2°) observer

360	0.000	0.000	0.000
380	0.040	0.000	0.187
400	−0.026	0.004	−0.120
420	2.114	0.041	10.065
440	7.323	0.411	36.235
460	6.815	1.281	39.090
480	1.843	2.797	16.753
500	−0.219	6.291	4.727
520	1.003	14.463	1.532
540	5.723	19.509	0.314
560	11.284	19.106	0.058
580	16.548	15.600	0.027
600	18.528	10.607	0.013
620	14.397	6.240	0.002
640	6.646	2.540	0.000
660	2.290	0.842	0.000
680	0.574	0.208	0.000
700	0.120	0.043	0.000
720	0.034	0.012	0.000
740	0.007	0.003	0.000
760	0.001	0.000	0.000
780	0.001	0.000	0.000
White point	95.047	100.000	108.883

Illuminant D_{65} 1964 (10°) observer

360	0.000	0.000	0.000
380	0.003	−0.001	0.025
400	0.056	0.013	0.199
420	2.951	0.280	13.768
440	7.227	1.042	36.808
460	6.578	2.534	37.827
480	1.278	4.872	14.226
500	−0.259	8.438	3.254
520	1.951	14.030	1.025
540	6.751	17.715	0.184
560	12.223	17.407	−0.013
580	16.779	14.210	0.004

Appendix 5 – *continued*

600	17.793	10.121	−0.001
620	13.135	5.971	0.000
640	5.859	2.399	0.000
660	1.901	0.741	0.000
680	0.469	0.184	0.000
700	0.088	0.034	0.000
720	0.023	0.009	0.000
740	0.005	0.002	0.000
760	0.001	0.000	0.000
780	0.000	0.000	0.000
White point	94.811	100.000	107.304

Illuminant F2 (cool white fluorescent lamp), 1931 (2°) observer

360	0.000	0.000	0.000
380	−0.015	−0.001	−0.075
400	0.126	0.006	0.604
420	0.723	0.016	3.459
440	7.638	0.413	37.775
460	2.320	0.518	13.826
480	0.931	1.364	8.340
500	−0.106	3.077	2.271
520	0.034	5.636	0.725
540	5.711	18.719	0.319
560	13.144	23.526	0.088
580	27.390	25.997	0.044
600	24.880	13.965	0.017
620	12.425	5.247	0.001
640	3.276	1.258	0.000
660	0.613	0.222	0.000
680	0.082	0.030	0.000
700	0.014	0.005	0.000
720	0.002	0.001	0.000
740	0.000	0.000	0.000
760	0.000	0.000	0.000
780	0.000	0.000	0.000
White point	99.186	100.000	67.393

Illuminant F2 (cool white fluorescent lamp), 1964 (10°) observer

360	0.000	0.000	0.000
380	−0.038	−0.005	−0.171
400	0.234	0.028	1.066
420	1.022	0.100	4.782
440	7.898	1.121	39.933
460	2.301	1.042	13.716
480	0.686	2.475	7.408
500	−0.133	4.279	1.613
520	0.444	5.769	0.511
540	6.953	17.713	0.191
560	14.911	22.281	−0.001

Appendix 5 – *continued*

580	28.878	24.639	0.002
600	24.810	13.883	0.000
620	11.708	5.211	0.000
640	3.014	1.241	0.000
660	0.516	0.197	0.000
680	0.073	0.030	0.000
700	0.010	0.004	0.000
720	0.001	0.001	0.000
740	0.000	0.000	0.000
760	0.000	0.000	0.000
780	0.000	0.000	0.000
White point	103.279	100.000	69.027

Illuminant F7 (broad-band daylight lamp (6500 K)), 1931 (2°) observer

360	0.000	0.000	0.000
380	−0.007	−0.001	−0.033
400	0.121	0.007	0.578
420	1.323	0.028	6.323
440	10.790	0.584	53.336
460	4.665	0.963	27.365
480	1.708	2.492	15.213
500	−0.218	5.611	4.189
520	0.379	11.237	1.309
540	7.709	23.952	0.351
560	10.453	18.318	0.071
580	18.791	17.848	0.030
600	17.996	10.198	0.013
620	13.114	5.650	0.001
640	5.970	2.291	0.000
660	1.965	0.720	0.000
680	0.204	0.074	0.000
700	0.073	0.026	0.000
720	0.003	0.001	0.000
740	0.003	0.001	0.000
760	0.000	0.000	0.000
780	0.000	0.000	0.000
White point	95.041	100.000	108.747

Illuminant F7 (broad-band daylight lamp (6500 K)), 1964 (10°) observer

360	0.000	0.000	0.000
380	−0.036	−0.005	−0.161
400	0.246	0.031	1.106
420	1.824	0.177	8.525
440	10.807	1.533	54.683
460	4.506	1.899	26.455
480	1.222	4.373	13.104
500	−0.261	7.596	2.884
520	1.147	11.062	0.890
540	9.029	21.938	0.199

Appendix 5 – continued

560	11.459	16.827	0.000
580	19.208	16.389	0.002
600	17.412	9.821	−0.001
620	12.049	5.451	0.000
640	5.311	2.182	0.000
660	1.641	0.638	0.000
680	0.169	0.067	0.000
700	0.055	0.021	0.000
720	0.001	0.000	0.000
740	0.002	0.001	0.000
760	0.000	0.000	0.000
780	0.000	0.000	0.000
White point	95.792	100.000	107.686

Illuminant F11 (narrow-band white fluorescent lamp similar to TL84), 1931 (2°) observer

360	0.000	0.000	0.000
380	−0.014	−0.001	−0.076
400	0.100	0.005	0.509
420	0.256	−0.001	1.093
440	8.207	0.419	40.877
460	1.559	0.623	9.228
480	0.600	0.507	8.258
500	1.524	7.107	4.371
520	−5.091	−14.004	−0.965
540	20.536	58.821	1.039
560	3.973	7.524	−0.034
580	9.894	9.370	0.032
600	24.253	13.848	0.011
620	37.637	17.208	0.009
640	−4.377	−2.270	−0.002
660	2.164	0.978	0.001
680	−0.411	−0.200	0.000
700	0.172	0.075	0.000
720	−0.025	−0.012	0.000
740	0.006	0.003	0.000
760	−0.001	−0.001	0.000
780	0.000	0.000	0.000
White point	100.962	100.000	64.350

Illuminant F11 (narrow-band white fluorescent lamp similar to TL84), 1964 (10°) observer

360	0.000	0.000	0.000
380	−0.029	−0.005	−0.142
400	0.181	0.026	0.869
420	0.414	0.019	1.729
440	8.515	1.220	43.348
460	1.544	0.977	9.002
480	0.319	1.693	7.470
500	1.673	8.341	3.484
520	−5.992	−13.547	−0.739

Appendix 5 – *continued*

540	24.601	55.948	0.625
560	4.494	7.060	−0.051
580	10.526	8.885	0.014
600	24.099	13.702	−0.004
620	36.033	17.112	0.001
640	−4.279	−2.247	0.000
660	2.026	0.952	0.000
680	−0.397	−0.198	0.000
700	0.155	0.072	0.000
720	−0.025	−0.013	0.000
740	0.006	0.003	0.000
760	−0.001	−0.001	0.000
780	0.000	0.000	0.000
White point	103.863	100.00	65.607

Appendix 6 Example of calculation of tristimulus values using weighting factors for illuminant C and CIE 1931 (2°) observer from Appendix 5

λ/nm	R_λ	$E_\lambda \bar{x}_\lambda$	$E_\lambda \bar{y}_\lambda$	$E_\lambda \bar{z}_\lambda$	$E_\lambda \bar{x}_\lambda R_\lambda$	$E_\lambda \bar{y}_\lambda R_\lambda$	$E_\lambda \bar{z}_\lambda R_\lambda$
360	20.8	0.000	0.000	0.000	0.000	0.000	0.000
380	22.7	0.066	0.000	0.311	1.498	0.000	7.060
400	25.6	−0.164	0.001	−0.777	−4.198	0.026	−19.891
420	32.5	2.373	0.044	11.296	77.123	1.430	367.120
440	41.0	8.595	0.491	42.561	352.395	20.131	1745.001
460	48.6	6.939	1.308	39.899	337.235	63.569	1939.091
480	57.0	2.045	3.062	18.451	116.565	174.534	1051.707
500	69.1	−0.217	6.596	4.728	−14.995	455.784	326.705
520	67.1	0.881	12.925	1.341	59.115	867.268	89.981
540	60.0	5.406	18.650	0.319	324.360	1119.000	19.140
560	48.5	11.842	20.143	0.059	574.337	976.936	2.862
580	38.4	17.169	16.095	0.028	659.290	618.048	1.075
600	29.9	18.383	10.537	0.013	549.652	315.056	0.389
620	25.9	14.348	6.211	0.002	371.613	160.865	0.052
640	25.5	7.148	2.743	0.000	182.274	69.947	0.000
660	22.9	2.484	0.911	0.000	56.884	20.862	0.000
680	19.7	0.600	0.218	0.000	11.820	4.295	0.000
700	30.8	0.136	0.049	0.000	4.189	1.509	0.000
720	45.6	0.031	0.011	0.000	1.414	0.502	0.000
740	52.7	0.006	0.002	0.000	0.316	0.105	0.000
760	57.9	0.002	0.001	0.000	0.116	0.058	0.000
780	60.2	0.000	0.000	0.000	0.000	0.000	0.000
Total		98.073	99.998	118.231	3661.002	4869.922	5530.291

$X = 3661.002/100 = 36.610$
$Y = 4869.922/100 = 48.699$
$Z = 5530.291/100 = 55.303$

Subject index

Abney's law of luminance additivity 483
abridged spectrophotometers 62
absorbance 34–5, 36
   of dye solutions 76–9
absorption coefficients (K) 34, 36–7, 39–40, 210
   additivity of 297–8
   of coatings 294–5, 359
      determination of 308–13, 313–15, 319–26, 328
      nonlinear concentration dependence of 326–7
   of coatings that are very brightly coloured 315–18, 318–19
   of dyed textiles (K/S_s) 2, 12–14
      concentration-dependent 236–9
      overall best-fit 235–6
   of semitransparent layers
      concentration dependence of 336–8
      determination of 334
   units 297
absorptivity 35, 40
accommodation 428
achromatic colours 125, 126, 128
achromatic vision 493
achromatopsia 500
action spectra (retinal ganglion cells) 442, 443
adaptation (visual)
   chromatic 116–19, 413, 480
   dark/light 476
   selective 455, 485
adapting field 476
additive colour mixing 85–6, 88–90
after-images 456, 484–5
age effects
   colour-vision defects 499, 500, 508
   flicker perception 388
   light absorption in crystalline lens 487, 501
amacrine cells (retinal) 431, 436, 437–8, 439, 441, 459
anomaloscopes 490, 500, 501–2, 511
anomalous trichromats 489–92, 493, 494–5, 496, 497, 499, 502–3, 511
anthracene, absorption and fluorescence transitions in 52
Aristotle 122
aromatic compounds, UV absorption by 50–51

artificial neural network theory, use in recipe formulation 282–5
assimilation effect 482
atomic absorption spectra 45–6
atomic emission spectra 44–5

bathochromic shift 50
Beer–Lambert law 35, 76
   deviations from 78
Beer's law 34, 35–6
Benham's top 486
Bezold–Brucke hue shift 482
bipolar cells (retinal) 431, 437, 438
black body (Planckian) radiators 8–11
'blind spot' 427, 429, 430
Bouguer's law *see* Lambert's law
brain, visual processing in 444–51
Brewster angle 31
brightness 389, 42, 166
   CIE definition of 140
   related to luminance 462–3
   SDC definition of 39, 165
bronzing 245, 339
build (of colorants) 170–71, 176
   of dyes 163, 164
   of pigments 167–8
Burnham–Clark–Munsell colour-memory test 509, 510
1,3-butadiene, UV absorption in 47, 48, 49

calibration panels
   for opaque layer database 307, 318
   preparation of 328–33
      by blending finished paints 328–9
      by full formulation 330–31
      by tinting base paint 329–30
calibration prints, for inks database 338–40
camouflage, breaking 504
carotenoids, UV absorption in 47, 49
cathode-ray tube displays 373–88, 402, 407, 411, 417, 418, 419
   calibration of 419–22
   cathode-ray tube construction 373–5
   effects of surround on colour appearance 412
   electron-beam characteristics 375–7
   raster scan and flicker in 386–8
   resolution of image in 379–80

cathode-ray tube displays *continued*
  shadow-mask colour displays 377–81
  see also under phosphors
cathodoluminescence 381
CE3000 spectrophotometer 71–2
chroma 128, 160, 166
  in CIELAB colour space 139–40
chromatic adaptation 116–19, 413, 480
chromatic colours 125, 126, 128
  full 127
chromaticity coordinates 105, 122, 477
chromaticity diagrams 91–3, 105–8, 126, 127, 477–9
chromaticity discrimination 477–9, 496–8
chromaticness, CIE definition of 140
CIE *see* Commission Internationale de l'Éclairage
CIE 1960 UCS diagram 136, 488
CIE 1976 UCS diagram 137
CIELAB colour space 136, 137–40, 403, 478
  colour differences in 111, 147–50, 153, 154, 186, 356
  Colour Map 179–81
  modifications of 112
  Pauli extension 137–8
CIELUV colour space 137, 138, 139–40, 403–4, 405
  colour differences in 111–12, 147, 150
City University colour-vision test 506–7
coatings
  reflection at air/coating interface 304–7
  see also paints; semitransparent layers
Coats Viyella colour physics system 271
colorimetry 81–120
  see also tristimulus colorimeter
colour-aptitude test 509, 510
colour atlases 83, 128–30
colour blindness 493
colour constancy 116–19, 413, 448, 451, 452, 457, 459, 461, 479–81
colour difference 82
  descriptors of components of 158–62, 165–6, 168–9
  evaluation 140–62, 289
  fundamental concepts of 142–6
  reliability of instrumental assessment of 142
  reliability of visual assessment of 141–2
  formulae 111–12, 146–58
  based on DIN colour-order system 181–2
  Christ formula 183–4
  CIE 1994 ($\Delta L^* \Delta C^*_{ab} \Delta H^*_{ab}$; CIE94) 156–8
  CMC(l:c) 150, 151–5, 158, 190
  Gall and Riedel formula 182–3
  JPC79 151–3
  M&S 150–51, 166
  SEK1 and SEK2 190–92
  see also under CIELAB colour space; CIELUV colour space
colour matching 27
  booth design and viewing arrangements for 28–9
  light sources for 12, 27–8
  tests for ability in 500, 501–3
  using coloured lights 84, 88–90
  see also colorimetry
colour-memory test 509, 510
colour mixing 84–8
colour-order systems 121–40
  see also colour specification
colour perception 57–8, 426–7, 448, 452, 461–2, 481–6
  methods of investigating 463–7
  see also colour constancy
colour-rendering index 26–7
colour solids 130
colour spaces 122, 142–4, 341–2
  Adams–Nickerson (ANLAB) 135, 136
  Adams chromatic value 132–4
  CIE 1964 ($U^*V^*W^*$) 136
  CIE xyY 131
  cube-root formulae for 135–6, 137, 147
  Hunter $L\alpha\beta$ 132
  Hunter Lab 134–5
  Scofield Lab 132
  see also CIELAB colour space; CIELUV colour space
colour specification 82–4
  CIE system 82, 83, 91, 98–102, 126, 131
    additions to 102–4
    perceptual non-uniformity of 110–12, 478–9
    standard illumination and viewing conditions for 100, 103–4
    standard light sources and illuminants for 99–100, 102–3, 106–7, 110
    standard observer for 94, 100, 103, 454, 487–8, 515–16, 517
    standard primaries for 98–9, 105
    standard of reflectance factor for 104
    units 101
    use in visual display characterisation 399–413
    usefulness and limitations of 108–10
  primaries for 88, 89, 98–9, 454
    arbitrarily chosen 90
    inadequacy of real 90–93
    unreal (imaginary) 91, 95
    using more than three 94

SUBJECT INDEX 527

colour temperatures 10–11
  of daylight 12
colour vision 89, 426–513
  chromaticity discrimination 477–9, 496–8
  defective 489–500
    acquired 499–500
    anomalous trichromacy 489–92
    dichromacy 492–3, 495, 496–8
    incidence of 498, 499, 500
    inherited 498–9
    monochromacy 493, 498, 499, 500
    tests for 500–12
      colour- (wavelength-)discrimination 502, 507–8
      colour-confusion 503–7, 509
      colour-matching 501–3
      task-related 508–10, 511–12
  individual differences in 486–9, 501
  luminance discrimination 468, 472–7
  mechanism of 430–52
    photopigment action 430–35
    postreceptoral retinal pathways 435–44
    postretinal processing 444–51
  models of 452–62
    opponent colour theory 132, 453, 455–6
    retinex theory 453–4, 456–8
    trichromatic theory 132–3, 453, 454–5
    zone theories 459
  'normal' 487–8, 503
  spectral purity discrimination 470–72, 495–6
  wavelength discrimination 43, 82, 460–61, 467–70
colour-weak individuals 503
colours
  achromatic and chromatic 125, 126
  attributes of 38–9, 121–2, 132, 462–3
    discrimination of 467–79
    in Munsell system 123, 126–9
    trace differences in 176
  elementary 126
  full chromatic 127
  importance of 81
  measurement of 57–80
    see also colorimetry
  naming 83, 464–5, 468
  production of 1–2
  unitary (unique) 126
  visual assessment of 130
COMIC system 216–17, 293
Commission Internationale de l'Éclairage (CIE)
  CIE 1960 UCS diagram 136, 488
  CIE 1976 UCS diagram 137
  colour definition 58

colour specification system *see under* colour specification
specification of colour-rendering properties 119–20
standard illuminants and standard sources 13–15
tint formula 202–3
whiteness formula 200–202, 203–4
computers, for computer colorant formulation 286
cone cells 428–9
  in anomalous trichromats 491–2
  coding action of 458, 460, 469
  cone–rod interaction 476
  distribution of 429, 460–61
  luminance range of action of 57, 473–4
  pathway of output of 438–9
  peak sensitivities of 433–4, 459–60
  photopigments in 430–32, 433–5, 486
  selective desensitisation of 485
  variations in 486
  vulnerability to damage 499
cone monochromats 493, 499
conjugated molecules, UV absorption in 47, 48, 49
contrast ratio 359
convergence principle 438–9
correction matrix 217, 221–2, 237, 267–8
  for use with fluorescent dyes 258–9
corresponding colours 117, 118
cost of colorants
  reduction by computer formulation 226–7, 271, 275–6, 277
  relative 174–5
critical flicker fusion frequency 387
critical pigment volume concentration 333
crystalline lens (of eye) 427, 428
  age-related light absorption in 487
cylindrical coordinates 123–5

Dalton, John 452
dark adaptation 473
daylight 11–13
  simulated 12, 14, 15, 17, 27–8, 102–3
DBH model 167, 173
depth (of colour) 166
  evaluation of 169–85
  evaluation of relative
    by $(K/S)_\lambda$ summation methods 179–81, 184
    by methods based on colour-order systems and colour spaces 181–5
    by single-wavelength methods 177–9
    by visual methods 175–7

depth (of colour) *continued*
  SDC definition of 38–9, 163, 165
  standard 183–4
    for fastness testing 173–4
deuteranomalous trichromats 491, 492, 495, 499, 502
deuteranopic dichromats 492, 494, 496, 499, 502, 503
deviant colour-normal individuals 503
dichromacy 492–3, 495, 496–8, 499, 502, 503
diffraction, colour production by 1
diffraction gratings 70–71
diode lasers 21, 23
dioptric power, of eye 427–8
dissimilar colorants 169
dullness 38–9, 42, 165–6
dyes
  build (build-up) of 163, 164
  components of colour and colour difference 163–7
  cost of
    reduction by computer formulation 226–7, 271, 275–6, 277
    relative 174–5
  dye–dye interactions 267, 271
  mixing 84–5, 87–8
  pH-sensitive 78–9
  photodegradation of 53–4
    see also light-fastness testing
  spectrophotometric analysis of solutions of 76–9
  see also recipe prediction for dyed textiles

Einstein, Albert 9
electroluminescence 21, 23
electromagnetic radiation 2, 3
  wave characterisation of 3–5
electromagnetic spectrum 5
elementary colours 126
energy content of radiation 6
energy-level diagrams 45, 46, 49, 51–2
excitation purity 107
eye 57
  role as transducer 426
  sensitivity of 5, 434
    colour-difference evaluation 130, 176
    colour discrimination 43, 82, 467
    effect of luminance (Purkinje shift) 402, 474
    see also V_λ curves
  structure and function of 427–30

f-numbers 428
Farnsworth–Munsell 100 hue colour-vision test 507–8
Farnsworth dichotomous D15 colour-vision test 506
fastness testing
  evaluation of results from 185–95
    colorimetric 189–93, 195, 196
    grey scales for use in 185–8
    reliability in 188–9
    visual 193–4
  standard depths for 173–4
  see also light-fastness testing
Fechner's colours 485–6
FFACs *see* fluorescent factors
FIAF computer program 182–3
field size metamerism 114
fluorescence 51–3, 72–6, 381
  quenching of 255, 257, 265
fluorescent dyes *see under* recipe prediction for dyed textiles
fluorescent factors (FFACs) 260–61, 264
fluorescent lamps/tubes 19–20
  'artificial daylight' 27–8
  colour-rendering properties 25, 26, 119
  for liquid-crystal display backlighting 395
fluorescent materials
  fastness testing of 193
  measurement of colour of 74–6
fluorescent whitening agents (FWAs) 74, 198–9
  colour assessment in presence of 339–40
    effect of triplet absorption in 64–5, 205, 206
    light sources for 12, 27–8, 60–61, 63, 75–6, 203–4, 252
    standard illuminants for 14
  effects in fastness testing 193
Forcius 122, 126
fovea 429
  'blue' photoreceptors absent from 401, 434, 460
Fraunhofer, Josef von 70
Fraunhofer lines 46
frequency 3–4
Fresnel's law 31
Freundlich isotherm 239–40
full chromatic colours 127
FWAs *see* fluorescent whitening agents

gamma function (CRT display) 377, 408–9, 422
ganglion cells (retinal) 438–9, 440, 441–3, 448–9, 460, 469, 473

gas discharge tubes 18–19
geometric metamerism 114, 489
gloss 109
    assessment of 28, 32–3
glutamate (retinal neurotransmitter) 435
Grassman's fourth rule of colorimetry 483
Grassman's law 88
grey scales 185–8, 253–4

Hardy spectrophotometer 69, 74
HBS model 166–7
Helmholtz–Kohlrausch effect 483, 484
Helmholtz, Herman von 89, 453, 464, 476, 489
Hering, Ewald 132, 453
Hertz, Heinrich 3
heterochromatic flicker photometry 484, 485
1,3,5-hexatriene, UV absorption in 47, 48, 49
hiding power of coatings 359–62
HMI lamp 18
holographic diffraction gratings 71
holographic filters 398
horizontal cells (retinal) 431, 435, 436–7, 438, 440, 442, 459
hue 38–9, 462–3
    in CIELAB colour space 139
    descriptors of differences in 158–62
    in Munsell system 126–7
    perception related to luminance level 482
    perception related to stimulus duration 485
    and wavelength position of light absorption 5, 39–40
hue circuits 126–7
Huematic colour-vision test 508
hydrogen
    atomic spectrum of 44–5
    UV absorption in 47
hypsochromic shift 53

illuminance 7
illuminant metamerism 59, 103, 113, 489
illuminants, standard 13–15, 99–100, 106–7, 514–15
    A 13–14, 64, 100
    B 14, 100, 102, 103
    C 14, 100, 102, 103
    D 14–15, 64, 102–3, 405–6
    for assessment of colour-rendering properties 119
incandescence 15
indicators (dyes) 78, 87

inducing field 476
influence matrix
    for pigmented materials 350, 353, 354, 357, 368, 369
    for textile dyeing 219, 220–21
infrared radiation 3, 5
inks see semitransparent layers
instrumental metamerism 417
Integ technique for depth measurement 179–81
interference, colour production by 1
irradiance 7
Ishihara colour-vision test 503–5, 506
isoconfusion lines 496–8

Jablonski diagrams 51–2
Judd polynomial 125–6, 131, 135, 401
just-noticeable differences 466–7

Kubelka–Munk analysis 36–8, 178–81
    applied to dyed textiles 209–14
        fibre blends 278–9
    applied to pigmented materials 294–8
        opaque layers 342
        semitransparent layers 299–304
    limitations of 298–9

Lambertian reflectors 299
Lambert's law 34–5, 36
Land, Edwin 453–4
Langmuir isotherm 239, 240
lasers 21, 22
    electronic transitions in He–Ne 46–7
lateral diffusion error 66
lateral geniculate nucleus 445, 469
least-squares analysis
    in curve fitting for dye recipe prediction 231–2, 236
    in optical coefficient analysis for opaque layers 318–27, 344, 345–6, 353
LEDs see light-emitting diodes
Leonardo da Vinci 122
light
    absorption of 1–2, 33–6, 51–3
        and hue 39–40
        see also Kubelka–Munk analysis
    emission of 1–2
    interactions with matter 29–38, 43–51
    mixing of coloured 85–7, 88–90, 454, 455, 483–4

light *continued*
  polarisation of 30, 31
  reflection of 1–2
    at surface 31
    diffuse 32–3
  refraction of 29, 30–31
  relation to colour 2–3, 444, 462
  scattering of 1–2, 31–2
    see also Kubelka–Munk analysis
  speed of 3
  wavelength range of visible 5
  see also infrared radiation; ultraviolet radiation
light adaptation 473
light-emitting diodes 21, 23
light-fastness testing 12–13, 54–6, 193–4
  evaluation of results from
    colorimetric 195, 196
    visual 193–4
  standard depths for 173–4
  standard light sources for 17, 55–6, 193
  wool standards for 55, 173, 193
light intensity, measurements of 6–8
light sources 15–23
  in colour-matching booths 27–8
  colour-rendering properties of 24, 25–7, 119–20
  control of UV content of 60–61, 64–5, 204
  daylight simulators 15
  illumination levels for 8
  luminous efficacy of 24–5
  for spectrophotometry 63–5
  for spectroradiometer calibration 416
  standard 13–15, 99–100
  for whiteness evaluation 203, 204
  see also spectral power distribution, *and under specific light sources*
lightness 125, 158, 166
lightness constancy 479–80
liquid crystals 389–90
liquid-crystal displays 389, 390–99, 417, 418, 419
  coloured 393–9
  double supertwist (D-STN) cell 392, 393, 394
  supertwist (STN) cell 392–3
  twisted-nematic (TN) cell 390–93
lumen, definition of 7–8
luminance 7, 415, 463
  additivity of 483, 484
  colour phenomena related to 482–3
  contrast of 439–41
  discrimination 468, 472–7
luminance factor 415
luminescence 2

luminosity
  perception of 483–4
  see also brightness
luminosity curves *see* V_λ curves
luminous efficacy 415
luminous efficiency 382, 415
luminous exitance 415
luminous flux 7, 415
luminous intensity 7, 415

MacAdam discrimination ellipses 144, 145, 478–9, 496, 497
MacAdam UCS diagram 131–2, 136
macula 429, 446
  pigment in 487, 496
Maxwell, James Clerk 3, 426, 453, 454, 489
Maxwell disc 86, 453
MBF/MBTF lamps 20
mercury, atomic emission spectrum 45
mercury-vapour lamps 18, 20
mesopic vision 428, 476
metamerism 43, 112–15, 116, 216, 488–9
  field size 114
  geometric 114, 489
  illuminant 59, 103, 113, 489
  instrumental 417
  minimising by computer formulation 226–33
  observer 93, 113–14, 488–9, 490
methanal, UV absorption in 48, 49
method of adjustment 465, 466
method of constant stimuli 465–6, 468, 502
method of limits 465, 468, 502
Microflash spectrophotometer 72, 73
monochromacy 493, 498, 499
  acquired 500
multi-layer perceptron 282
Munsell colour-order system 122–30

neural networks, use in recipe formulation 281–5
neurone action 435–7
neurotransmitters, retinal 435
neutral point (colour perception) 496, 497–8
neutral points (retinal ganglion cells) 443, 449, 460
Newton, Isaac
  hue circle 122, 126
  on nature of colour 2–3, 444, 462

observer metamerism 93, 113–14, 488–9, 490
opacity of layers 302–4
opponent colour theory (colour vision) 132, 453, 455–6

opponent responses
  of cortical cells 447–50
  of retinal cells 437, 438, 439–42, 448, 459
opsins 432, 435
optic chiasma 444, 445
optic radiations 445, 446
optical brighteners *see* fluorescent whitening agents
optimal colour stimuli 128

paints
  components of colour and colour difference 167–9
    determination of optical coefficients 308–13, 319–26
    calibration panels for 307, 318, 328–33
    estimates of error in 313–15
    for very bright colorants 315–18, 318–19
  hiding power of 359–62
  metallic 110, 114
  recipe prediction for 340–67
    colorimetric match prediction for 348–58, 364
    including fixed component 362–4
    recipe correction 364–7
    spectrophotometric curve matching for 343–8, 363
    theory applied to fibre blends 278–9
  spreading rate of 360–62
perfect reflecting diffuser 125, 138
period of radiation 4
pH-sensitive dyes 78–9
phosphorescence 381
phosphors
  for cathode-ray tube displays 381–3, 418
    ageing effects 382–3
    coating for 375
    geometrical distribution of 378–81
    lifetimes of 375–6
    luminance of 376
    P22 group 382–3, 384, 385, 402
    persistence of 375, 382, 387
    spectral characteristics of 383–6
  for fluorescent tubes 19, 20, 395
photochromatic interval 474
photochromism 194
  test for 56
photometry, defined 414
photopic vision 24–5, 428, 473–4, 487–8
  and colour vision defects 494
photopigments 430–35, 459–60
  anomalous 491–2, 494–5
  variations in 486

photoreceptors *see* cone cells; rod cells
pigmented materials *see* recipe prediction for pigmented materials
Pineo factor 234–5
pixel point spread function 375, 376
Planckian radiators 8–11
Planck's radiation expression 9–10
Planck's relationship (energy of radiation) 6
polarised light 30, 31
power of radiation 7
  see also spectral power distributions
prime colour lamp 19–20
prisms 69
protanomalous trichromats 490, 492, 495, 499, 502, 503
protanopic dichromats 492, 494, 496, 499, 502, 503
pseudo-isochromatic colour-vision tests 500, 503–6
psychophysical studies 463–4, 465–7
psychophysiology 463–4
Purkinje shift 402, 474

quantum concept of radiation 6, 9, 44–5
quantum yield (of photoreaction) 53
quartz halogen lamps 63
quenching of fluorescence 255
  dyes 257, 265

radiance 7, 414
radiance factors *see* spectral radiance factors
radiant efficiency 382, 414
radiant energy 414
radiant exitance 414
radiant exposure 7
radiant factor 414
radiant flux 7, 414
  measurement of distribution of 415–17
radiant intensity 7, 414
radio waves 3
radiometry, defined 414
Rayleigh colour-mixing equation 501, 503
recipe prediction for dyed textiles 209–91
  colorimetric curve matching 217–22
    algorithm for starting recipe 223–6
    four-dye prediction 227–30
    minimising metamerism by 227–9, 230, 232–3
  cost-effectiveness of systems for 288–90
  deviations from linearity in equations 233–49
    interpolation solutions 235–9
    physical chemistry solutions 239–45, 247–9
    piecewise linear solutions 245–7

recipe prediction for dyed textiles *continued*
  for fibre blends
    fibre mixtures 277
    loose-stock blends 277–81
  hardware for 286–8
  recipe correction 266–70, 289
    by matching test dyeing 268
    using correction factors for dye-uptake effects 268–9
    using correction matrix 217, 221–2, 237, 258–9, 267–8
    using deeper shade for standard 269–70
  reducing cost and metamerism by 226–33
  selection of dye combinations 271–7
    assessment of new dyes 276
    cost minimisation 271, 275–6, 277
    dye triangulation 272–7
    spectrophotometric curve matching 214–17, 225
    minimising metamerism by 230–32
    use with fibre blends 279–81
  starting recipe 223–6
  use of artificial neural networks in 281–5
  use of historical data in 270–71
  using fluorescent dyes 249–66
    correction matrix for 258–9
    two-monochromator method for 251, 259, 261
    using two-mode spectrophotometry 250, 259
recipe prediction for pigmented materials 292–372
  see also under coatings; paints; semitransparent layers
Redifon computer system 217, 293
reflectance 209
  calculated and measured values of 298
  calculation of tristimulus values from 95–8, 104–5, 524
  of dyed textiles 210–14, 234–5
  of fluorescent dyes 249–56
  of layers
    accounting for reflection at interface 304–7
    opaque layers 295–6, 342–3
      see also under calibration panels
    semitransparent layers 299–304, 334–6, 367–8
  measurement of 61–3
    for depth determinations 178
    spectrophotometers for 59–61, 62
    standards for 104
  prediction by Kubelka–Munk theory 294–9
  specular-included or -excluded 31, 67–8
  of very brightly coloured materials 315–18

refractive index 3
relative spectral luminosity functions *see* V_λ curves
repeatability of assessment 142
reproducibility of assessment 142
residual colour differences 172
retina 57, 428–30
  postreceptoral pathways in 435–44
  resolving power of 441, 460–61
retinex theory of colour vision 453–4, 456–8
rhodopsin 432, 433, 434, 474
Ritter, J W 3
robustness of recipe 358
rod cells 428–9
  cone–rod interaction 476
  pathway of output of 438–9
  photopigment in 430–33, 434–5, 474
  sensitivity at low luminance of 57, 401–2, 428–9, 473–4
rod monochromats 493, 499

S-potentials 437
sample preparation effects 285
sample presentation effects 285
saturation of colour
  CIE definition of 140
  perceived 481, 482–3
  related to spectral purity 462–3
saturation threshold, determination of 470–71
Saunderson equation 306, 336, 340
scaling-up 267, 271
scattering coefficients (S) 34, 36–7, 210
  additivity of 297–8
  of coatings 294–5, 359
    determination 308–13, 313–15, 319–26, 328
  of coatings that are very brightly coloured 315–18
  of semitransparent layers, determination 334
  units 297
Schuster–Kubelka–Munk theory *see* Kubelka–Munk analysis
scotopic vision 24, 25, 428, 434, 473–4, 475, 476
screen savers 382
selective adaptation studies 455, 485
self-quenching of fluorescence 255
semitransparent layers
  match prediction of 367–72
    colorimetric 368–70
    database calibration for 334–40
    opaque and transparent white for 370–72
    recipe correction 372
    spectrophotometric curve matching 368
  optical properties of 299–304

sensitivity of recipe 357–8
shade differences 172
similar colorants 169
Simplex iteration method (absorption coefficient calculation) 243, 244–6
simultaneous contrast phenomena 481–2
single-number shade passing (SNSP) 146, 151, 154
skin, damage by UV radiation 6, 13
Snell's law 30
SNSP see single-number shade passing
sodium
　atomic emission spectrum of 45
　lines in solar spectrum 46
sodium-vapour lamp 16, 18–19, 25, 26
Spectraflash 500 spectrophotometer 60–61
spectral analysers 61, 68–71
spectral luminous efficiency functions see V_λ curves
spectral power distributions 8, 9
　changes with temperature 8–11
　of fluorescent lamps 20
　of mercury-vapour lamps 18, 20
　of sodium-vapour lamp 19
　of standard illuminants and standard light sources 13–15, 64
　of xenon-arc lamp 17, 63–4
spectral purity 463
　discrimination 470–72, 495–6
　effect on perceived luminosity of 483
spectral radiance factors 73–4, 198–9, 250–56, 264–6
spectrophotometers
　for colour measurement 59–61
　　abridged 62
　　Datacolor Microflash instrument 72, 73
　　double-monochromator 74–5
　　dual-beam 60–61, 68, 72
　　geometry of instrument 65–8, 103–4
　　　for studies of fluorescent materials 74–5
　　Hardy instrument 69, 74
　　Macbeth CE3000 instrument 71–2
　　specifications for 79–80
　　Spectraflash 500 60–61
　　spectral analysers for 61, 68–71
　　standards for 62–3
　　Zeiss RFC instrument 69
　for computer colorant formulation 286, 307–8
　stability of 285
spectrophotometry
　reflectance see under reflectance
　transmission, applied to dye solutions 76–9, 177
spectroradiometers 415–17, 419

spectrum 2, 5
spectrum locus 105
spreading effect 482
spreading rate of coatings 360–62
standardisation of colorants 171–2
Stefan, Josef 9
stereoscopic vision 445
Stokes shift 73
strength (of colorant) 166–7, 168, 169–72, 174–5
　measurement of 40–42, 181
　SDC definition of 38–9
　uses of term 172
subtractive colour mixing 86–8
sun, spectrum of 46

telespectroradiometers 415–16
textural effects 109, 285–6
thermochromism 63
three-band fluorescent lamp 19–20, 26
TL84 lamp 20, 26, 150, 230
Tokyo Medical College colour-vision test 505
trace differences (colour attributes) 176
translucency 66–7
transmittance 34, 35
trichromacy 434, 454–5, 459, 461, 468–9
trichromatic theory of colour vision 132–3, 453, 454–5
triplet absorption 64, 205, 206
triplet state 53
tristimulus colorimeters 58–9, 93–4, 417–19
　for calibration of CRT displays 422–3
tristimulus values 58, 89–93, 94–5, 102
　calculation from measured reflectance values 95–8, 104–5, 524
　limitations of 109–10
　negative 91
　related to colour appearance 105
　weighting function tables 62
　see also chromaticity diagrams
tritanomalous trichromats 491, 492, 498, 499, 501
tritanopia, small-field 401
tritanopic dichromats 492, 494, 496, 498, 499, 501
tungsten–halogen lamp 16–17
　colour temperature of 11
tungsten-filament lamp 16, 27
　colour temperature of 11

UCS diagrams see uniform chromaticity scale diagrams
Ultralume lamp 20
ultraviolet radiation 3

ultraviolet radiation *continued*
  absorption in aromatic compounds 50–51
  absorption in simple molecules 47–9
  in daylight 12–13
  energy content of 6
  skin damage by 6, 13
uniform chromaticity scale (UCS) diagrams 131–2, 136–7
unitary (unique) colours 126

V_λ (relative spectral luminosity function) curves 8, 99, 414
  for mesopic vision 476
  for photopic vision 24–5, 442, 475, 476
    in colour defectives 494
    variation in 487–8
  for scotopic vision 24, 25, 475, 476
visual cortex 445, 446–51, 459
visual displays
  colorimetry of 399–413
    chromaticity of primaries 399–404
    colour appearance 411–13
    display characterisation 407–10
    white point setting 404–7, 420
  measurement and calibration of performance of 414–23
    calibration of display 419–23
    spectroradiometers for 415–17
    tricolorimeters for 417–19
    units for 414–15
  see also cathode-ray tube displays; liquid-crystal displays

visual purple *see* rhodopsin
visual thresholds, measurement of 465–6

wavelength 3, 4
  units for 4–5
wavenumber 4
Weber–Fechner law 472–3
Weber's law 125
white inks, opaque and transparent 370–72
white point, in visual displays 377, 404–7, 420
whiteness 296
  evaluation of 195–205
    by colorimetric methods 199–200
    by visual methods 199
    formulae for use in 200–204
    hue preferences in 197, 200–202
  evaluation of absolute 204
  perception of 197–9, 461
  of perfect reflecting diffuser 125
  standard for reflectance studies 308

X-linked genes 498
xenon-arc lamps 17, 55
  control of UV content of 60–61, 204
  for use in spectrophotometry 63–4
  for use in whiteness evaluation 204, 205

yarn, card winders for 285
yellowness, evaluation of 205–7
Young, Thomas 89, 452–3

Zeiss RFC3 spectrophotometer 69
zone theories of colour vision 459